Digital Signal Processing

————————————————————————————————————— *Second Edition*

Monson H. Hayes, Ph.D.
*Professor of Electrical and
Computer Engineering
Georgia Institute of Technology*

Schaum's Outline Series

New York   Chicago   San Francisco   Lisbon   London   Madrid
Mexico City   Milan   New Delhi   San Juan   Seoul
Singapore   Sydney   Toronto

MONSON H. HAYES is a professor of Electrical and Computer Engineering at the Georgia Institute of Technology in Atlanta, Georgia. He received his B.A. degree in Physics from the University of California, Berkeley, and his M.S.E.E. and Sc.D. degrees in Electrical Engineering and Computer Science from M.I.T. His research interests are in digital signal processing with applications in image and video processing. He has contributed more than 100 articles to journals and conference proceedings, and is the author of the textbook *Statistical Digital Signal Processing and Modeling*, John Wiley & Sons. He received the IEEE Senior Award for the author of a paper of exceptional merit from the ASSP Society of the IEEE in 1983, the Presidential Young Investigator Award for his "contributions to signal modeling including the development of algorithms for signal restoration from Fourier transform phase or magnitude".

1 2 3 4 5 6 7 8 9 10   CUS/CUS   1 9 8 7 6 5 4 3 2 1

ISBN   978-0-07-163509-7
MHID   0-07-163509-2

McGraw-Hill books are available at special quantity discounts to use as premiums and sales promotions or for use in corporate training programs. To contact a representative, please e-mail us at bulksales@mcgraw-hill.com.

For Sandy

Preface

Digital signal processing (DSP) is concerned with the representation of signals in digital form, and with the processing of these signals and the information that they carry. Although DSP, as we know it today, began to flourish in the 1960's, some of the important and powerful processing techniques that are in use today may be traced back to numerical algorithms that were proposed and studied centuries ago. Since the early 1970's, when the first DSP chips were introduced, the field of digital signal processing has evolved dramatically. With a tremendously rapid increase in the speed of DSP processors, along with a corresponding increase in their sophistication and computational power, digital signal processing has become an integral part of many commercial products and applications, and is becoming a commonplace term.

This book is concerned with the fundamentals of digital signal processing, and there are two ways that the reader may use this book to learn about DSP. First, it may be used as a supplement to any one of a number of excellent DSP textbooks by providing the reader with a rich source of worked problems and examples. Alternatively, it may be used as a self-study guide to DSP, using the method of *learning by example*. With either approach, this book has been written with the goal of providing the reader with a broad range of problems having different levels of difficulty. In addition to problems that may be considered *drill,* the reader will find more challenging problems that require some creativity in their solution, as well as problems that explore practical applications such as computing the payments on a home mortgage. When possible, a problem is worked in several different ways, or alternative methods of solution are suggested.

The nine chapters in this book cover what is typically considered to be the core material for an introductory course in DSP. The first chapter introduces the basics of digital signal processing, and lays the foundation for the material in the following chapters. The topics covered in this chapter include the description and characterization of discrete-type signals and systems, convolution, and linear constant coefficient difference equations. The second chapter considers the represention of discrete-time signals in the frequency domain. Specifically, we introduce the discrete-time Fourier transform (DTFT), develop a number of DTFT properties, and see how the DTFT may be used to solve difference equations and perform convolutions. Chapter 3 covers the important issues associated with sampling continuous-time signals. Of primary importance in this chapter is the sampling theorem, and the notion of aliasing. In Chapter 4, the z-transform is developed, which is the discrete-time equivalent of the Laplace transform for continuous-time signals. Then, in Chapter 5, we look at the system function, which is the z-transform of the unit sample response of a linear shift-invariant system, and introduce a number of different types of systems, such as allpass, linear phase, and minimum phase filters, and feedback systems.

The next two chapters are concerned with the Discrete Fourier Transform (DFT). In Chapter 6, we introduce the DFT, and develop a number of DFT properties. The key idea in this chapter is that multiplying the DFTs of two sequences corresponds to circular convolution in the time domain. Then, in Chapter 7, we develop a number of efficient algorithms for computing the DFT of a finite-length sequence. These algorithms are referred to, generically, as fast Fourier transforms (FFTs). Finally, the last two chapters consider the design and implementation of discrete-time systems. In Chapter 8 we look at different ways to implement a linear shift-invariant discrete-time system, and look at the sensitivity of these implementations to filter coefficient quantization. In addition, we analyze the propagation of round-off noise in fixed-point implementations of these systems. Then, in Chapter 9 we look at techniques for designing FIR and IIR linear shift-invariant filters. Although the primary focus is on the design of low-pass filters, techniques for designing other frequency selective filters, such as high-pass, bandpass, and bandstop filters are also considered.

It is hoped that this book will be a valuable tool in learning DSP. Feedback and comments are welcomed through the web site for this book, which may be found at

<code>http://www.ee.gatech.edu/users/mhayes/schaum</code>

Also available at this site will be important information, such as corrections or amplifications to problems in this book, additional reading and problems, and reader comments.

Contents

Chapter 1

Signals and Systems

1.1 INTRODUCTION

In this chapter we begin our study of digital signal processing by developing the notion of a discrete-time signal and a discrete-time system. We will concentrate on solving problems related to signal representations, signal manipulations, properties of signals, system classification, and system properties. First, in Sec. 1.2 we define precisely what is meant by a discrete-time signal and then develop some basic, yet important, operations that may be performed on these signals. Then, in Sec. 1.3 we consider discrete-time systems. Of special importance will be the notions of linearity, shift-invariance, causality, stability, and invertibility. It will be shown that for systems that are linear and shift-invariant, the input and output are related by a convolution sum. Properties of the convolution sum and methods for performing convolutions are then discussed in Sec. 1.4. Finally, in Sec. 1.5 we look at discrete-time systems that are described in terms of a difference equation.

1.2 DISCRETE-TIME SIGNALS

A discrete-time signal is an indexed sequence of real or complex numbers. Thus, a discrete-time signal is a function of an integer-valued variable, n, that is denoted by $x(n)$. Although the independent variable n need not necessarily represent "time" (n may, for example, correspond to a spatial coordinate or distance), $x(n)$ is generally referred to as a function of time. A discrete-time signal is undefined for noninteger values of n. Therefore, a real-valued signal $x(n)$ will be represented graphically in the form of a *lollipop* plot as shown in Fig. 1-1. In

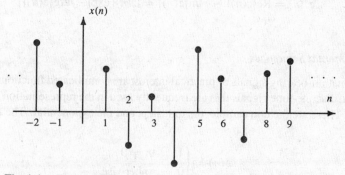

Fig. 1-1. The graphical representation of a discrete-time signal $x(n)$.

some problems and applications it is convenient to view $x(n)$ as a vector. Thus, the sequence values $x(0)$ to $x(N-1)$ may often be considered to be the elements of a column vector as follows:

$$\mathbf{x} = [x(0), \ x(1), \ldots, x(N-1)]^T$$

Discrete-time signals are often derived by sampling a continuous-time signal, such as speech, with an analog-to-digital (A/D) converter.[1] For example, a continuous-time signal $x_a(t)$ that is sampled at a rate of $f_s = 1/T_s$ samples per second produces the sampled signal $x(n)$, which is related to $x_a(t)$ as follows:

$$x(n) = x_a(nT_s)$$

Not all discrete-time signals, however, are obtained in this manner. Some signals may be considered to be naturally occurring discrete-time sequences because there is no physical analog-to-digital converter that is converting an

---

[1] Analog-to-digital conversion will be discussed in Chap. 3.

analog signal into a discrete-time signal. Examples of signals that fall into this category include daily stock market prices, population statistics, warehouse inventories, and the Wolfer sunspot numbers.[2]

1.2.1   Complex Sequences

In general, a discrete-time signal may be complex-valued. In fact, in a number of important applications such as digital communications, complex signals arise naturally. A complex signal may be expressed either in terms of its real and imaginary parts,

$$z(n) = a(n) + jb(n) = \text{Re}\{z(n)\} + j\text{Im}\{z(n)\}$$

or in polar form in terms of its magnitude and phase,

$$z(n) = |z(n)| \exp[j\arg\{z(n)\}]$$

The magnitude may be derived from the real and imaginary parts as follows:

$$|z(n)|^2 = \text{Re}^2\{z(n)\} + \text{Im}^2\{z(n)\}$$

whereas the phase may be found using

$$\arg\{z(n)\} = \tan^{-1}\frac{\text{Im}\{z(n)\}}{\text{Re}\{z(n)\}}$$

If $z(n)$ is a complex sequence, the *complex conjugate*, denoted by $z^*(n)$, is formed by changing the sign on the imaginary part of $z(n)$:

$$z^*(n) = \text{Re}\{z(n)\} - j\text{Im}\{z(n)\} = |z(n)| \exp[-j\arg\{z(n)\}]$$

1.2.2   Some Fundamental Sequences

Although most information-bearing signals of practical interest are complicated functions of time, there are three simple, yet important, discrete-time signals that are frequently used in the representation and description of more complicated signals. These are the unit sample, the unit step, and the exponential. The *unit sample*, denoted by $\delta(n)$, is defined by

$$\delta(n) = \begin{cases} 1 & n = 0 \\ 0 & \text{otherwise} \end{cases}$$

and plays the same role in discrete-time signal processing that the unit impulse plays in continuous-time signal processing. The *unit step*, denoted by $u(n)$, is defined by

$$u(n) = \begin{cases} 1 & n \geq 0 \\ 0 & \text{otherwise} \end{cases}$$

and is related to the unit sample by

$$u(n) = \sum_{k=-\infty}^{n} \delta(k)$$

Similarly, a unit sample may be written as a difference of two steps:

$$\delta(n) = u(n) - u(n-1)$$

---

[2]The Wolfer sunspot number R was introduced by Rudolf Wolf in 1848 as a measure of sunspot activity. Daily records are available back to 1818 and estimates of monthly means have been made since 1749. There has been much interest in studying the correlation between sunspot activity and terrestrial phenomena such as meteorological data and climatic variations.

Finally, an *exponential* sequence is defined by

$$x(n) = a^n$$

where a may be a real or complex number. Of particular interest is the exponential sequence that is formed when $a = e^{j\omega_0}$, where ω_0 is a real number. In this case, $x(n)$ is a complex exponential

$$e^{jn\omega_0} = \cos(n\omega_0) + j\sin(n\omega_0)$$

As we will see in the next chapter, complex exponentials are useful in the Fourier decomposition of signals.

1.2.3 Signal Duration

Discrete-time signals may be conveniently classified in terms of their duration or extent. For example, a discrete-time sequence is said to be a *finite-length sequence* if it is equal to zero for all values of n outside a finite interval $[N_1, N_2]$. Signals that are not finite in length, such as the unit step and the complex exponential, are said to be *infinite-length sequences*. Infinite-length sequences may further be classified as either being right-sided, left-sided, or two-sided. A *right-sided sequence* is any infinite-length sequence that is equal to zero for all values of $n < n_0$ for some integer n_0. The unit step is an example of a right-sided sequence. Similarly, an infinite-length sequence $x(n)$ is said to be *left-sided* if, for some integer n_0, $x(n) = 0$ for all $n > n_0$. An example of a left-sided sequence is

$$x(n) = u(n_0 - n) = \begin{cases} 1 & n \leq n_0 \\ 0 & n > n_0 \end{cases}$$

which is a time-reversed and delayed unit step. An infinite-length signal that is neither right-sided nor left-sided, such as the complex exponential, is referred to as a *two-sided sequence*.

1.2.4 Periodic and Aperiodic Sequences

A discrete-time signal may always be classified as either being *periodic* or *aperiodic*. A signal $x(n)$ is said to be periodic if, for some positive real integer N,

$$x(n) = x(n + N) \tag{1.1}$$

for all n. This is equivalent to saying that the sequence repeats itself every N samples. If a signal is periodic with period N, it is also periodic with period $2N$, period $3N$, and all other integer multiples of N. The *fundamental period*, which we will denote by N, is the smallest positive integer for which Eq. (*1.1*) is satisfied. If Eq. (*1.1*) is not satisfied for any integer N, $x(n)$ is said to be an aperiodic signal.

EXAMPLE 1.2.1 The signals

$$x_1(n) = a^n u(n) = \begin{cases} a^n & n \geq 0 \\ 0 & n < 0 \end{cases}$$

and

$$x_2(n) = \cos(n^2)$$

are not periodic, whereas the signal

$$x_3(n) = e^{j\pi n/8}$$

is periodic and has a fundamental period of $N = 16$.

If $x_1(n)$ is a sequence that is periodic with a period N_1, and $x_2(n)$ is another sequence that is periodic with a period N_2, the sum

$$x(n) = x_1(n) + x_2(n)$$

will always be periodic and the fundamental period is

$$N = \frac{N_1 N_2}{\gcd(N_1, N_2)} \tag{1.2}$$

where $\gcd(N_1, N_2)$ means *the greatest common divisor* of N_1 and N_2. The same is true for the product; that is,

$$x(n) = x_1(n)x_2(n)$$

will be periodic with a period N given by Eq. (*1.2*). However, the fundamental period may be smaller.

Given any sequence $x(n)$, a periodic signal may always be formed by replicating $x(n)$ as follows:

$$y(n) = \sum_{k=-\infty}^{\infty} x(n - kN)$$

where N is a positive integer. In this case, $y(n)$ will be periodic with period N.

1.2.5 Symmetric Sequences

A discrete-time signal will often possess some form of symmetry that may be exploited in solving problems. Two symmetries of interest are as follows:

Definition: A real-valued signal is said to be *even* if, for all n,

$$x(n) = x(-n)$$

whereas a signal is said to be *odd* if, for all n,

$$x(n) = -x(-n)$$

Any signal $x(n)$ may be decomposed into a sum of its even part, $x_e(n)$, and its odd part, $x_o(n)$, as follows:

$$x(n) = x_e(n) + x_o(n) \qquad (1.3)$$

To find the even part of $x(n)$ we form the sum

$$x_e(n) = \tfrac{1}{2}\{x(n) + x(-n)\}$$

whereas to find the odd part we take the difference

$$x_o(n) = \tfrac{1}{2}\{x(n) - x(-n)\}$$

For complex sequences the symmetries of interest are slightly different.

Definition: A complex signal is said to be *conjugate symmetric*[3] if, for all n,

$$x(n) = x^*(-n)$$

and a signal is said to be *conjugate antisymmetric* if, for all n,

$$x(n) = -x^*(-n)$$

Any complex signal may always be decomposed into a sum of a conjugate symmetric signal and a conjugate antisymmetric signal.

1.2.6 Signal Manipulations

In our study of discrete-time signals and systems we will be concerned with the manipulation of signals. These manipulations are generally compositions of a few basic signal transformations. These transformations may be classified either as those that are transformations of the independent variable n or those that are transformations of the amplitude of $x(n)$ (i.e., the dependent variable). In the following two subsections we will look briefly at these two classes of transformations and list those that are most commonly found in applications.

---

[3]A sequence that is conjugate symmetric is sometimes said to be hermitian.

Transformations of the Independent Variable

Sequences are often altered and manipulated by modifying the index n as follows:

$$y(n) = x(f(n))$$

where $f(n)$ is some function of n. If, for some value of n, $f(n)$ is not an integer, $y(n) = x(f(n))$ is undefined. Determining the effect of modifying the index n may always be accomplished using a simple tabular approach of listing, for each value of n, the value of $f(n)$ and then setting $y(n) = x(f(n))$. However, for many index transformations this is not necessary, and the sequence may be determined or plotted directly. The most common transformations include shifting, reversal, and scaling, which are defined below.

Shifting   This is the transformation defined by $f(n) = n - n_0$. If $y(n) = x(n - n_0)$, $x(n)$ is shifted to the right by n_0 samples if n_0 is positive (this is referred to as a delay), and it is shifted to the left by n_0 samples if n_0 is negative (referred to as an advance).

Reversal   This transformation is given by $f(n) = -n$ and simply involves "flipping" the signal $x(n)$ with respect to the index n.

Time Scaling   This transformation is defined by $f(n) = Mn$ or $f(n) = n/N$ where M and N are positive integers. In the case of $f(n) = Mn$, the sequence $x(Mn)$ is formed by taking every Mth sample of $x(n)$ (this operation is known as *down-sampling*). With $f(n) = n/N$ the sequence $y(n) = x(f(n))$ is defined as follows:

$$y(n) = \begin{cases} x\left(\dfrac{n}{N}\right) & n = 0, \pm N, \pm 2N, \cdots \\ 0 & \text{otherwise} \end{cases}$$

(this operation is known as *up-sampling*).

Examples of shifting, reversing, and time scaling a signal are illustrated in Fig. 1-2.

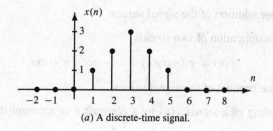

(a) A discrete-time signal.

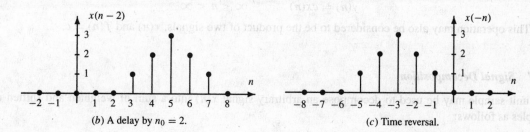

(b) A delay by $n_0 = 2$.   (c) Time reversal.

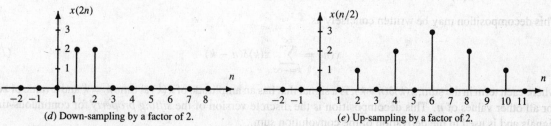

(d) Down-sampling by a factor of 2.   (e) Up-sampling by a factor of 2.

Fig. 1-2.   Illustration of the operations of shifting, reversal, and scaling of the independent variable n.

Shifting, reversal, and time-scaling operations are order-dependent. Therefore, one needs to be careful in evaluating compositions of these operations. For example, Fig. 1-3 shows two systems, one that consists of a delay followed by a reversal and one that is a reversal followed by a delay. As indicated, the outputs of these two systems are not the same.

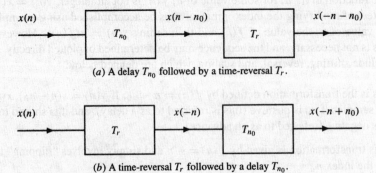

(a) A delay T_{n_0} followed by a time-reversal T_r.

(b) A time-reversal T_r followed by a delay T_{n_0}.

Fig. 1-3. Example illustrating that the operations of delay and reversal do not commute.

Addition, Multiplication, and Scaling

The most common types of amplitude transformations are addition, multiplication, and scaling. Performing these operations is straightforward and involves only pointwise operations on the signal.

Addition   The sum of two signals

$$y(n) = x_1(n) + x_2(n) \qquad -\infty < n < \infty$$

is formed by the pointwise addition of the signal values.

Multiplication   The multiplication of two signals

$$y(n) = x_1(n)x_2(n) \qquad -\infty < n < \infty$$

is formed by the pointwise product of the signal values.

Scaling   Amplitude scaling of a signal $x(n)$ by a constant c is accomplished by multiplying every signal value by c:

$$y(n) = cx(n) \qquad -\infty < n < \infty$$

This operation may also be considered to be the product of two signals, $x(n)$ and $f(n) = c$.

1.2.7   Signal Decomposition

The unit sample may be used to decompose an arbitrary signal $x(n)$ into a sum of weighted and shifted unit samples as follows:

$$x(n) = \cdots + x(-1)\delta(n+1) + x(0)\delta(n) + x(1)\delta(n-1) + x(2)\delta(n-2) + \cdots$$

This decomposition may be written concisely as

$$x(n) = \sum_{k=-\infty}^{\infty} x(k)\delta(n-k) \qquad\qquad (1.4)$$

where each term in the sum, $x(k)\delta(n-k)$, is a signal that has an amplitude of $x(k)$ at time $n = k$ and a value of zero for all other values of n. This decomposition is the discrete version of the *sifting property* for continuous-time signals and is used in the derivation of the convolution sum.

1.3   DISCRETE-TIME SYSTEMS

A discrete-time system is a mathematical operator or mapping that transforms one signal (the input) into another signal (the output) by means of a fixed set of rules or operations. The notation $T[\cdot]$ is used to represent a general system as shown in Fig. 1-4, in which an input signal $x(n)$ is transformed into an output signal $y(n)$ through the transformation $T[\cdot]$. The input-output properties of a system may be specified in any one of a number of different ways. The relationship between the input and output, for example, may be expressed in terms of a concise mathematical rule or function such as

$$y(n) = x^2(n)$$

or

$$y(n) = 0.5y(n-1) + x(n)$$

It is also possible, however, to describe a system in terms of an algorithm that provides a sequence of instructions or operations that is to be applied to the input signal, such as

$$y_1(n) = 0.5y_1(n-1) + 0.25x(n)$$
$$y_2(n) = 0.25y_2(n-1) + 0.5x(n)$$
$$y_3(n) = 0.4y_3(n-1) + 0.5x(n)$$
$$y(n) = y_1(n) + y_2(n) + y_3(n)$$

In some cases, a system may conveniently be specified in terms of a table that defines the set of all possible input-output signal pairs of interest.

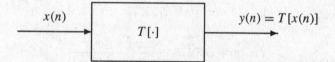

Fig. 1-4.   The representation of a discrete-time system as a transformation $T[\cdot]$ that maps an input signal $x(n)$ into an output signal $y(n)$.

Discrete-time systems may be classified in terms of the properties that they possess. The most common properties of interest include linearity, shift-invariance, causality, stability, and invertibility. These properties, along with a few others, are described in the following section.

1.3.1   System Properties

Memoryless System

The first property is concerned with whether or not a system has memory.

> **Definition:**   A system is said to be memoryless if the output at any time $n = n_0$ depends only on the input at time $n = n_0$.

In other words, a system is memoryless if, for any n_0, we are able to determine the value of $y(n_0)$ given only the value of $x(n_0)$.

EXAMPLE 1.3.1   The system

$$y(n) = x^2(n)$$

is memoryless because $y(n_0)$ depends only on the value of $x(n)$ at time n_0. The system

$$y(n) = x(n) + x(n-1)$$

on the other hand, is not memoryless because the output at time n_0 depends on the value of the input both at time n_0 and at time $n_0 - 1$.

Additivity

An additive system is one for which the response to a sum of inputs is equal to the sum of the inputs individually. Thus,

> **Definition:**   A system is said to be additive if
>
> $$T[x_1(n) + x_2(n)] = T[x_1(n)] + T[x_2(n)]$$

for any signals $x_1(n)$ and $x_2(n)$.

Homogeneity

A system is said to be homogeneous if scaling the input by a constant results in a scaling of the output by the same amount. Specifically,

> **Definition:**   A system is said to be homogeneous if
>
> $$T[cx(n)] = cT[x(n)]$$

for any *complex* constant c and for any input sequence $x(n)$.

EXAMPLE 1.3.2   The system defined by

$$y(n) = \frac{x^2(n)}{x(n-1)}$$

is not additive because

$$T[x_1(n) + x_2(n)] = \frac{(x_1(n) + x_2(n))^2}{x_1(n-1) + x_2(n-1)}$$

which is not the same as

$$T[x_1(n)] + T[x_2(n)] = \frac{x_1^2(n)}{x_1(n-1)} + \frac{x_2^2(n)}{x_2(n-1)}$$

This system is, however, homogeneous because, for an input $cx(n)$ the output is

$$T[cx(n)] = \frac{(cx(n))^2}{cx(n-1)} = c\frac{x^2(n)}{x(n-1)} = cT[x(n)]$$

On the other hand, the system defined by the equation

$$y(n) = x(n) + x^*(n-1)$$

is additive because

$$[x_1(n) + x_2(n)] + [x_1(n-1) + x_2(n-1)]^* = \big[x_1(n) + x_1^*(n-1)\big] + \big[x_2(n) + x_2^*(n-1)\big]$$

However, this system is not homogeneous because the response to $cx(n)$ is

$$T[cx(n)] = cx(n) + c^*x^*(n-1)$$

which is not the same as

$$cT[x(n)] = cx(n) + cx^*(n-1)$$

Linear Systems

A system that is both additive and homogeneous is said to be *linear*. Thus,

> **Definition:**   A system is said to be linear if
>
> $$T[a_1x_1(n) + a_2x_2(n)] = a_1T[x_1(n)] + a_2T[x_2(n)]$$

for any two inputs $x_1(n)$ and $x_2(n)$ and for any complex constants a_1 and a_2.

Linearity greatly simplifies the evaluation of the response of a system to a given input. For example, using the decomposition for $x(n)$ given in Eq. (1.4), and using the additivity property, it follows that the output $y(n)$ may be written as

$$y(n) = T[x(n)] = T\left[\sum_{k=-\infty}^{\infty} x(k)\delta(n-k)\right] = \sum_{k=-\infty}^{\infty} T[x(k)\delta(n-k)]$$

Because the coefficients $x(k)$ are constants, we may use the homogeneity property to write

$$y(n) = \sum_{k=-\infty}^{\infty} T[x(k)\delta(n-k)] = \sum_{k=-\infty}^{\infty} x(k)T[\delta(n-k)] \qquad (1.5)$$

If we define $h_k(n)$ to be the response of the system to a unit sample at time $n = k$,

$$h_k(n) = T[\delta(n-k)]$$

Eq. (1.5) becomes

$$y(n) = \sum_{k=-\infty}^{\infty} x(k)h_k(n) \qquad (1.6)$$

which is known as the *superposition summation*.

Shift-Invariance

If a system has the property that a shift (delay) in the input by n_0 results in a shift in the output by n_0, the system is said to be shift-invariant. More formally,

> **Definition:** Let $y(n)$ be the response of a system to an arbitrary input $x(n)$. The system is said to be shift-invariant if, for any delay n_0, the response to $x(n-n_0)$ is $y(n-n_0)$. A system that is not shift-invariant is said to be shift-varying.[4]

In effect, a system will be shift-invariant if its properties or characteristics do not change with time. To test for shift-invariance one needs to compare $y(n-n_0)$ to $T[x(n-n_0)]$. If they are the same for any input $x(n)$ and for all shifts n_0, the system is shift-invariant.

EXAMPLE 1.3.3   The system defined by

$$y(n) = x^2(n)$$

is shift-invariant, which may be shown as follows. If $y(n) = x^2(n)$ is the response of the system to $x(n)$, the response of the system to

$$x'(n) = x(n-n_0)$$

is

$$y'(n) = [x'(n)]^2 = x^2(n-n_0)$$

Because $y'(n) = y(n-n_0)$, the system is shift-invariant. However, the system described by the equation

$$y(n) = x(n) + x(-n)$$

is shift-varying. To see this, note that the system's response to the input $x(n) = \delta(n)$ is

$$y(n) = \delta(n) + \delta(-n) = 2\delta(n)$$

whereas the response to $x(n-1) = \delta(n-1)$ is

$$y'(n) = \delta(n-1) + \delta(-n-1)$$

Because this is not the same as $y(n-1) = 2\delta(n-1)$, the system is shift-varying.

---

[4]Some authors refer to this property as *time-invariance*. However, because n does not necessarily represent "time," shift-invariance is a bit more general.

Linear Shift-Invariant Systems

A system that is both linear and shift-invariant is referred to as a *linear shift-invariant* (LSI) system. If $h(n)$ is the response of an LSI system to the unit sample $\delta(n)$, its response to $\delta(n-k)$ will be $h(n-k)$. Therefore, in the superposition sum given in Eq. (*1.6*),

$$h_k(n) = h(n-k)$$

and it follows that

$$y(n) = \sum_{k=-\infty}^{\infty} x(k)h(n-k) \qquad (1.7)$$

Equation (*1.7*), which is known as the convolution sum, is written as

$$y(n) = x(n) * h(n)$$

where $*$ indicates the convolution operator. The sequence $h(n)$, referred to as the *unit sample response*, provides a complete characterization of an LSI system. In other words, the response of the system to *any* input $x(n)$ may be found once $h(n)$ is known.

Causality

A system property that is important for real-time applications is *causality*, which is defined as follows:

> **Definition:**   A system is said to be *causal* if, for any n_0, the response of the system at time n_0 depends only on the input up to time $n = n_0$.

For a causal system, changes in the output cannot precede changes in the input. Thus, if $x_1(n) = x_2(n)$ for $n \le n_0$, $y_1(n)$ must be equal to $y_2(n)$ for $n \le n_0$. Causal systems are therefore referred to as *nonanticipatory*. An LSI system will be causal if and only if $h(n)$ is equal to zero for $n < 0$.

EXAMPLE 1.3.4   The system described by the equation $y(n) = x(n) + x(n-1)$ is causal because the value of the output at any time $n = n_0$ depends only on the input $x(n)$ at time n_0 and at time $n_0 - 1$. The system described by $y(n) = x(n) + x(n+1)$, on the other hand, is noncausal because the output at time $n = n_0$ depends on the value of the input at time $n_0 + 1$.

Stability

In many applications, it is important for a system to have a response, $y(n)$, that is bounded in amplitude whenever the input is bounded. A system with this property is said to be *stable* in the bounded input-bounded output (BIBO) sense. Specifically,

> **Definition:**   A system is said to be stable in the bounded input-bounded output sense if, for any input that is bounded, $|x(n)| \le A < \infty$, the output will be bounded,
>
> $$|y(n)| \le B < \infty$$

For a linear shift-invariant system, stability is guaranteed if the unit sample response is absolutely summable:

$$\sum_{n=-\infty}^{\infty} |h(n)| < \infty \qquad (1.8)$$

EXAMPLE 1.3.5   An LSI system with unit sample response $h(n) = a^n u(n)$ will be stable whenever $|a| < 1$, because

$$\sum_{n=-\infty}^{\infty} |h(n)| = \sum_{n=0}^{\infty} |a|^n = \frac{1}{1 - |a|} \qquad |a| < 1$$

The system described by the equation $y(n) = nx(n)$, on the other hand, is not stable because the response to a unit step, $x(n) = u(n)$, is $y(n) = nu(n)$, which is unbounded.

Invertibility

A system property that is important in applications such as channel equalization and deconvolution is *invertibility*. A system is said to be *invertible* if the input to the system may be uniquely determined from the output. In order for a system to be invertible, it is necessary for distinct inputs to produce distinct outputs. In other words, given any two inputs $x_1(n)$ and $x_2(n)$ with $x_1(n) \neq x_2(n)$, it must be true that $y_1(n) \neq y_2(n)$.

EXAMPLE 1.3.6 The system defined by

$$y(n) = x(n)g(n)$$

is invertible if and only if $g(n) \neq 0$ for all n. In particular, given $y(n)$ with $g(n)$ nonzero for all n, $x(n)$ may be recovered from $y(n)$ as follows:

$$x(n) = \frac{y(n)}{g(n)}$$

1.4 CONVOLUTION

The relationship between the input to a linear shift-invariant system, $x(n)$, and the output, $y(n)$, is given by the convolution sum

$$x(n) * h(n) = \sum_{k=-\infty}^{\infty} x(k)h(n-k)$$

Because convolution is fundamental to the analysis and description of LSI systems, in this section we look at the mechanics of performing convolutions. We begin by listing some properties of convolution that may be used to simplify the evaluation of the convolution sum.

1.4.1 Convolution Properties

Convolution is a linear operator and, therefore, has a number of important properties including the commutative, associative, and distributive properties. The definitions and interpretations of these properties are summarized below.

Commutative Property

The commutative property states that the order in which two sequences are convolved is not important. Mathematically, the commutative property is

$$x(n) * h(n) = h(n) * x(n)$$

From a systems point of view, this property states that a system with a unit sample response $h(n)$ and input $x(n)$ behaves in exactly the same way as a system with unit sample response $x(n)$ and an input $h(n)$. This is illustrated in Fig. 1-5(a).

Associative Property

The convolution operator satisfies the associative property, which is

$$\{x(n) * h_1(n)\} * h_2(n) = x(n) * \{h_1(n) * h_2(n)\}$$

From a systems point of view, the associative property states that if two systems with unit sample responses $h_1(n)$ and $h_2(n)$ are connected in cascade as shown in Fig. 1-5(b), an equivalent system is one that has a unit sample response equal to the convolution of $h_1(n)$ and $h_2(n)$:

$$h_{eq}(n) = h_1(n) * h_2(n)$$

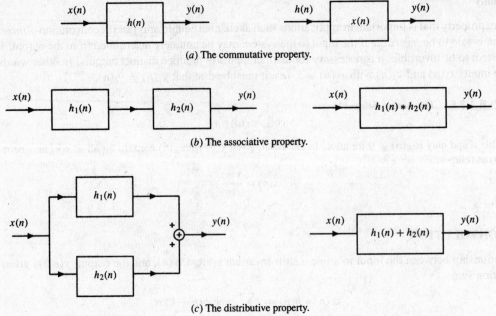

(a) The commutative property.

(b) The associative property.

(c) The distributive property.

Fig. 1-5. The interpretation of convolution properties from a systems point of view.

Distributive Property

The distributive property of the convolution operator states that

$$x(n) * \{h_1(n) + h_2(n)\} = x(n) * h_1(n) + x(n) * h_2(n)$$

From a systems point of view, this property asserts that if two systems with unit sample responses $h_1(n)$ and $h_2(n)$ are connected in parallel, as illustrated in Fig. 1-5(c), an equivalent system is one that has a unit sample response equal to the sum of $h_1(n)$ and $h_2(n)$:

$$h_{eq}(n) = h_1(n) + h_2(n)$$

1.4.2 Performing Convolutions

Having considered some of the properties of the convolution operator, we now look at the mechanics of performing convolutions. There are several different approaches that may be used, and the one that is the easiest will depend upon the form and type of sequences that are to be convolved.

Direct Evaluation

When the sequences that are being convolved may be described by simple closed-form mathematical expressions, the convolution is often most easily performed by directly evaluating the sum given in Eq. (1.7). In performing convolutions directly, it is usually necessary to evaluate finite or infinite sums involving terms of the form α^n or $n\alpha^n$. Listed in Table 1-1 are closed-form expressions for some of the more commonly encountered series.

EXAMPLE 1.4.1   Let us perform the convolution of the two signals

$$x(n) = a^n u(n) = \begin{cases} a^n & n \geq 0 \\ 0 & n < 0 \end{cases}$$

and

$$h(n) = u(n)$$

Table 1-1   Closed-form Expressions for Some Commonly Encountered Series

$$\sum_{n=0}^{N-1} a^n = \frac{1-a^N}{1-a} \qquad\qquad \sum_{n=0}^{\infty} a^n = \frac{1}{1-a} \qquad |a| < 1$$

$$\sum_{n=0}^{N-1} na^n = \frac{(N-1)a^{N+1} - Na^N + a}{(1-a)^2} \qquad\qquad \sum_{n=0}^{\infty} na^n = \frac{a}{(1-a)^2} \qquad |a| < 1$$

$$\sum_{n=0}^{N-1} n = \tfrac{1}{2}N(N-1) \qquad\qquad \sum_{n=0}^{N-1} n^2 = \tfrac{1}{6}N(N-1)(2N-1)$$

With the direct evaluation of the convolution sum we find

$$y(n) = x(n) * h(n) = \sum_{k=-\infty}^{\infty} x(k)h(n-k) = \sum_{k=-\infty}^{\infty} a^k u(k)u(n-k)$$

Because $u(k)$ is equal to zero for $k < 0$ and $u(n-k)$ is equal to zero for $k > n$, when $n < 0$, there are no nonzero terms in the sum and $y(n) = 0$. On the other hand, if $n \geq 0$,

$$y(n) = \sum_{k=0}^{n} a^k = \frac{1-a^{n+1}}{1-a}$$

Therefore,

$$y(n) = \frac{1-a^{n+1}}{1-a}u(n)$$

Graphical Approach

In addition to the direct method, convolutions may also be performed graphically. The steps involved in using the graphical approach are as follows:

1. Plot both sequences, $x(k)$ and $h(k)$, as functions of k.

2. Choose one of the sequences, say $h(k)$, and time-reverse it to form the sequence $h(-k)$.

3. Shift the time-reversed sequence by n. [*Note:* If $n > 0$, this corresponds to a shift to the right (delay), whereas if $n < 0$, this corresponds to a shift to the left (advance).]

4. Multiply the two sequences $x(k)$ and $h(n-k)$ and sum the product for all values of k. The resulting value will be equal to $y(n)$. This process is repeated for all possible shifts, n.

EXAMPLE 1.4.2   To illustrate the graphical approach to convolution, let us evaluate $y(n) = x(n)*h(n)$ where $x(n)$ and $h(n)$ are the sequences shown in Fig. 1-6 (a) and (b), respectively. To perform this convolution, we follow the steps listed above:

1. Because $x(k)$ and $h(k)$ are both plotted as a function of k in Fig. 1-6 (a) and (b), we next choose one of the sequences to reverse in time. In this example, we time-reverse $h(k)$, which is shown in Fig. 1-6 (c).

2. Forming the product, $x(k)h(-k)$, and summing over k, we find that $y(0) = 1$.

3. Shifting $h(k)$ to the right by one results in the sequence $h(1-k)$ shown in Fig. 1-6 (d). Forming the product, $x(k)h(1-k)$, and summing over k, we find that $y(1) = 3$.

4. Shifting $h(1-k)$ to the right again gives the sequence $h(2-k)$ shown in Fig. 1-6 (e). Forming the product, $x(k)h(2-k)$, and summing over k, we find that $y(2) = 6$.

5. Continuing in this manner, we find that $y(3) = 5$, $y(4) = 3$, and $y(n) = 0$ for $n > 4$.

6. We next take $h(-k)$ and shift it to the left by one as shown in Fig. 1-6 (f). Because the product, $x(k)h(-1-k)$, is equal to zero for all k, we find that $y(-1) = 0$. In fact, $y(n) = 0$ for all $n < 0$.

Figure 1-6 (g) shows the convolution for all n.

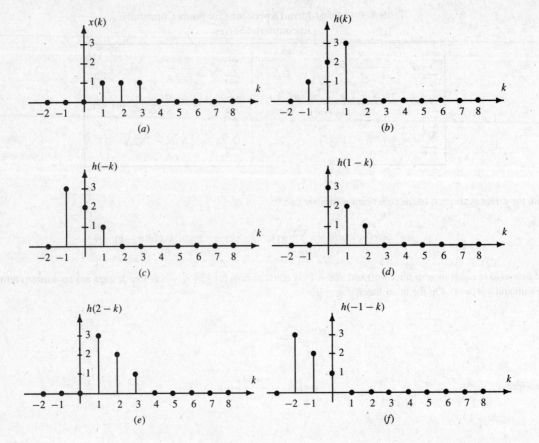

Fig. 1-6. The graphical approach to convolution.

A useful fact to remember in performing the convolution of two finite-length sequences is that if $x(n)$ is of length L_1 and $h(n)$ is of length L_2, $y(n) = x(n) * h(n)$ will be of length

$$L = L_1 + L_2 - 1$$

Furthermore, if the nonzero values of $x(n)$ are contained in the interval $[M_x, N_x]$ and the nonzero values of $h(n)$ are contained in the interval $[M_h, N_h]$, the nonzero values of $y(n)$ will be *confined* to the interval $[M_x + M_h, N_x + N_h]$.

EXAMPLE 1.4.3 Consider the convolution of the sequence

$$x(n) = \begin{cases} 1 & 10 \le n \le 20 \\ 0 & \text{otherwise} \end{cases}$$

with

$$h(n) = \begin{cases} n & -5 \le n \le 5 \\ 0 & \text{otherwise} \end{cases}$$

Because $x(n)$ is zero outside the interval $[10, 20]$, and $h(n)$ is zero outside the interval $[-5, 5]$, the nonzero values of the convolution, $y(n) = x(n) * h(n)$, will be contained in the interval $[5, 25]$.

Slide Rule Method

Another method for performing convolutions, which we call the *slide rule method*, is particularly convenient when both $x(n)$ and $h(n)$ are finite in length and short in duration. The steps involved in the slide rule method are as follows:

1. Write the values of $x(k)$ along the top of a piece of paper, and the values of $h(-k)$ along the top of another piece of paper as illustrated in Fig. 1-7.

2. Line up the two sequence values $x(0)$ and $h(0)$, multiply each pair of numbers, and add the products to form the value of $y(0)$.

3. Slide the paper with the time-reversed sequence $h(k)$ to the right by one, multiply each pair of numbers, sum the products to find the value $y(1)$, and repeat for all shifts to the right by $n > 0$. Do the same, shifting the time-reversed sequence to the left, to find the values of $y(n)$ for $n < 0$.

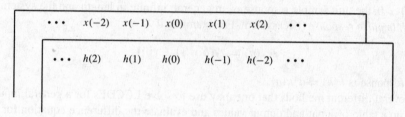

Fig. 1-7.   The slide rule approach to convolution.

In Chap. 2 we will see that another way to perform convolutions is to use the Fourier transform.

1.5   DIFFERENCE EQUATIONS

The convolution sum expresses the output of a linear shift-invariant system in terms of a linear combination of the input values $x(n)$. For example, a system that has a unit sample response $h(n) = \alpha^n u(n)$ is described by the equation

$$y(n) = \sum_{k=0}^{\infty} \alpha^k x(n-k) \qquad (1.9)$$

Although this equation allows one to compute the output $y(n)$ for an arbitrary input $x(n)$, from a computational point of view this representation is not very efficient. In some cases it may be possible to more efficiently express the output in terms of past values of the output in addition to the current and past values of the input. The previous system, for example, may be described more concisely as follows:

$$y(n) = \alpha y(n-1) + x(n) \qquad (1.10)$$

Equation (1.10) is a special case of what is known as a *linear constant coefficient difference equation*, or LCCDE. The general form of a LCCDE is

$$y(n) = \sum_{k=0}^{q} b(k)x(n-k) - \sum_{k=1}^{p} a(k)y(n-k) \qquad (1.11)$$

where the coefficients $a(k)$ and $b(k)$ are constants that define the system. If the difference equation has one or more terms $a(k)$ that are nonzero, the difference equation is said to be *recursive*. On the other hand, if all of the coefficients $a(k)$ are equal to zero, the difference equation is said to be *nonrecursive*. Thus, Eq. (1.10) is an example of a first-order recursive difference equation, whereas Eq. (1.9) is an infinite-order nonrecursive difference equation.

Difference equations provide a method for computing the response of a system, $y(n)$, to an arbitrary input $x(n)$. Before these equations may be solved, however, it is necessary to specify a set of *initial conditions*. For example, with an input $x(n)$ that begins at time $n = 0$, the solution to Eq. (1.11) at time $n = 0$ depends on the values of $y(-1), \ldots, y(-p)$. Therefore, these initial conditions must be specified before the solution for $n \geq 0$ may be found. When these initial conditions are zero, the system is said to be in *initial rest*.

For an LSI system that is described by a difference equation, the unit sample response, $h(n)$, is found by solving the difference equation for $x(n) = \delta(n)$ assuming initial rest. For a nonrecursive system, $a(k) = 0$, the difference equation becomes

$$y(n) = \sum_{k=0}^{q} b(k)x(n-k) \qquad (1.12)$$

and the output is simply a weighted sum of the current and past input values. As a result, the unit sample response is simply

$$h(n) = \sum_{k=0}^{q} b(k)\delta(n-k)$$

Thus, $h(n)$ is finite in length and the system is referred to as a *finite-length impulse response* (FIR) system. However, if $a(k) \neq 0$, the unit sample response is, in general, infinite in length and the system is referred to as an *infinite-length impulse response* (IIR) system. For example, if

$$y(n) = ay(n-1) + x(n)$$

the unit sample response is $h(n) = a^n u(n)$.

There are several different methods that one may use to solve LCCDEs for a general input $x(n)$. The first is to simply set up a table of input and output values and evaluate the difference equation for each value of n. This approach would be appropriate if only a few output values needed to be determined. Another approach is to use z-transforms. This approach will be discussed in Chap. 4. The third is the classical approach of finding the homogeneous and particular solutions, which we now describe.

Given an LCCDE, the general solution is a sum of two parts,

$$y(n) = y_h(n) + y_p(n)$$

where $y_h(n)$ is known as the *homogeneous solution* and $y_p(n)$ is the *particular solution*. The homogeneous solution is the response of the system to the initial conditions, assuming that the input $x(n) = 0$. The particular solution is the response of the system to the input $x(n)$, assuming zero initial conditions.

The homogeneous solution is found by solving the homogeneous difference equation

$$y(n) + \sum_{k=1}^{p} a(k)y(n-k) = 0 \qquad (1.13)$$

The solution to Eq. (*1.13*) may be found by assuming a solution of the form

$$y_h(n) = z^n$$

Substituting this solution into Eq. (*1.13*) we obtain the polynomial equation

$$z^n + \sum_{k=1}^{p} a(k)z^{n-k} = 0$$

or $\qquad z^{n-p}\{z^p + a(1)z^{p-1} + a(2)z^{p-2} + \cdots + a(p-1)z + a(p)\} = 0$

The polynomial in braces is called the *characteristic polynomial*. Because it is of degree p, it will have p roots, which may be either real or complex. If the coefficients $a(k)$ are real-valued, these roots will occur in complex-conjugate pairs (i.e., for each complex root z_i there will be another that is equal to z_i^*). If the p roots z_i are distinct, $z_i \neq z_k$ for $k \neq i$, the general solution to the homogeneous difference equation is

$$y_h(n) = \sum_{k=1}^{p} A_k z_k^n \qquad (1.14)$$

where the constants A_k are chosen to satisfy the initial conditions. For repeated roots, the solution must be modified as follows. If z_1 is a root of multiplicity m with the remaining $p - m$ roots distinct, the homogeneous solution becomes

$$y_h(n) = (A_1 + A_2 n + \cdots + A_m n^{m-1}) z_1^n + \sum_{k=m+1}^{p} A_k z_k^n \qquad (1.15)$$

For the particular solution, it is necessary to find the sequence $y_p(n)$ that satisfies the difference equation for the given $x(n)$. In general, this requires some creativity and insight. However, for many of the typical inputs that we are interested in, the solution will have the same form as the input. Table 1-2 lists the particular solution for some commonly encountered inputs. For example, if $x(n) = a^n u(n)$, the particular solution will be of the form

$$y_p(n) = Ca^n u(n)$$

provided a is not a root of the characteristic equation. The constant C is found by substituting the solution into the difference equation. Note that for $x(n) = C\delta(n)$ the particular solution is zero. Because $x(n) = 0$ for $n > 0$, the unit sample only affects the initial condition of $y(n)$.

**Table 1-2   The Particular Solution to an LCCDE
for Several Different Inputs**

Term in $x(n)$	Particular Solution
C	C_1
Cn	$C_1 n + C_2$
Ca^n	$C_1 a^n$
$C\cos(n\omega_0)$	$C_1 \cos(n\omega_0) + C_2 \sin(n\omega_0)$
$C\sin(n\omega_0)$	$C_1 \cos(n\omega_0) + C_2 \sin(n\omega_0)$
$Ca^n \cos(n\omega_0)$	$C_1 a^n \cos(n\omega_0) + C_2 a^n \sin(n\omega_0)$
$C\delta(n)$	None

EXAMPLE 1.5.1   Let us find the solution to the difference equation

$$y(n) - 0.25y(n - 2) = x(n) \qquad (1.16)$$

for $x(n) = u(n)$ assuming initial conditions of $y(-1) = 1$ and $y(-2) = 0$.

We begin by finding the particular solution. From Table 1-2 we see that for $x(n) = u(n)$

$$y_p(n) = C_1$$

Substituting this solution into the difference equation we find

$$C_1 - 0.25C_1 = 1$$

In order for this to hold, we must have

$$C_1 = \frac{1}{1 - 0.25} = \frac{4}{3}$$

To find the homogeneous solution, we set $y_h(n) = z^n$, which gives the characteristic polynomial

$$z^2 - 0.25 = 0$$

or

$$(z + 0.5)(z - 0.5) = 0$$

Therefore, the homogeneous solution has the form

$$y_h(n) = A_1(0.5)^n + A_2(-0.5)^n$$

Thus, the total solution is

$$y(n) = \tfrac{4}{3} + A_1(0.5)^n + A_2(-0.5)^n \qquad n \ge 0 \qquad (1.17)$$

The constants A_1 and A_2 must now be found so that the total solution satisfies the given initial conditions, $y(-1) = 1$ and $y(-2) = 0$. Because the solution given in Eq. (1.17) only applies for $n \geq 0$, we must derive an equivalent set of initial conditions for $y(0)$ and $y(1)$. Evaluating Eq. (1.16) at $n = 0$ and $n = 1$, we have

$$y(0) - 0.25y(-2) = x(0) = 1$$
$$y(1) - 0.25y(-1) = x(1) = 1$$

Substituting these *derived initial conditions* into Eq. (1.17) we have

$$y(0) = \tfrac{4}{3} + A_1 + A_2 = 1$$
$$y(1) = \tfrac{4}{3} + \tfrac{1}{2}A_1 - \tfrac{1}{2}A_2 = 1$$

Solving for A_1 and A_2 we find

$$A_1 = -\tfrac{1}{2} \qquad A_2 = \tfrac{1}{6}$$

Thus, the solution is

$$y(n) = \tfrac{4}{3} - (0.5)^{n+1} + \tfrac{1}{6}(-0.5)^n \qquad n \geq 0$$

Although we have focused thus far on linear difference equations with constant coefficients, not all systems and not all difference equations of interest are linear, and not all have constant coefficients. A system that computes a running average of a signal $x(n)$ over the interval $[0, n]$, for example, is defined by

$$y(n) = \frac{1}{n+1} \sum_{k=0}^{n} x(k) \qquad n \geq 0$$

This system may be represented by a difference equation that has time-varying coefficients:

$$y(n) = \frac{n}{n+1} y(n-1) + x(n) \qquad n \geq 0$$

Although more complicated and difficult to solve, nonlinear difference equations or difference equations with time-varying coefficients are important and arise frequently in many applications.

Solved Problems

Discrete-Time Signals

1.1 Determine whether or not the signals below are periodic and, for each signal that is periodic, determine the fundamental period.

(a) $x(n) = \cos(0.125\pi n)$

(b) $x(n) = \mathrm{Re}\{e^{jn\pi/12}\} + \mathrm{Im}\{e^{jn\pi/18}\}$

(c) $x(n) = \sin(\pi + 0.2n)$

(d) $x(n) = e^{j\frac{\pi}{16}n} \cos(n\pi/17)$

(a) Because $0.125\pi = \pi/8$, and

$$\cos\left(\frac{\pi}{8}n\right) = \cos\left(\frac{\pi}{8}(n+16)\right)$$

$x(n)$ is periodic with period $N = 16$.

(b) Here we have the sum of two periodic signals,

$$x(n) = \cos(n\pi/12) + \sin(n\pi/18)$$

with the period of the first signal being equal to $N_1 = 24$, and the period of the second, $N_2 = 36$. Therefore, the period of the sum is

$$N = \frac{N_1 N_2}{\gcd(N_1, N_2)} = \frac{(24)(36)}{\gcd(24, 36)} = \frac{(24)(36)}{12} = 72$$

(c) In order for this sequence to be periodic, we must be able to find a value for N such that

$$\sin(\pi + 0.2n) = \sin(\pi + 0.2(n + N))$$

The sine function is periodic with a period of 2π. Therefore, $0.2N$ must be an integer multiple of 2π. However, because π is an irrational number, no integer value of N exists that will make the equality true. Thus, this sequence is aperiodic.

(d) Here we have the product of two periodic sequences with periods $N_1 = 32$ and $N_2 = 34$. Therefore, the fundamental period is

$$N = \frac{(32)(34)}{\gcd(32, 34)} = \frac{(32)(34)}{2} = 544$$

1.2   Find the even and odd parts of the following signals:

(a) $x(n) = u(n)$

(b) $x(n) = \alpha^n u(n)$

The even part of a signal $x(n)$ is given by

$$x_e(n) = \tfrac{1}{2}[x(n) + x(-n)]$$

With $x(n) = u(n)$, we have

$$x_e(n) = \tfrac{1}{2}[u(n) + u(-n)] = \begin{cases} 1 & n = 0 \\ \tfrac{1}{2} & n \neq 0 \end{cases}$$

which may be written concisely as

$$x_e(n) = \tfrac{1}{2} + \tfrac{1}{2}\delta(n)$$

Therefore, the even part of the unit step is a sequence that has a constant value of $\tfrac{1}{2}$ for all n except at $n = 0$, where it has a value of 1.

The odd part of a signal $x(n)$ is given by the difference

$$x_o(n) = \tfrac{1}{2}[x(n) - x(-n)]$$

With $x(n) = u(n)$, this becomes

$$x_o(n) = \begin{cases} \tfrac{1}{2} & n > 0 \\ 0 & n = 0 \\ -\tfrac{1}{2} & n < 0 \end{cases}$$

or

$$x_o(n) = \tfrac{1}{2}\operatorname{sgn}(n)$$

where $\operatorname{sgn}(n)$ is the signum function.

With $x(n) = \alpha^n u(n)$, the even part is

$$x_e(n) = \tfrac{1}{2}[\alpha^n u(n) + \alpha^{-n} u(-n)] = \begin{cases} \tfrac{1}{2}\alpha^n & n > 0 \\ 1 & n = 0 \\ \tfrac{1}{2}\alpha^{-n} & n < 0 \end{cases}$$

or

$$x_e(n) = \tfrac{1}{2}\alpha^{|n|} + \tfrac{1}{2}\delta(n)$$

The odd part, on the other hand, is

$$x_o(n) = \tfrac{1}{2}[\alpha^n u(n) - \alpha^{-n} u(-n)] = \tfrac{1}{2}\alpha^{|n|}\operatorname{sgn}(n)$$

1.3   If $x_1(n)$ is even and $x_2(n)$ is odd, what is $y(n) = x_1(n) \cdot x_2(n)$?

If $y(n) = x_1(n) \cdot x_2(n)$,

$$y(-n) = x_1(-n) \cdot x_2(-n)$$

Because $x_1(n)$ is even, $x_1(n) = x_1(-n)$, and because $x_2(n)$ is odd, $x_2(n) = -x_2(-n)$. Therefore,

$$y(-n) = -x_1(n) \cdot x_2(n) = -y(n)$$

and it follows that $y(n)$ is odd.

1.4   If $x(n) = 0$ for $n < 0$, derive an expression for $x(n)$ in terms of its even part, $x_e(n)$, and, using this expression, find $x(n)$ when $x_e(n) = (0.9)^{|n|}u(n)$. Determine whether or not it is possible to derive a similar expression for $x(n)$ in terms of its odd part.

Because

$$x_e(n) = \tfrac{1}{2}[x(n) + x(-n)]$$

and

$$x_o(n) = \tfrac{1}{2}[x(n) - x(-n)]$$

note that when $x(n) = 0$ for $n < 0$,

$$x_e(n) = \tfrac{1}{2}x(n) \qquad n > 0$$

and

$$x_e(n) = x(n) \qquad n = 0$$

Therefore, $x(n)$ may be recovered from its even part as follows:

$$x(n) = \begin{cases} x_e(n) & n = 0 \\ 2x_e(n) & n > 0 \end{cases}$$

For example, with $x_e(n) = (0.9)^{|n|}u(n)$, we have

$$x(n) = \delta(n) + 2(0.9)^n u(n-1)$$

Unlike the case when only the even part of a sequence is known, if only the odd part is given, it is not possible to recover $x(n)$. The problem is in recovering the value of $x(0)$. Because $x_o(0)$ is always equal to zero, there is no information in the odd part of $x(n)$ about the value of $x(0)$. However, if we were given $x(0)$ along with the odd part, then, $x(n)$ could be recovered for all n.

1.5   If $x_e(n)$ is the conjugate symmetric part of a sequence $x(n)$, what symmetries do the real and imaginary parts of $x_e(n)$ possess?

The conjugate symmetric part of $x(n)$ is

$$x_e(n) = \tfrac{1}{2}[x(n) + x^*(-n)]$$

Expressing $x(n)$ in terms of its real and imaginary parts, we have

$$\begin{aligned} x_e(n) &= \tfrac{1}{2}[x_r(n) + jx_i(n) + \{x_r(-n) + jx_i(-n)\}^*] \\ &= \tfrac{1}{2}[x_r(n) + jx_i(n) + x_r(-n) - jx_i(-n)] \\ &= \tfrac{1}{2}[x_r(n) + x_r(-n)] + \tfrac{1}{2}j[x_i(n) - x_i(-n)] \end{aligned}$$

Therefore, the real part of $x_e(n)$ is even, and the imaginary part is odd.

1.6   Find the conjugate symmetric part of the sequence

$$x(n) = je^{jn\pi/4}$$

The conjugate symmetric part of $x(n)$ is

$$x_e(n) = \tfrac{1}{2}[x(n) + x^*(-n)] = \tfrac{1}{2}\big[je^{jn\pi/4} - je^{jn\pi/4}\big] = 0$$

Thus, this sequence is conjugate antisymmetric.

1.7   Given the sequence $x(n) = (6 - n)[u(n) - u(n - 6)]$, make a sketch of

(a) $y_1(n) = x(4 - n)$ (b) $y_2(n) = x(2n - 3)$

(c) $y_3(n) = x(8 - 3n)$ (d) $y_4(n) = x(n^2 - 2n + 1)$

(a) The sequence $x(n)$, illustrated in Fig. 1-8(a), is a linearly decreasing sequence that begins at index $n = 0$ and ends at index $n = 5$. The first sequence that is to be sketched, $y_1(n) = x(4 - n)$, is found by shifting $x(n)$ by four and time-reversing. Observe that at index $n = 4$, $y_1(n)$ is equal to $x(0)$. Therefore, $y_1(n)$ has a value of 6 at $n = 4$ and decreases linearly to the left (decreasing values of n) until $n = -1$, beyond which $y_1(n) = 0$. The sequence $y_1(n)$ is shown in Fig. 1-8(b).

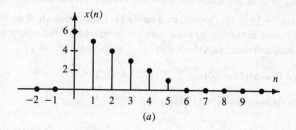

(a)

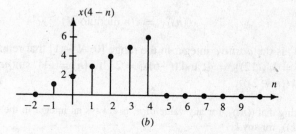

(b)

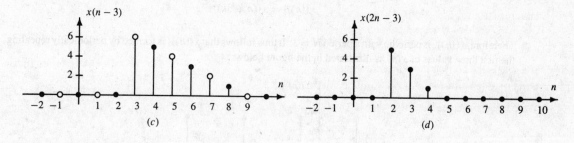

(c)   (d)

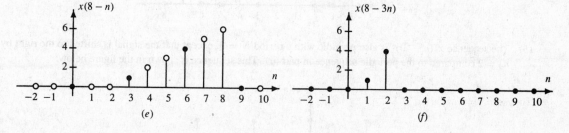

(e)   (f)

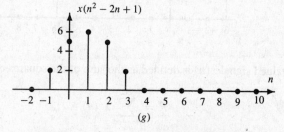

(g)

Fig. 1-8. Performing signal manipulations.

(b) The second sequence, $y_2(n) = x(2n - 3)$, is formed through the combination of time-shifting and downsampling. Therefore, $y_2(n)$ may be plotted by first shifting $x(n)$ to the right by three (delay) as shown in

Fig. 1-8(c). The sequence $y_2(n)$ is then formed by down-sampling by a factor of 2 (i.e., keeping only the even index terms as indicated by the solid circles in Fig. 1-8(c)). A sketch of $y_2(n)$ is shown in Fig. 1-8(d).

(c) The third sequence, $y_3(n) = x(8 - 3n)$, is formed through a combination of time-shifting, down-sampling, and time-reversal. To sketch $y_3(n)$ we begin by plotting $x(8 - n)$, which is formed by shifting $x(n)$ to the left by eight (advance) and reversing in time as shown in Fig. 1-8(e). Then, $y_3(n)$ is found by extracting every third sample of $x(8 - n)$, as indicated by the solid circles, which is plotted in Fig. 1-8(f).

(d) Finally, $y_4(n) = x(n^2 - 2n + 1)$ is formed by a nonlinear transformation of the time variable n. This sequence may be easily sketched by listing how the index n is mapped. First, note that if $n \geq 4$ or $n \leq -2$, then $n^2 - 2n + 1 \geq 9$ and, therefore, $y_4(n) = 0$. For $-1 \leq n \leq 3$ we have

$$y_4(-1) = y_4(3) = x(4) = 2 \qquad y_4(0) = y_4(2) = x(1) = 5 \qquad y_4(1) = x(0) = 6$$

The sequence $y_4(n)$ is sketched in Fig. 1-8(g).

1.8   The notation $x((n))_N$ is used to define the sequence that is formed as follows:

$$x((n))_N = x(n \text{ modulo } N)$$

where $(n \text{ modulo } N)$ is the *positive* integer in the range $[0, N - 1]$ that remains after dividing n by N. For example, $((3))_8 = 3$, $((12))_8 = 4$, and $((-6))_4 = 2$. If $x(n) = (\frac{1}{2})^n \sin(n\pi/2)u(n)$, make a sketch of (a) $x((n))_3$ and (b) $x((n - 2))_3$.

(a) We begin by noting that $((n))_3$, for any value of n, is always an integer in the range $[0, 2]$. In fact, because $((n))_3 = ((n + 3k))_3$ for any k,

$$x((n))_3 = x((n + 3k))_3$$

Therefore, $x((n))_3$ is periodic with a period $N = 3$. It thus follows that $x((n))_3$ is formed by periodically repeating the first three values of $x(n)$ as illustrated in the figure below:

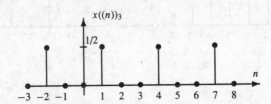

(b) The sequence $x((n - 2))_3$ is also periodic with a period $N = 3$, except that the signal is shifted to the right by $n_0 = 2$ compared to the periodic sequence in part (a). This sequence is shown in the figure below:

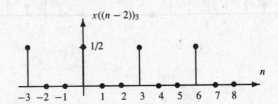

1.9   The power in a real-valued signal $x(n)$ is defined as the sum of the squares of the sequence values:

$$P = \sum_{n=-\infty}^{\infty} x^2(n)$$

Suppose that a sequence $x(n)$ has an even part $x_e(n)$ equal to

$$x_e(n) = \left(\tfrac{1}{2}\right)^{|n|}$$

If the power in $x(n)$ is $P = 5$, find the power in the odd part, $x_o(n)$, of $x(n)$.

This problem requires finding the relationship between the power in $x(n)$ and the power in the even and odd parts. By definition, $x(n) = x_e(n) + x_o(n)$. Therefore,

$$P = \sum_{n=-\infty}^{\infty} x^2(n) = \sum_{n=-\infty}^{\infty} [x_e(n) + x_o(n)]^2$$

$$= \sum_{n=-\infty}^{\infty} x_e^2(n) + \sum_{n=-\infty}^{\infty} x_o^2(n) + \sum_{n=-\infty}^{\infty} 2x_e(n)x_o(n)$$

Note that $x_e(n)x_o(n)$ is the product of an even sequence and an odd sequence and, therefore, the product is odd. Because the sum for all n of an odd sequence is equal to zero,

$$\sum_{n=-\infty}^{\infty} 2x_e(n)x_o(n) = 0$$

Thus, the power in $x(n)$ is

$$P = \sum_{n=-\infty}^{\infty} x_e^2(n) + \sum_{n=-\infty}^{\infty} x_o^2(n)$$

which says that the power in $x(n)$ is equal to the sum of the powers in its even and odd parts. Evaluating the power in the even part of $x(n)$, we find

$$P_e = \sum_{n=-\infty}^{\infty} \left(\tfrac{1}{2}\right)^{2|n|} = -1 + 2\sum_{n=0}^{\infty} \left(\tfrac{1}{2}\right)^{2n} = \tfrac{5}{3}$$

Therefore, with $P = 5$ we have

$$P_o = 5 - P_e = \tfrac{10}{3}$$

1.10   Consider the sequence

$$x(n) = \left(\tfrac{3}{2}\right)^n u(-n)$$

(a)   Find the numerical value of

$$A = \sum_{n=-\infty}^{\infty} x(n)$$

(b)   Compute the power in $x(n)$,

$$P = \sum_{n=-\infty}^{\infty} x^2(n)$$

(c)   If $x(n)$ is input to a time-varying system defined by $y(n) = nx(n)$, find the power in the output signal (i.e., evaluate the sum)

$$P = \sum_{n=-\infty}^{\infty} y^2(n)$$

(a)   This is a direct application of the geometric series

$$A = \sum_{n=-\infty}^{\infty} \left(\tfrac{3}{2}\right)^n u(-n) = \sum_{n=-\infty}^{0} \left(\tfrac{3}{2}\right)^n$$

With the substitution of $-n$ for n we have

$$A = \sum_{n=0}^{\infty} \left(\tfrac{3}{2}\right)^{-n} = \sum_{n=0}^{\infty} \left(\tfrac{2}{3}\right)^n$$

Therefore, it follows from the geometric series that

$$A = \frac{1}{1 - \tfrac{2}{3}} = 3$$

(b) To find the power in $x(n)$ we must evaluate the sum

$$P = \sum_{n=-\infty}^{\infty} x^2(n) = \sum_{n=-\infty}^{0} \left(\tfrac{3}{2}\right)^{2n}$$

Replacing n by $-n$ and using the geometric series, this sum becomes

$$P = \sum_{n=0}^{\infty} \left(\tfrac{3}{2}\right)^{-2n} = \sum_{n=0}^{\infty} \left(\tfrac{2}{3}\right)^{2n} = \frac{1}{1 - \tfrac{4}{9}} = \tfrac{9}{5}$$

(c) Finally, to find the power in $y(n) = nx(n)$ we must evaluate the sum

$$P = \sum_{n=-\infty}^{\infty} [nx(n)]^2 = \sum_{n=0}^{\infty} n^2 \left(\tfrac{4}{9}\right)^n \tag{1.18}$$

In Table 1-1 there is a closed-form expression for the sum

$$\sum_{n=0}^{\infty} n a^n = \frac{a}{(1-a)^2} \qquad |a| < 1 \tag{1.19}$$

but not for $\sum_{k=0}^{\infty} n^2 a^n$. However, we may derive a closed-form expression for this sum as follows.[5] Differentiating both sides of Eq. (1.19) with respect to a, we have

$$\sum_{n=0}^{\infty} n^2 a^{n-1} = \frac{d}{da} \frac{a}{(1-a)^2} = \frac{1+a}{(1-a)^3}$$

Therefore, we have the sum

$$\sum_{n=0}^{\infty} n^2 a^n = \frac{a(1+a)}{(1-a)^3}$$

Using this expression to evaluate Eq. (1.18), we find

$$P = \sum_{n=0}^{\infty} n^2 \left(\tfrac{4}{9}\right)^n = \frac{\left(\tfrac{4}{9}\right)\left(\tfrac{13}{9}\right)}{\left(\tfrac{5}{9}\right)^3} = \frac{468}{125}$$

1.11 Express the sequence

$$x(n) = \begin{cases} 1 & n = 0 \\ 2 & n = 1 \\ 3 & n = 2 \\ 0 & \text{else} \end{cases}$$

as a sum of scaled and shifted unit steps.

In this problem, we would like to perform a signal decomposition, expressing $x(n)$ as a sum of scaled and shifted unit steps. There are several ways to derive this decomposition. One way is to express $x(n)$ as a sum of weighted and shifted unit samples,

$$x(n) = \delta(n) + 2\delta(n-1) + 3\delta(n-2)$$

and use the fact that a unit sample may be written as the difference of two steps as follows:

$$\delta(n) = u(n) - u(n-1)$$

Therefore, $x(n) = u(n) - u(n-1) + 2[u(n-1) - u(n-2)] + 3[u(n-2) - u(n-3)]$

which gives the desired decomposition:

$$x(n) = u(n) + u(n-1) + u(n-2) - 3u(n-3)$$

---

[5]This method is very useful and should be remembered.

Another way to derive this decomposition more directly is as follows. First, we note that the decomposition should begin with a unit step, which generates a value of 1 at index $n = 0$. Because $x(n)$ increases to a value of 2 at $n = 1$, we must add a delayed unit step $u(n - 1)$. At $n = 2$, $x(n)$ again increases in amplitude by 1, so we add the delayed unit step $u(n - 2)$. At this point, we have

$$u(n) + u(n - 1) + u(n - 2) = \begin{cases} 1 & n = 0 \\ 2 & n = 1 \\ 3 & n \geq 2 \end{cases}$$

Thus, all that remains is to bring the sequence back to zero for $n \geq 3$. This may be done by subtracting the delayed unit step $3u(n - 3)$, which produces the same decomposition as before.

Discrete-Time Systems

1.12   For each of the systems below, $x(n)$ is the input and $y(n)$ is the output. Determine which systems are homogeneous, which systems are additive, and which are linear.

(a) $y(n) = \log(x(n))$

(b) $y(n) = 6x(n + 2) + 4x(n + 1) + 2x(n) + 1$

(c) $y(n) = 6x(n) + [x(n + 1)x(n - 1)]/x(n)$

(d) $y(n) = x(n)\sin(n\pi/2)$

(e) $y(n) = \text{Re}\{x(n)\}$

(f) $y(n) = \frac{1}{2}[x(n) + x^*(-n)]$

(a)   If the system is homogeneous,

$$y(n) = T[cx(n)] = cT[x(n)]$$

for any input $x(n)$ and for all complex constants c. The system $y(n) = \log(x(n))$ is not homogeneous because the response of the system to $x_1(n) = cx(n)$ is

$$y_1(n) = \log(x_1(n)) = \log(cx(n)) = \log c + \log(x(n))$$

which is not equal to $c\log(x(n))$. For the system to be additive, if $y_1(n)$ and $y_2(n)$ are the responses to the inputs $x_1(n)$ and $x_2(n)$, respectively, the response to $x(n) = x_1(n) + x_2(n)$ must be $y(n) = y_1(n) + y_2(n)$. For this system we have

$$T[x_1(n) + x_2(n)] = \log[x_1(n) + x_2(n)] \neq \log[x_1(n)] + \log[x_2(n)]$$

Therefore, the system is not additive. Finally, because the system is neither additive nor homogeneous, the system is nonlinear.

(b)   Note that if $y(n)$ is the response to $x(n)$,

$$y(n) = 6x(n + 2) + 4x(n + 1) + 2x(n) + 1$$

the response to $x_1(n) = cx(n)$ is

$$y_1(n) = 6x_1(n + 2) + 4x_1(n + 1) + 2x_1(n) + 1$$
$$= c\{6x(n + 2) + 4x(n + 1) + 2x(n)\} + 1$$

However, $$cy(n) = c\{6x(n + 2) + 4x(n + 1) + 2x(n) + 1\}$$

which is not the same as $y_1(n)$. Therefore, this system is not homogeneous. Similarly, note that the response to $x(n) = x_1(n) + x_2(n)$ is

$$y(n) = 6x(n + 2) + 4x(n + 1) + 2x(n) + 1$$
$$= 6\{x_1(n + 2) + x_2(n + 2)\} + 4\{x_1(n + 1) + x_2(n + 1)\} + 2\{x_1(n) + x_2(n)\} + 1$$
$$= y_1(n) + y_2(n) - 1$$

which is not equal to $y_1(n) + y_2(n)$. Therefore, this system is not additive and, as a result, is nonlinear.

(c) This system is homogeneous, because the response of the system to $x_1(n) = cx(n)$ is

$$y_1(n) = 6x_1(n) + \frac{x_1(n+1)x_1(n-1)}{x_1(n)}$$

$$= c\left[6x(n) + \frac{x(n+1)x(n-1)}{x(n)}\right] = cy(n)$$

The system is clearly, however, not additive and therefore is nonlinear.

(d) Let $y_1(n)$ and $y_2(n)$ be the responses of the system to the inputs $x_1(n)$ and $x_2(n)$, respectively. The response to the input

$$x(n) = ax_1(n) + bx_2(n)$$

is

$$y(n) = x(n)\sin\left(\frac{n\pi}{2}\right) = [ax_1(n) + bx_2(n)]\sin\left(\frac{n\pi}{2}\right)$$

$$= ax_1(n)\sin\left(\frac{n\pi}{2}\right) + bx_2(n)\sin\left(\frac{n\pi}{2}\right) = ay_1(n) + by_2(n)$$

(1.20)

Thus, it follows that this system is linear and, therefore, additive and homogeneous.

(e) Because the real part of the sum of two numbers is the sum of the real parts, if $y_1(n)$ is the response of the system to $x_1(n)$, and $y_2(n)$ is the response to $x_2(n)$, the response to $y(n) = y_1(n) + y_2(n)$ is

$$y(n) = \text{Re}\{x_1(n) + x_2(n)\} = \text{Re}\{x_1(n)\} + \text{Re}\{x_2(n)\} = y_1(n) + y_2(n)$$

Therefore the system is additive. It is not homogeneous, however, because

$$\text{Re}\{cx(n)\} \neq c\text{Re}\{x(n)\}$$

unless c is real. Thus, this system is nonlinear.

(f) For an input $x(n)$, this system produces an output that is the conjugate symmetric part of $x(n)$. If c is a complex constant, and if the input to the system is $x_1(n) = cx(n)$, the output is

$$y_1(n) = \tfrac{1}{2}\left[x_1(n) + x_1^*(-n)\right] = \tfrac{1}{2}[cx(n) + c^*x^*(-n)] \neq cy(n)$$

Therefore, this system is not homogeneous. This system is, however, additive because

$$T[x_1(n) + x_2(n)] = \tfrac{1}{2}\{[x_1(n) + x_2(n)] + [x_1(-n) + x_2(-n)]^*\}$$

$$= \tfrac{1}{2}\{[x_1(n) + x_1^*(-n)] + [x_2(n) + x_2^*(-n)]\}$$

$$= T[x_1(n)] + T[x_2(n)]$$

1.13 A linear system is one that is both homogeneous and additive.

(a) Give an example of a system that is homogeneous but not additive.

(b) Give an example of a system that is additive but not homogeneous.

There are many different systems that are either homogeneous or additive but not both. One example of a system that is homogeneous but not additive is the following:

$$y(n) = \frac{x(n-1)x(n)}{x(n+1)}$$

Specifically, note that if $x(n)$ is multiplied by a complex constant c, the output will be

$$y(n) = \frac{cx(n-1)\,cx(n)}{cx(n+1)} = c\frac{x(n-1)x(n)}{x(n+1)}$$

which is c times the response to $x(n)$. Therefore, the system is homogeneous. On the other hand, it should be clear that the system is not additive because, in general,

$$\frac{\{x_1(n-1) + x_2(n-1)\}\{x_1(n) + x_2(n)\}}{x_1(n+1) + x_2(n+1)} \neq \frac{x_1(n-1)x_1(n)}{x_1(n+1)} + \frac{x_2(n-1)x_2(n)}{x_2(n+1)}$$

An example of a system that is additive but not homogeneous is

$$y(n) = \text{Im}\{x(n)\}$$

Additivity follows from the fact that the imaginary part of a sum of complex numbers is equal to the sum of imaginary parts. This system is not homogeneous, however, because

$$y(n) = \text{Im}\{jx(n)\} \neq j\text{Im}\{x(n)\}$$

1.14 Determine whether or not each of the following systems is shift-invariant:

(a) $y(n) = x(n) + x(n-1) + x(n-2)$

(b) $y(n) = x(n)u(n)$

(c) $y(n) = \sum_{k=-\infty}^{n} x(k)$

(d) $y(n) = x(n^2)$

(e) $y(n) = x((n))_N$ (i.e., $y(n) = x(n \text{ modulo } N)$ as discussed in Prob. 1.8)

(f) $y(n) = x(-n)$

(a) Let $y(n)$ be the response of the system to an arbitrary input $x(n)$. To test for shift-invariance we want to compare the shifted response $y(n - n_0)$ with the response of the system to the shifted input $x(n - n_0)$. With

$$y(n) = x(n) + x(n-1) + x(n-2)$$

we have, for the shifted response,

$$y(n - n_0) = x(n - n_0) + x(n - n_0 - 1) + x(n - n_0 - 2)$$

Now, the response of the system to $x_1(n) = x(n - n_0)$ is

$$y_1(n) = x_1(n) + x_1(n-1) + x_1(n-2)$$
$$= x(n - n_0) + x(n - n_0 - 1) + x(n - n_0 - 2)$$

Because $y_1(n) = y(n - n_0)$, the system is shift-invariant.

(b) This system is a special case of a more general system that has an input-output description given by

$$y(n) = x(n)f(n)$$

where $f(n)$ is a shift-varying *gain*. Systems of this form are always shift-varying provided $f(n)$ is not a constant. To show this, assume that $f(n)$ is not constant and let n_1 and n_2 be two indices for which $f(n_1) \neq f(n_2)$. With an input $x_1(n) = \delta(n - n_1)$, note that the output $y_1(n)$ is

$$y_1(n) = f(n_1)\delta(n - n_1)$$

If, on the other hand, the input is $x_2(n) = \delta(n - n_2)$, the response is

$$y_2(n) = f(n_2)\delta(n - n_2)$$

Although $x_1(n)$ and $x_2(n)$ differ only by a shift, the responses $y_1(n)$ and $y_2(n)$ differ by a shift and a change in amplitude. Therefore, the system is shift-varying.

(c) Let

$$y(n) = \sum_{k=-\infty}^{n} x(k)$$

be the response of the system to an arbitrary input $x(n)$. The response of the system to the shifted input $x_1(n) = x(n - n_0)$ is

$$y_1(n) = \sum_{k=-\infty}^{n} x_1(k) = \sum_{k=-\infty}^{n} x(k - n_0) = \sum_{k=-\infty}^{n-n_0} x(k)$$

Because this is equal to $y(n - n_0)$, the system is shift-invariant.

(d) This system is shift-varying, which may be shown with a simple counterexample. Note that if $x(n) = \delta(n)$, the response will be $y(n) = \delta(n)$. However, if $x_1(n) = \delta(n-2)$, the response will be $y_1(n) = x_1(n^2) = \delta(n^2-2) = 0$, which is not equal to $y(n-2)$. Therefore, the system is shift-varying.

(e) With $y(n)$ the response to $x(n)$, note that for the input $x_1(n) = x(n - N)$, the output is

$$y_1(n) = x((n - N))_N = x((n))_N$$

which is the same as the response to $x(n)$. Because $y_1(n) \neq y(n-N)$, in general, this system is not shift-invariant.

(f) This system may easily be shown to be shift-varying with a counterexample. However, suppose we use the direct approach and let $x(n)$ be an input and $y(n) = x(-n)$ be the response. If we consider the shifted input, $x_1(n) = x(n - n_0)$, we find that the response is

$$y_1(n) = x_1(-n) = x(-n - n_0)$$

However, note that if we shift $y(n)$ by n_0,

$$y(n - n_0) = x(-(n - n_0)) = x(-n + n_0)$$

which is not equal to $y_1(n)$. Therefore, the system is shift-varying.

1.15 A linear discrete-time system is characterized by its response $h_k(n)$ to a delayed unit sample $\delta(n - k)$. For each linear system defined below, determine whether or not the system is shift-invariant.

(a) $h_k(n) = (n - k)u(n - k)$

(b) $h_k(n) = \delta(2n - k)$

(c) $h_k(n) = \begin{cases} \delta(n - k - 1) & k \text{ even} \\ 5u(n - k) & k \text{ odd} \end{cases}$

(a) Note that $h_k(n)$ is a function of $n - k$. This suggests that the system is shift-invariant. To verify this, let $y(n)$ be the response of the system to $x(n)$:

$$y(n) = \sum_{k=-\infty}^{\infty} h_k(n)x(k)$$

$$= \sum_{k=-\infty}^{\infty} (n - k)u(n - k)x(k) = \sum_{k=-\infty}^{n} (n - k)x(k) \qquad (1.21)$$

The response to a shifted input, $x(n - n_0)$, is

$$y_1(n) = \sum_{k=-\infty}^{\infty} x(k - n_0)h_k(n) = \sum_{k=-\infty}^{\infty} (n - k)u(n - k)x(k - n_0)$$

$$= \sum_{k=-\infty}^{n} (n - k)x(k - n_0)$$

With the substitution $l = k - n_0$ this becomes

$$y_1(n) = \sum_{l=-\infty}^{n-n_0} (n - n_0 - l)x(l)$$

From the expression for $y(n)$ given in Eq. (1.21), we see that

$$y(n - n_0) = \sum_{k=-\infty}^{n-n_0} (n - n_0 - k)x(k)$$

which is the same as $y_1(n)$. Therefore, this system is shift-invariant.

(b) For the second system, $h_k(n)$ is *not* a function of $n - k$. Therefore, we should expect this system to be shift-varying. Let us see if we can find an example that demonstrates that it is a shift-varying system. For the input $x(n) = \delta(n)$, the response is

$$y(n) = h_0(n) = \delta(2n) = \begin{cases} 1 & n = 0 \\ 0 & \text{else} \end{cases}$$

If we delay $x(n)$ by 1, the response to $x_1(n) = \delta(n - 1)$ is

$$y_1(n) = h_1(n) = \delta(2n - 1) = 0$$

Because $y_1(n) \neq y(n - 1)$, the system is shift-varying.

(c) Finally, for the last system, we see that although $h_k(n)$ is a function of $n - k$ for k even and a function of $(n - k)$ for k odd,

$$h_k(n) \neq h_{k-1}(n - 1)$$

In other words, the response of the system to $\delta(n - k - 1)$ is not equal to the response of the system to $\delta(n - k)$ delayed by 1. Therefore, this system is shift-varying.

1.16 Let $T[\cdot]$ be a linear system, not necessarily shift-invariant, that has a response $h_k(n)$ to the input $\delta(n - k)$. Derive a test in terms of $h_k(n)$ that allows one to determine whether or not the system is stable and whether or not the system is causal.

(a) The response of a linear system to an input $x(n)$ is

$$y(n) = \sum_{k=-\infty}^{\infty} h_k(n) x(k) \qquad (1.22)$$

Therefore, the output may be bounded as follows:

$$|y(n)| = \left| \sum_{k=-\infty}^{\infty} h_k(n) x(k) \right| \leq \sum_{k=-\infty}^{\infty} |h_k(n)| |x(k)|$$

If $x(n)$ is bounded, $|x(n)| \leq A < \infty$,

$$|y(n)| \leq A \sum_{k=-\infty}^{\infty} |h_k(n)|$$

As a result, if

$$\sum_{k=-\infty}^{\infty} |h_k(n)| \leq B < \infty \qquad \text{all } n \qquad (1.23)$$

the output will be bounded, and the system is stable. Equation (1.23) is a necessary condition for stability. To establish the sufficiency of this condition, we will show that if this summation is *not* finite, we can find a bounded input that will produce an unbounded output. Let us assume that $h_k(n)$ is bounded for all k and n [otherwise the system will be unstable, because the response to the bounded input $\delta(n - k)$ will be unbounded]. With $h_k(n)$ bounded for all k and n, suppose that the sum in Eq. (1.23) is unbounded for some n, say $n = n_0$. Let

$$x(n) = \text{sgn}\{h_n(n_0)\}$$

that is,

$$x(n) = \begin{cases} 1 & h_n(n_0) > 0 \\ 0 & h_n(n_0) = 0 \\ -1 & h_n(n_0) < 0 \end{cases}$$

For this input, the response at time $n = n_0$ is

$$y(n_0) = \sum_{k=-\infty}^{\infty} h_k(n_0) x(k) = \sum_{k=-\infty}^{\infty} h_k(n_0) \, \text{sgn}\{h_k(n_0)\} = \sum_{k=-\infty}^{\infty} |h_k(n_0)|$$

which, by assumption, is unbounded. Therefore, the system is unstable and we have established the sufficiency of the condition given in Eq. (1.23).

(b) Let us now consider causality. For an input $x(n)$, the response is as given in Eq. (1.22). In order for a system to be causal, the output $y(n)$ at time n_0 cannot depend on the input $x(n)$ for any $n > n_0$. Therefore, Eq. (1.22) must be of the form

$$y(n) = \sum_{k=-\infty}^{n} h_k(n)x(k)$$

This, however, will be true for any $x(n)$ if and only if

$$h_k(n) = 0 \qquad n < k$$

which is the desired test for causality.

1.17 Determine whether or not the systems defined in Prob. 1.15 are (a) stable and (b) causal.

(a) For the first system, $h_k(n) = (n - k)u(n - k)$, note that $h_k(n)$ grows linearly with n. Therefore, this system cannot be stable. For example, note that if $x(n) = \delta(n)$, the output will be

$$y(n) = h_0(n) = nu(n)$$

which is unbounded. Alternatively, we may use the test derived in Prob. 1.16 to check for stability. Because

$$\sum_{k=-\infty}^{\infty} |h_k(n)| = \sum_{k=-\infty}^{n} |n - k| = \sum_{k=0}^{\infty} |k| = \infty$$

this system is unstable. On the other hand, because $h_k(n) = 0$ for $n < k$, this system is causal.

(b) For the second system, $h_k(n) = \delta(2n - k)$, note that $h_k(n)$ has, at most, one nonzero value, and this nonzero value is equal to 1. Therefore,

$$\sum_{k=-\infty}^{\infty} |h_k(n)| \leq 1$$

for all n, and the system is stable. However, the system is not causal. To show this, note that if $x(n) = \delta(n - 2)$, the response is

$$y(n) = h_2(n) = \delta(2n - 2) = \delta(n - 1)$$

Because the system produces a response *before* the input occurs, it is noncausal.

(c) For the last system, note that

$$\sum_{k=-\infty}^{\infty} |h_k(n)| = \sum_{\substack{k=-\infty \\ k \text{ even}}}^{\infty} |h_k(n)| + \sum_{\substack{k=-\infty \\ k \text{ odd}}}^{\infty} |h_k(n)| \leq \sum_{\substack{k=-\infty \\ k \text{ odd}}}^{\infty} |h_k(n)|$$

$$= \sum_{\substack{k=-\infty \\ k \text{ odd}}}^{\infty} 5u(n - k) = \sum_{\substack{k=-\infty \\ k \text{ odd}}}^{n} 5$$

which is unbounded. Therefore, this system is unstable. Finally, because $h_k(n) = 0$ for $n < k$, the system is causal.

1.18 Consider a linear system that has a response to a delayed unit step given by

$$s_k(n) = k\delta(n - k)$$

That is, $s_k(n)$ is the response of the linear system to the input $x(n) = u(n - k)$. Find the response of this system to the input $x(n) = \delta(n - k)$, where k is an arbitrary integer, and determine whether or not this system is shift-invariant, stable, or causal.

Because this system is linear, we may find the response, $h_k(n)$, to the input $\delta(n - k)$ as follows. With $\delta(n - k) = u(n - k) - u(n - k - 1)$, using linearity it follows that

$$h_k(n) = s_k(n) - s_{k+1}(n) = k\delta(n - k) - (k + 1)\delta(n - k - 1)$$

which is shown below:

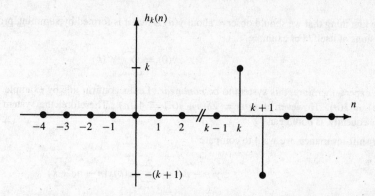

From this plot, we see that the system is not shift-invariant, because the response of the system to a unit sample changes in amplitude as the unit sample is advanced or delayed. However, because $h_k(n) = 0$ for $n < k$, the system is causal. Finally, because $h_k(n)$ is unbounded as a function of k, it follows that the system is unstable. In particular, note that the test for stability of a linear system derived in Prob. 1.16 requires that

$$\max_n \sum_{k=-\infty}^{\infty} |h_k(n)| \leq B < \infty$$

For this system,

$$\sum_{k=-\infty}^{\infty} |h_k(n)| = |2n|$$

Note that in evaluating this sum, we are summing over k. This is most easily performed by plotting $h_k(n)$ versus n as illustrated in the figure below.

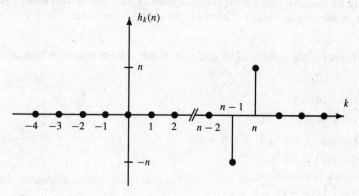

Because this sum cannot be bounded by a finite number B, this system is unstable. Because this system is unstable, we should be able to find a bounded input that produces an unbounded output. One such sequence is the following:

$$x(n) = \sum_{k=0}^{\infty} \delta(n - 2k)$$

The response is

$$y(n) = n(-1)^n u(n)$$

which is clearly unbounded.

1.19 Consider a system whose output $y(n)$ is related to the input $x(n)$ by

$$y(n) = \sum_{k=-\infty}^{\infty} x(k)x(n + k)$$

Determine whether or not the system is (a) linear, (b) shift-invariant, (c) stable, (d) causal.

(a) The first thing that we should observe about $y(n)$ is that it is formed by summing products of $x(n)$ with shifted versions of itself. For example,

$$y(0) = \sum_{k=-\infty}^{\infty} x^2(k)$$

We expect, therefore, this system to be *nonlinear*. Let us confirm this by example. Note that if $x(n) = \delta(n)$, $y(n) = \delta(n)$. However, if $x(n) = 2\delta(n)$, $y(n) = 4\delta(n)$. Therefore, the system is not homogeneous and, consequently, is nonlinear.

(b) For shift-invariance, we want to compare

$$y(n - n_0) = \sum_{k=-\infty}^{\infty} x(k)x(n - n_0 + k)$$

to the response of the system to $x_1(n) = x(n - n_0)$, which is

$$y_1(n) = \sum_{k=-\infty}^{\infty} x_1(k)x_1(n + k)$$

$$= \sum_{k=-\infty}^{\infty} x(k - n_0)x(n + k - n_0)$$

$$= \sum_{k'=-\infty}^{\infty} x(k')x(n + k')$$

where the last equality follows with the substitution $k' = k - n_0$. Because $y_1(n) \neq y(n - n_0)$, this system is not shift-invariant.

(c) For stability, note that if $x(n)$ is a unit step, $y(0)$ is unbounded. Therefore, this system is unstable.

(d) Finally, for causality, note that the output depends on the values of $x(n)$ for all n. For example, $y(0)$ is the sum of the squares of $x(k)$ for all k. Therefore, this system is not causal.

1.20 Given that $x(n)$ is the system input and $y(n)$ is the system output, which of the following systems are causal?

(a) $y(n) = x^2(n)u(n)$

(b) $y(n) = x(|n|)$

(c) $y(n) = x(n) + x(n - 3) + x(n - 10)$

(d) $y(n) = x(n) - x(n^2 - n)$

(e) $y(n) = \prod_{k=1}^{N} x(n - k)$

(f) $y(n) = \sum_{k=n}^{\infty} x(n - k)$

(a) The system $y(n) = x^2(n)u(n)$ is *memoryless* (i.e., the response of the system at time n depends only on the input at time n and on no other values of the input). Therefore, this system is causal.

(b) The system $y(n) = x(|n|)$ is an example of a noncausal system. This may be seen by looking at the output when $n < 0$. In particular, note that $y(-1) = x(1)$. Therefore, the output of the system at time $n = -1$ depends on the value of the input at a future time.

(c) For this system, in order to compute the output $y(n)$ at time n all we need to know is the value of the input $x(n)$ at times n, $n - 3$, and $n - 10$. Therefore, this system must be causal.

(d) This system is noncausal, which may be seen by evaluating $y(n)$ for $n < 0$. For example,

$$y(-1) = x(-1) - x(2)$$

Because $y(-1)$ depends on the value of $x(2)$, which occurs after time $n = -1$, this system is noncausal.

(e) The output of this system at time n is the product of the values of the input $x(n)$ at times $n-1, \ldots, n-N$. Therefore, because the output depends only on past values of the input signal, the system is causal.

(f) This system is not causal, which may be seen easily if we rewrite the system definition as follows:

$$y(n) = \sum_{k=n}^{\infty} x(n-k) = \sum_{l=-\infty}^{0} x(l)$$

Therefore, the input must be known for all $n \leq 0$ to determine the output at time n. For example, to find $y(-5)$ we must know $x(0), x(-1), x(-2), \ldots$. Thus, the system is noncausal.

1.21 Determine which of the following systems are stable:

(a) $y(n) = x^2(n)$

(b) $y(n) = e^{x(n)}/x(n-1)$

(c) $y(n) = \cos(x(n))$

(d) $y(n) = \sum_{k=-\infty}^{n} x(k)$

(e) $y(n) = \log(1 + |x(n)|)$

(f) $y(n) = x(n) * \cos(n\pi/8)$

(a) Let $x(n)$ be any bounded input with $|x(n)| < M$. Then it follows that the output, $y(n) = x^2(n)$, may be bounded by

$$|y(n)| = |x(n)|^2 < M^2$$

Therefore, this system is stable.

(b) This system is clearly not stable. For example, note that the response of the system to a unit sample $x(n) = \delta(n)$ is infinite for all values of n except $n = 1$.

(c) Because $|\cos(x)| \leq 1$ for all x, this system is stable.

(d) This system corresponds to a digital integrator and is unstable. Consider, for example, the step response of the system. With $x(n) = u(n)$ we have, for $n \geq 0$,

$$y(n) = \sum_{k=-\infty}^{n} u(k) = (n+1)$$

Although the input is bounded, $|x(n)| \leq 1$, the response of the system is unbounded.

(e) This system may be shown to be stable by using the following inequality:

$$\log(1+x) \leq x \qquad x \geq 0$$

Specifically, if $x(n)$ is bounded, $|x(n)| < M$,

$$|y(n)| = |\log(1 + |x(n)|)| \leq 1 + |x(n)| < 1 + M$$

Therefore, the output is bounded, and the system is stable.

(f) This system is not stable. This may be seen by considering the bounded input $x(n) = \cos(n\pi/8)$. Specifically, note that the output of the system at time $n = 0$ is

$$y(0) = \sum_{k=-\infty}^{\infty} x(k)h(-k) = \sum_{k=-\infty}^{\infty} \cos\left(\frac{n\pi}{8}\right)\cos\left(-\frac{n\pi}{8}\right) = \sum_{k=-\infty}^{\infty} \cos^2\left(\frac{n\pi}{8}\right)$$

which is unbounded. Alternatively, because the input-output relation is one of convolution, this is a linear shift-invariant system with a unit sample response

$$h(n) = \cos\left(\frac{n\pi}{8}\right)$$

Because a linear shift-invariant system will be stable only if

$$\sum_{n=-\infty}^{\infty} |h(n)| < \infty$$

we see that this system is not stable.

1.22   Determine which of the following systems are invertible:

(a) $y(n) = 2x(n)$

(b) $y(n) = nx(n)$

(c) $y(n) = x(n) - x(n-1)$

(d) $y(n) = \displaystyle\sum_{k=-\infty}^{n} x(k)$

(e) $y(n) = \text{Re}\{x(n)\}$

To test for invertibility, we may show that a system is invertible by designing an *inverse system* that uniquely recovers the input from the output, or we may show that a system is not invertible by finding two different inputs that produce the same output. Each system defined above will be tested for invertibility using one of these two methods.

(a)   This system is clearly invertible because, given the output $y(n)$, we may recover the input using $x(n) = 0.5y(n)$.

(b)   This system is not invertible, because the value of $x(n)$ at $n = 0$ cannot be recovered from $y(n)$. For example, the response of the system to $x(n)$ and to $x_1(n) = x(n) + \alpha\delta(n)$ will be the same for any α.

(c)   Due to the differencing between two successive input values, this system will not be invertible. For example, note that the inputs $x(n)$ and $x(n) + c$ will produce the same output for any value of c.

(d)   This system corresponds to an integrator and is an invertible system. To show that it is invertible, we may construct the inverse system, which is

$$x(n) = y(n) - y(n-1)$$

To show that this is the inverse system, note that

$$y(n) - y(n-1) = \sum_{k=-\infty}^{n} x(k) - \sum_{k=-\infty}^{n-1} x(k) = x(n)$$

(e)   Invertibility must hold for complex as well as real-valued signals. Therefore, this system is noninvertible because it discards the imaginary part of $x(n)$. One could state, however, that this system is invertible over the set of real-valued signals.

1.23   Consider the cascade of two systems, S_1 and S_2.

(a)   If both S_1 and S_2 are linear, shift-invariant, stable, and causal, will the cascade also be linear, shift-invariant, stable, and causal?

(b)   If both S_1 and S_2 are nonlinear, will the cascade be nonlinear?

(c)   If both S_1 and S_2 are shift-varying, will the cascade be shift-varying?

(a)   Linearity, shift-invariance, stability, and causality are easily shown to be preserved in a cascade. For example, the response of S_1 to the input $ax_1(n) + bx_2(n)$ will be $aw_1(n) + bw_2(n)$ due to the linearity of S_1. With this as the input to S_2, the response will be, again by linearity, $ay_1(n) + by_2(n)$. Therefore, if both S_1 and S_2 are linear, the cascade will be linear.

Similarly, for shift-invariance, if $x(n - n_0)$ is input to S_1, the response will be $w(n - n_0)$. In addition, because S_2 is shift-invariant, the response to $w(n - n_0)$ will be $y(n - n_0)$. Therefore, the response of the cascade to $x(n - n_0)$ is $y(n - n_0)$, and the cascade is shift-invariant.

To establish stability, note that with S_1 being stable, if $x(n)$ is bounded, the output $w(n)$ will be bounded. With $w(n)$ a bounded input to the stable system S_2, the response $y(n)$ will also be bounded. Therefore, the cascade is stable.

Finally, causality of the cascade follows by noting that if S_2 is causal, $y(n)$ at time $n = n_0$ depends only on $w(n)$ for $n \leq n_0$. With S_1 being causal, $w(n)$ for $n \leq n_0$ will depend only on the input $x(n)$ for $n \leq n_0$, and it follows that the cascade is causal.

(b) If S_1 and S_2 are nonlinear, it is not necessarily true that the cascade will be nonlinear because the second system may undo the nonlinearity of the first. For example, with

$$w(n) = S_1\{x(n)\} = \exp\{x(n)\}$$
$$y(n) = S_2\{w(n)\} = \log\{w(n)\}$$

although both S_1 and S_2 are nonlinear, the cascade is the identity system and, therefore, is linear.

(c) As in (b), if S_1 and S_2 are shift-varying, it is not necessarily true that the cascade will be shift-varying. For example, if the first system is a modulator,

$$w(n) = x(n) \cdot e^{jn\omega_0}$$

and the second is a demodulator,

$$y(n) = w(n) \cdot e^{-jn\omega_0}$$

the cascade is shift-invariant, even though a modulator and a demodulator are shift-varying. Another example is when S_1 is an up-sampler

$$w(n) = \begin{cases} x\left(\frac{n}{2}\right) & n = 0, \pm 2, \pm 4, \ldots \\ 0 & \text{else} \end{cases}$$

and S_2 is a down-sampler

$$y(n) = w(2n)$$

In this case, the cascade is shift-invariant, and $y(n) = x(n)$. However, if the order of the systems is reversed, the cascade will no longer be shift-invariant. Also, if a linear shift-invariant system, such as a unit delay, is inserted between the up-sampler and the down-sampler, the cascade of the three systems will, in general, be shift-varying.

Convolution

1.24 The first nonzero value of a finite-length sequence $x(n)$ occurs at index $n = -6$ and has a value $x(-6) = 3$, and the last nonzero value occurs at index $n = 24$ and has a value $x(24) = -4$. What is the index of the first nonzero value in the convolution

$$y(n) = x(n) * x(n)$$

and what is its value? What about the last nonzero value?

Because we are convolving two finite-length sequences, the index of the first nonzero value in the convolution is equal to the sum of the indices of the first nonzero values of the two sequences that are being convolved. In this case, the index is $n = -12$, and the value is

$$y(-12) = x^2(-6) = 9$$

Similarly, the index of the last nonzero value is at $n = 48$ and the value is

$$y(48) = x^2(24) = 16$$

1.25 The convolution of two finite-length sequences will be finite in length. Is it true that the convolution of a finite-length sequence with an infinite-length sequence will be infinite in length?

It is not necessarily true that the convolution of an infinite-length sequence with a finite-length sequence will be infinite in length. It may be either. Clearly, if $x(n) = \delta(n)$ and $h(n) = (0.5)^n u(n)$, the convolution will be an infinite-length sequence. However, it is possible for the finite-length sequence to *remove* the infinite-length tail of an infinite-length sequence. For example, note that

$$(0.5)^n u(n) - (0.5)^n u(n-1) = \delta(n)$$

Therefore, the convolution of $x(n) = \delta(n) - \frac{1}{2}\delta(n-1)$ with $h(n) = (0.5)^n u(n)$ will be finite in length:

$$\left[\delta(n) - \tfrac{1}{2}\delta(n-1)\right] * (0.5)^n u(n) = (0.5)^n u(n) - \tfrac{1}{2}(0.5)^{n-1} u(n-1) = \delta(n)$$

1.26   Find the convolution of the two finite-length sequences:

$$x(n) = 0.5n[u(n) - u(n-6)]$$

$$h(n) = 2\sin\left(\frac{n\pi}{2}\right)[u(n+3) - u(n-4)]$$

Shown in the figure below are the sequences $x(k)$ and $h(k)$.

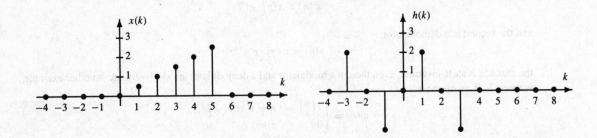

Because $h(n)$ is equal to zero outside the interval $[-3, 3]$, and $x(n)$ is zero outside the interval $[1, 5]$, the convolution $y(n) = x(n) * h(n)$ is zero outside the interval $[-2, 8]$.

One way to perform the convolution is to use the slide rule approach. Listing $x(k)$ and $h(-k)$ across two pieces of paper, aligning them at $k = 0$, we have the picture as shown below (the sequence $h(-k)$ is in front).

0	0	0	0	0.5	1	1.5	2	2.5
-2	0	2	0	-2	0	2	0	0

$$k = 0$$

Forming the sum of the products $x(k)h(-k)$, we obtain the value of $y(n)$ at time $n = 0$, which is $y(0) = 2$. Shifting $h(-k)$ to the left by one, multiplying and adding, we obtain the value of $y(n)$ at $n = -1$, which is $y(-1) = 2$. Shifting one more time to the left, forming the sum of products, we find $y(-2) = 1$, which is the last nonzero value of $y(n)$ for $n < 0$. Repeating the process by shifting $h(-k)$ to the right, we obtain the values of $y(n)$ for $n > 0$, which are

$$y(1) = 2 \qquad y(2) = 3 \qquad y(3) = -2 \qquad y(4) = -3$$

$$y(5) = 2 \qquad y(6) = 2 \qquad y(7) = -4 \qquad y(8) = -5$$

Another way to perform the convolution is to use the fact that

$$x(n) * \delta(n - n_0) = x(n - n_0)$$

Writing $h(n)$ as

$$h(n) = 2\delta(n+3) - 2\delta(n+1) + 2\delta(n-1) - 2\delta(n-3)$$

we may evaluate $y(n)$ as follows

$$y(n) = 2x(n + 3) - 2x(n + 1) + 2x(n - 1) - 2x(n - 3)$$

Making a table of these shifted sequences,

n	-2	-1	0	1	2	3	4	5	6	7	8
$2x(n + 3)$	1	2	3	4	5	0	0	0	0	0	0
$-2x(n + 1)$	0	0	-1	-2	-3	-4	-5	0	0	0	0
$2x(n - 1)$	0	0	0	0	1	2	3	4	5	0	0
$-2x(n - 3)$	0	0	0	0	0	0	-1	-2	-3	-4	-5
$y(n)$	1	2	2	2	3	-2	-3	2	2	-4	-5

and adding down the columns, we obtain the sequence $y(n)$.

1.27   Derive a closed-form expression for the convolution of $x(n)$ and $h(n)$ where

$$x(n) = \left(\tfrac{1}{6}\right)^{n-6} u(n)$$

$$h(n) = \left(\tfrac{1}{3}\right)^{n} u(n - 3)$$

Because both sequences are infinite in length, it is easier to evaluate the convolution sum directly:

$$y(n) = \sum_{k=-\infty}^{\infty} x(k)h(n - k)$$

Note that because $x(n) = 0$ for $n < 0$ and $h(n) = 0$ for $n < 3$, $y(n)$ will be equal to zero for $n < 3$. Substituting $x(n)$ and $h(n)$ into the convolution sum, we have

$$y(n) = \sum_{k=-\infty}^{\infty} \left(\tfrac{1}{6}\right)^{k-6} u(k)\left(\tfrac{1}{3}\right)^{n-k} u(n - k - 3)$$

Due to the step $u(k)$, the lower limit on the sum may be changed to $k = 0$, and because $u(n - k - 3)$ is zero for $k > n - 3$, the upper limit may be changed to $k = n - 3$. Thus, for $n \geq 3$ the convolution sum becomes

$$y(n) = \sum_{k=0}^{n-3} \left(\tfrac{1}{6}\right)^{k-6} \left(\tfrac{1}{3}\right)^{n-k} = 6^6 \left(\tfrac{1}{3}\right)^{n} \sum_{k=0}^{n-3} \left(\tfrac{3}{6}\right)^{k} \qquad n \geq 3$$

Using the geometric series to evaluate the sum, we have

$$y(n) = 6^6 \left(\tfrac{1}{3}\right)^{n} \frac{1 - \left(\tfrac{1}{2}\right)^{n-2}}{1 - \tfrac{1}{2}} = 2 \cdot 6^6 \left(\tfrac{1}{3}\right)^{n} \left[1 - 4\left(\tfrac{1}{2}\right)^{n}\right] \qquad n \geq 3$$

1.28   A linear shift-invariant system has a unit sample response

$$h(n) = u(-n - 1)$$

Find the output if the input is

$$x(n) = -n3^n u(-n)$$

Shown below are the sequences $x(n)$ and $h(n)$.

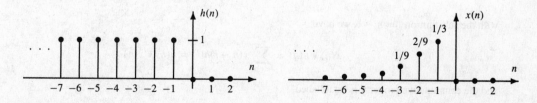

Because $x(n)$ is zero for $n > -1$, and $h(n)$ is equal to zero for $n > -1$, the convolution will be equal to zero for $n > -2$. Evaluating the convolution sum directly, we have

$$y(n) = \sum_{k=-\infty}^{\infty} x(k)h(n-k) = \sum_{k=-\infty}^{\infty} -k3^k u(-k)u(-(n-k)-1)$$

Because $u(-k) = 0$ for $k > 0$ and $u(-(n-k)-1) = 0$ for $k < n+1$, the convolution sum becomes

$$y(n) = \sum_{k=n+1}^{0} -k3^k \qquad n \le -2$$

With the change of variables $m = -k$, and using the series formulas given in Table 1-1, we have

$$y(n) = \sum_{m=0}^{-n-1} m\left(\tfrac{1}{3}\right)^m = \frac{(-n-1)\left(\tfrac{1}{3}\right)^{-n+1} + n\left(\tfrac{1}{3}\right)^{-n} + \tfrac{1}{3}}{\left(1 - \tfrac{1}{3}\right)^2}$$

$$= \tfrac{3}{4} + \tfrac{3}{4}(2n-1)\left(\tfrac{1}{3}\right)^{-n} \qquad n \le -2$$

Let us check this answer for a few values of n using graphical convolution. Time-reversing $x(k)$, we see that $h(k)$ and $x(-k)$ do not overlap for any k and, thus, $y(0) = 0$. In fact, it is not until we shift $x(-k)$ to the left by two that there is any overlap. With $x(-2-k)$ and $h(k)$ overlapping at one point, and the product being equal to $\tfrac{1}{3}$, it follows that $y(-2) = \tfrac{1}{3}$. Evaluating the expression above for $y(n)$ above at index $n = -2$, we obtain the same result. For $n = -3$, the sequences $x(-3-k)$ and $h(k)$ overlap at two points, and the sum of the products gives $y(-3) = \tfrac{1}{3} + \tfrac{2}{9} = \tfrac{5}{9}$, which, again, is the same as the expression above.

1.29   If the response of a linear shift-invariant system to a unit step (i.e., the step response) is

$$s(n) = n\left(\tfrac{1}{2}\right)^n u(n)$$

find the unit sample response, $h(n)$.

In this problem, we begin by noting that

$$\delta(n) = u(n) - u(n-1)$$

Therefore, the unit sample response, $h(n)$, is related to the step response, $s(n)$, as follows:

$$h(n) = s(n) - s(n-1)$$

Thus, given $s(n)$, we have

$$h(n) = s(n) - s(n-1)$$
$$= n\left(\tfrac{1}{2}\right)^n u(n) - (n-1)\left(\tfrac{1}{2}\right)^{n-1} u(n-1)$$
$$= \left[n\left(\tfrac{1}{2}\right)^n - 2(n-1)\left(\tfrac{1}{2}\right)^n\right] u(n-1)$$
$$= (2-n)\left(\tfrac{1}{2}\right)^n u(n-1)$$

1.30   Prove the commutative property of convolution

$$x(n) * h(n) = h(n) * x(n)$$

Proving the commutative property is straightforward and only involves a simple manipulation of the convolution sum. With the convolution of $x(n)$ with $h(n)$ given by

$$x(n) * h(n) = \sum_{k=-\infty}^{\infty} x(k)h(n-k)$$

with the substitution $l = n - k$, we have

$$x(n) * h(n) = \sum_{l=-\infty}^{\infty} x(n-l)h(l) = h(n) * x(n)$$

and the commutative property is established.

1.31   Prove the distributive property of convolution

$$h(n) * [x_1(n) + x_2(n)] = h(n) * x_1(n) + h(n) * x_2(n)$$

To prove the distributive property, we have

$$h(n) * [x_1(n) + x_2(n)] = \sum_{k=-\infty}^{\infty} h(k)[x_1(n-k) + x_2(n-k)]$$

Therefore,
$$h(n) * [x_1(n) + x_2(n)] = \sum_{k=-\infty}^{\infty} h(k)x_1(n-k) + \sum_{k=-\infty}^{\infty} h(k)x_2(n-k)$$

$$= h(n) * x_1(n) + h(n) * x_2(n) \tag{1.24}$$

and the property is established.

1.32   Let

$$h(n) = 3\left(\tfrac{1}{2}\right)^n u(n) - 2\left(\tfrac{1}{3}\right)^{n-1} u(n)$$

be the unit sample response of a linear shift-invariant system. If the input to this system is a unit step,

$$x(n) = \begin{cases} 1 & n \ge 0 \\ 0 & \text{else} \end{cases}$$

find $\lim_{n \to \infty} y(n)$ where $y(n) = h(n) * x(n)$.

With

$$y(n) = h(n) * x(n) = \sum_{k=-\infty}^{\infty} h(k)x(n-k)$$

if $x(n)$ is a unit step,

$$y(n) = \sum_{k=-\infty}^{\infty} h(k)u(n-k) = \sum_{k=-\infty}^{n} h(k)$$

Therefore,
$$\lim_{n \to \infty} y(n) = \sum_{k=-\infty}^{\infty} h(k)$$

Evaluating the sum, we have

$$\lim_{n \to \infty} y(n) = 3 \sum_{k=0}^{\infty} \left(\tfrac{1}{2}\right)^k - 2 \sum_{k=0}^{\infty} \left(\tfrac{1}{3}\right)^{k-1} = \frac{3}{1 - 1/2} - \frac{6}{1 - 1/3} = -3$$

1.33   Convolve

$$x(n) = (0.9)^n u(n)$$

with a ramp

$$h(n) = nu(n)$$

The convolution of $x(n)$ with $h(n)$ is

$$y(n) = x(n) * h(n) = \sum_{k=-\infty}^{\infty} x(k)h(n-k)$$

$$= \sum_{k=-\infty}^{\infty} [(0.9)^k u(k)][(n-k)u(n-k)]$$

Because $u(k)$ is zero for $k < 0$, and $u(n - k)$ is zero for $k > n$, this sum may be rewritten as follows:

$$y(n) = \sum_{k=0}^{n} (n - k)(0.9)^k \qquad n \geq 0$$

or

$$y(n) = n \sum_{k=0}^{n} (0.9)^k - \sum_{k=0}^{n} k(0.9)^k \qquad n \geq 0$$

Using the series given in Table 1-1, we have

$$y(n) = n \frac{1 - (0.9)^{n+1}}{1 - 0.9} - \frac{n(0.9)^{n+2} - (n + 1)(0.9)^{n+1} + 0.9}{(1 - 0.9)^2}$$

$$= 10n[1 - (0.9)^{n+1}] - 100[n(0.9)^{n+2} - (n + 1)(0.9)^{n+1} + 0.9] \qquad n \geq 0$$

which may be simplified to

$$y(n) = [10n - 90 + 90(0.9)^n]u(n)$$

1.34 Perform the convolution

$$y(n) = x(n) * h(n)$$

when

$$h(n) = \left(\tfrac{1}{2}\right)^n u(n)$$

and

$$x(n) = \left(\tfrac{1}{3}\right)^n [u(n) - u(n - 101)]$$

With

$$y(n) = x(n) * h(n) = \sum_{k=-\infty}^{\infty} x(k)h(n - k)$$

we begin by substituting $x(n)$ and $h(n)$ into the convolution sum

$$y(n) = \sum_{k=-\infty}^{\infty} \left(\tfrac{1}{3}\right)^k [u(k) - u(k - 101)]\left(\tfrac{1}{2}\right)^{n-k} u(n - k)$$

or

$$y(n) = \sum_{k=0}^{100} \left(\tfrac{1}{3}\right)^k \left(\tfrac{1}{2}\right)^{n-k} u(n - k)$$

To evaluate this sum, which depends on n, we consider three cases. First, for $n < 0$, the sum is equal to zero because $u(n - k) = 0$ for $0 \leq k \leq 100$. Therefore,

$$y(n) = 0 \qquad n < 0$$

Second, note that for $0 \leq n \leq 100$, the step $u(n - k)$ is only equal to 1 for $k \leq n$. Therefore,

$$y(n) = \sum_{k=0}^{n} \left(\tfrac{1}{3}\right)^k \left(\tfrac{1}{2}\right)^{n-k} = \left(\tfrac{1}{2}\right)^n \sum_{k=0}^{n} \left(\tfrac{2}{3}\right)^k$$

$$= \left(\tfrac{1}{2}\right)^n \frac{1 - \left(\tfrac{2}{3}\right)^{n+1}}{1 - \tfrac{2}{3}} = 3\left(\tfrac{1}{2}\right)^n \left[1 - \left(\tfrac{2}{3}\right)^{n+1}\right]$$

Finally, for $n \geq 100$, note that $u(n-k)$ is equal to 1 for all k in the range $0 \leq k \leq 100$. Therefore,

$$y(n) = \sum_{k=0}^{100} \left(\tfrac{1}{3}\right)^k \left(\tfrac{1}{2}\right)^{n-k} = \left(\tfrac{1}{2}\right)^n \sum_{k=0}^{100} \left(\tfrac{2}{3}\right)^k$$

$$= \left(\tfrac{1}{2}\right)^n \frac{1 - \left(\tfrac{2}{3}\right)^{101}}{1 - \tfrac{2}{3}} = 3\left(\tfrac{1}{2}\right)^n \left[1 - \left(\tfrac{2}{3}\right)^{101}\right]$$

In summary, we have

$$y(n) = \begin{cases} 0 & n < 0 \\ 3\left(\tfrac{1}{2}\right)^n \left[1 - \left(\tfrac{2}{3}\right)^{n+1}\right] & 0 \leq n \leq 100 \\ 3\left(\tfrac{1}{2}\right)^n \left[1 - \left(\tfrac{2}{3}\right)^{101}\right] & n \geq 100 \end{cases}$$

1.35   Let $h(n)$ be a truncated exponential

$$h(n) = \begin{cases} \alpha^n & 0 \leq n \leq 10 \\ 0 & \text{else} \end{cases}$$

and $x(n)$ a discrete pulse of the form

$$x(n) = \begin{cases} 1 & 0 \leq n \leq 5 \\ 0 & \text{else} \end{cases}$$

Find the convolution $y(n) = h(n) * x(n)$.

To find the convolution of these two finite-length sequences, we need to evaluate the sum

$$y(n) = h(n) * x(n) = \sum_{k=-\infty}^{\infty} h(k)x(n-k)$$

To evaluate this sum, it will be useful to make a plot of $h(k)$ and $x(n-k)$ as a function of k as shown in the following figure:

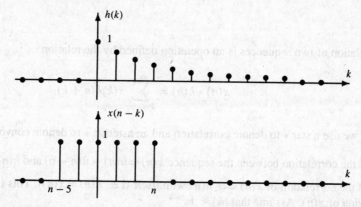

Note that the amount of overlap between $h(k)$ and $x(n-k)$ depends on the value of n. For example, if $n < 0$, there is no overlap, whereas for $0 \leq n \leq 5$, the two sequences overlap for $0 \leq k \leq n$. Therefore, in the following, we consider five separate cases.

Case 1 $n < 0$. When $n < 0$, there is no overlap between $h(k)$ and $x(n-k)$. Therefore, the product $h(k)x(n-k) = 0$ for all k, and $y(n) = 0$.

Case 2 $0 \leq n \leq 5$. For this case, the product $h(k)x(n-k)$ is nonzero only for k in the range $0 \leq k \leq n$. Therefore,

$$y(n) = \sum_{k=0}^{n} \alpha^k = \frac{1 - \alpha^{n+1}}{1 - \alpha}$$

Case 3 $6 \le n \le 10$. For $6 \le n \le 10$, all of the nonzero values of $x(n-k)$ are within the limits of the sum, and

$$y(n) = \sum_{k=n-5}^{n} \alpha^k = \sum_{k=0}^{5} \alpha^{k+(n-5)}$$

$$= \alpha^{n-5} \sum_{k=0}^{5} \alpha^k = \alpha^{n-5} \frac{1-\alpha^6}{1-\alpha}$$

Case 4 $11 \le n \le 15$. When n is in the range $11 \le n \le 15$, the sequences $h(k)$ and $x(n-k)$ overlap for $n-5 \le k \le 10$. Therefore,

$$y(n) = \sum_{k=n-5}^{10} \alpha^k = \sum_{k=0}^{15-n} \alpha^{k+(n-5)}$$

$$= \alpha^{n-5} \sum_{k=0}^{15-n} \alpha^k = \alpha^{n-5} \frac{1-\alpha^{16-n}}{1-\alpha}$$

Case 5 $n > 15$. Finally, for $n > 15$, there is again no overlap between $h(k)$ and $x(n-k)$, and the product $h(k)x(n-k)$ is equal to zero for all k. Therefore, $y(n) = 0$ for $n > 15$.

In summary, for the convolution we have

$$y(n) = \begin{cases} 0 & n < 0 \\ \dfrac{1-\alpha^{n+1}}{1-\alpha} & 0 \le n \le 5 \\ \alpha^{n-5} \dfrac{1-\alpha^6}{1-\alpha} & 6 \le n \le 10 \\ \alpha^{n-5} \dfrac{1-\alpha^{16-n}}{1-\alpha} & 11 \le n \le 15 \\ 0 & n > 15 \end{cases}$$

1.36   The correlation of two sequences is an operation defined by the relation

$$x(n) \star h(n) = \sum_{k=-\infty}^{\infty} x(k)h(n+k)$$

Note that we use a star $\star$ to denote correlation and an asterisk $*$ to denote convolution.

(a)   Find the correlation between the sequence $x(n) = u(n) - u(n-6)$ and $h(n) = u(n-2) - u(n-5)$.

(b)   Find the correlation of $x(n) = \alpha^n u(n)$ with itself (i.e., $h(n) = x(n)$). This is known as the *autocorrelation* of $x(n)$. Assume that $|\alpha| < 1$.

(a)   If we compare the expression for the correlation of $x(n)$ and $h(n)$ with the convolution

$$x(n) * h(n) = \sum_{k=-\infty}^{\infty} x(k)h(n-k)$$

we see that the only difference is that, in the case of convolution, $h(k)$ is time-reversed prior to shifting by n, whereas for correlation $h(k)$ is shifted without time-reversal. Therefore, with a graphical approach to compute the correlation, we simply need to plot $x(k)$ and $h(k)$, shift $h(k)$ by n (to the left if $n > 0$ and to the right if $n < 0$), multiply the two sequences $x(k)$ and $h(n+k)$, and sum the products. Shown in the figure below is a plot of $x(k)$ and $h(k)$.

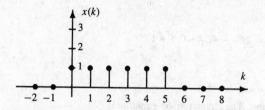 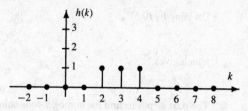

Denoting the correlation by $r_{xh}(n)$, it is clear that for $n = 0$ the correlation is equal to 3. In fact, this will be the value of $r_{xh}(n)$ for $-1 \leq n \leq 2$. For $n = 3$, $x(k)$ and $h(3 + k)$ only overlap at two points, and $r_{xh}(3) = 2$. Similarly, because $x(k)$ and $h(4 + k)$ only overlap at one point, $r_{xh}(4) = 1$. Finally, $r_{xh}(n) = 0$ for $n > 4$. Proceeding in a similar fashion for $n < 0$, we find that $r_{xh}(-2) = 2$, and $r_{xh}(-3) = 1$. The correlation is shown in the figure below.

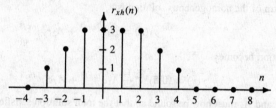

(b) Let $r_x(n)$ denote the autocorrelation of $x(n)$, and note that the autocorrelation is the convolution of $x(n)$ with $x(-n)$:

$$r_x(n) = x(n) \star x(n) = x(n) * x(-n) = \sum_{k=-\infty}^{\infty} x(k)x(n + k)$$

In addition observe that $r_x(n)$ is an even function of n:

$$r_x(-n) = \sum_{k=-\infty}^{\infty} x(k)x(-n + k) = \sum_{k'=-\infty}^{\infty} x(k' + n)x(k') = r_x(n)$$

Therefore, it is only necessary to find the values of $r_x(n)$ for $n \geq 0$. For $n \geq 0$, we have

$$r_x(n) = \sum_{k=-\infty}^{\infty} \alpha^k u(k)\alpha^{n+k} u(n + k) = \alpha^n \sum_{k=0}^{\infty} \alpha^{2k} = \frac{1}{1 - \alpha^2}\alpha^n \qquad n \geq 0$$

Using the symmetry of $r_x(n)$, we have, for $n < 0$,

$$r_x(n) = \frac{1}{1 - \alpha^2}\alpha^{-n} \qquad n \leq 0$$

Combining these two results together, we finally have

$$r_x(n) = \frac{1}{1 - \alpha^2}\alpha^{|n|}$$

Difference Equations

1.37 Consider a system described by the difference equation

$$y(n) = y(n - 1) - y(n - 2) + 0.5x(n) + 0.5x(n - 1)$$

Find the response of this system to the input

$$x(n) = (0.5)^n u(n)$$

with initial conditions $y(-1) = 0.75$ and $y(-2) = 0.25$.

The first step in solving this difference equation is to find the particular solution. With $x(n) = (0.5)^n u(n)$, we assume a solution of the form

$$y_p(n) = C_1(0.5)^n \qquad n \geq 0$$

Substituting this solution into the difference equation, we have

$$C_1(0.5)^n = C_1(0.5)^{n-1} - C_1(0.5)^{n-2} + 0.5(0.5)^n + 0.5(0.5)^{n-1} \qquad n \geq 0$$

Dividing by $(0.5)^n$,

$$C_1 = 2C_1 - 4C_1 + 0.5 + 1$$

which gives

$$C_1 = \tfrac{1}{2}$$

The next step is to find the homogeneous solution. The characteristic equation is

$$z^2 - z + 1 = 0$$

which has roots

$$z = \tfrac{1}{2}(1 \pm j\sqrt{3}) = e^{\pm j\pi/3}$$

Therefore, the form of the homogeneous solution is

$$y_h(n) = A_1 e^{jn\pi/3} + A_2 e^{-jn\pi/3}$$

and the total solution becomes

$$y(n) = (0.5)^{n+1} + A_1 e^{jn\pi/3} + A_2 e^{-jn\pi/3} \qquad n \geq 0 \qquad\qquad (1.25)$$

The constants A_1 and A_2 must now be found so that the total solution satisfies the given initial conditions, $y(-1) = 0.75$ and $y(-2) = 0.25$. Because the solution given in Eq. (1.25) is only applicable for $n \geq 0$, we must derive an equivalent set of initial conditions for $y(0)$ and $y(1)$. Evaluating the difference equation for $n = 0$ and $n = 1$, we have

$$y(0) = y(-1) - y(-2) + 0.5x(0) + 0.5x(-1) = 0.75 - 0.25 + 0.5 = 1$$

and

$$y(1) = y(0) - y(-1) + 0.5x(1) + 0.5x(0) = 1 - 0.75 + 0.25 + 0.5 = 1$$

Now, substituting these *derived initial conditions* into Eq. (1.25), we have

$$y(0) = 0.5 + A_1 + A_2 = 1$$
$$y(1) = 0.25 + A_1 e^{j\pi/3} + A_2 e^{-j\pi/3} = 1$$

Writing this pair of equations in the two unknowns A_1 and A_2 in matrix form,

$$\begin{bmatrix} 1 & 1 \\ e^{j\pi/3} & e^{-j\pi/3} \end{bmatrix} \begin{bmatrix} A_1 \\ A_2 \end{bmatrix} = \begin{bmatrix} 0.5 \\ 0.75 \end{bmatrix}$$

and solving, we find

$$\begin{bmatrix} A_1 \\ A_2 \end{bmatrix} = j\frac{\sqrt{3}}{3} \begin{bmatrix} \tfrac{1}{2}e^{-j\pi/3} - \tfrac{3}{4} \\ -\tfrac{1}{2}e^{j\pi/3} + \tfrac{3}{4} \end{bmatrix}$$

Substituting into Eq. (1.25) and simplifying, we find, after a bit of algebra,

$$y(n) = (0.5)^{n+1} + \frac{\sqrt{3}}{2}\sin\left(\frac{n\pi}{3}\right) - \frac{2\sqrt{3}}{2}\sin\left((n-1)\frac{\pi}{3}\right)$$

An important observation to make about this solution is that, because the difference equation has real coefficients, the roots of the characteristic polynomial are in complex-conjugate pairs. This ensures that the unit sample response is real. With a real-valued input $x(n)$, the response must be real and, therefore, it follows that A_2 will be the complex conjugate of A_1:

$$A_2 = A_1^*$$

1.38   A second-order recursive system is described by the LCCDE

$$y(n) = \tfrac{3}{4}y(n-1) - \tfrac{1}{8}y(n-2) + x(n) - x(n-1)$$

(a) Find the unit sample response $h(n)$ of this system.

(b) Find the system's response to the input $x(n) = u(n) - u(n-10)$ with zero initial conditions.

(c) Find the system's response to the input $x(n) = (\frac{1}{2})^n u(n)$ with zero initial conditions.

(a) To find the unit sample response, we must solve the difference equation with $x(n) = \delta(n)$ and initial rest conditions. The characteristic equation is

$$z^2 - \tfrac{3}{4}z + \tfrac{1}{8} = \left(z - \tfrac{1}{2}\right)\left(z - \tfrac{1}{4}\right)$$

Therefore, the homogeneous solution is

$$y_h(n) = A_1\left(\tfrac{1}{2}\right)^n + A_2\left(\tfrac{1}{4}\right)^n \qquad n \geq 0 \qquad\qquad (1.26)$$

Because the particular solution is zero when the system input is a unit sample, Eq. (1.26) represents the total solution.

To find the constants A_1 and A_2, we must derive the initial conditions at $n = 0$ and $n = 1$. With initial rest conditions, $y(-1) = y(-2) = 0$, it follows that

$$y(0) = \tfrac{3}{4}y(-1) - \tfrac{1}{8}y(-2) + x(0) - x(-1) = 1$$
$$y(1) = \tfrac{3}{4}y(0) - \tfrac{1}{8}y(-1) + x(1) - x(0) = \tfrac{3}{4} - 1 = -\tfrac{1}{4}$$

We may now write two equations in the two unknowns A_1 and A_2 by evaluating Eq. (1.26) at $n = 0$ and $n = 1$ as follows:

$$1 = A_1 + A_2$$
$$-\tfrac{1}{4} = \tfrac{1}{2}A_1 + \tfrac{1}{4}A_2$$

Solving for A_1 and A_2, we find

$$A_1 = -2 \qquad A_2 = 3$$

Thus,

$$y(n) = -2\left(\tfrac{1}{2}\right)^n + 3\left(\tfrac{1}{4}\right)^n \qquad n \geq 0$$

and the unit sample response is

$$h(n) = \left[-2\left(\tfrac{1}{2}\right)^n + 3\left(\tfrac{1}{4}\right)^n\right]u(n)$$

(b) To find the response of the system to $x(n) = u(n) - u(n - 10)$, we may proceed in one of two ways. First, we may perform the convolution of $h(n)$ with $x(n)$:

$$y(n) = x(n) * h(n) = \sum_{k=-\infty}^{\infty} x(k)h(n - k) = \sum_{k=0}^{9} h(n - k)$$

Alternatively, noting that the input is a sum of two steps, we may find the step response of the system, $s(n)$, and then using linearity, write the response as

$$y(n) = s(n) - s(n - 10)$$

Using this approach, we see from part (a) that the step response for $n \geq 0$ is

$$s(n) = h(n) * u(n) = \sum_{k=0}^{n} h(k) = \sum_{k=0}^{n}\left[-2\left(\tfrac{1}{2}\right)^k + 3\left(\tfrac{1}{4}\right)^k\right] \qquad n \geq 0$$

Evaluating the sums using the geometric series, we find

$$s(n) = \left[-2\frac{1 - \left(\tfrac{1}{2}\right)^{n+1}}{1 - \tfrac{1}{2}} + 3\frac{1 - \left(\tfrac{1}{4}\right)^{n+1}}{1 - \tfrac{1}{4}}\right]u(n) = \left[2\left(\tfrac{1}{2}\right)^n - \left(\tfrac{1}{4}\right)^n\right]u(n)$$

Thus, the desired solution is

$$y(n) = s(n) - s(n - 10) = \left[2\left(\tfrac{1}{2}\right)^n - \left(\tfrac{1}{4}\right)^n\right]u(n) - \left[2\left(\tfrac{1}{2}\right)^{n-10} - \left(\tfrac{1}{4}\right)^{n-10}\right]u(n - 10)$$

(c)   With $x(n) = (\frac{1}{2})^n u(n)$, note that $x(n)$ has the same form as one of the terms in the homogeneous solution. Therefore, the particular solution will not be of the form $y_p(n) = C(\frac{1}{2})^n$ as indicated in Table 1-2. If we were to substitute this particular solution into the difference equation, we would find that no value of C would work. As is the case when a root of the characteristic equation is of second order, the particular solution has the form

$$y_p(n) = Cn\left(\tfrac{1}{2}\right)^n$$

Substituting this into the difference equation, we have

$$Cn\left(\tfrac{1}{2}\right)^n = \tfrac{3}{4}C(n-1)\left(\tfrac{1}{2}\right)^{n-1} - \tfrac{1}{8}C(n-2)\left(\tfrac{1}{2}\right)^{n-2} + \left(\tfrac{1}{2}\right)^n - \left(\tfrac{1}{2}\right)^{n-1}$$

Dividing through by $(\frac{1}{2})^n$, we have

$$Cn = \tfrac{3}{2}C(n-1) - \tfrac{1}{2}C(n-2) - 1$$

Solving for C, we find that $C = -2$. Thus, the total solution is

$$y(n) = -2n\left(\tfrac{1}{2}\right)^n + A_1\left(\tfrac{1}{2}\right)^n + A_2\left(\tfrac{1}{4}\right)^n \qquad n \ge 0 \tag{1.27}$$

We now must solve for the constants A_1 and A_2. As we did in part (a), with zero initial conditions we find that $y(0) = 1$ and $y(1) = \frac{1}{4}$. Therefore, evaluating Eq. (1.27) at $n = 0$ and $n = 1$, we obtain the following two equations in the two unknowns A_1 and A_2:

$$1 = A_1 + A_2$$
$$\tfrac{1}{4} = -2\left(\tfrac{1}{2}\right) + \tfrac{1}{2}A_1 + \tfrac{1}{4}A_2$$

Solving for A_1 and A_2, we find that $A_1 = 4$ and $A_2 = 3$. Thus, the total solution becomes

$$y(n) = \left[-2n\left(\tfrac{1}{2}\right)^n + 4\left(\tfrac{1}{2}\right)^n - 3\left(\tfrac{1}{4}\right)^n\right]u(n)$$

1.39   A \$100,000 mortgage is to be paid off in *equal* monthly payments of d dollars. Interest, compounded monthly, is charged at the rate of 10 percent per annum on the unpaid balance [e.g., after the first month the total debt equals ($\$100,000 + \frac{0.10}{12}\$100,000$)]. Determine the amount of the payment, d, so that the mortgage is paid off in 30 years, and find the total amount of payments that are made over the 30-year period.

The total unpaid balance at the end of the nth month, in the absence of any additional loans or payments, is equal to the unpaid balance in the previous month plus the interest charged on the unpaid balance for the previous month. Therefore, with $y(n)$ the balance at the end of the nth month we have

$$y(n) = y(n-1) + \beta y(n-1)$$

where $\beta = \frac{0.10}{12}$ is the interest charged on the unpaid balance. In addition, the balance must be adjusted by the net amount of money leaving the bank into your pocket, which is simply the amount borrowed in the nth month minus the amount paid to the bank in the nth month. Thus

$$y(n) = y(n-1) + \beta y(n-1) + x_b(n) - x_p(n)$$

where $x_b(n)$ is the amount borrowed in the nth month and $x_p(n)$ is the amount paid in the nth month. Combining terms, we have

$$y(n) - vy(r-1) = x_b(n) - x_p(n) = x(n)$$

where $v = 1 + \beta = 1 + \frac{0.10}{12}$, and $x(n)$ is the net amount of money in the nth month that leaves the bank. Because a principal of p dollars is borrowed during month zero, and payments of d dollars begin with month 1, the *driving function*, $x(n)$, is

$$x(n) = x_b(n) - x_p(n) = p\delta(n) - du(n-1)$$

and the difference equation for $y(n)$ becomes

$$y(n) - vy(n-1) = p\delta(n) - du(n-1)$$

Because we are assuming zero initial conditions, $y(-1) = 0$, and because the input consists of a linear combination of a scaled unit sample and a scaled delayed step, the solution to the difference equation is simply

$$y(n) = ph(n) + ds(n-1)$$

where $h(n)$ and $s(n)$ are the unit sample and unit step response, respectively. To find the unit sample response, we write the difference equation in the form

$$y(n) = vy(n-1) + \delta(n)$$

The characteristic equation for this difference equation is

$$z - v = 0$$

and the homogeneous solution is

$$y(n) = Av^n \qquad n \geq 0$$

Because the input $x(n) = \delta(n)$ is equal to zero for $n > 0$, the particular solution is zero (all that the unit sample does is set the initial condition at $n = 0$). Evaluating the difference equation at $n = 0$, we have

$$y(0) = vy(-1) + 1 = 1$$

Therefore, it follows that $A = 1$ in the homogeneous solution, and that the unit sample response is

$$h(n) = v^n u(n)$$

The step response may now be found by convolving $h(n)$ with $u(n)$:

$$s(n) = h(n) * u(n) = \sum_{k=0}^{n} h(k) = \sum_{k=0}^{n} v^k = \frac{1 - v^{n+1}}{1 - v} \qquad n \geq 0$$

Thus, the total solution is

$$y(n) = ph(n) + ds(n-1) = pv^n u(n) - d\frac{1 - v^n}{1 - v} u(n-1)$$

We now want to find the value of d so that after 360 equal monthly payments the mortgage is paid off. In other words, we want to find d such that

$$y(360) = pv^{360} - d\frac{1 - v^{360}}{1 - v}$$
$$= \frac{1}{1 - v}[p(1 - v)v^{360} - d(1 - v^{360})] = 0$$

Solving for d, we have

$$d = \frac{p(1 - v)}{1 - v^{360}} v^{360}$$

With $v = \frac{12.1}{12}$ and $p = 100{,}000$, we have

$$d = 877.57$$

The total payment to the bank after 30 years is

$$C = (877.57)(360) = 315{,}925.20$$

1.40   Every second, each α particle within a reactor splits into eight β particles and each β particle splits into an α particle and two β particles. Schematically,

$$\alpha \longrightarrow 8\beta \qquad \beta \longrightarrow \alpha + 2\beta$$

Given that there is a single α particle in the reactor at time $n = 0$, find an expression for the total number of particles within the reactor at time n.

Let $\alpha(n)$ and $\beta(n)$ be the number of α particles and β particles within the reactor at time n. The behavior within the reactor may be described by the following pair of *coupled* difference equations:

$$\alpha(n+1) = \beta(n)$$

$$\beta(n+1) = 8\alpha(n) + 2\beta(n)$$

Before we can solve these difference equations, we must uncouple them. Therefore, let us derive a single difference equation for $\beta(n)$. From the first equation we see that $\alpha(n) = \beta(n-1)$. Substituting this relation into the second difference equation, we have

$$\beta(n+1) = 8\beta(n-1) + 2\beta(n)$$

or, equivalently,

$$\beta(n) = 2\beta(n-1) + 8\beta(n-2)$$

The characteristic equation for this difference equation is

$$z^2 - 2z - 8 = (z-4)(z+2) = 0$$

which gives the following homogeneous solution

$$\beta(n) = A_1(4)^n + A_2(-2)^n$$

Similarly, because $\alpha(n) = \beta(n-1)$, the solution for $\alpha(n)$ is

$$\alpha(n) = \frac{A_1}{4}4^n - \frac{A_2}{2}(-2)^n$$

With the initial conditions $\alpha(0) = 1$ and $\beta(0) = 0$, we may solve for A_1 and A_2 as follows:

$$\alpha(0) = \frac{A_1}{4} - \frac{A_2}{2} = 1$$

$$\beta(0) = A_1 + A_2 = 0$$

Therefore,

$$A_1 = \tfrac{4}{3} \qquad A_2 = -\tfrac{4}{3}$$

and the solutions for $\alpha(n)$ and $\beta(n)$ are

$$\alpha(n) = \tfrac{1}{3}(4)^n + \tfrac{2}{3}(-2)^n \qquad n \geq 0$$

$$\beta(n) = \tfrac{4}{3}(4)^n - \tfrac{4}{3}(-2)^n \qquad n \geq 0$$

Because we are interested in the total number of particles within the reactor at time n, with

$$y(n) = \alpha(n) + \beta(n)$$

we have

$$y(n) = \tfrac{1}{3}[5(4)^n + (-2)^{n+1}] \qquad n \geq 0$$

Supplementary Problems

Discrete-Time Signals

1.41 Find the period N of the sequence

$$x(n) = \cos\left(\frac{n\pi}{8}\right)\cos\left(\frac{2n\pi}{15}\right) + \sin\left(\frac{n\pi}{3}\right)\sin\left(\frac{n\pi}{4}\right)$$

1.42   The input to a linear shift-invariant system is periodic with period N.

(a)   Show that the output of the system is also periodic with period N.

(b)   If the system is linear but shift-varying, is the output guaranteed to be periodic?

(c)   If the system is nonlinear but shift-invariant, is the output guaranteed to be periodic?

1.43   If $x(n) = 0$ for $n < 0$, and the odd part is $x_o(n) = n(0.5)^{|n|}$, find $x(n)$ given that $x(0) = 1$.

1.44   Find the conjugate symmetric part of the sequence

$$x(n) = \left(\tfrac{1}{2} + j\tfrac{1}{2}\right)^n u(n)$$

1.45   If $x(n)$ is odd, what is $y(n) = x^2(n)$?

1.46   If $x(n) = 0$ for $n < 0$, P_e is the power in the even part of $x(n)$, and P_o is the power in the odd part, which of the following statements are true?

(a) $P_e \geq P_o$

(b) $P_o \geq P_e$

(c) $P_e = P_o$

(d)   None of the above are true.

1.47   Express the sequence

$$x(n) = \begin{cases} (-1)^n & -2 \leq n \leq 2 \\ 0 & \text{else} \end{cases}$$

as a sum of scaled and shifted unit steps.

1.48   Synthesize the triangular pulse

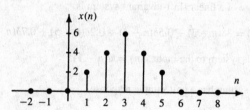

as a sum of scaled and shifted pulses,

$$p(n) = \begin{cases} 1 & 0 \leq n \leq 3 \\ 0 & \text{else} \end{cases}$$

Discrete-Time Systems

1.49   Listed below are several systems that relate the input $x(n)$ to the output $y(n)$. For each, determine whether the system is linear or nonlinear, shift-invariant or shift-varying, stable or unstable, causal or noncausal, and invertible or noninvertible.

(a) $y(n) = x(n) + x(-n)$

(b) $y(n) = \displaystyle\sum_{k=0}^{n} x(k)$

(c) $y(n) = \displaystyle\sum_{k=n-n_0}^{n+n_0} x(k)$

(d) $y(n) = \log\{x(n)\}$

(e) $y(n) = \text{median}\{x(n-1), x(n), x(n+1)\}$

1.50   Given below are the unit sample responses of several linear shift-invariant systems. For each system, determine the conditions on the parameter a in order for the system to be stable.

(a) $h(n) = a^n u(-n)$

(b) $h(n) = a^n \{u(n) - u(n - 100)\}$

(c) $h(n) = a^{|n|}$

1.51   Is it true that all memoryless systems are shift-invariant?

1.52   Consider the linear shift-invariant system described by the first-order linear constant coefficient difference equation

$$y(n) = ay(n - 1) + x(n)$$

Determine the conditions (if any) for which this system is stable.

1.53   Suppose that two systems, S_1 and S_2, are connected in parallel.

(a) If both S_1 and S_2 are linear, shift-invariant, stable, and causal, will the parallel connection always be linear, shift-invariant, stable, and causal?

(b) If both S_1 and S_2 are nonlinear, will the parallel connection necessarily be nonlinear?

(c) If both S_1 and S_2 are shift-varying, will the parallel connection necessarily be shift-varying?

Convolution

1.54   Find the convolution of the two sequences

$$x(n) = \delta(n - 2) - 2\delta(n - 4) + 3\delta(n - 6)$$
$$h(n) = 2\delta(n + 3) + \delta(n) + 2\delta(n - 2) + \delta(n - 3)$$

1.55   The unit sample response of a linear shift-invariant system is

$$h(n) = 3\delta(n - 3) + 0.5\delta(n - 4) + 0.2\delta(n - 5) + 0.7\delta(n - 6) - 0.8\delta(n - 7)$$

Find the response of this system to the input $x(n) = u(n - 1)$.

1.56   A linear shift-invariant system has a unit sample response

$$h(n) = u(-n)$$

Find the output if the input is

$$x(n) = \left(\tfrac{1}{3}\right)^n u(n)$$

1.57   The step response of a system is defined as the response of the system to a unit step $u(n)$.

(a) Let $s(n)$ be the step response of a linear shift-invariant system. Express $s(n)$ in terms of the unit sample response $h(n)$, and find $s(n)$ when $h(n) = u(n) - u(n - 6)$.

(b) Derive an expression for $h(n)$ in terms of $s(n)$ and find the unit sample response for a system whose step response is

$$s(n) = (-0.5)^n u(n)$$

1.58   The unit sample response of a linear shift-invariant system is shown below.

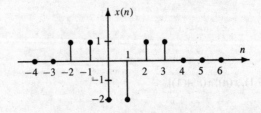

(a) Find the response of the system to the input $u(n-4)$.

(b) Repeat for $x(n) = (-1)^n u(n)$.

1.59   If $x(n) = (\frac{3}{4})^n u(n-2)$ and $h(n) = 2^n u(-n-5)$, find the convolution $y(n) = x(n) * h(n)$.

1.60   Given three sequences, $h(n)$, $g(n)$, and $r(n)$, express $g(n)$ in terms of $r(n)$ if

$$g(n) = \sum_{k=-\infty}^{\infty} h(n+k)h(5-k) \qquad r(n) = \sum_{k=-\infty}^{\infty} h(n-k)h(k)$$

1.61   Let $h(n) = a^n u(n)$ and $x(n) = b^n u(n)$. Find the convolution $y(n) = x(n) * h(n)$ assuming that $a \neq b$.

1.62   If $x(n) = a^n u(n)$, find the convolution $y(n) = x(n) * x(n)$.

1.63   The input to a linear shift-invariant system is the unit step, $x(n) = u(n)$, and the response is $y(n) = \delta(n)$. Find the unit sample response of this system.

1.64   If $h(n) = A\delta(n) + (\frac{1}{3})^n u(n)$ is the unit sample response of a linear shift-invariant system, and $s(n)$ is the step response (the response of the system to a unit step), find the value of the constant A so that $\lim_{n \to \infty} s(n) = 0$.

1.65   The unit sample response of a linear shift-invariant system is

$$h(n) = \left(\tfrac{1}{3}\right)^n u(n)$$

Find the response of the system to the complex exponential $x(n) = \exp(jn\pi/4)$.

1.66   Evaluate the convolution of the sequence $x(n) = n(\frac{1}{3})^n \cos(\pi n)$ with the unit step, $h(n) = u(n)$.

1.67   Let

$$x(n) = \begin{cases} n(0.5)^n & 0 \le n \le 5 \\ \dfrac{(0.25)^n}{n^2} & n < 0 \end{cases}$$

and $h(n) = e^{j\frac{\pi}{2}n} u(-n)$. If $y(n) = x(n) * h(n)$, what is the numerical value of $y(-2)$?

1.68   Given

$$x(n) = \begin{cases} \dfrac{n}{5} & 0 \le n \le 5 \\ 2 - \dfrac{n}{5} & 6 \le n \le 10 \\ 0 & \text{otherwise} \end{cases}$$

and $h(n) = \delta(n-2) + \delta(n-3) + \delta(n-4)$, at what value of n will the convolution $y(n) = x(n) * h(n)$ attain its maximum value, and what is this maximum value?

1.69   A linear system has a response $h_k(n) = \delta(2n-k)$ to the unit sample $\delta(n-k)$. Find the response of the system to the input $x(n) = u(n)$.

1.70   Consider the interconnection of three linear shift-invariant systems shown in the figure below.

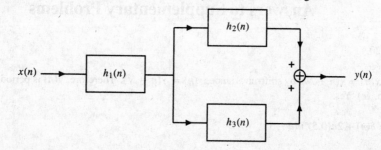

If $h_1(n) = u(n-2)$, $h_2(n) = nu(n)$ and $h_3(n) = \delta(n-2)$, find the unit sample response of the overall system.

Difference Equations

1.71   Consider the linear shift-invariant system described by the LCCDE

$$y(n) = -\tfrac{1}{2}y(n-1) + 2x(n)$$

Find the response of this system to the input

$$x(n) = \begin{cases} 2 & n = 0, 2, 4, 6, \ldots \\ 0 & \text{otherwise} \end{cases}$$

Hint: Write $x(n)$ as $\{1 + (-1)^n\}\, u(n)$ and use linearity.

1.72   Consider a system with input $x(n)$ and output $y(n)$ that satisfies the difference equation

$$y(n) = ny(n-1) + x(n)$$

If $x(n) = \delta(n)$, determine $y(n)$ for all n.

1.73   A linear shift-invariant system is described by the LCCDE

$$y(n) - 5y(n-1) + 6y(n-2) = x(n-1)$$

Find the step response of the system (i.e., the response to the input $x(n) = u(n)$).

1.74   A system is characterized by the difference equation

$$y(n) - 6y(n-1) + 8y(n-2) = 4x(n)$$

If the input is $x(n) = 2u(n) - 3nu(n)$, find the response of the system assuming initial conditions of $y(-1) = 2$ and $y(-2) = 1$.

1.75   Consider the system described by the difference equation

$$y(n) - y(n-1) + 0.25y(n-2) = x(n) - 0.25x(n-1)$$

(a)   Find the unit sample response of the system.

(b)   Find the response of the system to $x(n) = (0.25)^n u(n)$.

1.76   For a savings account that pays interest at the rate of 1 percent per month, if deposits are made on the first of each month at the rate of \$50 per month, how much money will there be in the account at the end of 1 year?

1.77   A savings account pays interest at the rate of 1 percent per month. With an initial deposit of \$50, how much will there be in the account after 10 years?

Answers to Supplementary Problems

1.41 $N = 120$.

1.42   (a) If $x(n) = x(n+N)$, by shift-invariance, $y(n) = y(n+N)$. Therefore, $y(n)$ is periodic with period N.
(b) No. (c) Yes.

1.43 $x(n) = \delta(n) + 2n(0.5)^n u(n-1)$.

1.44 $x_e(n) = \frac{1}{2}\left(\frac{\sqrt{2}}{2}\right)^{|n|} e^{jn\pi/4} + \frac{1}{2}\delta(n)$

1.45   Even.

1.46   (*a*) is true.

1.47 $x(n) = u(n+2) - 2u(n+1) + 2u(n) - 2u(n-1) + 2u(n-2) - u(n-3)$.

1.48 $x(n) = 2p(n-1) + 2p(n-2) + 2p(n-3) - 2p(n-4)$.

1.49   (*a*) Linear, shift-varying, stable, noncausal, noninvertible. (*b*) Linear, shift-varying, unstable, noncausal, invertible. (*c*) Linear, shift-invariant, stable, noncausal, invertible. (*d*) Nonlinear, shift-invariant, stable, causal, invertible. (*e*) Nonlinear, shift-invariant, stable, noncausal, noninvertible.

1.50   (*a*) $|a| > 1$. (*b*) Any finite a. (*c*) $|a| < 1$.

1.51   No. Consider the system $y(n) = x(n)\cos(n\pi/2)$.

1.52 $|a| < 1$.

1.53   (*a*) Yes. (*b*) No. (*c*) No.

1.54   The sequence values, beginning at index $n = -1$, are $y(n) = \{2, 0, -4, 1, 6, 0, 1, -1, -2, 6, 3\}$.

1.55 $y(n) = 3\delta(n-4) + 3.5\delta(n-5) + 3.7\delta(n-6) + 4.4\delta(n-7) + 3.6u(n-8)$.

1.56 $y(n) = \frac{3}{2}u(-n-1) + \frac{3}{2}\left(\frac{1}{3}\right)^n u(n)$.

1.57   (*a*) $s(n) = \sum_{k=-\infty}^{n} h(k)$. With $h(n) = u(n) - u(n-6)$ the step response is

$$s(n) = \begin{cases} 0 & n < 0 \\ (n+1) & 0 \le n \le 5 \\ 6 & 6 \le n \end{cases}$$

   (*b*) $h(n) = s(n) - s(n-1)$. If $s(n) = (-0.5)^n u(n)$, then $h(n) = \delta(n) + 3(-0.5)^n u(n-1)$.

1.58   (*a*) $y(n) = \delta(n-2) + 2\delta(n-3) - 2\delta(n-5) - \delta(n-6)$.

   (*b*) $y(n) = \delta(n+2) - 2\delta(n) + \delta(n-2)$.

1.59 $y(n) = \frac{1}{20}\left(\frac{3}{4}\right)^{n+5} u(n+2) + \frac{9}{40}(2)^n u(-n-3)$.

1.60 $g(n) = r(n+5)$.

1.61 $y(n) = \dfrac{b^{n+1} - a^{n+1}}{b-a} u(n)$.

1.62 $y(n) = (n+1) a^n u(n)$.

1.63 $h(n) = \delta(n) - \delta(n-1)$.

1.64 $A = -\frac{3}{2}$.

1.65 $y(n) = \dfrac{1}{1 - \frac{1}{3}e^{-j\pi/4}} e^{jn\pi/4}$.

1.66 $y(n) = \frac{1}{16}\big[(4n + 3)(-\frac{1}{3})^n - 3\big]u(n).$

1.67 $y(-2) = \frac{17}{4} - j\frac{119}{32}.$

1.68 $\max\{y(n)\} = \frac{13}{5}$, which occurs at index $n = 8$.

1.69 $y(n) = u(n).$

1.70 $y(n) = u(n - 4) + \frac{1}{2}(n - 2)(n - 1)\,u(n - 2).$

1.71 $y(n) = \big[4(-1)^n + \frac{4}{3} - \frac{4}{3}(-\frac{1}{2})^n\big]u(n).$

1.72 $y(n) = n!\,u(n).$

1.73 $y(n) = \big[\frac{1}{2} + (\frac{3}{2})(3)^n - 2(2^n)\big]u(n).$

1.74 $y(n) = \big[\frac{32}{3}[4^n - 1] - 4n + 12(2)^n\big]u(n).$

1.75   (a) $h(n) = \big[\frac{1}{2}n + 1\big]\big(\frac{1}{2}\big)^n u(n).$ (b) $y(n) = (n + 1)\big(\frac{1}{2}\big)^n u(n).$

1.76 $690.46.

1.77 $165.02.

Chapter 2

Fourier Analysis

2.1 INTRODUCTION

The Fourier representation of signals plays an extremely important role in both continuous-time and discrete-time signal processing. It provides a method for mapping signals into another "domain" in which to manipulate them. What makes the Fourier representation particularly useful is the property that the convolution operation is mapped to multiplication. In addition, the Fourier transform provides a different way to interpret signals and systems. In this chapter we will develop the discrete-time Fourier transform (i.e., a Fourier transform for discrete-time signals). We will show how complex exponentials are eigenfunctions of linear shift-invariant (LSI) systems and how this property leads to the notion of a frequency response representation of LSI systems. Finally, we will explore how the discrete-time Fourier transform may be used to solve linear constant-coefficient difference equations and perform convolutions.

2.2 FREQUENCY RESPONSE

Eigenfunctions of linear shift-invariant systems are sequences that, when input to the system, pass through with only a change in (complex) amplitude. That is to say, if the input is $x(n)$, the output is $y(n) = \lambda x(n)$, where λ, the eigenvalue, generally depends on the input $x(n)$.

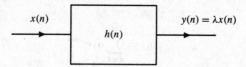

Signals of the form

$$x(n) = e^{jn\omega} \qquad -\infty < n < \infty$$

where ω is a constant, are eigenfunctions of LSI systems. This may be shown from the convolution sum:

$$y(n) = h(n) * x(n) = \sum_{k=-\infty}^{\infty} h(k)x(n-k)$$

$$= \sum_{k=-\infty}^{\infty} h(k)e^{j\omega(n-k)} = e^{jn\omega} \sum_{k=-\infty}^{\infty} h(k)e^{-jk\omega}$$

$$= H(e^{j\omega})e^{jn\omega}$$

Thus, the eigenvalue, which we denote by $H(e^{j\omega})$, is

$$H(e^{j\omega}) = \sum_{k=-\infty}^{\infty} h(k)e^{-jk\omega} \qquad (2.1)$$

Note that $H(e^{j\omega})$ is, in general, *complex-valued* and depends on the frequency ω of the complex exponential. Thus, it may be written in terms of its *real* and *imaginary* parts,

$$H(e^{j\omega}) = H_R(e^{j\omega}) + jH_I(e^{j\omega})$$

or in terms of its *magnitude* and *phase*,

$$H(e^{j\omega}) = |H(e^{j\omega})|e^{j\phi_h(\omega)}$$

55

where
$$|H(e^{j\omega})|^2 = H(e^{j\omega})H^*(e^{j\omega}) = H_R^2(e^{j\omega}) + H_I^2(e^{j\omega})$$

and
$$\phi_h(\omega) = \tan^{-1}\frac{H_I(e^{j\omega})}{H_R(e^{j\omega})}$$

Graphical representations of the frequency response are of great value in the analysis of LSI systems, and plots of the magnitude and phase are commonly used. However, another useful graphical representation is a plot of $20\log|H(e^{j\omega})|$ versus ω. The units on the log magnitude scale are decibels (abbreviated dB). Thus, 0 dB corresponds to a value of $|H(e^{j\omega})| = 1$, 20 dB is equivalent to $|H(e^{j\omega})| = 10$, -20 dB is equivalent to $|H(e^{j\omega})| = 0.1$, and so on. It is also useful to note that 6 dB corresponds approximately to $|H(e^{j\omega})| = 2$, and -6 dB is approximately $|H(e^{j\omega})| = 0.5$. One of the advantages of a log magnitude plot is that, because the logarithm expands the scale for small values of $|H(e^{j\omega})|$, it is useful in displaying the fine detail of the frequency response near zero.

A graphical representation that is often used instead of the phase is the *group delay*, which is defined as follows:

$$\tau_h(\omega) = -\frac{d\phi_h(\omega)}{d\omega}$$

In evaluating the group delay, the phase is taken to be a continuous and differentiable function of ω by adding integer multiples of π to the principal value of the phase (this is referred to as *unwrapping the phase*).

The function $H(e^{j\omega})$ is very useful and important in the characterization of LSI systems and is called the *frequency response*. The frequency response defines how a complex exponential is changed in (complex) amplitude when it is filtered by the system. The frequency response is particularly useful if we are able to decompose an input signal into a sum of complex exponentials. For example, the response of an LSI system to an input of the form

$$x(n) = \sum_{k=1}^{N}\alpha_k e^{jn\omega_k}$$

will be
$$y(n) = \sum_{k=1}^{N}\alpha_k H(e^{j\omega_k})e^{jn\omega_k}$$

where $H(e^{j\omega_k})$ is the frequency response of the system evaluated at frequency ω_k.

EXAMPLE 2.2.1 Let $x(n) = \cos(n\omega_0)$ be the input to a linear shift-invariant system with a real-valued unit sample response $h(n)$. If $x(n)$ is decomposed into a sum of two complex exponentials,

$$x(n) = \tfrac{1}{2}e^{jn\omega_0} + \tfrac{1}{2}e^{-jn\omega_0}$$

the response of the system may be written as

$$y(n) = \tfrac{1}{2}H(e^{j\omega_0})e^{jn\omega_0} + \tfrac{1}{2}H(e^{-j\omega_0})e^{-jn\omega_0}$$

Because $h(n)$ is real-valued, $H(e^{j\omega})$ is conjugate symmetric:

$$H(e^{-j\omega}) = H^*(e^{j\omega})$$

Therefore,
$$y(n) = \tfrac{1}{2}H(e^{j\omega_0})e^{jn\omega_0} + \tfrac{1}{2}H^*(e^{j\omega_0})e^{-jn\omega_0}$$

and it follows that

$$y(n) = \text{Re}\{H(e^{j\omega_0})e^{jn\omega_0}\} = |H(e^{j\omega_0})|\cos(n\omega_0 + \phi_h(\omega_0))$$

Periodicity

The frequency response is a complex-valued function of ω and is periodic with a period 2π. This is in sharp contrast with the frequency response of a linear time-invariant continuous-time system, which has a frequency response that is *not* periodic, in general. The reason for this periodicity stems from the fact that a discrete-time complex exponential of frequency ω_0 is the same as a complex exponential of frequency $\omega_0 + 2\pi$; that is,

$$x(n) = e^{jn\omega_0} = e^{jn(\omega_0 + 2\pi)}$$

Therefore, if the input to a linear shift-invariant system is $x(n) = e^{jn\omega_0}$, the response must be the same as the response to the signal $x(n) = e^{jn(\omega_0 + 2\pi)}$. This, in turn, requires that

$$H(e^{j\omega_0}) = H\left(e^{j(\omega_0 + 2\pi)}\right)$$

Symmetry

If $h(n)$ is real-valued, the frequency response is a *conjugate symmetric* function of frequency:

$$H(e^{-j\omega}) = H^*(e^{j\omega})$$

Conjugate symmetry of $H(e^{j\omega})$ implies that the real part is an even function of ω,

$$H_R(e^{j\omega}) = H_R(e^{-j\omega})$$

and that the imaginary part is odd,

$$H_I(e^{j\omega}) = -H_I(e^{-j\omega})$$

Conjugate symmetry also implies that the magnitude is even,

$$|H(e^{j\omega})| = |H(e^{-j\omega})|$$

and that the phase and group delay are odd,

$$\phi_h(\omega) = -\phi_h(-\omega) \qquad \tau_h(\omega) = -\tau_h(-\omega)$$

EXAMPLE 2.2.2   Consider the LSI system with unit sample response

$$h(n) = \alpha^n u(n)$$

where α is a real number with $|\alpha| < 1$. The frequency response is

$$H(e^{j\omega}) = \sum_{n=-\infty}^{\infty} h(n)e^{-jn\omega} = \sum_{n=0}^{\infty} \alpha^n e^{-jn\omega}$$

$$= \sum_{n=0}^{\infty} (\alpha e^{-j\omega})^n = \frac{1}{1 - \alpha e^{-j\omega}}$$

The squared magnitude of the frequency response is

$$|H(e^{j\omega})|^2 = H(e^{j\omega})H^*(e^{j\omega}) = \frac{1}{1 - \alpha e^{-j\omega}} \cdot \frac{1}{1 - \alpha e^{j\omega}} = \frac{1}{1 + \alpha^2 - 2\alpha \cos \omega}$$

and the phase is

$$\phi_h(\omega) = \tan^{-1} \frac{H_I(e^{j\omega})}{H_R(e^{j\omega})} = \tan^{-1} \frac{-\alpha \sin \omega}{1 - \alpha \cos \omega}$$

Finally, the group delay is found by differentiating the phase. The result is

$$\tau_h(\omega) = -\frac{\alpha^2 - \alpha \cos \omega}{1 + \alpha^2 - 2\alpha \cos \omega}$$

Inverting the Frequency Response

Given the frequency response of a linear shift-invariant system,

$$H(e^{j\omega}) = \sum_{n=-\infty}^{\infty} h(n)e^{-jn\omega}$$

the unit sample response may be recovered by integration:

$$h(n) = \frac{1}{2\pi} \int_{-\pi}^{\pi} H(e^{j\omega})e^{jn\omega}d\omega \qquad (2.2)$$

The integral may be taken over any period of length 2π.

EXAMPLE 2.2.3   For a system with a frequency response given by

$$H(e^{j\omega}) = \begin{cases} 1 & |\omega| \leq \omega_c \\ 0 & \omega_c < |\omega| \leq \pi \end{cases}$$

(this system is referred to as an ideal low-pass filter), the unit sample response is

$$h(n) = \frac{1}{2\pi} \int_{-\omega_c}^{\omega_c} e^{jn\omega}d\omega = \frac{1}{2j\pi n}[e^{jn\omega_c} - e^{-jn\omega_c}] = \frac{\sin n\omega_c}{\pi n}$$

Note that this system is noncausal (it is also unstable) and, therefore, unrealizable.

2.3 FILTERS

The term *digital filter*, or simply filter, is often used to refer to a discrete-time system. A digital filter is defined by J. F. Kaiser[1] as a "... computational process or algorithm by which a sampled signal or sequence of numbers (acting as the input) is transformed into a second sequence of numbers termed the output signal. The computational process may be that of lowpass filtering (smoothing), bandpass filtering, interpolation, the generation of derivatives, etc."

Filters may be characterized in terms of their system properties, such as linearity, shift-invariance, causality, stability, etc., and they may be classified in terms of the form of their frequency response. Some of these classifications are described below.

Linear Phase

A linear shift-invariant system is said to have linear phase if its frequency response is of the form

$$H(e^{j\omega}) = A(e^{j\omega})e^{-j\alpha\omega}$$

where α is a real number and $A(e^{j\omega})$ is a real-valued function of ω. Note that the phase of $H(e^{j\omega})$ is

$$\phi_h(\omega) = \begin{cases} -\alpha\omega & \text{when } A(e^{j\omega}) \geq 0 \\ -\alpha\omega + \pi & \text{when } A(e^{j\omega}) < 0 \end{cases}$$

Similarly, a filter is said to have *generalized linear phase* if the frequency response has the form

$$H(e^{j\omega}) = A(e^{j\omega})e^{-j(\alpha\omega-\beta)}$$

Thus, filters with linear phase or generalized linear phase have a constant group delay.

---

[1] *System Analysis by Digital Computer*, F. F. Kuo and J. F. Kaiser, Eds., John Wiley and Sons, New York, 1966.

Allpass

A system is said to be allpass filter if the frequency response magnitude is constant:

$$|H(e^{j\omega})| = c$$

An example of an allpass filter is the system that has a frequency response

$$H(e^{j\omega}) = \frac{e^{-j\omega} - \alpha}{1 - \alpha e^{-j\omega}}$$

where α is any real number with $|\alpha| < 1$. The unit sample response of this allpass filter is

$$h(n) = -\alpha\delta(n) + (1 - \alpha^2)\alpha^{n-1}u(n - 1)$$

Frequency Selective Filters

Many of the filters that are important in applications have piecewise constant frequency response magnitudes. These include the low-pass, high-pass, bandpass, and bandstop filters that are illustrated in Fig. 2-1. The intervals over which the frequency response magnitude is equal to 1 are called the *passbands*, and the intervals over which it is equal to 0 are called the *stopbands*. The frequencies that mark the edges of the passbands and stopbands are the *cutoff frequencies*.

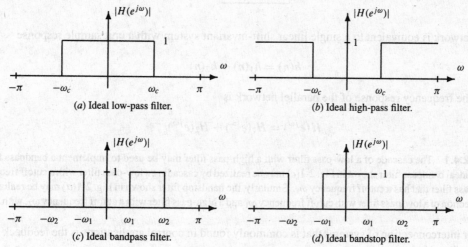

(a) Ideal low-pass filter.   (b) Ideal high-pass filter.

(c) Ideal bandpass filter.   (d) Ideal bandstop filter.

Fig. 2-1. Ideal filters.

2.4 INTERCONNECTION OF SYSTEMS

Filters are often interconnected to create systems that have desirable properties. Two common types of connections are series (cascade) and parallel. A cascade of two linear shift-invariant systems is shown in the figure below.

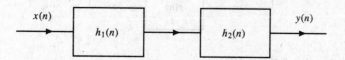

A cascade is equivalent to a single linear shift-invariant system with a unit sample response

$$h(n) = h_1(n) * h_2(n)$$

and a frequency response

$$H(e^{j\omega}) = H_1(e^{j\omega})H_2(e^{j\omega})$$

Note that the log magnitude of the cascade is the *sum* of the log magnitudes of the individual systems,

$$20\log|H(e^{j\omega})| = 20\log|H_1(e^{j\omega})| + 20\log|H_2(e^{j\omega})|$$

and the phase and group delay are additive,

$$\phi(\omega) = \phi_1(\omega) + \phi_2(\omega)$$
$$\tau(\omega) = \tau_1(\omega) + \tau_2(\omega)$$

A parallel connection of two linear shift-invariant systems is shown in the figure below.

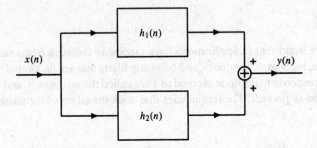

A parallel network is equivalent to a single linear shift-invariant system with a unit sample response

$$h(n) = h_1(n) + h_2(n)$$

Therefore, the frequency response of the parallel network is

$$H(e^{j\omega}) = H_1(e^{j\omega}) + H_2(e^{j\omega})$$

EXAMPLE 2.4.1 The cascade of a low-pass filter with a high-pass filter may be used to implement a bandpass filter. For example, the ideal bandpass filter shown in Fig. 2-1(c) may be realized by cascading a low-pass filter with a cutoff frequency ω_2 with a high-pass filter that has a cutoff frequency ω_1. Similarly, the bandstop filter shown in Fig. 2-1(d) may be realized with a parallel connection of a low-pass filter with cutoff frequency ω_1 and a high-pass filter with a cutoff frequency ω_2, with $\omega_2 > \omega_1$.

Another interconnection of systems that is commonly found in control applications is the feedback network shown in the figure below.

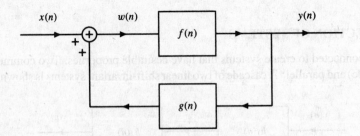

This network may be analyzed as follows. With

$$w(n) = x(n) + g(n) * y(n)$$

and

$$y(n) = f(n) * w(n)$$

we may use the Fourier analysis techniques described in the following section to show that the frequency response of this system, if it exists, is[2]

$$H(e^{j\omega}) = \frac{F(e^{j\omega})}{1 - F(e^{j\omega})G(e^{j\omega})}$$

2.5 THE DISCRETE-TIME FOURIER TRANSFORM

The frequency response of a linear shift-invariant system is found by multiplying $h(n)$ by a complex exponential, $e^{-jn\omega}$, and summing over n. The discrete-time Fourier transform (DTFT) of a sequence, $x(n)$, is defined in the same way,

$$X(e^{j\omega}) = \sum_{n=-\infty}^{\infty} x(n)e^{-jn\omega} \tag{2.3}$$

Thus, the frequency response of a linear shift-invariant system, $H(e^{j\omega})$, is the DTFT of the unit sample response, $h(n)$. In order for the DTFT of a sequence to exist, the summation in Eq. (2.3) must converge. This, in turn, requires that $x(n)$ be absolutely summable:

$$\sum_{n=-\infty}^{\infty} |x(n)| = S < \infty$$

EXAMPLE 2.5.1 The DTFT of the sequence

$$x_1(n) = \alpha^n u(n) \qquad |\alpha| < 1$$

is

$$X_1(e^{j\omega}) = \sum_{n=0}^{\infty} \alpha^n e^{-jn\omega} = \sum_{n=0}^{\infty} (\alpha e^{-j\omega})^n$$

Using the geometric series, this sum is

$$X_1(e^{j\omega}) = \frac{1}{1 - \alpha e^{-j\omega}}$$

provided $|\alpha| < 1$. Similarly, for the sequence

$$x_2(n) = -\alpha^n u(-n - 1) \qquad |\alpha| > 1$$

the DTFT is

$$X_2(e^{j\omega}) = \sum_{n=-\infty}^{\infty} x_2(n)e^{-jn\omega} = -\sum_{n=-\infty}^{-1} \alpha^n e^{-jn\omega}$$

Changing the limits on the sum, we have

$$X_2(e^{j\omega}) = -\sum_{n=1}^{\infty} \alpha^{-n} e^{jn\omega} = -\sum_{n=0}^{\infty} (\alpha^{-1} e^{j\omega})^n + 1$$

If $|\alpha| > 1$, this sum is

$$X_2(e^{j\omega}) = -\frac{1}{1 - \alpha^{-1}e^{j\omega}} + 1 = \frac{1}{1 - \alpha e^{-j\omega}}$$

Therefore, $x_1(n) = \alpha^n u(n)$ and $x_2(n) = -\alpha^n u(-n - 1)$ both have the same DTFT.

---

[2]It is possible that $g(n)$ will make the system unstable, in which case the DTFT of $h(n)$ will not exist. Feedback systems are typically analyzed using z-transforms.

Given $X(e^{j\omega})$, the sequence $x(n)$ may be recovered using the inverse DTFT,

$$x(n) = \frac{1}{2\pi} \int_{-\pi}^{\pi} X(e^{j\omega}) e^{jn\omega} d\omega \qquad (2.4)$$

The inverse DTFT may be viewed as a decomposition of $x(n)$ into a linear combination of all complex exponentials that have frequencies in the range $-\pi < \omega \leq \pi$. Table 2-1 contains a list of some useful DTFT pairs.

Table 2-1   Some Common DTFT Pairs

Sequence	Discrete-Time Fourier Transform
$\delta(n)$	1
$\delta(n - n_0)$	$e^{-jn_0\omega}$
1	$2\pi\delta(\omega)$
$e^{jn\omega_0}$	$2\pi\delta(\omega - \omega_0)$
$a^n u(n), \quad |a| < 1$	$\dfrac{1}{1 - ae^{-j\omega}}$
$-a^n u(-n - 1), \quad |a| > 1$	$\dfrac{1}{1 - ae^{-j\omega}}$
$(n + 1)a^n u(n), \quad |a| < 1$	$\dfrac{1}{(1 - ae^{-j\omega})^2}$
$\cos n\omega_0$	$\pi\delta(\omega + \omega_0) + \pi\delta(\omega - \omega_0)$

EXAMPLE 2.5.2   Suppose $X(e^{j\omega})$ consists of an impulse at frequency $\omega = \omega_0$:

$$X(e^{j\omega}) = \delta(\omega - \omega_0)$$

Using the inverse DTFT, we have

$$x(n) = \frac{1}{2\pi} \int_{-\pi}^{\pi} X(e^{j\omega}) e^{jn\omega} d\omega = \frac{1}{2\pi} e^{jn\omega_0}$$

Note that although $x(n)$ is not absolutely summable, by allowing the DTFT to contain impulses, we may consider the DTFT of sequences that contain complex exponentials. As another example, if

$$X(e^{j\omega}) = \pi\delta(\omega - \omega_0) + \pi\delta(\omega + \omega_0)$$

computing the inverse DTFT, we find

$$x(n) = \tfrac{1}{2} e^{jn\omega_0} + \tfrac{1}{2} e^{-jn\omega_0} = \cos(n\omega_0)$$

2.6   DTFT PROPERTIES

There are a number of properties of the DTFT that may be used to simplify the evaluation of the DTFT and its inverse. Some of these properties are described below. A summary of the DTFT properties appears in Table 2-2.

Periodicity

The discrete-time Fourier transform is periodic in ω with a period of 2π:

$$X(e^{j\omega}) = X(e^{j(\omega+2\pi)})$$

This property follows directly from the definition of the DTFT and the periodicity of the complex exponentials:

$$X(e^{j(\omega+2\pi)}) = \sum_{n=-\infty}^{\infty} x(n)e^{-jn(\omega+2\pi)} = \sum_{n=-\infty}^{\infty} x(n)e^{-jn\omega}e^{-j2\pi n}$$

$$= \sum_{n=-\infty}^{\infty} x(n)e^{-jn\omega} = X(e^{j\omega})$$

Table 2-2　Properties of the DTFT

Property	Sequence	Discrete-Time Fourier Transform
Linearity	$ax(n) + by(n)$	$aX(e^{j\omega}) + bY(e^{j\omega})$
Shift	$x(n - n_0)$	$e^{-jn_0\omega}X(e^{j\omega})$
Time-reversal	$x(-n)$	$X(e^{-j\omega})$
Modulation	$e^{jn\omega_0}x(n)$	$X(e^{j(\omega-\omega_0)})$
Convolution	$x(n) * y(n)$	$X(e^{j\omega})Y(e^{j\omega})$
Conjugation	$x^*(n)$	$X^*(e^{-j\omega})$
Derivative	$nx(n)$	$j\dfrac{dX(e^{j\omega})}{d\omega}$
Multiplication	$x(n)y(n)$	$\dfrac{1}{2\pi}\displaystyle\int_{-\pi}^{\pi} X(e^{j\theta})Y(e^{j(\omega-\theta)})d\theta$

Note: Given the DTFTs $X(e^{j\omega})$ and $Y(e^{j\omega})$ of $x(n)$ and $y(n)$, this table lists the DTFTs of sequences that are formed from $x(n)$ and $y(n)$.

Symmetry

The DTFT often has some symmetries that may be exploited to simplify the evaluation of the DTFT or the inverse DTFT. These properties are listed in the table below.

$x(n)$	$X(e^{j\omega})$
Real and even	Real and even
Real and odd	Imaginary and odd
Imaginary and even	Imaginary and even
Imaginary and odd	Real and odd

Note that these properties may be combined. For example, if $x(n)$ is conjugate symmetric, its real part is even and its imaginary part is odd. Therefore, it follows that $X(e^{j\omega})$ is real-valued. Similarly, note that if $x(n)$ is real, the real part of $X(e^{j\omega})$ is even and the imaginary part is odd. Thus, $X(e^{j\omega})$ is conjugate symmetric.

Linearity

The discrete-time Fourier transform is a *linear* operator. That is to say, if $X_1(e^{j\omega})$ is the DTFT of $x_1(n)$, and $X_2(e^{j\omega})$ is the DTFT of $x_2(n)$,

$$ax_1(n) + bx_2(n) \stackrel{DTFT}{\Longleftrightarrow} aX_1(e^{j\omega}) + bX_2(e^{j\omega})$$

Shifting Property

Shifting a sequence in time results in the multiplication of the DTFT by a complex exponential (linear phase term):

$$x(n - n_0) \stackrel{DTFT}{\Longleftrightarrow} e^{-jn_0\omega}X(e^{j\omega})$$

Time-Reversal

Time-reversing a sequence results in a frequency reversal of the DTFT:

$$x(-n) \stackrel{DTFT}{\Longleftrightarrow} X(e^{-j\omega})$$

Modulation

Multiplying a sequence by a complex exponential results in a shift in frequency of the DTFT:

$$e^{jn\omega_0}x(n) \overset{DTFT}{\Longleftrightarrow} X\left(e^{j(\omega-\omega_0)}\right)$$

Thus, modulating a sequence by a cosine of frequency ω_0 shifts the spectrum up and down in frequency by ω_0:

$$x(n)\cos n\omega_0 \overset{DTFT}{\Longleftrightarrow} \tfrac{1}{2}X\left(e^{j(\omega-\omega_0)}\right) + \tfrac{1}{2}X\left(e^{j(\omega+\omega_0)}\right)$$

Convolution Theorem

Perhaps the most important result in linear systems theory is that convolution in the time domain is equivalent to multiplication in the frequency domain. Specifically, this theorem says that the DTFT of a sequence that is formed by convolving two sequences, $x(n)$ and $h(n)$, is the product of the DTFTs of $x(n)$ and $h(n)$:

$$h(n) * x(n) \overset{DTFT}{\Longleftrightarrow} H(e^{j\omega})X(e^{j\omega})$$

Multiplication (Periodic Convolution) Theorem

As with the time-shift and modulation properties, there is a dual to the convolution theorem that states that multiplication in the time domain corresponds to (periodic) convolution in the frequency domain:

$$x(n)y(n) \overset{DTFT}{\Longleftrightarrow} \frac{1}{2\pi}\int_{-\pi}^{\pi} X(e^{j\theta})Y\left(e^{j(\omega-\theta)}\right)d\theta$$

Parseval's Theorem

A corollary to the multiplication theorem is *Parseval's theorem*, which is

$$\sum_{n=-\infty}^{\infty} |x(n)|^2 = \frac{1}{2\pi}\int_{-\pi}^{\pi} |X(e^{j\omega})|^2 d\omega$$

Parseval's theorem is referred to as the *conservation of energy theorem*, because it states that the DTFT operator preserves energy when going from the time domain into the frequency domain.

2.7 APPLICATIONS

In this section, we present some applications of the DTFT in discrete-time signal analysis. These include finding the frequency response of an LSI system that is described by a difference equation, performing convolutions, solving difference equations that have zero initial conditions, and designing inverse systems.

2.7.1 LSI Systems and LCCDEs

An important *subclass* of LSI systems contains those whose input, $x(n)$, and output, $y(n)$, are related by a linear constant coefficient difference equation (LCCDE):

$$y(n) = -\sum_{k=1}^{p} a(k)y(n-k) + \sum_{k=0}^{q} b(k)x(n-k) \tag{2.5}$$

The linearity and shift properties of the DTFT may be used to express this difference equation in the frequency domain as follows:

$$Y(e^{j\omega}) = -\sum_{k=1}^{p} a(k)e^{-jk\omega}Y(e^{j\omega}) + \sum_{k=0}^{q} b(k)e^{-jk\omega}X(e^{j\omega})$$

or

$$Y(e^{j\omega})\left[1 + \sum_{k=1}^{p} a(k)e^{-jk\omega}\right] = X(e^{j\omega})\sum_{k=0}^{q} b(k)e^{-jk\omega}$$

Therefore, the frequency response of this system is

$$H(e^{j\omega}) = \frac{Y(e^{j\omega})}{X(e^{j\omega})} = \frac{\displaystyle\sum_{k=0}^{q} b(k)e^{-jk\omega}}{1 + \displaystyle\sum_{k=1}^{p} a(k)e^{-jk\omega}} \qquad (2.6)$$

EXAMPLE 2.7.1   Consider the linear shift-invariant system characterized by the second-order linear constant coefficient difference equation

$$y(n) = 1.3433y(n-1) - 0.9025y(n-2) + x(n) - 1.4142x(n-1) + x(n-2)$$

The frequency response may be found by inspection without solving the difference equation for $h(n)$ as follows:

$$H(e^{j\omega}) = \frac{1 - 1.4142e^{-j\omega} + e^{-2j\omega}}{1 - 1.3433e^{-j\omega} + 0.9025e^{-2j\omega}}$$

Note that this problem may also be worked in the reverse direction. For example, given a frequency response function such as

$$H(e^{j\omega}) = \frac{1 + e^{-2j\omega}}{2 - e^{-j\omega} + 0.5e^{-2j\omega}}$$

a difference equation may be easily found that will implement this system. First, dividing numerator and denominator by 2 and rewriting the frequency response as follows,

$$H(e^{j\omega}) = \frac{0.5 + 0.5e^{-2j\omega}}{1 - 0.5e^{-j\omega} + 0.25e^{-2j\omega}}$$

we see that a difference equation for this system is

$$y(n) = 0.5y(n-1) - 0.25y(n-2) + 0.5x(n) + 0.5x(n-2)$$

2.7.2   Performing Convolutions

Because the DTFT maps convolution in the time domain into multiplication in the frequency domain, the DTFT provides an alternative to performing convolutions in the time domain. The following example illustrates the procedure.

EXAMPLE 2.7.2   If the unit sample response of an LSI system is

$$h(n) = \alpha^n u(n)$$

let us find the response of the system to the input $x(n) = \beta^n u(n)$, where $|\alpha| < 1$, $|\beta| < 1$, and $\alpha \neq \beta$. Because the output of the system is the convolution of $x(n)$ with $h(n)$,

$$y(n) = h(n) * x(n)$$

the DTFT of $y(n)$ is

$$Y(e^{j\omega}) = H(e^{j\omega})X(e^{j\omega}) = \frac{1}{1 - \alpha e^{-j\omega}}\frac{1}{1 - \beta e^{-j\omega}}$$

Therefore, all that is required is to find the inverse DTFT of $Y(e^{j\omega})$. This may be done easily by expanding $Y(e^{j\omega})$ as follows:

$$Y(e^{j\omega}) = \frac{1}{(1 - \alpha e^{-j\omega})(1 - \beta e^{-j\omega})} = \frac{A}{1 - \alpha e^{-j\omega}} + \frac{B}{1 - \beta e^{-j\omega}}$$

where A and B are constants that are to be determined. Expressing the right-hand side of this expansion over a common denominator,

$$\frac{1}{(1 - \alpha e^{-j\omega})(1 - \beta e^{-j\omega})} = \frac{(A + B) - (A\beta + B\alpha)e^{-j\omega}}{(1 - \alpha e^{-j\omega})(1 - \beta e^{-j\omega})}$$

and equating coefficients, the constants A and B may be found by solving the pair of equations

$$A + B = 1$$
$$A\beta + B\alpha = 0$$

The result is

$$A = \frac{\alpha}{\alpha - \beta} \qquad B = -\frac{\beta}{\alpha - \beta}$$

Therefore,

$$Y(e^{j\omega}) = \frac{\alpha/(\alpha - \beta)}{1 - \alpha e^{-j\omega}} - \frac{\beta/(\alpha - \beta)}{1 - \beta e^{-j\omega}}$$

and it follows that the inverse DTFT is

$$y(n) = \left[\frac{\alpha}{\alpha - \beta}\alpha^n - \frac{\beta}{\alpha - \beta}\beta^n \right] u(n)$$

2.7.3 Solving Difference Equations

In Chap. 1 we looked at methods for solving difference equations in the "time domain." The DTFT may be used to solve difference equations in the "frequency domain" provided that the initial conditions are zero. The procedure is simply to transform the difference equation into the frequency domain by taking the DTFT of each term in the equation, solving for the desired term, and finding the inverse DTFT.

EXAMPLE 2.7.3 Let us solve the following LCCDE for $y(n)$ assuming zero initial conditions,

$$y(n) - 0.25y(n - 1) = x(n) - x(n - 2)$$

for $x(n) = \delta(n)$. We begin by taking the DTFT of each term in the difference equation:

$$Y(e^{j\omega}) - 0.25e^{-j\omega}Y(e^{j\omega}) = X(e^{j\omega}) - e^{-2j\omega}X(e^{j\omega})$$

Because the DTFT of $x(n)$ is $X(e^{j\omega}) = 1$,

$$Y(e^{j\omega}) = \frac{1 - e^{-2j\omega}}{1 - 0.25e^{-j\omega}} = \frac{1}{1 - 0.25e^{-j\omega}} - \frac{e^{-2j\omega}}{1 - 0.25e^{-j\omega}}$$

Using the DTFT pair

$$(0.25)^n u(n) \overset{DTFT}{\Longleftrightarrow} \frac{1}{1 - 0.25e^{-j\omega}}$$

the inverse DTFT of $Y(e^{j\omega})$ may be easily found using the linearity and shift properties,

$$y(n) = (0.25)^n u(n) - (0.25)^{n-2} u(n - 2)$$

2.7.4 Inverse Systems

The inverse of a system with unit sample response $h(n)$ is a system that has a unit sample response $g(n)$ such that

$$h(n) * g(n) = \delta(n)$$

In terms of the frequency response, it is easy to see that, if the inverse of $H(e^{j\omega})$ exists, it is equal to

$$G(e^{j\omega}) = \frac{1}{H(e^{j\omega})}$$

Care must be exercised, however, because not all systems are invertible or, if the inverse exists, it may be noncausal. For example, the ideal low-pass filter in Example 2.2.3 does not have an inverse, and the inverse of the system

$$H(e^{j\omega}) = 1 - 2e^{-j\omega}$$

is

$$G(e^{j\omega}) = \frac{1}{1 - 2e^{-j\omega}}$$

which corresponds to a system that has a noncausal unit sample response

$$g(n) = -2^{-n}u(-n - 1)$$

EXAMPLE 2.7.4   If the frequency response of an LSI system is

$$H(e^{e^{j\omega}}) = \frac{1 - \frac{1}{4}e^{-j\omega}}{1 + \frac{1}{2}e^{-j\omega}}$$

the inverse system is

$$G(e^{j\omega}) = \frac{1}{H(e^{j\omega})} = \frac{1 + \frac{1}{2}e^{-j\omega}}{1 - \frac{1}{4}e^{-j\omega}}$$

which has a unit sample response

$$g(n) = (0.25)^n u(n) + 0.5(0.25)^{n-1} u(n - 1)$$

Solved Problems

Frequency Response

2.1   Let $h(n)$ be the unit sample response of an LSI system. Find the frequency response when

(a) $h(n) = \delta(n) + 6\delta(n - 1) + 3\delta(n - 2)$

(b) $h(n) = \left(\frac{1}{3}\right)^{n+2} u(n - 2)$.

(a) This system has a unit sample response that is finite in length. Therefore, the frequency response is a polynomial in $e^{j\omega}$, with the coefficients of the polynomial equal to the values of $h(n)$:

$$H(e^{j\omega}) = 1 + 6e^{-j\omega} + 3e^{-2j\omega}$$

This may be shown more formally by writing

$$H(e^{j\omega}) = \sum_{n=-\infty}^{\infty} h(n)e^{-jn\omega} = \sum_{n=-\infty}^{\infty} [\delta(n) + 6\delta(n - 1) + 3\delta(n - 2)]e^{-jn\omega}$$

Because

$$\sum_{n=-\infty}^{\infty} \delta(n - n_0)e^{-jn\omega} = e^{-jn_0\omega}$$

then

$$H(e^{j\omega}) = 1 + 6e^{-j\omega} + 3e^{-2j\omega}$$

(b) For the second system, the frequency response is

$$H(e^{j\omega}) = \sum_{n=-\infty}^{\infty} h(n)e^{-jn\omega} = \sum_{n=2}^{\infty} \left(\tfrac{1}{3}\right)^{n+2} e^{-jn\omega}$$

Changing the limits on the sum so that it begins with $n = 0$, we have

$$H(e^{j\omega}) = \sum_{n=0}^{\infty} \left(\tfrac{1}{3}\right)^{n+4} e^{-j(n+2)\omega} = \left(\tfrac{1}{3}\right)^4 e^{-2j\omega} \sum_{n=0}^{\infty} \left(\tfrac{1}{3} e^{-j\omega}\right)^n$$

Using the geometric series, we find

$$H(e^{j\omega}) = \left(\tfrac{1}{3}\right)^4 \frac{e^{-2j\omega}}{1 - \tfrac{1}{3}e^{-j\omega}}$$

2.2 An Lth-order moving average filter is a linear shift-invariant system that, for an input $x(n)$, produces the output

$$y(n) = \frac{1}{L+1} \sum_{k=0}^{L} x(n-k)$$

Find the frequency response of this system.

If the input to the moving average filter is $x(n) = \delta(n)$, the response, by definition, will be the unit sample response, $h(n)$. Therefore,

$$h(n) = \frac{1}{L+1} \sum_{k=0}^{L} \delta(n-k)$$

and

$$H(e^{j\omega}) = \frac{1}{L+1} \sum_{k=0}^{L} e^{-jk\omega}$$

Using the geometric series, we have

$$H(e^{j\omega}) = \frac{1}{L+1} \frac{1 - e^{-j(L+1)\omega}}{1 - e^{-j\omega}}$$

Factoring out a term $e^{-j(L+1)\omega/2}$ from the numerator, and a term $e^{-j\omega/2}$ from the denominator, we have

$$H(e^{j\omega}) = \frac{1}{L+1} e^{-jL\omega/2} \frac{e^{j(L+1)\omega/2} - e^{-j(L+1)\omega/2}}{e^{j\omega/2} - e^{-j\omega/2}}$$

or

$$H(e^{j\omega}) = \frac{1}{L+1} e^{-jL\omega/2} \frac{\sin(L+1)\omega/2}{\sin \omega/2}$$

2.3 The input to a linear shift-invariant system is

$$x(n) = 2\cos\left(\frac{n\pi}{4}\right) + 3\sin\left(\frac{3n\pi}{4} + \frac{\pi}{8}\right)$$

Find the output if the unit sample response of the system is

$$h(n) = 2\frac{\sin[(n-1)\pi/2]}{(n-1)\pi}$$

This problem may be solved using the eigenfunction property of LSI systems. Specifically, as we saw in Example 2.2.1, if the input to an LSI system is $x(n) = \cos(n\omega_0)$, the response will be

$$y(n) = |H(e^{j\omega_0})| \cos(n\omega_0 + \phi_h(\omega_0))$$

FOURIER ANALYSIS

Therefore, we need to find the frequency response of the system. In Example 2.2.3, it was shown that the unit sample response of an ideal low-pass filter,

$$H_1(e^{j\omega}) = \begin{cases} 1 & |\omega| \le \omega_c \\ 0 & \omega_c < |\omega| \le \pi \end{cases}$$

is

$$h_1(n) = \frac{\sin n\omega_c}{\pi n}$$

Because $h(n) = 2h_1(n-1)$ with $\omega_c = \pi/2$, an expression may be derived for $H(e^{j\omega})$ in terms of $H_1(e^{j\omega})$ as follows:

$$H(e^{j\omega}) = \sum_{n=-\infty}^{\infty} h(n)e^{-jn\omega} = \sum_{n=-\infty}^{\infty} 2h_1(n-1)e^{-jn\omega}$$

$$= 2\sum_{n=-\infty}^{\infty} h_1(n)e^{-j(n+1)\omega} = 2e^{-j\omega}\sum_{n=-\infty}^{\infty} h_1(n)e^{-jn\omega} = 2e^{-j\omega}H_1(e^{j\omega})$$

Therefore,

$$H(e^{j\omega}) = \begin{cases} 2e^{-j\omega} & |\omega| \le \dfrac{\pi}{2} \\ 0 & \dfrac{\pi}{2} < |\omega| \le \pi \end{cases}$$

Because $|H(e^{j\omega})| = 0$ at $\omega = 3\pi/4$, the sinusoid in $x(n)$ is *filtered out*, and the output is simply

$$y(n) = 2|H(e^{j\pi/4})|\cos\left(\frac{n\pi}{4} + \phi_h\left(\frac{\pi}{4}\right)\right)$$

$$= 4\cos\left(\frac{n\pi}{4} + \frac{\pi}{4}\right) = 4\cos\left[(n-1)\frac{\pi}{4}\right]$$

2.4 Find the magnitude, phase, and group delay of a system that has a unit sample response

$$h(n) = \delta(n) - \alpha\delta(n-1)$$

where α is real.

The frequency response of this system is

$$H(e^{j\omega}) = 1 - \alpha e^{-j\omega} = 1 - \alpha\cos\omega + j\alpha\sin\omega$$

Therefore, the magnitude squared is

$$|H(e^{j\omega})|^2 = H(e^{j\omega})H^*(e^{j\omega}) = (1 - \alpha e^{-j\omega})\cdot(1 - \alpha e^{j\omega}) = 1 + \alpha^2 - 2\alpha\cos\omega$$

The phase, on the other hand, is

$$\phi_h(\omega) = \tan^{-1}\frac{H_I(e^{j\omega})}{H_R(e^{j\omega})} = \tan^{-1}\frac{\alpha\sin\omega}{1 - \alpha\cos\omega}$$

Finally, the group delay may be found by differentiating the phase (see Prob. 2.19). Alternatively, we may note that because this system is the inverse of the one considered in Example 2.2.2, the phase and the group delay are simply the negative of those found in the example. Therefore, we have

$$\tau_h(\omega) = \frac{\alpha^2 - \alpha\cos\omega}{1 + \alpha^2 - 2\alpha\cos\omega}$$

2.5 A 90° phase shifter is a system with a frequency response

$$H(e^{j\omega}) = \begin{cases} -j & 0 < \omega < \pi \\ j & -\pi < \omega < 0 \end{cases}$$

Note that the magnitude is constant for all ω, and the phase is $-\pi/2$ for $0 < \omega < \pi$ and $\pi/2$ for $-\pi < \omega < 0$. Find the unit sample response of this system.

The unit sample response may be found by integration:

$$h(n) = \frac{1}{2\pi} \int_{-\pi}^{\pi} H(e^{j\omega}) e^{jn\omega} d\omega = \frac{1}{2\pi} \int_{-\pi}^{0} j e^{jn\omega} d\omega - \frac{1}{2\pi} \int_{0}^{\pi} j e^{jn\omega} d\omega$$

$$= \frac{1}{2\pi n} e^{jn\omega} \Big|_{-\pi}^{0} - \frac{1}{2\pi n} e^{jn\omega} \Big|_{0}^{\pi} = \frac{1}{2\pi n}[1 - e^{-jn\pi}] - \frac{1}{2\pi n}[e^{jn\pi} - 1]$$

$$= \frac{1}{\pi n}[1 - e^{jn\pi}] = \frac{1}{\pi n}[1 - (-1)^n]$$

Therefore, we have

$$h(n) = \begin{cases} \dfrac{2}{n\pi} & n \text{ odd} \\ 0 & n \text{ even} \end{cases}$$

which may also be expressed as

$$h(n) = \begin{cases} \dfrac{2}{\pi} \dfrac{\sin^2(n\pi/2)}{n} & n \neq 0 \\ 0 & n = 0 \end{cases}$$

Filters

2.6   Let $h(n)$ be the unit sample response of a low-pass filter with a cutoff frequency ω_c.

(a) What type of filter has a unit sample response $g(n) = (-1)^n h(n)$?

(b) If a filter with a unit sample response $h(n)$ is implemented with a difference equation of the form

$$y(n) = \sum_{k=1}^{p} a(k)y(n-k) + \sum_{k=0}^{q} b(k)x(n-k) \tag{2.7}$$

how should this difference equation be modified to implement the system that has a unit sample response $g(n) = (-1)^n h(n)$?

(a) Given that $g(n) = (-1)^n h(n)$, the frequency response $G(e^{j\omega})$ is related to the frequency response of the low-pass filter, $H(e^{j\omega})$, as follows:

$$G(e^{j\omega}) = \sum_{n=-\infty}^{\infty} g(n)e^{-jn\omega} = \sum_{n=-\infty}^{\infty} (-1)^n h(n)e^{-jn\omega}$$

$$= \sum_{n=-\infty}^{\infty} h(n)e^{-jn(\omega-\pi)} = H\left(e^{j(\omega-\pi)}\right)$$

Therefore, $G(e^{j\omega})$ is formed by shifting $H(e^{j\omega})$ in frequency by π. Thus, if the passband of the low-pass filter is $|\omega| \leq \omega_c$, the passband of $G(e^{j\omega})$ will be $\pi - \omega_c < |\omega| \leq \pi$. As a result, it follows that $g(n)$ is the unit sample response of a high-pass filter.

(b) If a filter with a unit sample response $h(n)$ may be realized by the difference equation given in Eq. (2.7), the frequency response of the filter is

$$H(e^{j\omega}) = \frac{\displaystyle\sum_{k=0}^{q} b(k)e^{-jk\omega}}{1 - \displaystyle\sum_{k=1}^{p} a(k)e^{-jk\omega}}$$

Multiplying $h(n)$ by $(-1)^n$ produces a system with a frequency response

$$G(e^{j\omega}) = H\left(e^{j(\omega-\pi)}\right) = \frac{\displaystyle\sum_{k=0}^{q} b(k)e^{-jk(\omega-\pi)}}{1 - \displaystyle\sum_{k=1}^{p} a(k)e^{-jk(\omega-\pi)}}$$

Because $e^{jk\pi} = (-1)^k$,

$$G(e^{j\omega}) = \frac{\sum_{k=0}^{q} (-1)^k b(k) e^{-jk\omega}}{1 - \sum_{k=1}^{p} (-1)^k a(k) e^{-jk\omega}}$$

and the difference equation becomes

$$y(n) = \sum_{k=1}^{p} (-1)^k a(k) y(n-k) + \sum_{k=0}^{q} (-1)^k b(k) x(n-k)$$

That is, the coefficients $a(k)$ and $b(k)$ for k odd are negated.

2.7 Let $H(e^{j\omega})$ be the frequency response of an ideal low-pass filter with a cutoff frequency ω_c as shown in the figure below.

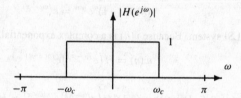

Assume that the phase is linear, $\phi_h(\omega) = -n_0\omega$. Determine whether or not it is possible to find an input $x(n)$ and a cutoff frequency $\omega_c < \pi$ that will produce the output

$$y(n) = \begin{cases} 1 & n = 0, 1, \ldots, 20 \\ 0 & \text{otherwise} \end{cases}$$

If $X(e^{j\omega})$ is the DTFT of $x(n)$, the output of the low-pass filter will have a DTFT

$$Y(e^{j\omega}) = H(e^{j\omega})X(e^{j\omega})$$

Therefore, $Y(e^{j\omega})$ must be equal to zero for $\omega_c \le |\omega| \le \pi$. However, the DTFT of $y(n)$ is

$$Y(e^{j\omega}) = \sum_{n=0}^{20} e^{-jn\omega} = \frac{1 - e^{-21j\omega}}{1 - e^{-j\omega}}$$

which is not zero for $\omega_c \le |\omega| \le \pi$. Therefore, there is no value for $\omega_c < \pi$, and no input $x(n)$ that will generate the given output $y(n)$.

2.8 Let $h(n)$ be the unit sample response of an ideal low-pass filter with a cutoff frequency $\omega_c = \pi/4$. Shown in the figure below is a linear shift-invariant system that is formed from a cascade of a low-pass filter and two modulators. Find the frequency response of the overall system relating the input $x(n)$ to the output $y(n)$.

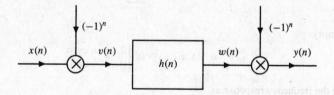

There are two ways that we may use to find the frequency response of this system. The first is to note that because the input to the low-pass filter is $(-1)^n x(n)$, the output of the low-pass filter is

$$w(n) = h(n) * [(-1)^n x(n)] = \sum_{k=-\infty}^{\infty} h(k)(-1)^{n-k} x(n-k)$$

Therefore,
$$y(n) = (-1)^n w(n) = (-1)^n \left\{ \sum_{k=-\infty}^{\infty} h(k)(-1)^{n-k} x(n-k) \right\}$$

Bringing the term $(-1)^n$ inside the summation, and using the fact that $(-1)^{n-k} = (-1)^{k-n}$, we have

$$y(n) = \sum_{k=-\infty}^{\infty} h(k)(-1)^n(-1)^{k-n} x(n-k) = \sum_{k=-\infty}^{\infty} (-1)^k h(k) x(n-k) = [(-1)^n h(n)] * x(n)$$

Thus, the unit sample response of the overall system is $(-1)^n h(n)$, and the frequency response is

$$\text{DTFT}\{(-1)^n h(n)\} = H\left(e^{j(\omega-\pi)}\right) = \begin{cases} 1 & \dfrac{3\pi}{4} \leq |\omega| \leq \pi \\ 0 & \text{otherwise} \end{cases}$$

Another way to determine the frequency response is to find the response of the system to a complex exponential, $x(n) = e^{jn\omega}$. Modulating by $(-1)^n = e^{-jn\pi}$ produces the sequence

$$v(n) = (-1)^n e^{jn\omega} = e^{jn(\omega-\pi)}$$

which is the input to the LSI system. Because $v(n)$ is a complex exponential, the response of the system to $v(n)$ is

$$w(n) = H\left(e^{j(\omega-\pi)}\right) e^{jn(\omega-\pi)}$$

Finally, with
$$y(n) = (-1)^n w(n) = (-1)^n H\left(e^{j(\omega-\pi)}\right) e^{jn(\omega-\pi)} = H\left(e^{j(\omega-\pi)}\right) e^{jn\omega}$$

it follows that the frequency response of the overall system is $H\left(e^{j(\omega-\pi)}\right)$ as we found before.

2.9 If $h(n)$ is the unit sample response of an ideal low-pass filter with a cutoff frequency $\omega_c = \pi/4$, find the frequency response of the filter that has a unit sample response $g(n) = h(2n)$.

To find the frequency response of this system, we may work the problem in one of two ways. The first is to note that because the unit sample response of an ideal low-pass filter with a cutoff frequency $\omega_c = \pi/4$ is

$$h(n) = \frac{\sin(n\pi/4)}{n\pi}$$

then
$$h(2n) = \frac{\sin(2n\pi/4)}{2n\pi} = \frac{1}{2} \frac{\sin(n\pi/2)}{n\pi}$$

which is the unit sample response of a low-pass filter with a magnitude of $\frac{1}{2}$ and a cutoff frequency $\omega = \pi/2$. The second way to work this problem is to find the frequency response of the system that has a unit sample response $g(n) = h(2n)$, given that $H(e^{j\omega})$ is the frequency response of a system with a unit sample response $h(n)$. Although more difficult than the first approach, this will give a general expression for the frequency response $G(e^{j\omega})$ in terms of $H(e^{j\omega})$ that may be applied to any system. To find the frequency response, we must evaluate the sum

$$G(e^{j\omega}) = \sum_{n=-\infty}^{\infty} h(2n) e^{-jn\omega}$$

Using the identity
$$1 + (-1)^n = \begin{cases} 2 & n \text{ even} \\ 0 & n \text{ odd} \end{cases}$$

we may write the frequency response as

$$G(e^{j\omega}) = \frac{1}{2} \sum_{n=-\infty}^{\infty} [1 + (-1)^n] h(n) e^{-jn\omega/2}$$

$$= \frac{1}{2} \sum_{n=-\infty}^{\infty} h(n) e^{-jn\omega/2} + \frac{1}{2} \sum_{n=-\infty}^{\infty} (-1)^n h(n) e^{-jn\omega/2}$$

In terms of $H(e^{j\omega})$, the first term may be written as

$$\frac{1}{2} \sum_{n=-\infty}^{\infty} h(n)e^{-jn\omega/2} = \frac{1}{2}H(e^{j\omega/2})$$

whereas the second term is

$$\frac{1}{2} \sum_{n=-\infty}^{\infty} (-1)^n h(n)e^{-jn\omega/2} = \frac{1}{2} \sum_{n=-\infty}^{\infty} h(n)e^{-jn(\omega+2\pi)/2} = \frac{1}{2}H(e^{j(\omega+2\pi)/2})$$

Therefore, $$G(e^{j\omega}) = \frac{1}{2}H(e^{j\omega/2}) + \frac{1}{2}H(e^{j(\omega+2\pi)/2})$$

With $H(e^{j\omega})$ the frequency response of a low-pass filter with a cutoff frequency $\omega_c = \pi/4$, this gives the same result as before.

2.10   Consider the high-pass filter that has a cutoff frequency $\omega_c = 3\pi/4$ as shown in the following figure:

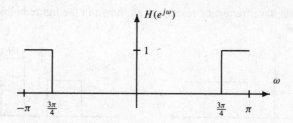

(a) Find the unit sample response, $h(n)$.

(b) A new system is defined so that its unit sample response is $h_1(n) = h(2n)$. Sketch the frequency response, $H_1(e^{j\omega})$, of this system.

(a) The unit sample response may be found two different ways. The first is to use the inverse DTFT formula and perform the integration. The second approach is to use the modulation property and note that if

$$H_{lp}(e^{j\omega}) = \begin{cases} 1 & \text{for } |\omega| \le \dfrac{\pi}{4} \\ 0 & \text{otherwise} \end{cases}$$

$H(e^{j\omega})$ may be written as

$$H(e^{j\omega}) = H_{lp}(e^{j(\omega-\pi)})$$

Therefore, it follows from the modulation property that

$$h(n) = e^{jn\pi} h_{lp}(n) = (-1)^n h_{lp}(n)$$

With $$h_{lp}(n) = \frac{\sin(n\pi/4)}{n\pi}$$

we have $$h(n) = (-1)^n \frac{\sin(n\pi/4)}{n\pi}$$

(b) The frequency response of the system that has a unit sample response $h_1(n) = h(2n)$ may be found by evaluating the discrete-time Fourier transform sum directly:

$$H_1(e^{j\omega}) = \sum_{n=-\infty}^{\infty} h_1(n)e^{-jn\omega} = \sum_{n=-\infty}^{\infty} h(2n)e^{-jn\omega} = \sum_{n \text{ even}} h(n)e^{-jn\omega/2}$$

However, an easier approach is to note that

$$h_1(n) = h(2n) = (-1)^{2n} \frac{\sin(2n\pi/4)}{2n\pi} = \frac{\sin(n\pi/2)}{2n\pi}$$

which is a low-pass filter with a cutoff frequency of $\pi/2$ and a gain of $\frac{1}{2}$. A plot of $H_1(e^{j\omega})$ is shown in the following figure:

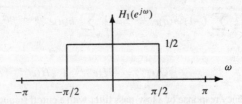

Interconnection of Systems

2.11 The ideal filters that have frequency responses as shown in the figure below are connected in cascade.

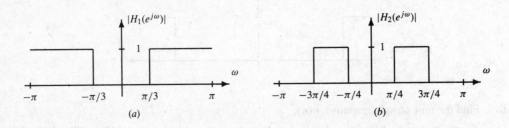

For an arbitrary input $x(n)$, find the range of frequencies that can be present in the output $y(n)$. Repeat for the case in which the two systems are connected in parallel.

If these two filters are connected in cascade, the frequency response of the cascade is

$$H(e^{j\omega}) = H_1(e^{j\omega})H_2(e^{j\omega})$$

Therefore, any frequencies in the output, $y(n)$, must be passed by *both* filters. Because the passband for $H_1(e^{j\omega})$ is $|\omega| > \pi/3$, and the passband for $H_2(e^{j\omega})$ is $\pi/4 < |\omega| < 3\pi/4$, the passband for the cascade (the frequencies for which both $|H_1(e^{j\omega})|$ and $|H_2(e^{j\omega})|$ are equal to 1) is

$$\frac{\pi}{3} \le |\omega| \le \frac{3\pi}{4}$$

With a parallel connection, the overall frequency response is

$$H(e^{j\omega}) = H_1(e^{j\omega}) + H_2(e^{j\omega})$$

Therefore, the frequencies that are contained in the output are those that are passed by *either* filter, or

$$|\omega| > \frac{\pi}{4}$$

2.12 Consider the following interconnection of linear shift-invariant systems:

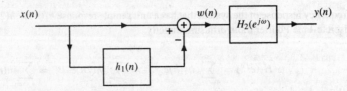

where $\qquad h_1(n) = \delta(n-1) \quad$ and $\quad H_2(e^{j\omega}) = \begin{cases} 1 & |\omega| \leq \dfrac{\pi}{2} \\ 0 & \dfrac{\pi}{2} < |\omega| \leq \pi \end{cases}$

Find the frequency response and the unit sample response of this system.

To find the unit sample response, let $x(n) = \delta(n)$. The output of the adder is then

$$w(n) = \delta(n) - \delta(n-1)$$

Because $w(n)$ is input to an LSI system with a unit sample response $h_2(n)$,

$$y(n) = h_2(n) - h_2(n-1)$$

where $\qquad h_2(n) = \dfrac{1}{2\pi} \int_{-\pi}^{\pi} H_2(e^{j\omega}) e^{jn\omega} d\omega = \dfrac{1}{2\pi} \int_{-\pi/2}^{\pi/2} e^{jn\omega} d\omega = \dfrac{\sin(n\pi/2)}{n\pi}$

Therefore, the unit sample response of the overall system is

$$h(n) = \frac{\sin(n\pi/2)}{n\pi} - \frac{\sin[(n-1)\pi/2]}{(n-1)\pi}$$

To find the frequency response, note that

$$W(e^{j\omega}) = 1 - e^{-j\omega}$$

Therefore, $\qquad H(e^{j\omega}) = W(e^{j\omega})H_2(e^{j\omega}) = [1 - e^{-j\omega}]H_2(e^{j\omega}) = \begin{cases} 1 - e^{-j\omega} & |\omega| \leq \dfrac{\pi}{2} \\ 0 & \dfrac{\pi}{2} < |\omega| \leq \pi \end{cases}$

2.13   Consider the interconnection of LSI systems shown in the following figure:

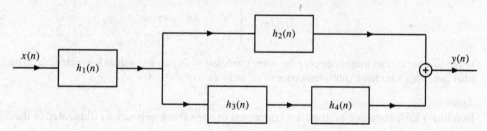

(*a*)   Express the frequency response of the overall system in terms of $H_1(e^{j\omega})$, $H_2(e^{j\omega})$, $H_3(e^{j\omega})$, and $H_4(e^{j\omega})$.

(*b*)   Find the frequency response if

$$h_1(n) = \delta(n) + 2\delta(n-2) + \delta(n-4)$$
$$h_2(n) = h_3(n) = (0.2)^n u(n)$$
$$h_4(n) = \delta(n-2)$$

(*a*)   Because $h_2(n)$ is in parallel with the cascade of $h_3(n)$ and $h_4(n)$, the frequency response of the parallel network is

$$G(e^{j\omega}) = H_2(e^{j\omega}) + H_3(e^{j\omega})H_4(e^{j\omega})$$

With $h_1(n)$ being in cascade with $g(n)$, the overall frequency response becomes

$$H(e^{j\omega}) = H_1(e^{j\omega})[H_2(e^{j\omega}) + H_3(e^{j\omega})H_4(e^{j\omega})]$$

(b) The frequency responses of the systems in this interconnection are

$$H_1(e^{j\omega}) = 1 + 2e^{-j2\omega} + e^{-j4\omega} = (1 + e^{-j2\omega})^2$$

$$H_2(e^{j\omega}) = H_3(e^{j\omega}) = \frac{1}{1 - 0.2e^{-j\omega}}$$

$$H_4(e^{j\omega}) = e^{-j2\omega}$$

Therefore,
$$H(e^{j\omega}) = H_1(e^{j\omega})[H_2(e^{j\omega}) + H_3(e^{j\omega})H_4(e^{j\omega})]$$
$$= H_1(e^{j\omega})H_2(e^{j\omega})[1 + H_4(e^{j\omega})]$$
$$= \frac{(1 + e^{-j2\omega})^3}{1 - 0.2e^{-j\omega}}$$

2.14 Suppose that the frequency response of a linear shift-invariant system is piecewise constant as shown in the following figure:

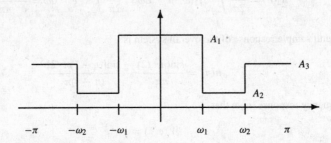

Describe how this filter may be implemented as a parallel connection of low-pass filters.

This filter may be viewed as a summation of a low-pass filter, a bandpass filter, and a high-pass filter. Because both a bandpass filter and a high-pass filter may be synthesized using a parallel connection of low-pass filters, we may proceed as follows. First, we put an allpass filter $H_3(e^{j\omega}) = A_3$ in parallel with a low-pass filter with a cutoff frequency ω_2 and a gain of $A_2 - A_3$. This parallel network has a frequency response

$$H(e^{j\omega}) = \begin{cases} A_2 & |\omega| < \omega_2 \\ A_3 & \omega_2 \le |\omega| < \pi \end{cases}$$

To produce the correct magnitude over the lower band, $|\omega| < \omega_1$, we add a third low-pass filter in parallel with the other two. This filter has a cutoff frequency of ω_1 and a gain of $A_1 - A_2$.

2.15 Two linear shift-invariant systems are connected in a feedback network as illustrated in the figure below.

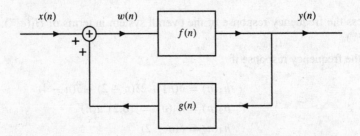

Assuming that the overall system is stable, so that $H(e^{j\omega})$ exists, show that the frequency response of this feedback network is

$$H(e^{j\omega}) = \frac{Y(e^{j\omega})}{X(e^{j\omega})} = \frac{F(e^{j\omega})}{1 - F(e^{j\omega})G(e^{j\omega})}$$

To analyze this network, we begin by noting that

$$w(n) = x(n) + g(n) * y(n)$$

which, in the frequency domain, becomes

$$W(e^{j\omega}) = X(e^{j\omega}) + G(e^{j\omega})Y(e^{j\omega})$$

Because

$$Y(e^{j\omega}) = F(e^{j\omega})W(e^{j\omega})$$

then

$$Y(e^{j\omega}) = F(e^{j\omega})[X(e^{j\omega}) + G(e^{j\omega})Y(e^{j\omega})]$$

Solving for $Y(e^{j\omega})$ yields

$$Y(e^{j\omega}) = \frac{F(e^{j\omega})}{1 - F(e^{j\omega})G(e^{j\omega})}X(e^{j\omega})$$

Therefore, the frequency response is

$$H(e^{j\omega}) = \frac{Y(e^{j\omega})}{X(e^{j\omega})} = \frac{F(e^{j\omega})}{1 - F(e^{j\omega})G(e^{j\omega})}$$

The Discrete-Time Fourier Transform

2.16 A linear shift-invariant system is described by the LCCDE

$$y(n) = 0.5y(n - 1) + bx(n)$$

Find the value of b so that $|H(e^{j\omega})|$ is equal to 1 at $\omega = 0$, and find the *half-power point* (i.e., the frequency at which $|H(e^{j\omega})|^2$ is equal to one-half of its peak value, which occurs at $\omega = 0$).

The frequency response of the system described by this difference equation is

$$H(e^{j\omega}) = \frac{b}{1 - 0.5e^{-j\omega}}$$

Because

$$|H(e^{j\omega})|^2 = \frac{b^2}{(1 - 0.5e^{-j\omega})(1 - 0.5e^{j\omega})} = \frac{b^2}{1.25 - \cos\omega}$$

$|H(e^{j\omega})|$ will be equal to 1 at $\omega = 0$ if

$$\frac{b^2}{1.25 - 1} = 1$$

This will be true when $b = \pm 0.5$.

To find the half-power point, we want to find the frequency for which

$$|H(e^{j\omega})|^2 = \frac{0.25}{1.25 - \cos\omega} = 0.5$$

This occurs when

$$\cos\omega = 0.75$$

or $\omega = 0.23\pi$.

2.17 Consider the system defined by the difference equation

$$y(n) = ay(n - 1) + bx(n) + x(n - 1)$$

where a and b are real, and $|a| < 1$. Find the relationship between a and b that must exist if the frequency response is to have a constant magnitude for all ω, that is,

$$|H(e^{j\omega})| = 1$$

Assuming that this relationship is satisfied, find the output of the system when $a = \frac{1}{2}$ and

$$x(n) = \left(\tfrac{1}{2}\right)^n u(n)$$

The frequency response of the LSI system described by this difference equation is

$$H(e^{j\omega}) = \frac{b + e^{-j\omega}}{1 - ae^{-j\omega}}$$

The squared magnitude is

$$|H(e^{j\omega})|^2 = \frac{(b + e^{-j\omega})(b + e^{j\omega})}{(1 - ae^{-j\omega})(1 - ae^{j\omega})} = \frac{1 + b^2 + 2b\cos\omega}{1 + a^2 - 2a\cos\omega}$$

Therefore, it follows that $|H(e^{j\omega})|^2 = 1$ if and only if $b = -a$.
With $a = \frac{1}{2}$ and $b = -\frac{1}{2}$, if $x(n) = \left(\tfrac{1}{2}\right)^n u(n)$, $Y(e^{j\omega})$ is given by

$$Y(e^{j\omega}) = H(e^{j\omega})X(e^{j\omega}) = \frac{-\frac{1}{2} + e^{-j\omega}}{1 - \frac{1}{2}e^{-j\omega}} \cdot \frac{1}{1 - \frac{1}{2}e^{-j\omega}} = \frac{-\frac{1}{2} + e^{-j\omega}}{\left(1 - \frac{1}{2}e^{-j\omega}\right)^2}$$

Using the DTFT pair

$$(n+1)a^n u(n) \overset{DTFT}{\Longleftrightarrow} \frac{1}{(1 - ae^{-j\omega})^2}$$

given in Table 2-1, and using the linearity and delay properties of the DTFT, we have

$$y(n) = -\tfrac{1}{2}(n+1)\left(\tfrac{1}{2}\right)^n u(n) + n\left(\tfrac{1}{2}\right)^{n-1} u(n-1)$$

What we observe from this example is that although $|H(e^{j\omega})| = 1$, the nonlinear phase has a significant effect on the values of the input sequence.

2.18 Show that the group delay of a linear shift-invariant system with a frequency response $H(e^{j\omega})$ may be expressed as

$$\tau_h(\omega) = \frac{H_R(e^{j\omega})G_R(e^{j\omega}) + H_I(e^{j\omega})G_I(e^{j\omega})}{|H(e^{j\omega})|^2}$$

where $H_R(e^{j\omega})$ and $H_I(e^{j\omega})$ are the real and imaginary parts of $H(e^{j\omega})$, respectively, and $G_R(e^{j\omega})$ and $G_I(e^{j\omega})$ are the real and imaginary parts of the DTFT of $nh(n)$.

In terms of magnitude and phase, the frequency response is

$$H(e^{j\omega}) = |H(e^{j\omega})|e^{j\phi_h(\omega)}$$

Note that if we take the logarithm of $H(e^{j\omega})$, we have an explicit expression for the phase

$$\ln H(e^{j\omega}) = \ln|H(e^{j\omega})| + j\phi_h(\omega)$$

Differentiating with respect to ω, we have

$$\frac{d}{d\omega}\ln H(e^{j\omega}) = \frac{1}{H(e^{j\omega})}\frac{d}{d\omega}H(e^{j\omega}) = \frac{d}{d\omega}\ln|H(e^{j\omega})| + j\frac{d}{d\omega}\phi_h(\omega)$$

Equating the imaginary parts of both sides of this equation yields

$$\frac{d}{d\omega}\phi_h(\omega) = \text{Im}\left\{\frac{1}{H(e^{j\omega})}\frac{d}{d\omega}H(e^{j\omega})\right\}$$

If we define

$$\frac{d}{d\omega}H(e^{j\omega}) = H_R'(e^{j\omega}) + jH_I'(e^{j\omega})$$

where $H_R'(e^{j\omega})$ is the derivative of the real part of $H(e^{j\omega})$ and $H_I'(e^{j\omega})$ is the derivative of the imaginary part, the group delay may be written as

$$\tau_h(\omega) = -\frac{d}{d\omega}\phi_h(\omega) = -\text{Im}\left\{\frac{H_R'(e^{j\omega}) + jH_I'(e^{j\omega})}{H(e^{j\omega})}\right\}$$

Multiplying the numerator and denominator by $H^*(e^{j\omega}) = H_R(e^{j\omega}) - jH_I(e^{j\omega})$ yields

$$\tau_h(\omega) = -\text{Im}\left\{\frac{\left[H_R'(e^{j\omega}) + jH_I'(e^{j\omega})\right]\left[H_R(e^{j\omega}) - jH_I(e^{j\omega})\right]}{|H(e^{j\omega})|^2}\right\}$$

$$= \frac{H_I(e^{j\omega})H_R'(e^{j\omega}) - H_R(e^{j\omega})H_I'(e^{j\omega})}{|H(e^{j\omega})|^2}$$

Finally, recall that if $H(e^{j\omega})$ is the DTFT of $h(n)$, the DTFT of $g(n) = nh(n)$ is

$$G(e^{j\omega}) = G_R(e^{j\omega}) + jG_I(e^{j\omega}) = j\frac{d}{d\omega}H(e^{j\omega}) = -H_I'(e^{j\omega}) + jH_R'(e^{j\omega})$$

where $G_R(e^{j\omega})$ is the real part of the DTFT of $nh(n)$, and $G_I(e^{j\omega})$ is the imaginary part. Therefore, $H_R'(e^{j\omega}) = G_I(e^{j\omega})$ and $H_I'(e^{j\omega}) = -G_R(e^{j\omega})$. Expressed in terms of $G_R(e^{j\omega})$ and $G_I(e^{j\omega})$, the group delay becomes

$$\tau_h(\omega) = \frac{H_R(e^{j\omega})G_R(e^{j\omega}) + H_I(e^{j\omega})G_I(e^{j\omega})}{|H(e^{j\omega})|^2}$$

Note that this expression for the group delay is convenient for digital evaluation, because it only requires computing the DTFT of $h(n)$ and $nh(n)$, and no derivatives.

2.19 Find the group delay for each of the following systems, where α is a real number:

(a) $H_1(e^{j\omega}) = 1 - \alpha e^{-j\omega}$

(b) $H_2(e^{j\omega}) = \dfrac{1}{1 - \alpha e^{-j\omega}}$

(c) $H_3(e^{j\omega}) = \dfrac{1}{1 - 2\alpha\cos\theta e^{-j\omega} + \alpha^2 e^{-j2\omega}}$

(a) For the first system, the frequency response is

$$H_1(e^{j\omega}) = 1 - \alpha\cos\omega + j\alpha\sin\omega$$

Therefore, the phase is

$$\phi_1(\omega) = \tan^{-1}\frac{\alpha\sin\omega}{1 - \alpha\cos\omega}$$

Because

$$\frac{d}{dx}\tan^{-1}u = \frac{1}{1 + u^2}\frac{du}{dx}$$

the group delay is

$$\tau_1(\omega) = -\frac{d}{d\omega}\phi_1(\omega) = -\frac{1}{1 + \left(\frac{\alpha\sin\omega}{1-\alpha\cos\omega}\right)^2}\frac{d}{d\omega}\left(\frac{\alpha\sin\omega}{1 - \alpha\cos\omega}\right)$$

Therefore,

$$\tau_1(\omega) = -\frac{1}{1 + \left(\frac{\alpha\sin\omega}{1-\alpha\cos\omega}\right)^2}\frac{(1 - \alpha\cos\omega)\alpha\cos\omega - (\alpha\sin\omega)^2}{(1 - \alpha\cos\omega)^2}$$

which, after simplification, becomes

$$\tau_1(\omega) = -\frac{(1 - \alpha \cos \omega)\alpha \cos \omega - (\alpha \sin \omega)^2}{(1 - \alpha \cos \omega)^2 + (\alpha \sin \omega)^2} = \frac{\alpha^2 - \alpha \cos \omega}{1 + \alpha^2 - 2\alpha \cos \omega}$$

Another way to solve this problem is to use the expression for the group delay derived in Prob. 2.18. With

$$H_1(e^{j\omega}) = 1 - \alpha \cos \omega + j\alpha \sin \omega$$

we see that

$$H_R(e^{j\omega}) = 1 - \alpha \cos \omega \qquad H_I(e^{j\omega}) = \alpha \sin \omega$$

Because the unit sample response is

$$h(n) = \delta(n) - \alpha\delta(n - 1)$$

then

$$g(n) = nh(n) = -\alpha\delta(n - 1)$$

and

$$G(e^{j\omega}) = -\alpha e^{-j\omega} = -\alpha \cos \omega + j\alpha \sin \omega$$

Therefore, the group delay is

$$\begin{aligned}
\tau_1(\omega) &= \frac{H_R(e^{j\omega})G_R(e^{j\omega}) + H_I(e^{j\omega})G_I(e^{j\omega})}{|H(e^{j\omega})|^2} \\
&= \frac{-\alpha \cos \omega(1 - \alpha \cos \omega) + (\alpha \sin \omega)^2}{(1 - \alpha \cos \omega)^2 + (\alpha \sin \omega)^2} = \frac{\alpha^2 - \alpha \cos \omega}{1 + \alpha^2 - 2\alpha \cos \omega}
\end{aligned}$$

which is the same as before.

(b) Having found the group delay for $H_1(e^{j\omega}) = 1 - \alpha e^{-j\omega}$, we may easily derive the group delay for $H_2(e^{j\omega})$, which is the inverse of $H_1(e^{j\omega})$:

$$H_2(e^{j\omega}) = \frac{1}{1 - \alpha e^{-j\omega}} = \frac{1}{H_1(e^{j\omega})}$$

Specifically, because

$$H_2(e^{j\omega}) = \frac{1}{H_2(e^{j\omega})}$$

$\phi_2(\omega) = -\phi_1(\omega)$ and, therefore,

$$\tau_2(\omega) = -\tau_1(\omega) = -\frac{\alpha^2 - \alpha \cos \omega}{1 + \alpha^2 - 2\alpha \cos \omega}$$

(c) For the last system, $H_3(e^{j\omega})$ may be factored as follows:

$$H_3(e^{j\omega}) = \frac{1}{1 - 2\alpha \cos \theta e^{-j\omega} + \alpha^2 e^{-j2\omega}} = \frac{1}{1 - \alpha e^{j\theta} e^{-j\omega}} \frac{1}{1 - \alpha e^{-j\theta} e^{-j\omega}}$$

The group delay of $H_3(e^{j\omega})$ is thus the sum of the group delays of these two factors. Furthermore, the group delay of each factor may be found straightforwardly by differentiating the phase. However, the group delay of these terms may also be found from $\tau_2(\omega)$ in part (b) if we use the modulation property of the DTFT. Specifically, recall that if $X(e^{j\omega})$ is the DTFT of $x(n)$, the DTFT of $e^{jn\theta}x(n)$ is

$$e^{jn\theta}x(n) \overset{DTFT}{\Longleftrightarrow} X(e^{j(\omega-\theta)}) = |X(e^{j(\omega-\theta)})|e^{j\phi(\omega-\theta)}$$

Therefore, if the group delay of $x(n)$ is $\tau(\omega)$, the group delay of $e^{jn\theta}x(n)$ will be $\tau(\omega - \theta)$. In part (b), we found that the group delay of $H(e^{j\omega}) = 1/(1 - \alpha e^{-j\omega})$ is

$$\tau(\omega) = -\frac{\alpha^2 - \alpha \cos \omega}{1 + \alpha^2 - 2\alpha \cos \omega}$$

Thus, it follows from the modulation property that the group delay of $H(e^{j\omega}) = 1/(1 - \alpha e^{-j(\omega-\theta)})$ is

$$\tau(\omega) = -\frac{\alpha^2 - \alpha \cos(\omega - \theta)}{1 + \alpha^2 - 2\alpha \cos(\omega - \theta)}$$

and that the group delay of $H(e^{j\omega}) = 1/(1 - \alpha e^{-j(\omega+\theta)})$ is

$$\tau(\omega) = -\frac{\alpha^2 - \alpha \cos(\omega + \theta)}{1 + \alpha^2 - 2\alpha \cos(\omega + \theta)}$$

Therefore, the group delay of $H_3(e^{j\omega})$ is the sum of these:

$$\tau_3(\omega) = -\frac{\alpha^2 - \alpha \cos(\omega - \theta)}{1 + \alpha^2 - 2\alpha \cos(\omega - \theta)} - \frac{\alpha^2 - \alpha \cos(\omega + \theta)}{1 + \alpha^2 - 2\alpha \cos(\omega + \theta)}$$

2.20   Find the DTFT of each of the following sequences:

(a) $x_1(n) = \left(\frac{1}{2}\right)^n u(n + 3)$

(b) $x_2(n) = \alpha^n \sin(n\omega_0)\, u(n)$

(c) $x_3(n) = \begin{cases} \left(\frac{1}{2}\right)^n & n = 0, 2, 4, \ldots \\ 0 & \text{otherwise} \end{cases}$

(a)   For the first sequence, the DTFT may be evaluated directly as follows:

$$X_1(e^{j\omega}) = \sum_{n=-3}^{\infty} \left(\frac{1}{2}\right)^n e^{-jn\omega} = \sum_{n=-3}^{\infty} \left(\frac{1}{2}e^{-j\omega}\right)^n$$

$$= \left(\frac{1}{2}e^{-j\omega}\right)^{-3} \sum_{n=0}^{\infty} \left(\frac{1}{2}e^{-j\omega}\right)^n = \frac{8e^{3j\omega}}{1 - \frac{1}{2}e^{-j\omega}}$$

(b)   The best way to find the DTFT of $x_2(n)$ is to express the sinusoid as a sum of two complex exponentials as follows:

$$x_2(n) = \frac{1}{2j}[\alpha^n e^{jn\omega_0} - \alpha^n e^{-jn\omega_0}]u(n)$$

The DTFT of the first term is

$$\frac{1}{2j} \sum_{n=0}^{\infty} \alpha^n e^{jn\omega_0} e^{-jn\omega} = \frac{1}{2j} \sum_{n=0}^{\infty} \left(\alpha e^{-j(\omega-\omega_0)}\right)^n = \frac{1}{2j} \frac{1}{1 - \alpha e^{-j(\omega-\omega_0)}}$$

Similarly, for the second term we have

$$-\frac{1}{2j} \sum_{n=0}^{\infty} \alpha^n e^{-jn\omega_0} e^{-jn\omega} = -\frac{1}{2j} \frac{1}{1 - \alpha e^{-j(\omega+\omega_0)}}$$

Therefore,

$$X_2(e^{j\omega}) = \frac{1}{2j}\left[\frac{1}{1 - \alpha e^{-j(\omega-\omega_0)}} - \frac{1}{1 - \alpha e^{-j(\omega+\omega_0)}}\right] = \frac{(\alpha \sin \omega_0)e^{-j\omega}}{1 - (2\alpha \cos \omega_0)e^{-j\omega} + \alpha^2 e^{-2j\omega}}$$

(c)   Finally, for $x_3(n)$, we have

$$X_3(e^{j\omega}) = \sum_{n=-\infty}^{\infty} x_3(n)e^{-jn\omega} = \sum_{n=0,2,4,\ldots}^{\infty} \left(\frac{1}{2}\right)^n e^{-jn\omega}$$

Therefore, $$X_3(e^{j\omega}) = \sum_{n=0}^{\infty} \left(\frac{1}{2}\right)^{2n} e^{-2jn\omega} = \sum_{n=0}^{\infty} \left(\frac{1}{4}e^{-2j\omega}\right)^n = \frac{1}{1 - \frac{1}{4}e^{-2j\omega}}$$

2.21 Because the DTFT of the output of a linear shift-invariant filter with frequency response $H(e^{j\omega})$ is

$$Y(e^{j\omega}) = X(e^{j\omega})H(e^{j\omega})$$

where $X(e^{j\omega})$ is the DTFT of the input, it follows that an LSI system cannot produce frequencies in the output that are not present in the input. Therefore, if a system introduces *new* frequencies, the system must be nonlinear and/or shift-varying. For each of the following systems, find the frequencies that are present in the output when $x(n) = \cos(n\omega_0)$:

(a) $y(n) = x^2(n)$

(b) $y(n) = x(n)\cos(n\pi/4)$

(c) $y(n) = x(2n)$

(a) With $x(n) = \cos(n\omega_0)$, the output of the square-law device is

$$y(n) = \cos^2(n\omega_0)$$

Using the trigonometric identity

$$\cos^2 A = \tfrac{1}{2} + \tfrac{1}{2}\cos(2A)$$

it follows that

$$y(n) = \tfrac{1}{2} + \tfrac{1}{2}\cos(2n\omega_0)$$

Therefore, although the only frequencies present in the input are $\omega = \pm\omega_0$, the frequencies in the output are $\omega = 0,\ \pm 2\omega_0$. Because this system is nonlinear, it creates frequencies in the output that are not in the input.

(b) For the modulator, the output is

$$y(n) = x(n)\cos\left(\frac{n\pi}{4}\right) = \cos(n\omega_0)\cos\left(\frac{n\pi}{4}\right)$$

Using the trigonometric identity

$$2\cos A \cos B = \cos(A+B) + \cos(A-B)$$

it follows that

$$y(n) = \tfrac{1}{2}\cos\left(n\omega_0 + \frac{n\pi}{4}\right) + \tfrac{1}{2}\cos\left(n\omega_0 - \frac{n\pi}{4}\right)$$

Therefore, the frequencies in the output are $\omega = \omega_0 \pm \pi/4$, which are different from those in the input. This is because the modulator is a shift-varying system.

(c) The last system, called a down-sampler, produces the output

$$y(n) = x(2n) = \cos(2n\omega_0)$$

thus creating frequencies in the output that are not present in the input. The down-sampler is a shift-varying system.

2.22 For each of the following pairs of signals, $x(n)$ and $y(n)$, determine whether or not there is a linear shift-invariant system that has the given response, $y(n)$, to the given input, $x(n)$. If such a system exists, determine whether or not the system is unique, and find the frequency response of an LSI system with the desired behavior. If no such LSI system exits, explain why.

(a) $x(n) = \left(\tfrac{1}{2}\right)^n u(n)$, $y(n) = \left(\tfrac{1}{4}\right)^n u(n)$

(b) $x(n) = e^{jn\pi/4}$, $y(n) = 0.5e^{jn\pi/4}$

(c) $x(n) = \dfrac{\sin(n\pi/4)}{n\pi}$, $y(n) = \dfrac{\sin(n\pi/2)}{n\pi}$

(d) $x(n) = u(n)$, $y(n) = \delta(n)$

(*a*) For the first input-output pair, we have

$$X(e^{j\omega}) = \frac{1}{1 - \frac{1}{2}e^{-j\omega}} \qquad Y(e^{j\omega}) = \frac{1}{1 - \frac{1}{4}e^{-j\omega}}$$

Because $X(e^{j\omega})$ is nonzero for all ω, the system that produces the response $y(n)$ is unique and is given by

$$H(e^{j\omega}) = \frac{Y(e^{j\omega})}{X(e^{j\omega})} = \frac{1 - \frac{1}{2}e^{-j\omega}}{1 - \frac{1}{4}e^{-j\omega}}$$

(*b*) For the second system, note that the input is a complex exponential with a frequency $\omega = \pi/4$. Therefore, if the system is LSI, the output must be a complex exponential of exactly the same frequency, that is,

$$y(n) = H(e^{j\pi/4})e^{jn\pi/4}$$

Because the output is

$$y(n) = 0.5\, e^{jn\pi/4}$$

any LSI system with

$$H(e^{j\pi/4}) = 0.5$$

will produce the given response. Thus, the system is not unique. One possible system is the low-pass filter

$$H(e^{j\omega}) = \begin{cases} 0.5 & |\omega| < \dfrac{\pi}{2} \\ 0 & \text{otherwise} \end{cases}$$

(*c*) For the third system, recall that an ideal low-pass filter with a cutoff frequency ω_c has a unit sample response given by (see Example 2.2.3)

$$h(n) = \frac{\sin n\omega_c}{\pi n}$$

Therefore, the DTFT of the input $x(n)$ is

$$X(e^{j\omega}) = \begin{cases} 1 & |\omega| < \dfrac{\pi}{4} \\ 0 & \text{otherwise} \end{cases}$$

and the DTFT of the output $y(n)$ is

$$Y(e^{j\omega}) = \begin{cases} 1 & |\omega| < \dfrac{\pi}{2} \\ 0 & \text{otherwise} \end{cases}$$

Because $X(e^{j\omega}) = 0$ for $|\omega| > \pi/4$, if the system is to be linear and shift-invariant, $Y(e^{j\omega})$ must be equal to zero for $|\omega| > \pi/4$ (an LSI system cannot produce new frequencies). Because this is not the case, no LSI system will produce the given input-output pair.

(*d*) For the last system, we are given $x(n) = u(n)$ and $y(n) = \delta(n)$. Therefore,

$$X(e^{j\omega}) = \frac{1}{1 - e^{-j\omega}} \quad \text{and} \quad Y(e^{j\omega}) = 1$$

As in part (*a*), there is a unique LSI system that produces this input-output pair, and the frequency response of this system is

$$H(e^{j\omega}) = \frac{Y(e^{j\omega})}{X(e^{j\omega})} = 1 - e^{-j\omega}$$

2.23 Find the DTFT of the two-sided sequence

$$x(n) = \left(\tfrac{1}{4}\right)^{|n|}$$

Note that we may write $x(n)$ as the sum of a left-sided sequence and a right-sided sequence as follows:

$$x(n) = \left(\tfrac{1}{4}\right)^n u(n) + \left(\tfrac{1}{4}\right)^{-n} u(-n) - \delta(n)$$

where the last term is included to remove the extra term that is introduced at $n = 0$ by the two exponential sequences. The DTFT of the first term is

$$\left(\tfrac{1}{4}\right)^n u(n) \overset{DTFT}{\Longleftrightarrow} \frac{1}{1 - \tfrac{1}{4}e^{-j\omega}}$$

and, using the time-reversal property, it follows that the DTFT of the second term is

$$\left(\tfrac{1}{4}\right)^{-n} u(-n) \overset{DTFT}{\Longleftrightarrow} \frac{1}{1 - \tfrac{1}{4}e^{j\omega}}$$

Therefore,

$$X(e^{j\omega}) = \frac{1}{1 - \tfrac{1}{4}e^{-j\omega}} + \frac{1}{1 - \tfrac{1}{4}e^{j\omega}} - 1$$

$$= \frac{\tfrac{15}{16}}{\left(1 - \tfrac{1}{4}e^{-j\omega}\right)\left(1 - \tfrac{1}{4}e^{j\omega}\right)} = \frac{\tfrac{15}{16}}{\tfrac{17}{16} - \tfrac{1}{2}\cos\omega}$$

2.24 Use the orthogonality of the complex exponentials

$$\frac{1}{2\pi}\int_{-\pi}^{\pi} e^{jk\omega}e^{-jl\omega}d\omega = \begin{cases} 1 & k = l \\ 0 & k \neq l \end{cases}$$

to show that $x(n)$ may be recovered from $X(e^{j\omega})$ as follows:

$$x(n) = \frac{1}{2\pi}\int_{-\pi}^{\pi} X(e^{j\omega})e^{jn\omega}d\omega$$

Given a sequence $x(n)$, the DTFT is defined by

$$X(e^{j\omega}) = \sum_{k=-\infty}^{\infty} x(k)e^{-jk\omega}$$

To recover $x(n)$ from $X(e^{j\omega})$, it is necessary to "filter out" all of the terms in the sum except one (i.e., we must isolate a single term in the sum). This may be done by multiplying both sides of the equation by a complex exponential, $e^{jn\omega}$:

$$X(e^{j\omega})e^{jn\omega} = \sum_{k=-\infty}^{\infty} x(k)e^{-jk\omega}e^{jn\omega}$$

and integrating from $-\pi$ to π,

$$\int_{-\pi}^{\pi} X(e^{j\omega})e^{jn\omega}d\omega = \int_{-\pi}^{\pi}\left[\sum_{k=-\infty}^{\infty} x(k)e^{-jk\omega}\right]e^{jn\omega}d\omega$$

Interchanging the order of the integral and the sum on the right gives

$$\int_{-\pi}^{\pi} X(e^{j\omega})e^{jn\omega}d\omega = \sum_{k=-\infty}^{\infty} x(k)\int_{-\pi}^{\pi} e^{-jk\omega}e^{jn\omega}d\omega$$

Using the orthogonality of the complex exponentials, it follows that the integral is zero when $k \neq n$, and it is equal to 2π when $k = n$. Therefore,

$$\int_{-\pi}^{\pi} X(e^{j\omega})e^{jn\omega}d\omega = 2\pi x(n)$$

Dividing both sides by 2π gives the desired result.

2.25 Find the inverse DTFT of $X(e^{j\omega})$ shown in the figure below:

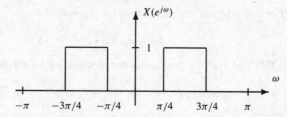

Because $X(e^{j\omega})$ is a piecewise constant function of ω, finding the inverse DTFT may be easily accomplished by integration. Using the inverse DTFT, we have

$$x(n) = \frac{1}{2\pi} \int_{-\pi}^{\pi} X(e^{j\omega})e^{jn\omega}d\omega$$

$$= \frac{1}{2\pi} \int_{\pi/4}^{3\pi/4} e^{jn\omega}d\omega + \frac{1}{2\pi} \int_{-3\pi/4}^{-\pi/4} e^{jn\omega}d\omega$$

$$= \frac{1}{2\pi jn} e^{jn\omega}\Big|_{\pi/4}^{3\pi/4} + \frac{1}{2\pi jn} e^{jn\omega}\Big|_{-3\pi/4}^{-\pi/4}$$

$$= \frac{1}{2\pi jn}[e^{j3n\pi/4} - e^{jn\pi/4}] + \frac{1}{2\pi jn}[e^{-jn\pi/4} - e^{-j3n\pi/4}]$$

Rearranging the terms, we have

$$x(n) = \frac{1}{2\pi jn}[e^{j3n\pi/4} - e^{-j3n\pi/4}] - \frac{1}{2\pi jn}[e^{jn\pi/4} - e^{-jn\pi/4}] = \frac{\sin(3\pi n/4)}{\pi n} - \frac{\sin(\pi n/4)}{\pi n}$$

which is the desired result.

It is interesting to note that $x(n)$ is expressed as the difference of two sequences, with the first being an ideal low-pass filter with a cutoff frequency of $3\pi/4$, and the second an ideal low-pass filter with a cutoff frequency of $\pi/4$. This is a consequence of the fact that $X(e^{j\omega})$ may be expressed as

$$X(e^{j\omega}) = X_1(e^{j\omega}) - X_2(e^{j\omega})$$

where

$$X_1(e^{j\omega}) = \begin{cases} 1 & |\omega| < \dfrac{3\pi}{4} \\ 0 & \text{otherwise} \end{cases}$$

and

$$X_2(e^{j\omega}) = \begin{cases} 1 & |\omega| < \dfrac{\pi}{4} \\ 0 & \text{otherwise} \end{cases}$$

Another way to evaluate the inverse DTFT is to observe that $X(e^{j\omega})$ may be written as

$$X(e^{j\omega}) = X_2\big(e^{j(\omega+\frac{\pi}{2})}\big) + X_2\big(e^{j(\omega-\frac{\pi}{2})}\big)$$

where $X_2(e^{j\omega})$ is the ideal low-pass filter defined above. Thus, $X(e^{j\omega})$ may be viewed as a modulated low-pass filter:

$$x(n) = 2x_2(n)\cos\left(\frac{n\pi}{2}\right)$$

With

$$x_2(n) = \frac{\sin(n\pi/4)}{n\pi}$$

$x(n)$ may also be written as

$$x(n) = 2\frac{\sin(n\pi/4)\cos(n\pi/2)}{n\pi}$$

This may be shown to be equivalent to the previous representation for $x(n)$ by using the trigonometric identity

$$2\sin A \sin B = \sin(A+B) + \sin(A-B)$$

2.26 Find the inverse DTFT of $X(e^{j\omega}) = \cos^2 \omega$.

Recall that the DTFT of a delayed unit sample is a complex exponential:

$$\delta(n - n_0) \stackrel{DTFT}{\Longleftrightarrow} e^{-jn_0\omega}$$

Therefore, the inverse DTFT of $X(e^{j\omega}) = \cos^2 \omega$ may be found easily if we expand it in terms of complex exponentials:

$$X(e^{j\omega}) = \left(\tfrac{1}{2}e^{j\omega} + \tfrac{1}{2}e^{-j\omega}\right)^2 = \tfrac{1}{2} + \tfrac{1}{4}e^{j2\omega} + \tfrac{1}{4}e^{-j2\omega}$$

Thus, it follows that $x(n)$ is

$$x(n) = \tfrac{1}{2}\delta(n) + \tfrac{1}{4}\delta(n + 2) + \tfrac{1}{4}\delta(n - 2)$$

2.27 If $h(n)$ is the unit sample response of a *real* and *causal* linear shift-invariant system, show that the system is completely specified by the real part of its frequency response:

$$H_R(e^{j\omega}) = \text{Re}\{H(e^{j\omega})\}$$

In other words, show that $H(e^{j\omega})$ may be uniquely recovered from its real part.

Recall from the symmetry properties of the DTFT that if $h(n)$ is real, $H(e^{j\omega})$ is conjugate symmetric. Therefore, if $H(e^{j\omega})$ is written in terms of its real and imaginary parts,

$$H(e^{j\omega}) = H_R(e^{j\omega}) + jH_I(e^{j\omega})$$

then the real part, $H_R(e^{j\omega})$, is the DTFT of the *even part* of $h(n)$:

$$h_e(n) = \tfrac{1}{2}[h(n) + h(-n)] \stackrel{DTFT}{\Longleftrightarrow} H_R(e^{j\omega})$$

Therefore, given $H_R(e^{j\omega})$, or $h_e(n)$, the question is how to recover $h(n)$. Note that if $h(n)$ is *causal*, $h(n) = 0$ for $n < 0$, and

$$h_e(n) = \begin{cases} \tfrac{1}{2}h(n) & n > 0 \\ h(0) & n = 0 \\ \tfrac{1}{2}h(-n) & n < 0 \end{cases}$$

As a result, $h(n)$ may be recovered from $h_e(n)$ as follows:

$$h(n) = \begin{cases} 2h_e(n) & n > 0 \\ h_e(0) & n = 0 \\ 0 & n < 0 \end{cases}$$

2.28 If $h(n)$ is real and causal, and if

$$H_R(e^{j\omega}) = \text{Re}\{H(e^{j\omega})\} = 1 + \alpha \cos 2\omega$$

find $h(n)$.

Because the real part of $H(e^{j\omega})$ is

$$H_R(e^{j\omega}) = 1 + \alpha \cos 2\omega = 1 + \tfrac{1}{2}\alpha e^{j2\omega} + \tfrac{1}{2}\alpha e^{-j2\omega}$$

the even part of $h(n)$, which is the inverse DTFT of $H_R(e^{j\omega})$, is

$$h_e(n) = \delta(n) + \tfrac{1}{2}\alpha\delta(n + 2) + \tfrac{1}{2}\alpha\delta(n - 2)$$

With $h(n)$ a causal sequence, it follows from the results of Prob. 2.27 that

$$h(n) = \begin{cases} 2h_e(n) & n > 0 \\ h_e(0) & n = 0 \\ 0 & n < 0 \end{cases}$$

which gives

$$h(n) = \delta(n) + \alpha\delta(n - 2)$$

DTFT Properties

2.29　Show that if $X(e^{j\omega})$ is real and even, $x(n)$ is real and even.

For $x(n)$ we have

$$x(n) = \frac{1}{2\pi}\int_{-\pi}^{\pi} X(e^{j\omega})e^{jn\omega}d\omega = \frac{1}{2\pi}\int_{-\pi}^{\pi} X(e^{j\omega})\cos(n\omega)d\omega - j\frac{1}{2\pi}\int_{-\pi}^{\pi} X(e^{j\omega})\sin(n\omega)d\omega$$

If $X(e^{j\omega})$ is real and even, then $X(e^{j\omega})\sin(n\omega)$ is real and odd. Therefore, when integrated from $-\pi$ to π, the integral is zero. Thus, $x(n)$ may be written as

$$x(n) = \frac{1}{2\pi}\int_{-\pi}^{\pi} X(e^{j\omega})\cos(n\omega)d\omega$$

and it follows that $x(n)$ is real. Finally, because $X(e^{j\omega})\cos(n\omega)$ is real and even, $x(n)$ is real and even, that is,

$$x(-n) = \frac{1}{2\pi}\int_{-\pi}^{\pi} X(e^{j\omega})\cos(-n\omega)d\omega = \frac{1}{2\pi}\int_{-\pi}^{\pi} X(e^{j\omega})\cos(n\omega)d\omega = x(n)$$

2.30　Prove the convolution theorem.

There are several ways to prove the convolution theorem. One way is by a straightforward manipulation of the DTFT sum. Specifically, if $y(n) = h(n) * x(n)$,

$$y(n) = \sum_{l=-\infty}^{\infty} h(l)x(n-l)$$

and the DTFT of $y(n)$ is

$$Y(e^{j\omega}) = \sum_{n=-\infty}^{\infty}\left[\sum_{l=-\infty}^{\infty} h(l)x(n-l)\right]e^{-jn\omega} = \sum_{l=-\infty}^{\infty} h(l)\left[\sum_{n=-\infty}^{\infty} x(n-l)e^{-jn\omega}\right]$$

Note that the expression in brackets is the DTFT of $x(n-l)$. Using the delay property of the DTFT, this is equal to $X(e^{j\omega})e^{-jl\omega}$, and the right side of this equation becomes

$$Y(e^{j\omega}) = \sum_{l=-\infty}^{\infty} h(l)X(e^{j\omega})e^{-jl\omega}$$

Factoring out $X(e^{j\omega})$ from the sum, which does not depend on l, we have

$$Y(e^{j\omega}) = X(e^{j\omega})\sum_{l=-\infty}^{\infty} h(l)e^{-jl\omega} = X(e^{j\omega})H(e^{j\omega})$$

which proves the theorem.

Another way to prove the convolution theorem is to consider the following cascade of two LSI systems, one with a unit sample response of $h(n)$ and the other with a unit sample response of $x(n)$:

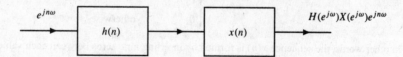

If the input to this cascade is a complex exponential, $e^{jn\omega}$, the output of the first system is $H(e^{j\omega})e^{jn\omega}$. Because this complex exponential is the input to the second system, the output is $H(e^{j\omega})X(e^{j\omega})e^{jn\omega}$. Therefore, $H(e^{j\omega})X(e^{j\omega})$ is the frequency response of the cascade, and because the unit sample response of the cascade is the convolution $h(n) * x(n)$, we have the DTFT pair

$$h(n) * x(n) \stackrel{DTFT}{\Longleftrightarrow} H(e^{j\omega})X(e^{j\omega})$$

which establishes the convolution theorem.

2.31 Derive the *up-sampling* property of the DTFT, which states that if $X(e^{j\omega})$ is the DTFT of $x(n)$, the DTFT of

$$y(n) = \begin{cases} x\left(\dfrac{n}{L}\right) & n = 0, \pm L, \pm 2L, \ldots \\ 0 & \text{otherwise} \end{cases}$$

is

$$Y(e^{j\omega}) = X(e^{jL\omega})$$

From the definition of the DTFT, we have

$$Y(e^{j\omega}) = \sum_{n=-\infty}^{\infty} y(n)e^{-jn\omega}$$

Because $y(n)$ is equal to zero except when n is an integer multiple of L,

$$Y(e^{j\omega}) = \sum_{n=-\infty}^{\infty} y(nL)e^{-jnL\omega} = \sum_{n=-\infty}^{\infty} x(n)e^{-jnL\omega}$$

or

$$Y(e^{j\omega}) = X(e^{jL\omega})$$

Thus, $Y(e^{j\omega})$ is formed by scaling $X(e^{j\omega})$ in frequency.

2.32 Find the inverse DTFT of

$$X(e^{j\omega}) = \frac{1}{1 - \frac{1}{3}e^{-j10\omega}}$$

For this problem, the direct approach of performing the integration

$$x(n) = \frac{1}{2\pi} \int_{-\pi}^{\pi} X(e^{j\omega})e^{jn\omega}d\omega$$

is not easy. However, a simple approach is to recall that the inverse DTFT of

$$Y(e^{j\omega}) = \frac{1}{1 - \frac{1}{3}e^{-j\omega}}$$

is

$$y(n) = \left(\tfrac{1}{3}\right)^n u(n)$$

and to note that $Y(e^{j\omega})$ is related to $X(e^{j\omega})$ by scaling in frequency,

$$X(e^{j\omega}) = Y(e^{j10\omega})$$

Therefore, it follows from the up-sampling property in Prob. 2.31 that

$$x(n) = \begin{cases} \left(\tfrac{1}{3}\right)^{n/10} & n = 0, \pm 10, \pm 20, \ldots \\ 0 & \text{otherwise} \end{cases}$$

In other words, the sequence $x(n)$ is formed by inserting nine zeros between each value of $y(n)$.

2.33 Let $x(n)$ be a sequence with a DTFT $X(e^{j\omega})$. For each of the following sequences that are formed from $x(n)$, express the DTFT in terms of $X(e^{j\omega})$:

(a) $x^*(-n)$

(b) $x(n) * x^*(-n)$

(c) $x(2n + 1)$

(a) The DTFT of $x^*(-n)$ is

$$\text{DTFT}\{x^*(-n)\} = \sum_{n=-\infty}^{\infty} x^*(-n)e^{-jn\omega} = \sum_{n=-\infty}^{\infty} x^*(n)e^{jn\omega}$$

Bringing the conjugate outside, we have

$$\text{DTFT}\{x^*(-n)\} = \left[\sum_{n=-\infty}^{\infty} x(n)e^{-jn\omega}\right]^* = X^*(e^{j\omega})$$

which leads to the DTFT pair

$$x^*(-n) \overset{DTFT}{\Longleftrightarrow} X^*(e^{j\omega})$$

(b) For $y(n) = x(n) * x^*(-n)$, note that because $y(n)$ is the convolution of two sequences, the DTFT of $y(n)$ is the product of the DTFTs of $x(n)$ and $x^*(-n)$. As shown in part (a), the DTFT of $x^*(-n)$ is $X^*(e^{j\omega})$. Therefore, we have the DTFT pair

$$x(n) * x^*(-n) \overset{DTFT}{\Longleftrightarrow} X(e^{j\omega})X^*(e^{j\omega}) = |X(e^{j\omega})|^2$$

(c) For $x(2n + 1)$ we have

$$\text{DTFT}\{x(2n + 1)\} = \sum_{n=-\infty}^{\infty} x(2n + 1)e^{-jn\omega} = \sum_{n \text{ odd}} x(n)e^{-jn\omega}$$

To evaluate this sum, a "trick" is to use the identity

$$1 - (-1)^n = \begin{cases} 2 & n \text{ odd} \\ 0 & n \text{ even} \end{cases}$$

This allows us to write the DTFT as follows:

$$\text{DTFT}\{x(2n + 1)\} = \sum_{n \text{ odd}} x(n)e^{-jn\omega} = \frac{1}{2}\sum_{n=-\infty}^{\infty} [1 - (-1)^n]x(n)e^{-jn\omega}$$

$$= \frac{1}{2}\sum_{n=-\infty}^{\infty} x(n)e^{-jn\omega} - \frac{1}{2}\sum_{n=-\infty}^{\infty} (-1)^n x(n)e^{-jn\omega}$$

Because the first sum is simply $X(e^{j\omega})$, and the second is the DTFT of the modulated signal

$$(-1)^n x(n) = e^{jn\pi} x(n)$$

then

$$\text{DTFT}\{x(2n + 1)\} = \frac{1}{2}\left[X(e^{j\omega}) - X\left(e^{j(\omega-\pi)}\right)\right]$$

2.34 Let $x(n)$ be the sequence

$$x(n) = \delta(n + 1) - \delta(n) + 2\delta(n - 1) + 3\delta(n - 2)$$

which has a DTFT

$$X(e^{j\omega}) = X_R(e^{j\omega}) + jX_I(e^{j\omega})$$

where $X_R(e^{j\omega})$ and $X_I(e^{j\omega})$ are the real part and the imaginary part of $X(e^{j\omega})$, respectively. Find the sequence $y(n)$ that has a DTFT given by

$$Y(e^{j\omega}) = X_I(e^{j\omega}) + jX_R(e^{j\omega})e^{j2\omega}$$

The key to solving this problem is to recall that if $x(n)$ is real, and if $X(e^{j\omega})$ is written in terms of its real and imaginary parts, $X_R(e^{j\omega})$ is the DTFT of the *even part* of $x(n)$, and $X_I(e^{j\omega})$ is the DTFT of the *odd part*:

$$x_e(n) = \frac{1}{2}[x(n) + x(-n)] \overset{DTFT}{\Longleftrightarrow} X_R(e^{j\omega})$$

$$x_o(n) = \frac{1}{2}[x(n) - x(-n)] \overset{DTFT}{\Longleftrightarrow} jX_I(e^{j\omega})$$

Therefore, the DTFT of $-jx_o(n)$ is $X_I(e^{j\omega})$,

$$-jx_o(n) \overset{DTFT}{\Longleftrightarrow} X_I(e^{j\omega})$$

and the DTFT of $jx_e(n+2)$ is

$$jx_e(n+2) \overset{DTFT}{\Longleftrightarrow} jX_R(e^{j\omega})e^{j2\omega}$$

Thus,

$$jx_e(n+2) - jx_o(n) \overset{DTFT}{\Longleftrightarrow} Y(e^{j\omega}) = X_I(e^{j\omega}) + jX_R(e^{j\omega})e^{j2\omega}$$

and it follows that

$$y(n) = jx_e(n+2) - jx_o(n)$$

Finally, with $x_e(n)$ and $x_o(n)$ as tabulated below,

n	-2	-1	0	1	2
$x_e(n)$	3/2	3/2	-1	3/2	3/2
$x_o(n)$	$-3/2$	$-1/2$	0	$1/2$	3/2

it follows that $y(n)$, which is formed from these two sequences, is as shown below:

n	-4	-3	-2	-1	0	1	2
$y(n)$	$3j/2$	$3j/2$	$j/2$	$2j$	$3j/2$	$-j/2$	$-3j/2$

2.35 Let $x(n)$ be the sequence

$$x(n) = 2\delta(n+2) - \delta(n+1) + 3\delta(n) - \delta(n-1) + 2\delta(n-2)$$

Evaluate the following quantities without explicitly finding $X(e^{j\omega})$:

(a) $X(e^{j\omega})|_{\omega=0}$

(b) $\phi_x(\omega)$

(c) $\int_{-\pi}^{\pi} X(e^{j\omega})d\omega$

(d) $X(e^{j\omega})|_{\omega=\pi}$

(e) $\int_{-\pi}^{\pi} |X(e^{j\omega})|^2 d\omega$

(a) Because the DTFT of $x(n)$ is

$$X(e^{j\omega}) = \sum_{n=-\infty}^{\infty} x(n)e^{-jn\omega}$$

note that if we evaluate $X(e^{j\omega})$ at $\omega=0$, we have

$$X(e^{j\omega})|_{\omega=0} = \sum_{n=-\infty}^{\infty} x(n)$$

which is simply the sum of the values of $x(n)$. Therefore, for the given sequence it follows that

$$X(e^{j\omega})|_{\omega=0} = 5$$

(b) To evaluate the phase, note that because $x(n)$ is real and even, $X(e^{j\omega})$ is real and even and, therefore, the phase is equal zero or π for all ω.

(c) From the inverse DTFT,

$$x(n) = \frac{1}{2\pi} \int_{-\pi}^{\pi} X(e^{j\omega})e^{jn\omega}d\omega$$

note that when $n = 0$:

$$x(0) = \frac{1}{2\pi} \int_{-\pi}^{\pi} X(e^{j\omega})d\omega$$

Therefore, it follows that

$$\int_{-\pi}^{\pi} X(e^{j\omega})d\omega = 2\pi x(0) = 6\pi$$

(d) Evaluating the DTFT of $x(n)$ at $\omega = \pi$, we have

$$X(e^{j\omega})|_{\omega=\pi} = \sum_{n=-\infty}^{\infty} x(n)e^{-jn\pi} = \sum_{n=-\infty}^{\infty} (-1)^n x(n)$$

which, for the given values of $x(n)$, evaluates to

$$X(e^{j\omega})|_{\omega=\pi} = 9$$

(e) From Parseval's theorem, we know that

$$\frac{1}{2\pi} \int_{-\pi}^{\pi} |X(e^{j\omega})|^2 d\omega = \sum_{n=-\infty}^{\infty} |x(n)|^2$$

Therefore,

$$\int_{-\pi}^{\pi} |X(e^{j\omega})|^2 d\omega = 2\pi \sum_{n=-\infty}^{\infty} |x(n)|^2 = 38\pi$$

2.36 The *center of gravity* of a sequence $x(n)$ is defined by

$$c = \frac{\displaystyle\sum_{n=-\infty}^{\infty} nx(n)}{\displaystyle\sum_{n=-\infty}^{\infty} x(n)}$$

and is used as a measure of the *time delay* of a sequence. Find an expression for c in terms of the DTFT of $x(n)$, and find the value of c for the sequence $x(n)$ that has a DTFT as shown in the figure below.

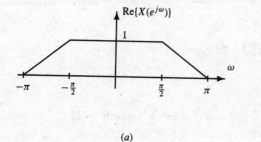

(a)

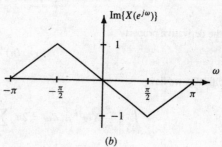

(b)

To find the value of c in terms of $X(e^{j\omega})$, first note that the denominator is simply the value of $X(e^{j\omega})$ evaluated at $\omega = 0$:

$$\sum_{n=-\infty}^{\infty} x(n) = X(e^{j\omega})|_{\omega=0}$$

For the numerator, recall the DTFT pair

$$nx(n) \overset{DTFT}{\Longleftrightarrow} j\frac{d}{d\omega}X(e^{j\omega})$$

Therefore,

$$\sum_{n=-\infty}^{\infty} nx(n) = j\frac{d}{d\omega}X(e^{j\omega})\Big|_{\omega=0}$$

and c may be evaluated in terms of $X(e^{j\omega})$ as follows:

$$c = j\frac{\dfrac{d}{d\omega}X(e^{j\omega})\Big|_{\omega=0}}{X(e^{j\omega})|_{\omega=0}}$$

For the DTFT that is given, we see that

$$X(e^{j\omega})|_{\omega=0} = 1$$

and

$$\frac{d}{d\omega}X(e^{j\omega})\Big|_{\omega=0} = \frac{d}{d\omega}\text{Re}\{X(e^{j\omega})\}_{\omega=0} + j\frac{d}{d\omega}\text{Im}\{X(e^{j\omega})\}_{\omega=0} = -j\frac{2}{\pi}$$

Therefore,

$$c = \frac{2}{\pi}$$

2.37 For the sequence $x(n)$ plotted in the figure below,

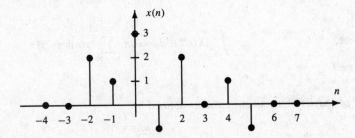

evaluate the integral

$$\int_{-\pi}^{\pi}\left|\frac{d}{d\omega}X(e^{j\omega})\right|^2 d\omega$$

This integral is easy to evaluate if we use Parseval's theorem

$$\frac{1}{2\pi}\int_{-\pi}^{\pi}|W(e^{j\omega})|^2 d\omega = \sum_{n=-\infty}^{\infty}|w(n)|^2$$

and the derivative property

$$nx(n) \overset{DTFT}{\Longleftrightarrow} j\frac{d}{d\omega}X(e^{j\omega})$$

Specifically, we have

$$\int_{-\pi}^{\pi}\left|\frac{d}{d\omega}X(e^{j\omega})\right|^2 d\omega = 2\pi\sum_{n=-\infty}^{\infty}|-jnx(n)|^2 = 2\pi\sum_{n=-\infty}^{\infty}|nx(n)|^2 = 75$$

Applications

2.38 A linear shift-invariant system has a frequency response

$$H(e^{j\omega}) = e^{j\omega}\frac{1}{1.1 + \cos\omega}$$

Find an LCCDE that relates the input to the output.

To convert $H(e^{j\omega})$ into a difference equation, we must first express $H(e^{j\omega})$ in terms of complex exponentials. Expanding the cosine into a sum of two complex exponentials, we have

$$H(e^{j\omega}) = \frac{e^{j\omega}}{1.1 + \frac{1}{2}e^{-j\omega} + \frac{1}{2}e^{j\omega}}$$

Multiplying numerator and denominator by $2e^{-j\omega}$ gives

$$H(e^{j\omega}) = \frac{Y(e^{j\omega})}{X(e^{j\omega})} = \frac{2}{1 + 2.2e^{-j\omega} + e^{-2j\omega}}$$

Cross-multiplying, we have

$$[1 + 2.2e^{-j\omega} + e^{-2j\omega}]Y(e^{j\omega}) = 2X(e^{j\omega})$$

which leads to the following difference equation when we take the inverse DTFT of each term:

$$y(n) + 2.2y(n-1) + y(n-2) = 2x(n)$$

2.39 Find the frequency response of a linear shift-invariant system whose input and output satisfy the difference equation

$$y(n) - 0.5y(n-1) = x(n) + 2x(n-1) + x(n-2)$$

To find the frequency response, we begin by finding the DTFT of each term in the difference equation

$$(1 - 0.5e^{-j\omega})Y(e^{j\omega}) = (1 + 2e^{-j\omega} + e^{-j2\omega})X(e^{j\omega})$$

Because $H(e^{j\omega}) = Y(e^{j\omega})/X(e^{j\omega})$, we have

$$H(e^{j\omega}) = \frac{1 + 2e^{-j\omega} + e^{-2j\omega}}{1 - 0.5e^{-j\omega}}$$

2.40 Write a difference equation to implement a system with a frequency response

$$H(e^{j\omega}) = \frac{1 - 0.5e^{-j\omega} + e^{-3j\omega}}{1 + 0.5e^{-j\omega} + 0.75e^{-2j\omega}}$$

With

$$H(e^{j\omega}) = \frac{Y(e^{j\omega})}{X(e^{j\omega})} = \frac{1 - 0.5e^{-j\omega} + e^{-3j\omega}}{1 + 0.5e^{-j\omega} + 0.75e^{-2j\omega}}$$

after cross-multiplying, we have

$$[1 + 0.5e^{-j\omega} + 0.75e^{-2j\omega}]Y(e^{j\omega}) = [1 - 0.5e^{-j\omega} + e^{-3j\omega}]X(e^{j\omega})$$

Taking the inverse DTFT of each term gives the desired difference equation

$$y(n) + 0.5y(n-1) + 0.75y(n-2) = x(n) - 0.5x(n-1) + x(n-3)$$

2.41 Find a difference equation to realize a linear shift-invariant system that has a frequency response

$$H(e^{j\omega}) = \tan \omega$$

To find a difference equation for $H(e^{j\omega})$, we must first express $\tan \omega$ in terms of complex exponentials:

$$\tan \omega = \frac{\sin \omega}{\cos \omega} = \frac{1}{j} \frac{e^{j\omega} - e^{-j\omega}}{e^{j\omega} + e^{-j\omega}}$$

With $H(e^{j\omega}) = Y(e^{j\omega})/X(e^{j\omega})$ we have, after cross-multiplying,

$$j[e^{j\omega} + e^{-j\omega}]Y(e^{j\omega}) = [e^{j\omega} - e^{-j\omega}]X(e^{j\omega})$$

Inverse transforming, we obtain the following difference equation:

$$jy(n+1) + jy(n-1) = x(n+1) - x(n-1)$$

By introducing a delay and dividing by j, this difference equation may be written in the more standard form

$$y(n) = -y(n-2) - jx(n) + jx(n-2)$$

2.42   Find a difference equation to implement a filter that has a unit sample response

$$h(n) = \left(\tfrac{1}{4}\right)^n \cos\left(\frac{n\pi}{3}\right) u(n)$$

To find a difference equation for this system, we must first find the frequency response $H(e^{j\omega})$. Expressing $h(n)$ in terms of complex exponentials,

$$h(n) = \tfrac{1}{2}\left(\tfrac{1}{4}\right)^n e^{jn\pi/3} u(n) + \tfrac{1}{2}\left(\tfrac{1}{4}\right)^n e^{-jn\pi/3} u(n)$$

it follows that the frequency response is

$$
\begin{aligned}
H(e^{j\omega}) &= \frac{1}{2}\frac{1}{1 - \tfrac{1}{4}e^{j\pi/3}e^{-j\omega}} + \frac{1}{2}\frac{1}{1 - \tfrac{1}{4}e^{-j\pi/3}e^{-j\omega}} \\
&= \frac{1}{2}\frac{1 - \tfrac{1}{4}e^{j\pi/3}e^{-j\omega} + 1 - \tfrac{1}{4}e^{-j\pi/3}e^{-j\omega}}{\left(1 - \tfrac{1}{4}e^{j\pi/3}e^{-j\omega}\right)\left(1 - \tfrac{1}{4}e^{-j\pi/3}e^{-j\omega}\right)} \\
&= \frac{1}{2}\frac{2 - \tfrac{1}{2}\cos(\pi/3)e^{-j\omega}}{1 - \tfrac{1}{2}\cos(\pi/3)e^{-j\omega} + \tfrac{1}{16}e^{-2j\omega}}
\end{aligned}
$$

Therefore, the difference equation for this system is

$$y(n) = \tfrac{1}{2}\left(\cos\frac{\pi}{3}\right)y(n-1) - \tfrac{1}{16}y(n-2) + x(n) - \tfrac{1}{4}\left(\cos\frac{\pi}{3}\right)x(n-1)$$

2.43   A system with input $x(n)$ and output $y(n)$ is described by the following set of coupled linear constant coefficient difference equations:

$$
\begin{aligned}
y(n) &= \tfrac{1}{2}y(n-1) + 2v(n) + v(n-1) \\
v(n) &= \tfrac{1}{3}v(n-1) + w(n-1) \\
w(n) &= \tfrac{1}{2}x(n) + 2x(n-1)
\end{aligned}
$$

Find a single linear constant coefficient difference equation that describes this system, and find the frequency response $H(e^{j\omega})$.

To find the frequency response for this system of difference equations, we first express each equation in the frequency domain:

$$
\begin{aligned}
\left[1 - \tfrac{1}{2}e^{-j\omega}\right]Y(e^{j\omega}) &= [2 + e^{-j\omega}]V(e^{j\omega}) \\
\left[1 - \tfrac{1}{3}e^{-j\omega}\right]V(e^{j\omega}) &= e^{-j\omega}W(e^{j\omega}) \\
W(e^{j\omega}) &= \left[\tfrac{1}{2} + 2e^{-j\omega}\right]X(e^{j\omega})
\end{aligned}
$$

Using the last two equations to express $V(e^{j\omega})$ in terms of $X(e^{j\omega})$, we have

$$V(e^{j\omega}) = \frac{e^{-j\omega}\left[\tfrac{1}{2} + 2e^{-j\omega}\right]}{1 - \tfrac{1}{3}e^{-j\omega}}X(e^{j\omega})$$

Substituting this expression for $V(e^{j\omega})$ into the first equation and solving for $Y(e^{j\omega})$ gives

$$Y(e^{j\omega}) = \frac{2 + e^{-j\omega}}{1 - \tfrac{1}{2}e^{-j\omega}} \cdot \frac{\tfrac{1}{2} + 2e^{-j\omega}}{1 - \tfrac{1}{3}e^{-j\omega}}e^{-j\omega}X(e^{j\omega})$$

Therefore, the frequency response is

$$H(e^{j\omega}) = \frac{Y(e^{j\omega})}{X(e^{j\omega})} = \frac{e^{-j\omega} + \frac{9}{2}e^{-2j\omega} + 2e^{-3j\omega}}{1 - \frac{5}{6}e^{-j\omega} + \frac{1}{6}e^{-j2\omega}}$$

Cross-multiplying, we have

$$Y(e^{j\omega})\left[1 - \tfrac{5}{6}e^{-j\omega} + \tfrac{1}{6}e^{-j2\omega}\right] = X(e^{j\omega})\left[e^{-j\omega} + \tfrac{9}{2}e^{-2j\omega} + 2e^{-3j\omega}\right]$$

and taking the inverse DTFT of each term gives the difference equation for the system:

$$y(n) - \tfrac{5}{6}y(n-1) + \tfrac{1}{6}y(n-2) = x(n-1) + \tfrac{9}{2}x(n-2) + 2x(n-3)$$

2.44 A linear shift-invariant system with input $x(n)$ and output $v(n)$ is described by the difference equation

$$v(n) = x(n) + \alpha x(n-1)$$

This system is cascaded with another system with input $v(n)$ and output $y(n)$ that is described by the difference equation

$$y(n) = \tfrac{1}{7}y(n-1) + v(n)$$

What value of α will guarantee that $y(n) = x(n)$?

Substituting the first equation into the second, we obtain a single difference equation that describes the overall system, that is,

$$y(n) = \tfrac{1}{7}y(n-1) + x(n) + \alpha x(n-1)$$

Taking the DTFT of both sides of the equation, we have

$$Y(e^{j\omega}) = \tfrac{1}{7}e^{-j\omega}Y(e^{j\omega}) + X(e^{j\omega}) + \alpha e^{-j\omega}X(e^{j\omega})$$

If $y(n) = x(n)$, $Y(e^{j\omega}) = X(e^{j\omega})$, and it is clear that this will be true if and only if $\alpha = -\tfrac{1}{7}$.

2.45 Find the input $x(n)$ that will produce a response, $y(n) = \delta(n)$, for a system described by the LCCDE

$$y(n) - \tfrac{1}{4}y(n-1) = x(n) - \tfrac{1}{8}x(n-2)$$

This problem is easily solved if we express this difference equation in the frequency domain. Specifically, we have

$$Y(e^{j\omega}) - \tfrac{1}{4}e^{-j\omega}Y(e^{j\omega}) = X(e^{j\omega}) - \tfrac{1}{8}e^{-j2\omega}X(e^{j\omega})$$

If we want the output to be $y(n) = \delta(n)$, $Y(e^{j\omega}) = 1$, and we have

$$1 - \tfrac{1}{4}e^{-j\omega} = X(e^{j\omega}) - \tfrac{1}{8}e^{-j2\omega}X(e^{j\omega})$$

Solving for $X(e^{j\omega})$ gives

$$X(e^{j\omega}) = \frac{1 - \tfrac{1}{4}e^{-j\omega}}{1 - \tfrac{1}{8}e^{-j2\omega}}$$

To find the inverse DTFT of $X(e^{j\omega})$, recall that

$$\left(\tfrac{1}{8}\right)^n u(n) \overset{DTFT}{\Longleftrightarrow} \frac{1}{1 - \tfrac{1}{8}e^{-j\omega}}$$

Therefore, the inverse DTFT of

$$W(e^{j\omega}) = \frac{1}{1 - \tfrac{1}{8}e^{-j2\omega}}$$

is the sequence

$$w(n) = \begin{cases} \left(\tfrac{1}{8}\right)^{n/2} & n = 0, 2, 4, \ldots \\ 0 & \text{otherwise} \end{cases}$$

and $x(n)$ is given by

$$x(n) = \begin{cases} \left(\tfrac{1}{8}\right)^{n/2} & n = 0, 2, 4, \ldots \\ -\tfrac{1}{4}\left(\tfrac{1}{8}\right)^{(n-1)/2} & n = 1, 3, 5, \ldots \end{cases}$$

Supplementary Problems

Frequency Response

2.46   Consider a linear shift-invariant system with a unit sample response

$$h(n) = \delta(n) + \delta(n - 1)$$

Find the output of the system when the input is

$$x(n) = 1 + 5\cos\left(\frac{n\pi}{10}\right)$$

2.47   If the unit sample response of a linear shift-invariant system is

$$h(n) = a^n u(n)$$

with $|a| < 1$, find the response of the system to the input $x(n) = 1$. Repeat for $x(n) = (-1)^n$.

2.48   Find the frequency response of the system that has a unit sample response

$$h(n) = \begin{cases} \cos(n\omega_0) & 0 \le n \le N - 1 \\ 0 & \text{otherwise} \end{cases}$$

2.49   The input to a linear shift-invariant system is

$$x(n) = \cos\left(\frac{n\pi}{8}\right) + \cos\left(\frac{3n\pi}{4}\right)$$

Find the output when the unit sample response is

(a) $h(n) = \dfrac{\sin(7n\pi/8)}{7n\pi/8}$

(b) $h(n) = \dfrac{\sin(n\pi/2)}{n} * e^{j\frac{\pi}{4}n}\dfrac{\sin(n\pi/4)}{n}$

2.50   The input to a linear shift-invariant system is

$$x(n) = n\left(\tfrac{1}{2}\right)^n u(n)$$

and the output is

$$y(n) = \left(\tfrac{1}{3}\right)^{n-2} u(n - 2) - \tfrac{1}{2}\left(\tfrac{1}{3}\right)^{n-3} u(n - 3)$$

Find the frequency response $H(e^{j\omega})$.

2.51   Find the frequency response of the system described by the LCCDE

$$y(n) = \tfrac{1}{3}y(n - 10) + x(n) + \tfrac{1}{4}x(n - 10)$$

2.52   Find the group delay of the system that has a frequency response

$$H(e^{j\omega}) = \frac{1 - \frac{1}{2}e^{-j\omega}}{1 + \frac{1}{4}e^{-j\omega}}$$

Filters

2.53   What is the unit sample response of an ideal bandstop filter with a lower cutoff frequency of ω_1 and an upper cutoff frequency of ω_2?

2.54   If $h(n)$ is the unit sample response of an ideal low-pass filter with a gain of one and a cutoff frequency $\omega_c = \pi/8$, what is $g(n) = \cos(n\pi/2)h(n)$?

The Interconnection of Systems

2.55   What type of filter has a unit sample response

$$h(n) = \delta(n) - \frac{\sin(n\pi/3)}{n\pi}$$

2.56   What is the magnitude of the frequency response of the cascade of the following two systems?

$$H_1(e^{j\omega}) = \frac{e^{-j\omega} - 0.5}{1 - 0.5e^{-j\omega}} \qquad h_2(n) = \delta(n) - \frac{\sin(n\pi/4)}{n\pi}$$

The Discrete-Time Fourier Transform

2.57   Find the DTFT of the sequence

$$x(n) = \begin{cases} N + 1 - |n| & |n| \leq N \\ 0 & \text{else} \end{cases}$$

2.58   For each of the following systems, find the frequencies that are present in the output when the DTFT of the input $x(n)$ is

$$X(e^{j\omega}) = \begin{cases} 1 & |\omega| \leq \dfrac{\pi}{3} \\ 0 & \dfrac{\pi}{3} < |\omega| \leq \pi \end{cases}$$

(a) $y(n) = x^2(n)$

(b) $y(n) = x(n)\cos(n\pi/2)$

(c) $y(n) = x(2n)$

2.59   Let $x(n) = e^{jn\pi/4}u(n)$ and $y(n) = 0.5e^{jn\pi/4}u(n)$. Determine whether or not there is a linear shift-invariant system that has the response, $y(n)$, to the input $x(n)$. If such a system exists, determine whether or not the system is unique, and find the frequency response of an LSI system with the desired behavior. If no such LSI system exists, explain why.

2.60   Find the inverse DTFT of $X(e^{j\omega})$ illustrated in the figure below.

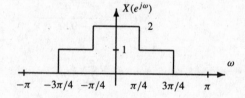

2.61   Find the inverse DTFT of $X(e^{j\omega})$ illustrated in the figure below.

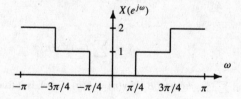

2.62   Find the inverse DTFT of $X(e^{j\omega}) = \cos 2\omega + j \sin \omega$.

2.63   Find the inverse DTFT of

$$X(e^{j\omega}) = \begin{cases} e^{j2\omega} & \dfrac{\pi}{3} < |\omega| < \dfrac{2\pi}{3} \\ 0 & \text{otherwise} \end{cases}$$

2.64   Find the DTFT of

$$x(n) = \frac{\sin(n\pi/2)}{n\pi} \frac{\sin(n\pi/4)}{n\pi}$$

2.65   If $x(n)$ is real and causal, and

$$\text{Re}\{X(e^{j\omega})\} = \frac{1}{\frac{5}{4} - \cos \omega}$$

find $x(n)$.

DTFT Properties

2.66   Let $x(n)$ be a sequence with a DTFT $X(e^{j\omega})$. For each of the following sequences that are related to $x(n)$, express the DTFT in terms of $X(e^{j\omega})$:

(a) $x^*(n)$

(b) $x(n) - x(n-2)$

(c) $x(2n)$

(d) $x(n) * x(n-1)$

2.67   If the DTFT of $x(n) = (\frac{1}{5})^n u(n+2)$ is $X(e^{j\omega})$, find the sequence that has a DTFT given by $Y(e^{j\omega}) = X(e^{j2\omega})$.

2.68   Let $x(n)$ be the sequence

$$x(n) = 2\delta(n+3) - 2\delta(n+1) + \delta(n-1) + 3\delta(n-2)$$

If the DTFT of $x(n)$ is expressed in terms of its real and imaginary parts as follows,

$$X(e^{j\omega}) = X_R(e^{j\omega}) + jX_I(e^{j\omega})$$

find the sequence $y(n)$ that has a DTFT given by

$$X(e^{j\omega}) = X_R(e^{j\omega}) + jX_I(e^{j\omega})e^{-j2\omega}$$

2.69   The DTFT of a sequence $x(n)$ is

$$X(e^{j\omega}) = \frac{3}{(1 - 0.8e^{-j\omega})^5}$$

Evaluate the sum

$$S = \sum_{n=-\infty}^{\infty} x(n)$$

2.70   The DTFT of a sequence $x(n)$ is

$$X(e^{j\omega}) = \cos^3(3\omega)$$

Evaluate the sum

$$S = \sum_{n=-\infty}^{\infty} (-1)^n x(n)$$

2.71   Let $\sum_{n=-\infty}^{\infty} x(n) = A$ and $\sum_{n=-\infty}^{\infty} h(n) = B$. If $y(n) = h(n) * x(n)$, is it true that $\sum_{n=-\infty}^{\infty} y(n) = A \cdot B$?

2.72   Evaluate the following integral:

$$\int_{-\pi}^{\pi} \frac{e^{j\omega}}{1 - 0.3e^{-j\omega}} d\omega$$

2.73   Using the *center of gravity* (see problem 2.36), find the time delay of the sequence

$$x(n) = \alpha^n u(n)$$

Applications of the DTFT

2.74 A causal linear shift-invariant system is defined by the difference equation

$$2y(n) - y(n-2) = x(n-1) + 3x(n-2) + 2x(n-3)$$

Find the frequency response, $H(e^{j\omega})$.

2.75 The frequency response of a linear shift-invariant system is

$$H(e^{j\omega}) = e^{j\omega} \frac{1}{2 + e^{-2j\omega}}$$

Find an LCCDE that relates the input to the output.

2.76 Find the inverse of the system that has a unit sample response $h(n) = n(-\frac{1}{6})^n u(n-3)$.

Answers to Supplementary Problems

2.46 $y(n) = 2 + 9.88 \cos[(n - \frac{1}{2})\pi/10]$.

2.47 $y(n) = \dfrac{1}{1-a}$ and $y(n) = \dfrac{1}{1+a}$.

2.48 $H(e^{j\omega}) = \frac{1}{2} e^{-j\frac{N-1}{2}(\omega - \omega_0)} \dfrac{\sin\frac{N}{2}(\omega - \omega_0)}{\sin\frac{1}{2}(\omega - \omega_0)} + \frac{1}{2} e^{-j\frac{N-1}{2}(\omega + \omega_0)} \dfrac{\sin\frac{N}{2}(\omega + \omega_0)}{\sin\frac{1}{2}(\omega + \omega_0)}$.

2.49 (a) $y(n) = \frac{8}{7} x(n)$. (b) $y(n) = \frac{1}{2}\pi^2 e^{jn\pi/8}$.

2.50 $H(e^{j\omega}) = \dfrac{2e^{-j\omega}(1 - \frac{1}{2}e^{-j\omega})^3}{(1 - \frac{1}{3}e^{-j\omega})}$.

2.51 $H(e^{j\omega}) = \dfrac{1 + \frac{1}{4}e^{-j10\omega}}{1 - \frac{1}{3}e^{-j10\omega}}$.

2.52 $\tau(\omega) = \dfrac{1 - 2\cos\omega}{5 - 4\cos\omega} - \dfrac{1 + 4\cos\omega}{17 + 8\cos\omega}$.

2.53 $h(n) = h_1(n) + h_2(n)$ where $h_1(n) = \dfrac{\sin(n\omega_1)}{n\omega_1}$ is an ideal low-pass filter with a cutoff frequency of ω_1, and $h_2(n) = \delta(n) - \dfrac{\sin(n\omega_2)}{n\omega_2}$ is an ideal high-pass filter with a cutoff frequency of ω_2.

2.54 A bandpass filter with a lower cutoff frequency of $\omega_1 = 3\pi/8$, an upper cutoff frequency of $\omega_2 = 5\pi/8$, and a gain of one half.

2.55 A high-pass filter with a cutoff frequency $\omega_c = \pi/3$.

2.56 $|H(e^{j\omega})| = 1$ for $|\omega| > \frac{\pi}{4}$ and $|H(e^{j\omega})| = 0$ otherwise.

2.57 $X(e^{j\omega}) = \dfrac{\sin^2 \frac{(N+1)}{2}\omega}{\sin^2 \frac{\omega}{2}}$.

2.58   (a) $|\omega| < \dfrac{2\pi}{3}$. (b) $\dfrac{\pi}{6} < |\omega| < \dfrac{5\pi}{6}$. (c) $|\omega| < \dfrac{2\pi}{3}$.

2.59   Unique, $h(n) = \frac{1}{2}\delta(n)$.

2.60 $x(n) = \dfrac{\sin\frac{n\pi}{4}}{n\pi} + \dfrac{\sin\frac{3n\pi}{4}}{n\pi}$.

2.61 $x(n) = 2\delta(n) - \dfrac{\sin\frac{3n\pi}{4}}{n\pi} - \dfrac{\sin\frac{n\pi}{4}}{n\pi}$.

2.62 $x(n) = \frac{1}{2}\delta(n+2) + \frac{1}{2}\delta(n+1) - \frac{1}{2}\delta(n-1) + \frac{1}{2}\delta(n-2)$.

2.63 $2\cos\frac{(n+2)\pi}{2} \cdot \dfrac{\sin\frac{(n+2)\pi}{6}}{(n+2)\pi}$.

2.64   The DTFT is constant with an amplitude of $\frac{1}{4}$ for $|\omega| < \dfrac{\pi}{4}$, and it decreases linearly to zero at $\omega = \pm\dfrac{3\pi}{4}$.

2.65 $x(n) = \frac{4}{3}\delta(n) + \frac{8}{3}\left(\frac{1}{2}\right)^n u(n-1)$.

2.66   (a) $X^*(e^{-j\omega})$. (b) $(1 - e^{-j2\omega})X(e^{j\omega})$. (c) $\frac{1}{2}X(e^{j\omega}) + \frac{1}{2}X(e^{j(\omega+\pi)})$. (d) $e^{-j\omega}X^2(e^{j\omega})$.

2.67 $y(n) = \begin{cases} \left(\frac{1}{5}\right)^{n/2} & n, = -4, -2, 0, 2, \ldots, \\ 0 & \text{otherwise} \end{cases}$

2.68   Beginning with index $n = -3$, the sequence values are $[1, \frac{3}{2}, \frac{1}{2}, -\frac{3}{2}, -2, \frac{3}{2}, \frac{5}{2}, \frac{3}{2}, -1]$.

2.69 $3 \cdot 5^5$.

2.70 -1.

2.71   Yes.

2.72 0.6π.

2.73 $\dfrac{\alpha}{1-\alpha}$.

2.74 $H(e^{j\omega}) = \frac{1}{2}e^{-j\omega}\dfrac{1 + 3e^{-j\omega} + 2e^{-j2\omega}}{1 - \frac{1}{2}e^{-j2\omega}}$.

2.75 $y(n) = \frac{1}{2}x(n+1) - \frac{1}{2}y(n-2)$.

2.76 $-72e^{3j\omega}\dfrac{(1 + \frac{1}{6}e^{-j\omega})^2}{(1 + \frac{1}{9}e^{-j\omega})}$.

Chapter 3

Sampling

3.1 INTRODUCTION

Most discrete-time signals come from *sampling* a continuous-time signal, such as speech and audio signals, radar and sonar data, and seismic and biological signals. The process of converting these signals into digital form is called *analog-to-digital* (A/D) conversion. The reverse process of reconstructing an analog signal from its samples is known as *digital-to-analog* (D/A) conversion. This chapter examines the issues related to A/D and D/A conversion. Fundamental to this discussion is the *sampling theorem*, which gives precise conditions under which an analog signal may be uniquely represented in terms of its samples.

3.2 ANALOG-TO-DIGITAL CONVERSION

An A/D converter transforms an analog signal into a digital sequence. The input to the A/D converter, $x_a(t)$, is a real-valued function of a continuous variable, t. Thus, for each value of t, the function $x_a(t)$ may be any real number. The output of the A/D is a *bit stream* that corresponds to a discrete-time sequence, $x(n)$, with an amplitude that is quantized, for each value of n, to one of a finite number of possible values. The components of an A/D converter are shown in Fig. 3-1. The first is the sampler, which is sometimes referred to as a *continuous-to-discrete* (C/D) converter, or *ideal A/D converter*. The sampler converts the continuous-time signal $x_a(t)$ into a discrete-time sequence $x(n)$ by extracting the values of $x_a(t)$ at integer multiples of the sampling period, T_s,

$$x(n) = x_a(nT_s)$$

Because the samples $x_a(nT_s)$ have a continuous range of possible amplitudes, the second component of the A/D converter is the quantizer, which maps the continuous amplitude into a discrete set of amplitudes. For a uniform quantizer, the quantization process is defined by the number of bits and the quantization interval Δ. The last component is the encoder, which takes the digital signal $\hat{x}(n)$ and produces a sequence of binary codewords.

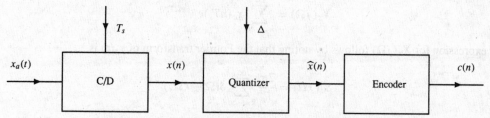

Fig. 3-1.   The components of an analog-to-digital converter.

3.2.1   Periodic Sampling

Typically, discrete-time signals are formed by *periodically sampling* a continuous-time signal

$$x(n) = x_a(nT_s) \tag{3.1}$$

The sample spacing T_s is the sampling period, and $f_s = 1/T_s$ is the sampling frequency in samples per second. A convenient way to view this sampling process is illustrated in Fig. 3-2(a). First, the continuous-time signal is multiplied by a periodic sequence of impulses,

$$s_a(t) = \sum_{n=-\infty}^{\infty} \delta(t - nT_s)$$

to form the *sampled signal*

$$x_s(t) = x_a(t)s_a(t) = \sum_{n=-\infty}^{\infty} x_a(nT_s)\delta(t - nT_s)$$

Then, the sampled signal is converted into a discrete-time signal by mapping the impulses that are spaced in time by T_s into a sequence $x(n)$ where the sample values are indexed by the integer variable n:

$$x(n) = x_a(nT_s)$$

This process is illustrated in Fig. 3-2(b).

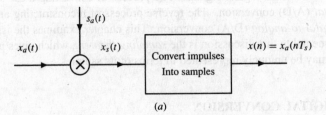

(a)

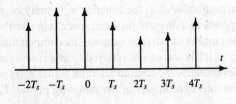
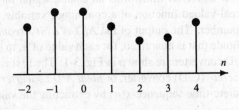

(b)

Fig. 3-2. Continuous-to-discrete conversion. (a) A model that consists of multiplying $x_a(t)$ by a sequence of impulses, followed by a system that converts impulses into samples. (b) An example that illustrates the conversion process.

The effect of the C/D converter may be analyzed in the frequency domain as follows. Because the Fourier transform of $\delta(t - nT_s)$ is $e^{-jn\Omega T_s}$, the Fourier transform of the sampled signal $x_s(t)$ is

$$X_s(j\Omega) = \sum_{n=-\infty}^{\infty} x_a(nT_s)e^{-jn\Omega T_s} \tag{3.2}$$

Another expression for $X_s(j\Omega)$ follows by noting that the Fourier transform of $s_a(t)$ is

$$S_a(j\Omega) = \frac{2\pi}{T_s} \sum_{k=-\infty}^{\infty} \delta(\Omega - k\Omega_s)$$

where $\Omega_s = 2\pi/T_s$ is the sampling frequency in radians per second. Therefore,

$$X_s(j\Omega) = \frac{1}{2\pi}X_a(j\Omega) * S_a(j\Omega) = \frac{1}{T_s} \sum_{k=-\infty}^{\infty} X_a(j\Omega - jk\Omega_s)$$

Finally, the discrete-time Fourier transform of $x(n)$ is

$$X(e^{j\omega}) = \sum_{n=-\infty}^{\infty} x(n)e^{-jn\omega} = \sum_{n=-\infty}^{\infty} x_a(nT_s)e^{-jn\omega} \tag{3.3}$$

Comparing Eq. (3.3) with Eq. (3.2), it follows that

$$X(e^{j\omega}) = X_s(j\Omega)|_{\Omega=\omega/T_s} = \frac{1}{T_s} \sum_{k=-\infty}^{\infty} X_a\left(j\frac{\omega}{T_s} - j\frac{2\pi k}{T_s}\right) \tag{3.4}$$

Thus, $X(e^{j\omega})$ is a frequency-scaled version of $X_s(j\Omega)$, with the scaling defined by

$$\omega = \Omega T_s$$

This scaling, which makes $X(e^{j\omega})$ periodic with a period of 2π, is a consequence of the time-scaling that occurs when $x_s(t)$ is converted to $x(n)$.

EXAMPLE 3.2.1   Suppose that $x_a(t)$ is strictly bandlimited so that $X_a(j\Omega) = 0$ for $|\Omega| > \Omega_0$ as shown in the figure below.

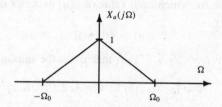

If $x_a(t)$ is sampled with a sampling frequency $\Omega_s \geq 2\Omega_0$, the Fourier transform of $x_s(t)$ is formed by periodically replicating $X_a(j\Omega)$ as illustrated in the figure below.

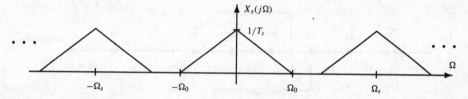

However, if $\Omega_s < 2\Omega_0$, the shifted spectra $X_a(j\Omega - jk\Omega_s)$ overlap, and when these spectra are summed to form $X_s(j\Omega)$, the result is as shown in the figure below.

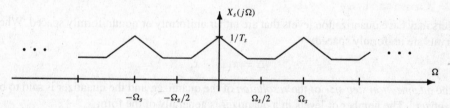

This overlapping of spectral components is called *aliasing*. When aliasing occurs, the frequency content of $x_a(t)$ is corrupted, and $X_a(j\Omega)$ cannot be recovered from $X_s(j\Omega)$.

As illustrated in Example 3.2.1, if $x_a(t)$ is strictly bandlimited so that the highest frequency in $x_a(t)$ is Ω_0, and if the sampling frequency is greater than $2\Omega_0$,

$$\Omega_s \geq 2\Omega_0$$

no aliasing occurs, and $x_a(t)$ may be uniquely recovered from its samples $x_a(nT_s)$ with a low-pass filter. The following is a statement of the famous Nyquist sampling theorem:

Sampling Theorem: If $x_a(t)$ is strictly bandlimited,

$$X_a(j\Omega) = 0 \qquad |\Omega| > \Omega_0$$

then $x_a(t)$ may be uniquely recovered from its samples $x_a(nT_s)$ if

$$\Omega_s = \frac{2\pi}{T_s} \geq 2\Omega_0$$

The frequency Ω_0 is called the *Nyquist frequency*, and the minimum sampling frequency, $\Omega_s = 2\Omega_0$, is called the *Nyquist rate*.

Because the signals that are found in physical systems will never be strictly bandlimited, an analog *anti-aliasing* filter is typically used to filter the signal prior to sampling in order to minimize the amount of energy above the Nyquist frequency and to reduce the amount of aliasing that occurs in the A/D converter.

3.2.2   *Quantization and Encoding*

A quantizer is a nonlinear and noninvertible system that transforms an input sequence $x(n)$ that has a continuous range of amplitudes into a sequence for which each value of $x(n)$ assumes one of a finite number of possible values. This operation is denoted by

$$\hat{x}(n) = Q[x(n)]$$

The quantizer has $L + 1$ *decision levels* $x_1, x_2, \ldots, x_{L+1}$ that divide the amplitude range for $x(n)$ into L *intervals*

$$I_k = [x_k, x_{k+1}] \qquad k = 1, 2, \ldots, L$$

For an input $x(n)$ that falls within interval I_k, the quantizer assigns a value within this interval, $\hat{x}_k$, to $x(n)$. This process is illustrated in Fig. 3-3.

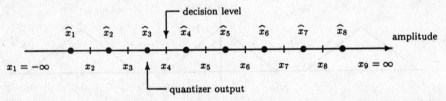

Fig. 3-3.   A quantizer with nine decision levels that divide the input amplitudes into eight quantization intervals and eight possible quantizer outputs, $\hat{x}_k$.

Quantizers may have quantization levels that are either uniformly or nonuniformly spaced. When the quantization intervals are uniformly spaced,

$$\Delta = x_{k+1} - x_k$$

Δ is called the *quantization step size* or the *resolution* of the quantizer, and the quantizer is said to be a *uniform* or *linear* quantizer.[1] The number of levels in a quantizer is generally of the form

$$L = 2^{B+1}$$

in order to make the most efficient use of a $(B + 1)$-bit binary code word. A 3-bit uniform quantizer in which the quantizer output is *rounded* to the nearest quantization level is illustrated in Fig. 3-4. With $L = 2^{B+1}$ quantization levels and a step size Δ, the *range* of the quantizer is

$$R = 2^{B+1} \cdot \Delta$$

Therefore, if the quantizer input is bounded,

$$|x(n)| \le X_{\max}$$

the range of possible input values may be covered with a step size

$$\Delta = \frac{X_{\max}}{2^B}$$

With rounding, the quantization error

$$e(n) = Q[x(n)] - x(n)$$

---

[1] In some applications, such as speech coding, the quantizer levels are *adaptive* (i.e., they change with time).

will be bounded by

$$-\frac{\Delta}{2} < e(n) < \frac{\Delta}{2}$$

However, if $|x(n)|$ exceeds $X_{\max}$, then $x(n)$ will be *clipped*, and the quantization error could be very large.

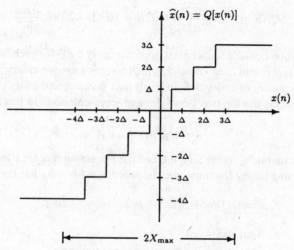

Fig. 3-4.   A 3-bit uniform quantizer.

A useful model for the quantization process is given in Fig. 3-5. Here, the quantization error is assumed to be an additive noise source. Because the quantization error is typically not known, the quantization error is described statistically. It is generally assumed that $e(n)$ is a sequence of random variables where

1. The statistics of $e(n)$ do not change with time (the quantization noise is a stationary random process).

2. The quantization noise $e(n)$ is a sequence of *uncorrelated* random variables.

3. The quantization noise $e(n)$ is *uncorrelated* with the quantizer input $x(n)$.

4. The probability density function of $e(n)$ is uniformly distributed over the range of values of the quantization error.

Although it is easy to find cases in which these assumptions do not hold (e.g., if $x(n)$ is a constant), they are generally valid for rapidly varying signals with fine quantization (Δ small).

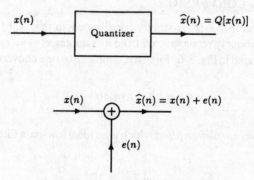

Fig. 3-5.   A quantization noise model.

With rounding, the quantization noise is uniformly distributed over the interval $[-\Delta/2, \Delta/2]$, and the quantization noise power (the variance) is

$$\sigma_e^2 = \frac{\Delta^2}{12}$$

With a step size

$$\Delta = \frac{X_{\max}}{2^B}$$

and a signal power σ_x^2, the signal-to-quantization noise ratio, in decibels (dB), is

$$\text{SQNR} = 10\log\frac{\sigma_x^2}{\sigma_e^2} = 6.02B + 10.81 - 20\log\frac{X_{\max}}{\sigma_x} \qquad (3.5)$$

Thus, the signal-to-quantization noise ratio increases approximately 6 dB for each bit.

The output of the quantizer is sent to an *encoder*, which assigns a unique binary number *(codeword)* to each quantization level. Any assignment of codewords to levels may be used, and many coding schemes exist. Most digital signal processing systems use the two's-complement representation. In this system, with a $(B + 1)$ bit codeword,

$$c = [b_0, b_1, \ldots, b_B]$$

the leftmost or most significant bit, b_0, is the sign bit, and the remaining bits are used to represent either binary integers or fractions. Assuming binary fractions, the codeword $b_0 b_1 b_2 \cdots b_B$ has the value

$$x = (-1)b_0 + b_1 2^{-1} + b_2 2^{-2} + \cdots + b_B 2^{-B}$$

An example is given below for a 3-bit codeword.

Binary Symbol	Numeric Value
0 1 1	$\frac{3}{4}$
0 1 0	$\frac{1}{2}$
0 0 1	$\frac{1}{4}$
0 0 0	0
1 1 1	$-\frac{1}{4}$
1 1 0	$-\frac{1}{2}$
1 0 1	$-\frac{3}{4}$
1 0 0	-1

3.3 DIGITAL-TO-ANALOG CONVERSION

As stated in the sampling theorem, if $x_a(t)$ is strictly bandlimited so that $X_a(j\Omega) = 0$ for $|\Omega| > \Omega_0$, and if $T_s < \pi / \Omega_0$, then $x_a(t)$ may be uniquely reconstructed from its samples $x(n) = x_a(nT_s)$. The reconstruction process involves two steps, as illustrated in Fig. 3-6. First, the samples $x(n)$ are converted into a sequence of impulses,

$$x_s(t) = \sum_{n=-\infty}^{\infty} x(n)\delta(t - nT_s)$$

and then $x_a(t)$ is filtered with a *reconstruction filter*, which is an ideal low-pass filter that has a frequency response given by

$$H_r(j\Omega) = \begin{cases} T_s & |\Omega| \le \dfrac{\pi}{T_s} \\ 0 & |\Omega| > \dfrac{\pi}{T_s} \end{cases}$$

This system is called an *ideal discrete-to-continuous* (D/C) converter. Because the impulse response of the reconstruction filter is

$$h_r(t) = \frac{\sin(\pi t / T_s)}{\pi t / T_s}$$

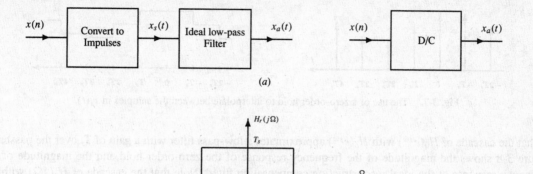

Fig. 3-6. (a) A discrete-to-continuous converter with an ideal low-pass reconstruction filter. (b) The frequency response of the ideal reconstruction filter.

the output of the filter is

$$x_a(t) = \sum_{n=-\infty}^{\infty} x(n)h_r(t - nT_s) = \sum_{n=-\infty}^{\infty} x(n)\frac{\sin \pi(t - nT_s)/T_s}{\pi(t - nT_s)/T_s} \tag{3.6}$$

This *interpolation formula* shows how $x_a(t)$ is reconstructed from its samples $x(n) = x_a(nT_s)$. In the frequency domain, the interpolation formula becomes

$$X_a(j\Omega) = \sum_{n=-\infty}^{\infty} x(n)H_r(j\Omega)e^{-jn\Omega T_s}$$

$$= H_r(j\Omega) \sum_{n=-\infty}^{\infty} x(n)e^{-jn\Omega T_s} = H_r(j\Omega)X(e^{j\Omega T_s}) \tag{3.7}$$

which is equivalent to

$$X_a(j\Omega) = \begin{cases} T_s X(e^{j\Omega T_s}) & |\Omega| < \dfrac{\pi}{T_s} \\ 0 & \text{otherwise} \end{cases} \tag{3.8}$$

Thus, $X(e^{j\omega})$ is frequency scaled ($\omega = \Omega T_s$), and then the low-pass filter removes all frequencies in the periodic spectrum $X(e^{j\Omega T_s})$ above the cutoff frequency $\Omega_c = \pi/T_s$.

Because it is not possible to implement an ideal low-pass filter, many D/A converters use a *zero-order hold* for the reconstruction filter. The impulse response of a zero-order hold is

$$h_0(t) = \begin{cases} 1 & 0 \le t \le T_s \\ 0 & \text{otherwise} \end{cases}$$

and the frequency response is

$$H_0(j\Omega) = e^{-j\Omega T_s/2}\frac{\sin(\Omega T_s/2)}{\Omega/2}$$

After a sequence of samples $x_a(nT_s)$ has been converted to impulses, the zero-order hold produces the staircase approximation to $x_a(t)$ shown in Fig. 3-7. With a zero-order hold, it is common to postprocess the output with a *reconstruction compensation filter* that approximates the frequency response

$$H_c(j\Omega) = \begin{cases} \dfrac{\Omega T_s/2}{\sin(\Omega T_s/2)}e^{j\Omega T_s/2} & |\Omega| \le \dfrac{\pi}{T_s} \\ 0 & |\Omega| > \dfrac{\pi}{T_s} \end{cases}$$

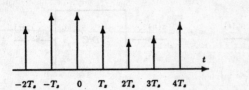
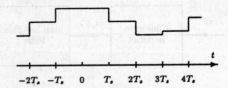

Fig. 3-7. The use of a zero-order hold to interpolate between the samples in $x_s(t)$.

so that the cascade of $H_0(e^{j\omega})$ with $H_c(e^{j\omega})$ approximates a low-pass filter with a gain of T_s over the passband. Figure 3-8 shows the magnitude of the frequency response of the zero-order hold and the magnitude of the frequency response of the ideal reconstruction compensation filter. Note that the cascade of $H_c(j\Omega)$ with the zero-order hold is an ideal low-pass filter.

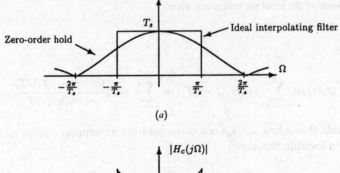

(a)

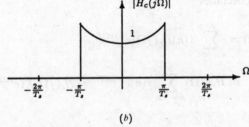

(b)

Fig. 3-8. (a) The magnitude of the frequency response of a zero-order hold compared to the ideal reconstruction filter. (b) The ideal reconstruction compensation filter.

3.4 DISCRETE-TIME PROCESSING OF ANALOG SIGNALS

One of the important applications of A/D and D/A converters is the processing of analog signals with a discrete-time system. In the ideal case, the overall system, shown in Fig. 3-9, consists of the cascade of a C/D converter, a discrete-time system, and a D/C converter. Thus, we are assuming that the sampled signal is not quantized and that an ideal low-pass filter is used for the reconstruction filter in the D/C converter. Because the input $x_a(t)$ and the output $y_a(t)$ are analog signals, the overall system corresponds to a continuous-time system. To analyze this system, note that the C/D converter produces the discrete-time signal $x(n)$, which has a DTFT given by

$$X(e^{j\omega}) = \frac{1}{T_s} \sum_{k=-\infty}^{\infty} X_a\left(j\frac{\omega}{T_s} - j\frac{2\pi k}{T_s}\right)$$

If the discrete-time system is linear and shift-invariant with a frequency response $H(e^{j\omega})$,

$$Y(e^{j\omega}) = H(e^{j\omega})X(e^{j\omega}) = H(e^{j\omega})\frac{1}{T_s} \sum_{k=-\infty}^{\infty} X_a\left(j\frac{\omega}{T_s} - j\frac{2\pi k}{T_s}\right)$$

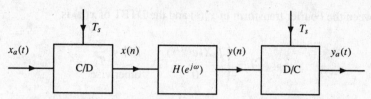

Fig. 3-9. Processing an analog signal using a discrete-time system.

Finally, the D/C converter produces the continuous-time signal $y_a(t)$ from the samples $y(n)$ as follows:

$$y_a(t) = \sum_{n=-\infty}^{\infty} y(n) \frac{\sin \pi(t - nT_s)/T_s}{\pi(t - nT_s)/T_s}$$

Either using Eq. (3.7) or by taking the DTFT directly, in the frequency domain this relationship becomes

$$Y_a(j\Omega) = H_r(j\Omega)Y(e^{j\Omega T_s}) = H_r(j\Omega)H(e^{j\Omega T_s})X(e^{j\Omega T_s})$$

or

$$Y_a(j\Omega) = H_r(j\Omega)H(e^{j\Omega T_s})\frac{1}{T_s} \sum_{k=-\infty}^{\infty} X_a\left(j\Omega - j\frac{2\pi k}{T_s}\right)$$

If $x_a(t)$ is bandlimited with $X_a(j\Omega) = 0$ for $|\Omega| > \pi/T_s$, the low-pass filter $H_r(j\Omega)$ eliminates all terms in the sum except the first one, and

$$Y_a(j\Omega) = \begin{cases} H(e^{j\Omega T_s})X_a(j\Omega) & |\Omega| \le \dfrac{\pi}{T_s} \\ 0 & |\Omega| > \dfrac{\pi}{T_s} \end{cases}$$

Therefore, the overall system behaves as a linear time-invariant continuous-time system with an effective frequency response

$$H_a(j\Omega) = \begin{cases} H(e^{j\Omega T_s}) & |\Omega| \le \dfrac{\pi}{T_s} \\ 0 & \text{otherwise} \end{cases} \tag{3.9}$$

Just as a continuous-time system may be implemented in terms of a discrete-time system, it is also possible to implement a discrete-time system in terms of a continuous-time system as illustrated Fig. 3-10. The signal $x_a(t)$ is related to the sequence values $x(n)$ as follows:

$$x_a(t) = \sum_{n=-\infty}^{\infty} x(n) \frac{\sin \pi(t - nT_s)/T_s}{\pi(t - nT_s)/T_s}$$

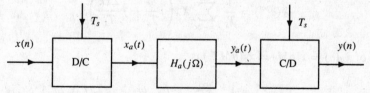

Fig. 3-10. Processing a discrete-time signal using a continuous-time system.

Because $x_a(t)$ is bandlimited, $y_a(t)$ is also bandlimited and may be represented in terms of its samples as follows:

$$y_a(t) = \sum_{n=-\infty}^{\infty} y(n) \frac{\sin \pi(t - nT_s)/T_s}{\pi(t - nT_s)/T_s}$$

The relationship between the Fourier transform of $x_a(t)$ and the DTFT of $x(n)$ is

$$X_a(j\Omega) = \begin{cases} T_s X(e^{j\Omega T_s}) & |\Omega| < \dfrac{\pi}{T_s} \\ 0 & \text{otherwise} \end{cases}$$

and the relationship between the Fourier transforms of $x_a(t)$ and $y_a(t)$ is

$$Y_a(j\Omega) = \begin{cases} H_a(j\Omega)X_a(j\Omega) & |\Omega| < \dfrac{\pi}{T_s} \\ 0 & \text{otherwise} \end{cases}$$

Therefore,

$$Y(e^{j\omega}) = \frac{1}{T_s} Y_a\left(\frac{j\omega}{T_s}\right) \qquad |\omega| < \pi$$

and the frequency response of the equivalent discrete-time system is

$$H(e^{j\omega}) = H_a\left(\frac{j\omega}{T_s}\right) \qquad |\omega| < \pi \tag{3.10}$$

3.5 SAMPLE RATE CONVERSION

In many practical applications of digital signal processing, one is faced with the problem of changing the sampling rate of a signal. The process of converting a signal from one rate to another is called *sample rate conversion*. There are two ways that sample rate conversion may be done. First, the sampled signal may be converted back into an analog signal and then resampled. Alternatively, the signal may be *resampled* in the digital domain. This approach has the advantage of not introducing additional distortion in passing the signal through an additional D/A and A/D converter. In this section, we describe how sample rate conversion may be performed digitally.

3.5.1 Sample Rate Reduction by an Integer Factor

Suppose that we would like to reduce the sampling rate by an integer factor, M. With a new sampling period $T_s' = MT_s$, the resampled signal is

$$x_d(n) = x_a(nT_s') = x_a(nMT_s) = x(nM)$$

Therefore, reducing the sampling rate by an integer factor M may be accomplished by taking every Mth sample of $x(n)$. The system for performing this operation, called a *down-sampler*, is shown in Fig. 3-11(a). Down-sampling generally results in aliasing. Specifically, recall that the DTFT of $x(n) = x_a(nT_s)$ is

$$X(e^{j\omega}) = \frac{1}{T_s} \sum_{k=-\infty}^{\infty} X_a\left(j\frac{\omega}{T_s} - j\frac{2\pi k}{T_s}\right)$$

Similarly, the DTFT of $x_d(n) = x(nM) = x_a(nMT_s)$ is

$$X_d(e^{j\omega}) = \frac{1}{MT_s} \sum_{r=-\infty}^{\infty} X_a\left(j\frac{\omega}{MT_s} - j\frac{2\pi r}{MT_s}\right)$$

Note that the summation index r in the expression for $X_d(e^{j\omega})$ may be expressed as

$$r = i + kM$$

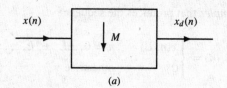

(a)

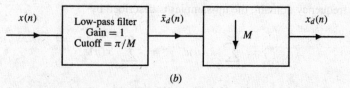

(b)

Fig. 3-11. *(a)* Down-sampling by an integer factor M. *(b)* Decimation by a factor of M, where $H(e^{j\omega})$ is a low-pass filter with a cutoff frequency $\omega_c = \pi/M$.

where $-\infty < k < \infty$ and $0 \le i \le M - 1$. Therefore, $X_d(e^{j\omega})$ may be expressed as

$$X_d(e^{j\omega}) = \frac{1}{M} \sum_{i=0}^{M-1} \left[\frac{1}{T_s} \sum_{k=-\infty}^{\infty} X_a\left(j\frac{\omega}{MT_s} - j\frac{2\pi k}{T_s} - j\frac{2\pi i}{MT_s} \right) \right]$$

The term inside the square brackets is

$$X\left(e^{j(\omega-2\pi i)/M}\right) = \frac{1}{T_s} \sum_{k=-\infty}^{\infty} X_a\left(j\frac{(\omega-2\pi i)}{MT_s} - j\frac{2\pi k}{T_s} \right)$$

Thus, the relationship between $X(e^{j\omega})$ and $X_d(e^{j\omega})$ is

$$X_d(e^{j\omega}) = \frac{1}{M} \sum_{k=0}^{M-1} X\left(e^{j(\omega-2\pi k)/M}\right) \tag{3.11}$$

Therefore, in order to prevent aliasing, $x(n)$ should be filtered prior to down-sampling with a low-pass filter that has a cutoff frequency $\omega_c = \pi/M$. The cascade of a low-pass filter with a down-sampler illustrated in Fig. 3-11(*b*) is called a *decimator*.

3.5.2　Sample Rate Increase by an Integer Factor

Suppose that we would like to increase the sampling rate by an integer factor L. If $x_a(t)$ is sampled with a sampling frequency $f_s = 1/T_s$, then

$$x(n) = x_a(nT_s)$$

To increase the sampling rate by an integer factor L, it is necessary to extract the samples

$$x_i(n) = x_a\left(\frac{nT_s}{L} \right)$$

from $x(n)$. The samples of $x_i(n)$ for values of n that are integer multiples of L are easily extracted from $x(n)$ as follows:

$$x_i(nL) = x(n)$$

Shown in Fig. 3-12(a) is an *up-sampler* that produces the sequence

$$\tilde{x}_i(n) = \begin{cases} x(n/L) & n = 0, \pm L, \pm 2L, \ldots \\ 0 & \text{otherwise} \end{cases}$$

In other words, the up-sampler expands the time scale by a factor of L by inserting $L - 1$ zeros between each sample of $x(n)$. In the frequency domain, the up-sampler is described by

$$\tilde{X}_i(e^{j\omega}) = \sum_{n=-\infty}^{\infty} \tilde{x}_i(n)e^{-jn\omega} = \sum_{n=-\infty}^{\infty} x(n)e^{-jnL\omega}$$

or

$$\tilde{X}_i(e^{j\omega}) = X(e^{jL\omega}) \qquad (3.12)$$

Therefore, $X(e^{j\omega})$ is simply scaled in frequency. After up-sampling, it is necessary to remove the frequency scaled *images* of $X_a(j\Omega)$, except those that are at integer multiples of 2π. This is accomplished by filtering $\tilde{x}_i(n)$

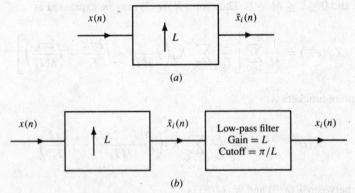

(a)

(b)

Fig. 3-12. (a) Up-sampling by an integer factor L. (b) Interpolation by a factor of L.

with a low-pass filter that has a cutoff frequency of π/L and a gain of L. In the time domain, the low-pass filter interpolates between the samples at integer multiples of L as shown in Fig. 3-13. The cascade of an up-sampler with a low-pass filter shown in Fig. 3-12(b) is called an *interpolator*. The interpolation process in the frequency domain is illustrated in Fig. 3-14.

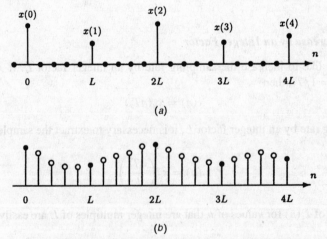

(a)

(b)

Fig. 3-13. (a) The output of the up-sampler. (b) The interpolation between the samples $\tilde{x}_i(n)$ that is performed by the low-pass filter.

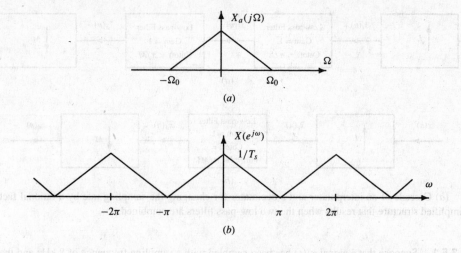

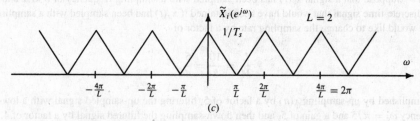

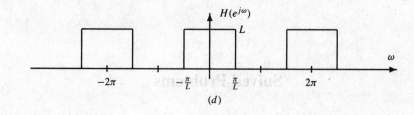

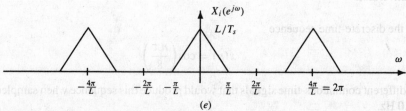

Fig. 3-14. Frequency domain illustration of the process of interpolation. (*a*) The continuous-time signal. (*b*) The DTFT of the sampled signal $x(n) = x_a(nT_s)$. (*c*) The DTFT of the up-sampler output. (*d*) The ideal low-pass filter to perform the interpolation. (*e*) The DTFT of the interpolated signal.

3.5.3 Sample Rate Conversion by a Rational Factor

The cascade of a decimator that reduces the sampling rate by a factor of M with an interpolator that increases the sampling rate by vital factor of L results in a system that changes the sampling rate by a rational factor of L/M. This cascade is illustrated in Fig. 3-15(*a*). Because the cascade of two low-pass filters with cutoff frequencies π/M and π/L is equivalent to a single low-pass filter with a cutoff frequency

$$\omega_c = \min\left\{ \frac{\pi}{M}, \frac{\pi}{L} \right\}$$

the sample rate converter may be simplified as illustrated in Fig. 3-15(*b*).

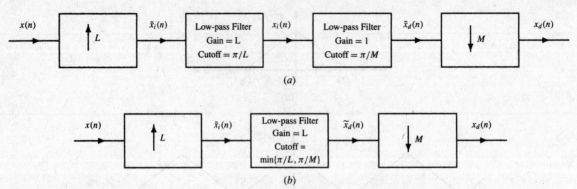

Fig. 3-15. (a) Cascade of an interpolator and a decimator for changing the sampling rate by a rational factor L/M. (b) A simplified structure that results when the two low-pass filters are combined.

EXAMPLE 3.5.1 Suppose that a signal $x_a(t)$ has been sampled with a sampling frequency of 8 kHz and that we would like to derive the discrete-time signal that would have been obtained if $x_a(t)$ had been sampled with a sampling frequency of 10 kHz. Thus, we would like to change the sampling rate by a factor of

$$\frac{L}{M} = \frac{10}{8} = \frac{5}{4}$$

This may be accomplished by up-sampling $x(n)$ by a factor of 5, filtering the up-sampled signal with a low-pass filter that has a cutoff frequency $\omega_c = \pi/5$ and a gain of 5, and then down-sampling the filtered signal by a factor of 4.

Solved Problems

A/D and D/A Conversion

3.1   Consider the discrete-time sequence

$$x(n) = \cos\left(\frac{n\pi}{8}\right)$$

Find two different continuous-time signals that would produce this sequence when sampled at a frequency of $f_s = 10$ Hz.

A continuous-time sinusoid

$$x_a(t) = \cos(\Omega_0 t) = \cos(2\pi f_0 t)$$

that is sampled with a sampling frequency of f_s results in the discrete-time sequence

$$x(n) = x_a(nT_s) = \cos\left(2\pi \frac{f_0}{f_s} n\right)$$

However, note that for any integer k,

$$\cos\left(2\pi \frac{f_0}{f_s} n\right) = \cos\left(2\pi \frac{f_0 + kf_s}{f_s} n\right)$$

Therefore, any sinusoid with a frequency

$$f = f_0 + kf_s$$

will produce the same sequence when sampled with a sampling frequency f_s. With $x(n) = \cos(n\pi/8)$, we want

$$2\pi \frac{f_0}{f_s} = \frac{\pi}{8}$$

or

$$f_0 = \tfrac{1}{16} f_s = 625 \text{ Hz}$$

Therefore, two signals that produce the given sequence are

$$x_1(t) = \cos(1250\pi t)$$

and

$$x_2(t) = \cos(21250\pi t)$$

3.2   If the Nyquist rate for $x_a(t)$ is Ω_s, what is the Nyquist rate for each of the following signals that are derived from $x_a(t)$?

(a) $\dfrac{dx_a(t)}{dt}$

(b) $x_a(2t)$

(c) $x_a^2(t)$

(d) $x_a(t)\cos(\Omega_0 t)$

(a) The Nyquist rate is equal to twice the highest frequency in $x_a(t)$. If

$$y_a(t) = \frac{dx_a(t)}{dt}$$

then

$$Y_a(j\Omega) = j\Omega X_a(j\Omega)$$

Thus, if $X_a(j\Omega) = 0$ for $|\Omega| > \Omega_0$, the same will be true for $Y_a(j\Omega)$. Therefore, the Nyquist frequency is not changed by differentiation.

(b) The signal $y_a(t) = x_a(2t)$ is formed from $x_a(t)$ by *compressing* the time axis by a factor of 2. This results in an *expansion* of the frequency axis by a factor of 2. Specifically, note that

$$Y_a(j\Omega) = \int_{-\infty}^{\infty} y_a(t)e^{-j\Omega t}\, dt = \int_{-\infty}^{\infty} x_a(2t)e^{-j\Omega t}\, dt$$

$$= \int_{-\infty}^{\infty} \tfrac{1}{2}x_a(\tau)e^{-j\Omega \tau/2}\, d\tau = \tfrac{1}{2}X_a\!\left(\frac{j\Omega}{2}\right)$$

Consequently, if the Nyquist frequency for $x_a(t)$ is Ω_s, the Nyquist frequency for $y_a(t)$ will be $2\Omega_s$.

(c) When two signals are multiplied, their Fourier transforms are convolved. Therefore, if

$$y_a(t) = x_a^2(t)$$

then

$$Y_a(j\Omega) = \frac{1}{2\pi} X_a(j\Omega) * X_a(j\Omega)$$

Thus, the highest frequency in $y_a(t)$ will be twice that of $x_a(t)$, and the Nyquist frequency will be $2\Omega_s$.

(d) Modulating a signal by $\cos(\Omega_0 t)$ shifts the spectrum of $x_a(t)$ up and down by Ω_0. Therefore, the Nyquist frequency for $y_a(t) = \cos(\Omega_0 t)x_a(t)$ will be $\Omega_s + 2\Omega_0$.

3.3   Let $h_a(t)$ be the impulse response of a causal continuous-time filter with a system function

$$H_a(s) = \frac{s+a}{(s+a)^2 + b^2}$$

Thus, $H_a(s)$ has a zero at $s = -a$ and a pair of poles at $s = -a \pm jb$. By sampling $h_a(t)$ we form a discrete-time filter with a unit sample response

$$h(n) = h_a(nT_s)$$

Find the frequency response $H(e^{j\omega})$ of the discrete-time filter.

To find the frequency response $H(e^{j\omega})$, it is necessary to find the impulse response of the analog filter, $h_a(t)$, sample the impulse response,

$$h(n) = h_a(nT_s)$$

and then find the discrete-time Fourier transform,

$$H(e^{j\omega}) = \sum_{n=-\infty}^{\infty} h(n)e^{-jn\omega}$$

To find the impulse response, we first perform a partial fraction expansion of $H_a(s)$ as follows:

$$H_a(s) = \frac{A}{s+(a+jb)} + \frac{B}{s+(a-jb)} \tag{3.13}$$

The constant A is

$$A = [(s+a+jb)H_a(s)]_{s=-a-jb} = \frac{s+a}{s+(a-jb)}\bigg|_{s=-a-jb} = \tfrac{1}{2}$$

Similarly, for B we have

$$B = [(s+a-jb)H_a(s)]_{s=-a+jb} = \frac{s+a}{s+(a+jb)}\bigg|_{s=-a+jb} = \tfrac{1}{2}$$

Therefore, $$H_a(s) = \frac{\frac{1}{2}}{s+(a+jb)} + \frac{\frac{1}{2}}{s+(a-jb)}$$

Another way to find the constants A and B would be to write Eq. (3.13) over a common denominator,

$$H_a(s) = \frac{s+a}{(s+a)^2 + b^2} = \frac{A(s+a-jb) + B(s+a+jb)}{(s+a)^2 + b^2}$$

and equate the polynomial coefficients in the numerators of $H_a(s)$:

$$A + B = 1$$
$$A(a-jb) + B(a+jb) = a$$

Solving these two equations for A and B gives the same result as before. From the partial fraction expansion of $H_a(s)$, the impulse response may be found using the Laplace transform pair

$$e^{-\alpha t}u(t) \Leftrightarrow \frac{1}{s+\alpha}$$

Specifically, we have

$$h_a(t) = \tfrac{1}{2}e^{(-a-jb)t}u(t) + \tfrac{1}{2}e^{(-a+jb)t}u(t) = e^{-at}\cos(bt)u(t)$$

Sampling $h_a(t)$, we have

$$h(n) = h_a(nT_s) = e^{-anT_s} \cos(bnT_s)u(n)$$

Finally, for the frequency response we have

$$H(e^{j\omega}) = \sum_{n=-\infty}^{\infty} h(n)e^{-jn\omega} = \sum_{n=0}^{\infty} e^{-anT_s} \cos(bnT_s)e^{-jn\omega}$$

$$= \sum_{n=0}^{\infty} \frac{1}{2} e^{(-a-jb)nT_s} e^{-jn\omega} + \sum_{n=0}^{\infty} \frac{1}{2} e^{(-a+jb)nT_s} e^{-jn\omega}$$

$$= \sum_{n=0}^{\infty} \frac{1}{2} (e^{-aT_s})^n e^{-jn(\omega+bT_s)} + \sum_{n=0}^{\infty} \frac{1}{2} (e^{-aT_s})^n e^{-jn(\omega-bT_s)}$$

Note that in order for these sums to converge, and for the frequency response to exist, it is necessary that

$$|e^{-aT_s}| < 1$$

or, because $T_s > 0$, we must have $a > 0$. In other words, the poles of $H_a(s)$ must lie in the left-half s-plane or, equivalently, $h_a(t)$ must be a stable filter. With $a > 0$ we have

$$H(e^{j\omega}) = \frac{\frac{1}{2}}{1 - e^{(-a-jb)T_s}e^{-j\omega}} + \frac{\frac{1}{2}}{1 - e^{(-a+jb)T_s}e^{-j\omega}}$$

which, after combining over a common denominator and simplifying, gives

$$H(e^{j\omega}) = \frac{1 - e^{-aT_s}\cos(bT_s)e^{-j\omega}}{1 - 2e^{-aT_s}\cos(bT_s)e^{-j\omega} + e^{-2aT_s}e^{-j2\omega}}$$

3.4   A continuous-time filter has a system function

$$H_a(s) = \frac{1}{s+1}$$

If $h_a(t)$ is sampled to form a discrete-time system with a unit sample response

$$h(n) = h_a(nT_s)$$

find the value for T_s so that $H(e^{j\omega})$ at $\omega = \pi/2$ is down 6 dB from its maximum value at $\omega = 0$, that is,

$$10 \log \frac{|H(e^{j\pi/2})|^2}{|H(e^{j0})|^2} = -6$$

The impulse response of the continuous-time system is

$$h_a(t) = e^{-t}u(t)$$

When sampled with a sampling period T_s, the resulting unit sample response is

$$h(n) = h_a(nT_s) = e^{-nT_s}u(n)$$

and the frequency response is

$$H(e^{j\omega}) = \sum_{n=0}^{\infty} e^{-nT_s}e^{-jn\omega} = \sum_{n=0}^{\infty} e^{-(T_s+j\omega)n} = \frac{1}{1 - e^{-T_s}e^{-j\omega}}$$

With

$$|H(e^{j0})|^2 = \frac{1}{(1 - e^{-T_s})^2}$$

and

$$|H(e^{j\pi/2})|^2 = \frac{1}{1 + e^{-2T_s}}$$

it follows that we want

$$10 \log \frac{|H(e^{j\pi/2})|^2}{|H(e^{j0})|^2} = 10 \log \frac{(1 - e^{-T_s})^2}{1 + e^{-2T_s}} = -6$$

or

$$\frac{(1 - e^{-T_s})^2}{1 + e^{-2T_s}} = 10^{-0.6} = 0.2512$$

Thus, we have

$$1 - 2e^{-T_s} + e^{-2T_s} = 0.2512 [1 + e^{-2T_s}]$$

or

$$0.7488e^{-2T_s} - 2e^{-T_s} + 0.7488 = 0$$

which is a quadratic equation in e^{-T_s}. Solving for the roots of this quadratic equation, we find

$$e^{-T_s} = \frac{1}{2(0.7488)} \left[2 \pm \sqrt{4 - 4(0.7488)^2} \right] = \frac{1}{0.7488} [1 \pm 0.6628] = 2.2206, \ 0.4503$$

Taking the natural logarithm, and selecting the positive value for T_s, we have

$$T_s = 0.7978$$

3.5   A continuous-time signal $x_a(t)$ is bandlimited with $X_a(j\Omega) = 0$ for $|\Omega| > \Omega_0$. If $x_a(t)$ is sampled with a sampling frequency $\Omega_s \geq 2\Omega_0$, how is the energy in $x(n)$,

$$E_d = \sum_{n=-\infty}^{\infty} |x(n)|^2$$

related to the energy in $x_a(t)$,

$$E_a = \int_{-\infty}^{\infty} |x_a(t)|^2 \, dt$$

and the sampling period T_s?

Using Parseval's theorem, the energy in the analog signal $x_a(t)$ may be expressed in the frequency domain as follows:

$$E_a = \int_{-\infty}^{\infty} |x_a(t)|^2 \, dt = \frac{1}{2\pi} \int_{-\infty}^{\infty} |X_a(j\Omega)|^2 \, d\Omega$$

Because $x_a(t)$ is bandlimited with $X_a(j\Omega) = 0$ for $|\Omega| > \Omega_0$,

$$E_a = \frac{1}{2\pi} \int_{-\Omega_0}^{\Omega_0} |X_a(j\Omega)|^2 \, d\Omega$$

Sampling $x_a(t)$ at or above the Nyquist rate results in a sequence $x(n)$ with a discrete-time Fourier transform

$$X(e^{j\omega}) = \begin{cases} \dfrac{1}{T_s} X_a\left(\dfrac{j\omega}{T_s}\right) & |\omega| \leq \Omega_0 T_s \\ 0 & \Omega_0 T_s < |\omega| \leq \pi \end{cases}$$

Therefore, the energy in $x(n)$, using Parseval's theorem, is

$$E_d = \sum_{n=-\infty}^{\infty} |x(n)|^2 = \frac{1}{2\pi} \int_{-\pi}^{\pi} |X(e^{j\omega})|^2 \, d\omega$$

$$= \frac{1}{2\pi} \int_{-\Omega_0 T_s}^{\Omega_0 T_s} \frac{1}{T_s^2} \left| X_a\left(\frac{j\omega}{T_s}\right) \right|^2 d\omega$$

$$= \frac{1}{2\pi T_s} \int_{-\Omega_0}^{\Omega_0} |X_a(ju)|^2 \, du = \frac{1}{T_s} E_a$$

and we have

$$E_d = \frac{1}{T_s} E_a$$

As a check on this result, suppose that $x_a(t)$ is a bandlimited signal with a spectrum shown in the figure below.

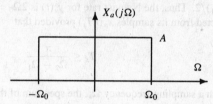

The energy in $x_a(t)$ is

$$E_a = \frac{1}{2\pi} A^2 \cdot 2\Omega_0 = \frac{A^2 \Omega_0}{\pi}$$

When sampled with a sampling frequency $\Omega_s \geq 2\Omega_0$, the DTFT of the sampled signal is as shown in the following figure:

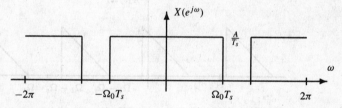

Therefore, the energy in $x(n)$ is

$$E_d = \frac{1}{2\pi} \left(\frac{A}{T_s}\right)^2 \cdot 2 \, \Omega_0 T_s = \frac{A^2 \Omega_0}{\pi T_s} = \frac{1}{T_s} E_a$$

3.6   A complex bandpass analog signal $x_a(t)$ has a Fourier transform that is nonzero over the frequency range $[\Omega_1, \Omega_2]$ as shown in the figure below.

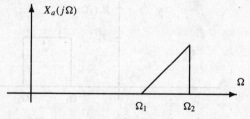

The signal is sampled to produce the sequence $x(n) = x_a(nT_s)$.

(a)   What is the smallest sampling frequency that can be used so that $x_a(t)$ may be recovered from its samples $x(n)$?

(b) For this minimum sampling frequency, find the interpolation formula for $x_a(t)$ in terms of $x(n)$.

(a) Because the highest frequency in $x_a(t)$ is Ω_2, the Nyquist rate is $2\Omega_2$. However, note that if $x_a(t)$ is modulated with a complex exponential of frequency $(\Omega_2 + \Omega_1)/2$,

$$y_a(t) = x_a(t)e^{-j(\Omega_2+\Omega_1)t/2}$$

then $y_a(t)$ is a (complex) low-pass signal with a spectrum shown in the following figure:

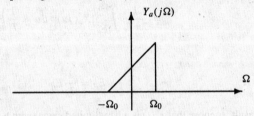

where $\Omega_0 = (\Omega_2 - \Omega_1)/2$. Thus, the Nyquist rate for $y_a(t)$ is $2\Omega_0 = \Omega_2 - \Omega_1$, which suggests that $x_a(t)$ may be uniquely reconstructed from its samples $x_a(nT_s)$ provided that

$$T_s \leq \frac{\pi}{\Omega_2 - \Omega_1}$$

If $x_a(t)$ is sampled with a sampling frequency Ω_s, the spectrum of the sampled signal is

$$X_s(j\Omega) = \frac{1}{T_s} \sum_{k=-\infty}^{\infty} X_a(j\Omega - jk\Omega_s)$$

as illustrated below.

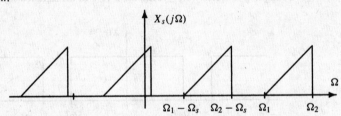

In order for there to be no interference between the shifted spectra, it is necessary that

$$\Omega_2 - \Omega_s \leq \Omega_1$$

or

$$\Omega_s \geq \Omega_2 - \Omega_1$$

If this condition is satisfied, $x_a(t)$ may be uniquely reconstructed from $x_s(t)$ using a bandpass filter with a frequency response as shown below.

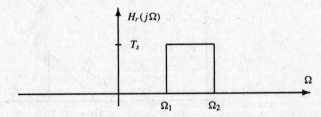

(b) With a sampling frequency $\Omega_s = \Omega_2 - \Omega_1$, the reconstruction filter is a complex bandpass filter with an impulse response

$$h_a(t) = T_s \frac{\sin(\Omega_s t/2)}{\pi t} e^{-j(\Omega_2+\Omega_1)t/2}$$

Therefore, the output of the reconstruction filter, which produces the complex bandpass signal $x_a(t)$, is

$$x_a(t) = \sum_{n=-\infty}^{\infty} x(n)h_r(t-nT_s) = T_s \sum_{n=-\infty}^{\infty} x(n) \frac{\sin \Omega_s(t-nT_s)/2}{\pi(t-nT_s)} e^{-j(\Omega_2+\Omega_1)(t-nT_s)/2}$$

3.7 Given a real-valued bandpass signal $x_a(t)$ with $X_a(f) = 0$ for $|f| < f_1$ and $|f| > f_2$, the Nyquist sampling theorem says that the minimum sampling frequency is $f_s = 2f_2$. However, in some cases, the signal may be sampled at a lower rate.

(a) Suppose that $f_1 = 8$ kHz and $f_2 = 10$ kHz. Make a sketch of the discrete-time Fourier transform of $x(n) = x_a(nT_s)$ if $f_s = 1/T_s = 4$ kHz.

(b) Define the bandwidth of the bandpass signal to be

$$B = f_2 - f_1$$

and the center frequency to be

$$f_c = \frac{f_2 + f_1}{2}$$

Show that if $f_c > B/2$ and f_2 is an integer multiple of the bandwidth B, no aliasing will occur if $x_a(t)$ is sampled at a sampling frequency $f_s = 2B$.

(c) Repeat part (b) for the case in which f_2 is not an integer multiple of the bandwidth B.

(a) Let $x_a(t)$ have a spectrum as shown in the figure below.

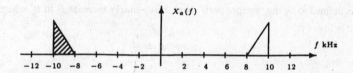

The spectrum of the sampled signal

$$x_s(t) = \sum_{n=-\infty}^{\infty} x_a(nT_s)\delta(t-nT_s)$$

is

$$X_s(f) = \frac{1}{T_s} \sum_{k=-\infty}^{\infty} X_a(f-kf_s)$$

which is formed by shifting $X_a(f)$ by integer multiples of the sampling frequency and summing. With $f_s = 4$ kHz, we have the spectrum sketched below.

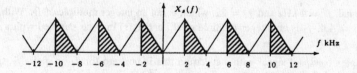

Note that $X_a(f)$ is not aliased. Therefore, with the appropriate processing of $x_s(t)$, the signal $x_a(t)$ may be recovered from its samples. Finally, the DTFT of the discrete-time sequence $x(n) = x_a(nT_s)$ is

$$X(e^{j\omega}) = X_s\left(\frac{j\omega}{T_s}\right)$$

which is sketched below.

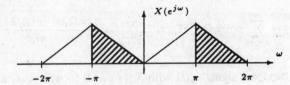

(b) If f_2 is an integer multiple of B, we may express f_1 and f_2 as follows:

$$f_1 = (l-1)B \qquad f_2 = lB$$

With a sampling frequency of $f_s = 2B$, the sampled signal has a spectrum

$$X_s(f) = \frac{1}{T_s} \sum_{k=-\infty}^{\infty} X_a(f - 2kB)$$

Because $X_a(f)$ is nonzero only for $(l-1)B < |f| < lB$, there is only one term in the sum that contributes to $X_s(f)$ in the frequency range $0 < f < B$ and only one term that contributes to the frequency range $-B < f < 0$ (draw a picture as in part (a) to see this clearly). Therefore, there is no aliasing, and $x_a(t)$ may be sampled without aliasing if a sampling frequency $f_s = 2B$.

(c) If f_2 is not an integer multiple of B, we may always increase B until this is the case. Specifically, let

$$k = \left\lfloor \frac{f_2}{B} \right\rfloor$$

where $\lfloor \cdot \rfloor$ is defined to be the "integer part." Now, if we simply increase B to B' where

$$k = \frac{f_2}{B'}$$

we have the case described in part (b) where f_2 is an integer multiple of the bandwidth. Thus, $x_a(t)$ may be sampled without aliasing a sampling frequency of

$$f_s = 2B' = \frac{2f_2}{\lfloor f_2/B \rfloor}$$

3.8 Determine the minimum sampling frequency for each of the following bandpass signals:

(a) $x_a(t)$ is real with $X_a(f)$ nonzero only for 9 kHz $< |f| <$ 12 kHz.

(b) $x_a(t)$ is real with $X_a(f)$ nonzero only for 18 kHz $< |f| <$ 22 kHz.

(c) $x_a(t)$ is complex with $X_a(f)$ nonzero only for 30 kHz $< f <$ 35 kHz.

(a) For this signal, the bandwidth is $B = f_2 - f_1 = 3$ kHz, and $f_2 = 12 = 4B$ is an integer multiple of B. Therefore, the minimum sampling frequency is $f_s = 2B = 6$ kHz.

(b) For this signal, $B = 4$ kHz and $f_2 = 22$, which is not an integer multiple of B. With $\lfloor f_2/B \rfloor = 5$, if we let $B' = f_2/5 = 4.4$, f_2 is an integer multiple of B', and $x_a(t)$ may be sampled with a sampling frequency of $f_s = 2B' = 8.8$ kHz.

(c) For a complex bandpass signal with a spectrum that is nonzero for $f_1 < f < f_2$, the minimum sampling frequency is $f_s = f_2 - f_1$. Thus, for this signal, $f_s = 5$ kHz.

3.9 How many bits are needed in an A/D converter if we want a signal-to-quantization noise ratio of at least 90 dB? Assume that $x_a(t)$ is gaussian with a variance σ_x^2, and that the range of the quantizer extends from $-3\sigma_x$ to $3\sigma_x$; that is, $X_{max} = 3\sigma_x$ (with this value for X_{max}, only about one out of every 1000 samples will exceed the quantizer range).

For a $(B + 1)$-bit quantizer, the signal-to-quantization noise ratio is

$$SQNR = 6.02B + 10.81 - 20 \log \frac{X_{max}}{\sigma_x}$$

With $X_{max} = 3\sigma_x$ this becomes

$$SQNR = 6.02B + 10.81 - 20 \log 3 = 6.02B + 10.81 - 9.54 = 6.02B + 1.27$$

If we want a signal-to-quantization noise ratio of 90 dB, we require

$$B = \frac{90 - 1.27}{6.02} = 14.74$$

or $B + 1 = 16$ bits.

3.10   An image is to be sampled with a signal-to-quantization noise ratio of at least 80 dB. Unlike many other signals, the image samples are nonnegative. Assume that the sampling device is calibrated so that the sampled image intensities fall within the range from 0 to 1. How many bits are needed to achieve the desired signal-to-quantization noise ratio?

For a bipolar signal with amplitudes that fall within the range $[-X_{max}, X_{max}]$, the signal-to-quantization noise ratio is

$$SQNR = 6.02B + 10.81 - 20 \log \frac{X_{max}}{\sigma_x}$$

For a nonnegative signal that is confined to the interval $[0, 1]$, the signal-to-quantization noise ratio is equivalent to the bipolar case if we set $X_{max} = 0.5$. If we assume that the intensities of the image are uniformly distributed over the interval $[0, 1]$,

$$\sigma_x^2 = \tfrac{1}{12}$$

Therefore, $$SQNR = 6.02B + 10.81 - 20 \log \frac{\sqrt{12}}{2} = 6.02B + 6.03$$

and for a signal-to-quantization noise ratio of 80 dB, we require

$$B = \frac{80 - 6.03}{6.02} = 12.29$$

or $B + 1 = 14$ bits.

3.11   Suppose that we have a set of unquantized samples, $x(n)$, that are nonnegative for all n. A method for quantizing $x(n)$ that is often used in speech processing is as follows. First, we form the sequence

$$y(n) = \log[x(n)]$$

Then $y(n)$ is quantized with a $(B + 1)$-bit uniform quantizer,

$$\hat{y}(n) = Q[y(n)] = y(n) + e(n)$$

The quantized signal samples are then obtained by exponentiating $\hat{y}(n)$,

$$\hat{x}(n) = \exp\{\hat{y}(n)\}$$

Show that if $e(n)$ is small, the signal-to-quantization noise ratio is independent of the signal power.

With $$\hat{y}(n) = Q[y(n)] = y(n) + e(n) = \log[x(n)] + e(n)$$

we have, for $\hat{x}(n)$,

$$\hat{x}(n) = \exp\{\log[x(n)] + e(n)\} = x(n) \cdot \exp\{e(n)\}$$

If $e(n) \ll 1$, we may use the expansion

$$\exp\{e(n)\} \approx 1 + e(n)$$

to write

$$\hat{x}(n) = x(n)[1 + e(n)] = x(n) + f(n)$$

where $f(n) = x(n)e(n)$ is a (signal-dependent) quantization noise. If we assume that the quantization noise $e(n)$ is statistically independent of $x(n)$,

$$E\{f^2(n)\} = E\{x^2(n)\} \cdot E\{e^2(n)\}$$

and the signal-to-quantization noise ratio is

$$\text{SQNR} = 10 \log \frac{E\{x^2(n)\}}{E\{f^2(n)\}} = -10 \log E\{e^2(n)\}$$

which is independent of the signal power.

Discrete-Time Processing of Analog Signals

3.12   A continuous-time signal $x_a(t)$ is to be filtered to remove frequency components in the range $5\,\text{kHz} \le f \le 10\,\text{kHz}$. The maximum frequency present in $x_a(t)$ is 20 kHz. The filtering is to be done by sampling $x_a(t)$, filtering the sampled signal, and reconstructing an analog signal using an ideal D/C converter. Find the minimum sampling frequency that may be used to avoid aliasing, and for this minimum sampling rate, find the frequency response of the ideal digital filter $H(e^{j\omega})$ that will remove the desired frequencies from $x_a(t)$.

Because the highest frequency in $x_a(t)$ is 20 kHz, the minimum sampling frequency to avoid aliasing is $f_s = 40\,\text{kHz}$. The relationship between the continuous frequency variable Ω and the discrete frequency variable ω is given by

$$\omega = \Omega T_s$$

or

$$\omega = 2\pi \frac{f}{f_s}$$

Therefore, the frequency range $5\,\text{kHz} \le f \le 10\,\text{kHz}$ corresponds to a digital frequency range

$$\frac{\pi}{4} \le \omega \le \frac{\pi}{2}$$

and the desired digital filter is a bandstop filter that has a frequency response as illustrated in the figure below.

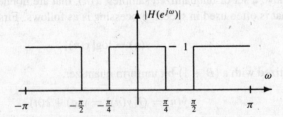

3.13   A major problem in the recording of electrocardiograms (ECGs) is the appearance of unwanted 60-Hz interference in the output. The causes of this power line interference include magnetic induction, displacement currents in the leads on the body of the patient, and equipment interconnections. Assume that the bandwidth of the signal of interest is 1 kHz, that is,

$$X_a(f) = 0 \qquad |f| > 1000\,\text{Hz}$$

The analog signal is converted into a discrete-time signal with an ideal A/D converter operating using a sampling frequency f_s. The resulting signal $x(n) = x_a(nT_S)$ is then processed with a discrete-time system that is described by the difference equation

$$y(n) = x(n) + ax(n-1) + bx(n-2)$$

The filtered signal, $y(n)$, is then converted back into an analog signal using an ideal D/A converter. Design a system for removing the 60-Hz interference by specifying values for f_s, a, and b so that a 60-Hz signal of the form

$$w_a(t) = A\sin(120\pi t)$$

will not appear in the output of the D/A converter.

The signal that is to have the 60-Hz noise removed is bandlimited to 1000 Hz. Therefore, in order to avoid aliasing when the signal is sampled, we require a sampling frequency

$$f_s \geq 2000$$

Using the minimum rate of 2000 Hz, note that a 60-Hz signal $w_a(t) = \sin(120\pi t)$ becomes

$$w(n) = w_a(nT_s) = \sin\left(\frac{120\pi n}{2000}\right) = \sin(n\omega_0)$$

where $\omega_0 = 0.06\pi$. Recall that complex exponentials are eigenfunctions of linear shift-invariant systems. Therefore, if the input to an LSI system is $x(n) = e^{jn\omega_0}$, the output is

$$y(n) = H(e^{j\omega_0})e^{jn\omega_0}$$

Because

$$w(n) = \frac{e^{jn\omega_0} - e^{-jn\omega_0}}{2j}$$

$w(n)$ will be removed from $x(n)$ if we design a filter so that $H(e^{j\omega})$ is equal to zero at $\omega = \pm\omega_0$. Because $H(e^{j\omega})$ is a second-order filter with a frequency response

$$H(e^{j\omega}) = 1 + ae^{-j\omega} + be^{-j2\omega}$$

it may be factored as follows:

$$H(e^{j\omega}) = (1 - \alpha e^{-j\omega})(1 - \beta e^{-j\omega})$$

Therefore, $H(e^{j\omega})$ will be zero for $\omega = \pm\omega_0$ if $\alpha = e^{j\omega_0}$ and $\beta = e^{-j\omega_0}$. In this case, we have

$$H(e^{j\omega}) = 1 - 2(\cos\omega_0)e^{-j\omega} + e^{-j2\omega}$$

Thus, our requirements are that

$$a = -2\cos\omega_0 = -2\cos(0.06\pi) \qquad b = 1$$

and $f_s = 2000$.

3.14   The following system is used to process an analog signal with a discrete-time system.

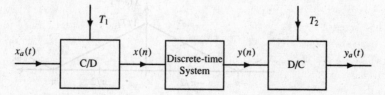

Suppose that $x_a(t)$ is bandlimited with $X_a(f) = 0$ for $|f| > 5$ kHz as shown in the figure below,

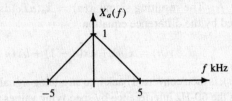

and that the discrete-time system is an ideal low-pass filter with a cutoff frequency of $\pi/2$.

(a) Find the Fourier transform of $y_a(t)$ if the sampling frequencies are $f_1 = f_2 = 10$ kHz.

(b) Repeat for $f_1 = 20$ kHz and $f_2 = 10$ kHz.

(c) Repeat for $f_1 = 10$ kHz and $f_2 = 20$ kHz.

(a) When the sampling frequencies of the C/D and D/C converters are the same, and $x_a(t)$ is bandlimited with $X_a(j\Omega) = 0$ for $|\Omega| > \pi/T_1$, this system is equivalent to an analog filter with a frequency response

$$H_a(j\Omega) = \begin{cases} H(e^{j\Omega T_1}) & |\Omega| < \dfrac{\pi}{T_1} \\[2mm] 0 & \text{else} \end{cases}$$

Therefore, if $H(e^{j\omega})$ is a low-pass filter with a cutoff frequency $\pi/2$, the cutoff frequency of $H_a(j\Omega)$, denoted by Ω_0, is given by

$$\Omega_0 T_1 = \frac{\pi}{2}$$

or

$$2\pi f_0 \cdot T_1 = \frac{\pi}{2}$$

Thus, $f_0 = \frac{1}{4}f_1 = 2500$ Hz

(b) When the sampling frequencies of the C/D and D/C are different, it is best to plot the spectrum of the signals as they progress through the system. With $X_a(f)$ as shown above, the discrete-time Fourier transform of $x(n)$ is

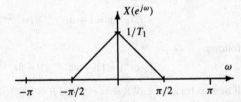

Because the cutoff frequency of the discrete-time low-pass filter is $\pi/2$, $y(n) = x(n)$, and the output of the D/C converter is as plotted below.

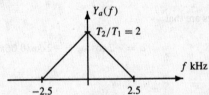

(c) With $f_1 = 10$ kHz, we are sampling $x_a(t)$ at the Nyquist rate, and the spectrum of $x(n)$ is

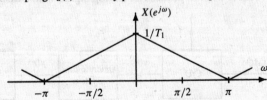

and the output of the low-pass filter is as shown below.

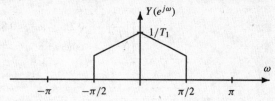

Therefore, the spectrum of $y_a(t)$ is as follows:

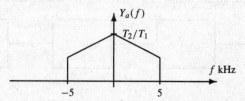

3.15 Consider the system in Fig. 3-9 for implementing a continuous-time system in terms of a discrete-time system. Assume that the input to the C/D converter is bandlimited to $\Omega_0 = \Omega_s/2$ and that the unit sample response of the discrete-time system is

$$h(n) = \delta(n) - 0.9\delta(n-1)$$

Find the overall frequency response of this system.

Assuming bandlimited inputs with $X_a(j\Omega) = 0$ for $|\Omega| > \Omega_s/2$, the output $Y_a(j\Omega)$ is related to the input $X_a(j\Omega)$ as follows:

$$Y_a(j\Omega) = H_a(j\Omega)X(j\Omega)$$

where

$$H_a(j\Omega) = \begin{cases} H(e^{j\Omega T_s}) & |\Omega| < \dfrac{\pi}{T_s} \\ 0 & \text{otherwise} \end{cases}$$

Because the frequency response of the discrete-time system is

$$H(e^{j\omega}) = 1 - 0.9e^{-j\omega}$$

then

$$H_a(j\Omega) = \begin{cases} 1 - 0.9e^{-j\Omega T_s} & |\Omega| < \dfrac{\pi}{T_s} \\ 0 & \text{otherwise} \end{cases}$$

3.16 Consider the system shown in Fig. 3-9 for implementing a continuous-time system in terms of a discrete-time system. Assuming that the input signals $x_a(t)$ are bandlimited so that $X_a(f) = 0$ for $|f| > 10$ kHz, find the discrete-time system that produces the output

$$Y_a(f) = \begin{cases} |f|X_a(f) & 2000 \leq |f| \leq 8000 \\ 0 & \text{otherwise} \end{cases}$$

For bandlimited inputs, the system in Fig. 3-9 is a linear shift-invariant system with an effective frequency response equal to

$$H_a(j\Omega) = \begin{cases} H(e^{j\Omega T_s}) & |\Omega| < \dfrac{\pi}{T_s} \\ 0 & \text{otherwise} \end{cases}$$

The system that we would like to realize has a frequency response

$$H_a(j\Omega) = \begin{cases} |\Omega| & 4000\pi \leq |\Omega| \leq 16000\pi \\ 0 & \text{otherwise} \end{cases}$$

If we assume a sampling frequency $f_s = 20$ kHz, the frequency response of the discrete-time system should be

$$H(e^{j\omega}) = \begin{cases} \left| \dfrac{\omega}{T_s} \right| & 0.2\pi \leq |\omega| \leq 0.8\pi \\ 0 & \text{otherwise} \end{cases}$$

where $T_s = 1/20000$.

3.17   Diagrammed in the figure below is a hybrid digital-analog network.

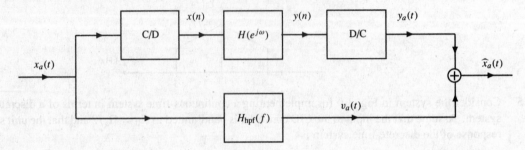

The discrete-time system $H(e^{j\omega})$ is a low-pass filter

$$H(e^{j\omega}) = \begin{cases} A & |\omega| \leq \omega_0 \\ 0 & \text{else} \end{cases}$$

and the analog system $H_{\text{hpf}}(f)$ is a high-pass filter with a frequency response as shown below.

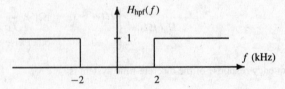

The input $x_a(t)$ is bandlimited to 4 kHz, and the sampling frequencies of the ideal C/D and D/C converters are 10 kHz. Find the value for A and ω_0 that will result in perfect reconstruction of $x_a(t)$,

$$\hat{x}_a(t) = x_a(t)$$

Because $x_a(t)$ is bandlimited to 4 kHz, the upper branch of this hybrid system acts as an ideal analog low-pass filter with a frequency response

$$H_{\text{lpf}}(f) = \begin{cases} A & |f| \leq \dfrac{\omega_0}{2\pi T_s} \\ 0 & |f| > \dfrac{\omega_0}{2\pi T_s} \end{cases}$$

Because the analog network is a high-pass filter with a cutoff frequency of 4 kHz, and

$$\hat{X}_a(f) = Y_a(f) + V_a(f)$$

$\hat{x}_a(t)$ will be equal to $x_a(t)$ provided that $A = 1$ and

$$\frac{\omega_0}{2\pi T_s} = 2000$$

or $\omega_0 = 0.4\pi$.

3.18   A digital sequence $x(n)$ is to be transmitted across a linear time-invariant bandlimited channel as illustrated in the figure below.

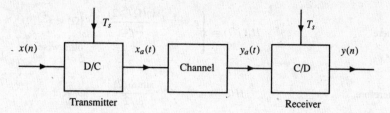

Transmitter   Receiver

The transmitter is a D/C converter, and the receiver simply samples the received waveform $y_a(t)$:

$$y(n) = y_a(nT_s)$$

Assume that the channel may be modeled as an ideal low-pass filter with a cutoff frequency of 4 kHz:

$$G_a(j\Omega) = \begin{cases} 1 & |\Omega| \leq 2\pi(4000) \\ 0 & |\Omega| > 2\pi(4000) \end{cases}$$

(a) Assuming an ideal C/D and D/C, and perfect synchronization between the transmitter and receiver, what values of T_s (if any) will guarantee that $y(n) = x(n)$?

(b) Suppose that the D/C is nonideal. Specifically, suppose that $x(n)$ is first converted to an impulse train and then a zero-order hold is used to perform the "interpolation" between the sample values. In other words, the impulse response of the interpolating filter is a pulse of duration T_s:

$$h_a(t) = \begin{cases} 1 & 0 \leq t \leq T_s \\ 0 & \text{otherwise} \end{cases}$$

Because the received sequence $y(n)$ will no longer be equal to $x(n)$, in order to improve the performance of the receiver, the received samples are processed with a digital filter as shown below.

$y(n)$ 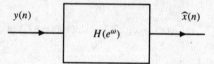 $\widehat{x}(n)$

$H(e^\omega)$

Find the frequency response of the filter that should be used to filter $y(n)$.

(a) The output of the D/C converter is a bandlimited signal $x_a(t)$ with a Fourier transform that is equal to zero for $|f| > f_s/2$. Because $x_a(t)$ is passed through a bandlimited channel that rejects all frequencies greater than 4 kHz, in order for there to be no distortion at the receiver, it is necessary that

$$\frac{f_s}{2} < 4000$$

or $f_s < 8000$

Thus, the C/D and D/C converters must operate at a rate less than 8 kHz.

(b) In order to get the maximum amount of data through the channel per unit of time, we will let T_s be the minimum sampling period,

$$T_s = \frac{1}{8000}$$

When the reconstruction filter in the D/C converter is a zero-order hold, the frequency response of the discrete-time system that relates the input sequence $x(n)$ to the reconstructed sequence $y(n)$ is

$$H(e^{j\omega}) = H_a\left(\frac{j\omega}{T_s}\right) \qquad |\omega| < \pi$$

where

$$H_a(j\Omega) = \begin{cases} e^{-j\Omega T_s/2}\dfrac{\sin(\Omega T_s/2)}{\Omega/2} & |\Omega| < \dfrac{\pi}{T_s} \\ 0 & \text{otherwise} \end{cases}$$

Therefore,

$$H(e^{j\omega}) = T_s e^{-j\omega/2}\frac{\sin(\omega/2)}{\omega/2} \qquad |\omega| < \pi$$

and the discrete-time filter for processing $y(n)$ to remove the distortion introduced by the zero-order hold should approximate the response

$$G(e^{j\omega}) = \frac{\omega/2}{\sin(\omega/2)} e^{j\omega/2} \qquad |\omega| < \pi$$

3.19 Consider the following system for processing a continuous-time signal with a discrete-time system:

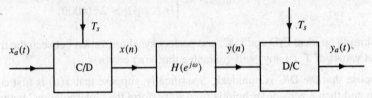

The frequency response of the discrete-time filter is

$$H(e^{j\omega}) = \frac{2\left(\frac{1}{3} - e^{-j\omega}\right)}{1 - \frac{1}{3}e^{-j\omega}}$$

If $f_s = 2$ kHz and $x_a(t) = \sin(1000\pi t)$, find the output $y_a(t)$.

Sampling $x_a(t) = \sin(1000\pi t)$ with a sampling frequency $f_s = 2000$ produces the discrete-time sequence

$$x(n) = x_a(nT_s) = \sin(1000\pi n T_s) = \sin\left(\frac{n\pi}{2}\right)$$

This sequence is then filtered with the discrete-time filter

$$H(e^{j\omega}) = \frac{2\left(\frac{1}{3} - e^{-j\omega}\right)}{1 - \frac{1}{3}e^{-j\omega}}$$

Because $x(n)$ is a sinusoid, the response is

$$y(n) = A\sin\left(\frac{n\pi}{2} + \phi\right)$$

where A and ϕ are the magnitude and phase, respectively, of the frequency response at $\omega = \pi/2$. With

$$|H(e^{j\omega})|^2 = 4\frac{\left(\frac{1}{3} - e^{-j\omega}\right)}{\left(1 - \frac{1}{3}e^{-j\omega}\right)} \cdot \frac{\left(\frac{1}{3} - e^{j\omega}\right)}{\left(1 - \frac{1}{3}e^{j\omega}\right)} = 4\frac{\frac{10}{9} - \frac{2}{3}\cos\omega}{\frac{10}{9} - \frac{2}{3}\cos\omega} = 4$$

it follows that $|H(e^{j\omega})| = 2$. We may evaluate the phase as follows:

$$H(e^{j\omega}) = 2\frac{\frac{1}{3} - e^{-j\omega}}{1 - \frac{1}{3}e^{-j\omega}} \cdot \frac{1 - \frac{1}{3}e^{j\omega}}{1 - \frac{1}{3}e^{j\omega}}$$

$$= 2\frac{\frac{2}{3} - \frac{1}{9}e^{j\omega} - e^{-j\omega}}{\left|1 - \frac{1}{3}e^{-j\omega}\right|^2} = 2\frac{\frac{2}{3} - \frac{10}{9}\cos\omega + j\frac{8}{9}\sin\omega}{\left|1 - \frac{1}{3}e^{-j\omega}\right|^2}$$

Therefore,
$$\phi_h(\omega) = \tan^{-1}\frac{\frac{8}{9}\sin\omega}{\frac{2}{3}-\frac{10}{9}\cos\omega}$$

which, when evaluated at $\omega = \pi/2$, gives

$$\phi_h(\omega)|_{\omega=\pi/2} = \tan^{-1}\frac{8/9}{2/3} = \tan^{-1}\frac{4}{3} = 0.2952\pi$$

Thus,
$$y(n) = 2\sin\left(\frac{\pi}{2}[n + 0.5903]\right)$$

3.20 Consider the following system consisting of an ideal D/C converter, a linear time-invariant filter, and an ideal C/D converter.

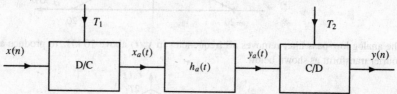

The continuous-time system $h_a(t)$ is an ideal low-pass filter with a frequency response

$$H_a(f) = \begin{cases} 1 & |f| \leq 10 \text{ kHz} \\ 0 & \text{otherwise} \end{cases}$$

(a) If $T_1 = T_2 = 10^{-4}$, find an expression relating the output $y(n)$ to the input $x(n)$.

(b) If $T_1 = \left(\frac{1}{4}\right) \times 10^{-4}$ and $T_2 = 10^{-4}$, find $y(n)$ when

$$x(n) = \left[\frac{\sin(n\pi/2)}{n\pi/2}\right]^2$$

(a) When $T_1 = T_2$, this system behaves as a linear shift-invariant discrete-time system with a frequency response

$$H(e^{j\omega}) = H_a\left(\frac{j\omega}{T_1}\right) \qquad |\omega| < \pi$$

Because $H_a(j\Omega) = 1$ for $|\Omega| < 2\pi \cdot 10^4$,

$$H(e^{j\omega}) = 1 \qquad |\omega| < \pi$$

and
$$h(n) = \delta(n)$$

Therefore, $y(n) = x(n)$.

Another way to analyze this system is to note that the output of the D/C converter, $x_a(t)$, is bandlimited to $f = 5$ kHz. Because $H_a(f)$ is an ideal low-pass filter with a cutoff frequency 10 kHz, $y_a(t) = x_a(t)$. Therefore, this system is equivalent to the one shown below.

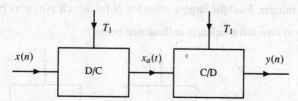

Because an ideal D/C converter followed by an ideal D/C converter is the identity system, $y(n) = x(n)$.

(b) When $T_1 \neq T_2$, this system is, in general, no longer a linear shift-invariant system. However, we may analyze this system in the frequency domain as follows. First, note that the DTFT of $x(n)$ is as illustrated in the following figure:

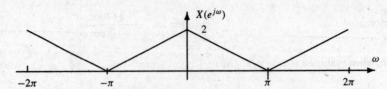

Thus, the output of the D/C converter is a bandlimited signal that has a Fourier transform as shown in the following figure:

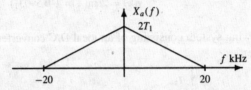

The analog low-pass filter removes all frequencies in $x_a(t)$ above 10 kHz to produce a signal $y_a(t)$ that has a Fourier transform as shown below.

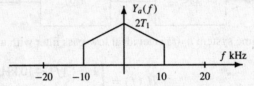

Because the highest frequency in $y_a(t)$ is 10 kHz, the Nyquist rate is 20 kHz. However, the sampling frequency of the C/D converter is 10 kHz, so $y_a(t)$ will be aliased. The DTFT of $y(n)$ is related to $Y_a(j\Omega)$ as follows:

$$Y(e^{j\omega}) = \frac{1}{T_2} \sum_{k=-\infty}^{\infty} Y_a\left(j\frac{\omega}{T_2} - j\frac{2\pi k}{T_2}\right)$$

Summing the shifted and scaled transforms yields

$$Y(e^{j\omega}) = \tfrac{3}{4} \qquad |\omega| < \pi$$

Therefore, $y(n) = \tfrac{3}{4}\delta(n)$

Sample Rate Conversion

3.21 Suppose that a discrete-time sequence $x(n)$ is bandlimited so that

$$X(e^{j\omega}) = 0 \qquad 0.3\pi < |\omega| < \pi$$

This sequence is then sampled to form the sequence

$$y(n) = x(nN)$$

where N is an integer. Find the largest value for N for which $x(n)$ may be uniquely recovered from $y(n)$.

The easiest way to view this problem is as illustrated below.

Converting $x(n)$ into a continuous-time signal with an ideal D/C converter with a sampling frequency f_s produces a continuous-time signal $x_a(t)$ that is bandlimited to $f_0 = 0.3 \cdot f_s/2$. Therefore, $x_a(t)$ may be sampled, without aliasing, if we use a sampling frequency $f_s' \geq 2 f_0 = 0.3 f_s$, or

$$T_s' < \frac{T_s}{0.3} = 3.33\bar{3} T_s$$

Therefore, if $T_s' = 3T_s$,

$$y(n) = x_a(3nT_s) = x(3n)$$

and $x(n)$ may be uniquely recovered from $y(n)$. Thus, $N = 3$.

3.22   Consider the following system:

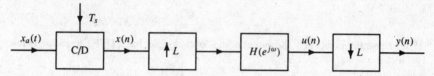

Assume that $X_a(f) = 0$ for $|f| > 1/T_s$ and that

$$H(e^{j\omega}) = \begin{cases} e^{-j\omega} & |\omega| \leq \dfrac{\pi}{L} \\ 0 & \dfrac{\pi}{L} < |\omega| \leq \pi \end{cases}$$

How is the output of the discrete-time system, $y(n)$, related to the input signal $x_a(t)$?

In this system, the bandlimited signal $x_a(t)$ is sampled, without aliasing, to produce the sampled signal $x(n) = x_a(nT_s)$. Up-sampling $x(n)$ by a factor of L, and filtering with an ideal low-pass filter with a cutoff frequency $\omega_c = \pi/L$, produces the signal

$$w(n) = x_a\left(\frac{nT_s}{L}\right)$$

that is, a signal that is sampled with a sampling frequency Lf_s. However, because the low-pass filter has linear phase with a group delay of one sample, the interpolated up-sampled signal is delayed by 1. Therefore, the output of the low-pass filter is

$$u(n) = w(n-1) = x_a\left([n-1]\frac{T_s}{L}\right)$$

Down-sampling by L then produces the output

$$y(n) = u(Ln) = w(Ln-1) = x_a\left(nT_s - \frac{T_s}{L}\right)$$

Thus, $y(n)$ corresponds to samples of $x_a(t - t_0)$ where $t_0 = T_s/L$.

3.23   Consider the system shown in the figure below.

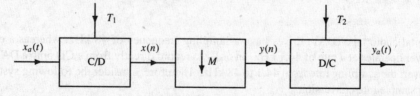

Assume that the input is bandlimited, $X_a(j\Omega) = 0$ for $|\Omega| > 2\pi \cdot 1000$.

(a)   What constraints must be placed on M, T_1, and T_2 in order for $y_a(t)$ to be equal to $x_a(t)$?

(b)   If $f_1 = f_2 = 20$ kHz and $M = 4$, find an expression for $y_a(t)$ in terms of $x_a(t)$.

(a) Suppose that $x_a(t)$ has a Fourier transform as shown in the figure below.

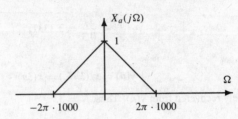

Because $y(n) = x(Mn) = x_a(nMT_1)$, in order to prevent $x(n)$ from being aliased, it is necessary that

$$MT_1 < \frac{1}{2000}$$

If this constraint is satisfied, the output of the down-sampler has a DTFT as shown below.

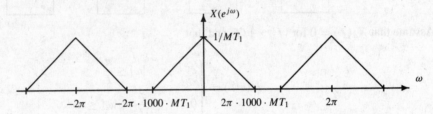

Going through the D/C converter produces the signal $y_a(t)$, which has the Fourier transform shown below.

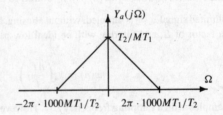

Therefore, in order for $y_a(t)$ to be equal to $x_a(t)$, we require that

1. $MT_1 \leq 1/2000$ in order to avoid aliasing.
2. $T_2 = MT_1$ to prevent frequency scaling.

(b) With $T_1 = T_2 = 1/20000$ and $M = 4$, note that

$$MT_1 = \frac{1}{5000} < \frac{1}{2000}$$

Therefore, there is no aliasing. Thus, as we see from the figure above,

$$Y_a(j\Omega) = \tfrac{1}{4} X_a\left(\frac{j\Omega}{4}\right)$$

or $$y_a(t) = x_a(4t)$$

3.24 Digital audio tape (DAT) drives have a sampling frequency of 48 kHz, whereas a compact disk (CD) player operates at a rate of 44.1 kHz. In order to record directly from a CD onto a DAT, it is necessary to convert the sampling rate from 44.1 to 48 kHz. Therefore, consider the following system for performing this sample rate conversion:

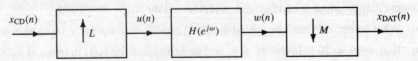

Find the smallest possible values for L and M and find the appropriate filter $H(e^{j\omega})$ to perform this conversion.

Given that $48000 = 2^7 \cdot 3 \cdot 5^3$ and $44100 = 2^2 \cdot 3^2 \cdot 5^2 \cdot 7^2$, to change the sampling rate we require

$$\frac{L}{M} = \frac{2^7 \cdot 3 \cdot 5^3}{2^2 \cdot 3^2 \cdot 5^2 \cdot 7^2} = \frac{2^5 \cdot 5}{3 \cdot 7^2} = \frac{160}{147}$$

Therefore, if we up-sample by $L = 160$ and then down-sample by $M = 147$, we achieve the desired sample rate conversion. The low-pass filter that we require is one that has a cutoff frequency

$$\omega_c = \min\left(\frac{\pi}{L}, \frac{\pi}{M}\right) = \frac{\pi}{160}$$

and the gain of the filter should be equal to $L = 160$.

3.25   Suppose that we would like to slow a segment of speech to one-half its normal speed. The speech signal $s_a(t)$ is assumed to have no energy outside of 5 kHz, and is sampled at a rate of 10 kHz, yielding the sequence

$$s(n) = s_a(nT_s)$$

The following system is proposed to create the slowed-down speech signal.

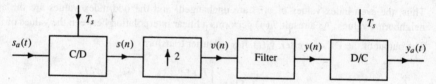

Assume that $S_a(j\Omega)$ is as shown in the following figure:

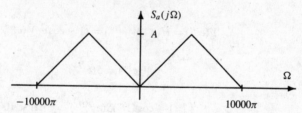

(a)   Find the spectrum of $v(n)$.

(b)   Suppose that the discrete-time filter is described by the difference equation

$$y(n) = v(n) + \tfrac{1}{2}[v(n-1) + v(n+1)]$$

Find the frequency response of the filter and describe its effect on $v(n)$.

(c)   What is $Y_a(j\Omega)$ in terms of $X_a(j\Omega)$? Does $y_a(t)$ correspond to slowed-down speech?

(a)   Since $s_a(t)$ is sampled at the Nyquist rate, the DTFT of the sampled speech signal, $s(n)$, is as follows:

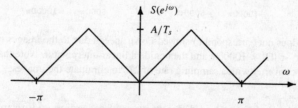

Up-sampling by a factor of 2 scales the frequency axis of $S(e^{j\omega})$ by a factor of two as shown below.

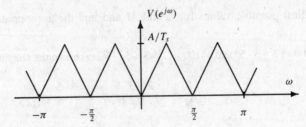

(b) The unit sample response of the discrete-time filter is

$$h(n) = \tfrac{1}{2}\delta(n+1) + \delta(n) + \tfrac{1}{2}\delta(n-1)$$

which has a frequency response

$$H(e^{j\omega}) = 1 + \cos\omega$$

To see the effect of this filter on $v(n)$, note that due to the up-sampling, $v(n) = 0$ for n odd. Therefore, with

$$y(n) = v(n) + \tfrac{1}{2}v(n-1) + \tfrac{1}{2}v(n+1)$$

it follows that

$$y(n) = \begin{cases} v(n) & n \text{ odd} \\ \tfrac{1}{2}v(n-1) + \tfrac{1}{2}v(n+1) & n \text{ even} \end{cases}$$

Thus, the even-index values of $v(n)$ are unchanged, and the odd-index values are the average of the two neighboring values. As a result, $h(n)$ performs a linear interpolation between the values of $v(n)$.

(c) The output of the DC converter, $y_a(t)$, has a Fourier transform

$$Y_a(j\Omega) = \begin{cases} T_s Y(e^{j\Omega T_s}) & |\Omega| < \pi/T_s \\ 0 & \text{otherwise} \end{cases}$$

Since

$$Y(e^{j\omega}) = H(e^{j\omega})V(e^{j\omega}) = (1 + \cos\omega)V(e^{j\omega})$$

and

$$V(e^{j\omega}) = S(e^{j2\omega})$$

then

$$Y_a(j\Omega) = \begin{cases} T_s(1 + \cos\Omega T_s)S(e^{j2\Omega T_s}) & |\Omega| < 10000\pi \\ 0 & \text{otherwise} \end{cases}$$

which is the product of $(1 + \cos\Omega T_s)$ and $T_s S(e^{j2\Omega T_s})$ as illustrated below.

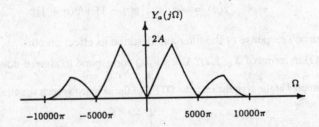

Thus, $y_a(t)$ does not correspond to slowed-down speech due to the images of $s_a(t)$ that occur in the frequency range $5000\pi < |\Omega| < 10000\pi$ and the nonideal linear interpolator. Note that a better approximation would be to use a DC converter with a sampling rate of $2T_s$ to eliminate the images.

3.26 Shown in the figure below are two different ways of cascading an up-sampler with a down-sampler.

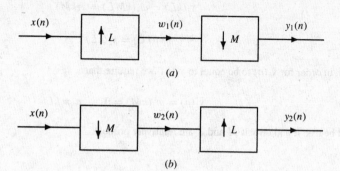

(a)   If $M = L$, show that the two systems are not identical.

(b)   Under what conditions will the two systems be identical?

(a)   In the first system, which consists of an up-sampler followed by a down-sampler, note that $w_1(n)$ is a sequence that is formed by inserting $L - 1$ zeros between each value of $x(n)$. The down-sampler then extracts every Lth value of $w_1(n)$, thus producing the output

$$y_1(n) = x(n)$$

In the second system, however, the down-sampler extracts every Lth sample of $x(n)$ and discards the rest. The up-sampler then inserts $L - 1$ zeros between each value of $w_2(n)$. Thus,

$$y_2(n) = \begin{cases} x\left(\dfrac{nM}{L}\right) & n = 0, \pm L, \pm 2L, \dots \\ 0 & \text{else} \end{cases}$$

Therefore, the two systems are not the same.

(b)   In order to analyze these systems when $L \neq M$, note that $y_2(n)$ in the second system has the form shown in the following figure:

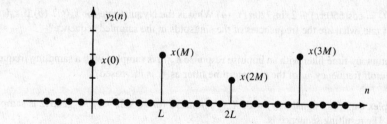

On the other hand, the sequence $w_1(n)$ in the first system is as shown below.

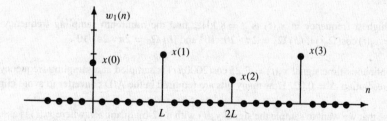

Note that $y_1(n)$ is formed by extracting every Mth value of $w_1(n)$,

$$y_1(n) = w_1(nM)$$

Clearly, $$y_1(kL) = w_1(kML) = x(kM)$$

so $$y_1(kL) = y_2(kL)$$

However, in order for $y_1(n)$ to be equal to $y_2(n)$, we require that

$$y_1(n) = w_1(nM) = 0 \qquad n \neq kL$$

This will be true if and only if M and L are relatively prime.

Supplementary Problems

A/D and D/A Conversion

3.27 Find two different continuous-time signals that will produce the sequence

$$x(n) = \cos(0.15n\pi)$$

when sampled with a sampling frequency of 8 kHz.

3.28 If the Nyquist rate for $x_a(t)$ is Ω_s, find the Nyquist rate for (a) $x^2(2t)$, (b) $x(t/3)$, (c) $x(t) * x(t)$.

3.29 A continuous-time signal $x_a(t)$ is known to be uniquely recoverable from its samples $x_a(nT_s)$ when $T_s = 1$ ms. What is the highest frequency in $X_a(f)$?

3.30 Suppose that $x_a(t)$ is bandlimited to 8 kHz (that is, $X_a(f) = 0$ for $|f| > 8000$). (a) What is the Nyquist rate for $x_a(t)$? (b) What is the Nyquist rate for $x_a(t)\cos(2\pi \cdot 1000t)$?

3.31 Let $x_a(t) = \cos(650\pi t) + 2\sin(720\pi t)$. (a) What is the Nyquist rate for $x_a(t)$? (b) If $x_a(t)$ is sampled at twice the Nyquist rate, what are the frequencies of the sinusoids in the sampled sequence?

3.32 If a continuous-time filter with an impulse response $h_a(t)$ is sampled with a sampling frequency of f_s, what happens to the cutoff frequency ω_c of the discrete-time filter as f_s is increased?

3.33 A complex bandpass signal $x_a(t)$ with $X_a(f)$ nonzero for 10 kHz $< f <$ 12 kHz is sampled at a sampling rate of 2 kHz. The resulting sequence is

$$x(n) = \delta(n)$$

What is $x_a(t)$?

3.34 If the highest frequency in $x_a(t)$ is $f = 8$ kHz, find the minimum sampling frequency for the bandpass signal $y_a(t) = x_a(t)\cos(\Omega_0 t)$ if (a) $\Omega_0 = 2\pi \cdot 20 \cdot 10^3$ and (b) $\Omega_0 = 2\pi \cdot 24 \cdot 10^3$.

3.35 The continuous-time signal $x_a(t) = 7.25\cos(2000\pi t)$ is sampled at a sampling frequency of 8 kHz and quantized with a resolution $\Delta = 0.02$. How many bits are required in the A/D converter to avoid clipping $x_a(t)$?

3.36 Suppose that we want to sample the signal $x_a(t)$ with a 12-bit quantizer, where $x_a(t)$ is assumed to be gaussian with a variance σ_x^2. What is the signal-to-quantization noise ratio if we want the range of the quantizer to extend from $-3\sigma_x$ to $3\sigma_x$?

3.37 Suppose that an analog waveform is sampled with a sampling frequency of 10 kHz and that $x_a(t)$ contains a strong 60-Hz interference signal. If the only information in $x_a(t)$ of interest is in the frequency band above

60 Hz, the interference may be eliminated with a discrete-time high-pass filter that has a frequency response of the form

$$H(e^{j\omega}) = \begin{cases} 0 & |\omega| < \omega_c \\ 1 & \omega_c \leq \omega \leq \pi \end{cases}$$

What is the smallest cutoff frequency ω_c that may be used and still eliminate the 60-Hz interference?

3.38   *True or False*: If $x(n)$ has a discrete-time Fourier transform that is equal to zero for $\pi/4 < |\omega| < \pi$,

$$x(n) = \sum_{k=-\infty}^{\infty} x(4k)\frac{\sin[\pi(n-4k)/4]}{\pi(n-4k)/4}$$

Discrete-Time Processing of Analog Signals

3.39   The system shown in Fig. 3-9 may be used to process an analog signal with a discrete-time system. Assume that $x_a(t)$ is bandlimited with $X_a(f) = 0$ for $|f| > 10$ kHz as shown in the figure below.

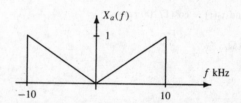

If the discrete-time system is an ideal low-pass filter with a cutoff frequency of $\pi/4$, find the Fourier transform of $y_a(t)$ when (*a*) $f_1 = 20$ kHz and $f_2 = 10$ kHz and (*b*) $f_1 = 10$ kHz and $f_2 = 20$ kHz.

3.40   For bandlimited input signals, the system shown in Fig. 3-10 is a linear time-invariant continuous-time system. If

$$y(n) = \tfrac{1}{2}y(n-1) + x(n)$$

find the frequency response of the equivalent continuous-time system.

3.41   For bandlimited input signals, the system shown in Fig. 3-10 is a linear time-invariant continuous-time system. If the overall system is to be a differentiator,

$$y_a(t) = \frac{d}{dt}x_a(t)$$

how should the frequency response of the discrete-time system be defined?

Sample Rate Conversion

3.42   The up-sampler and down-sampler are components that are found in interpolators and decimators, respectively. Are these systems linear? Are they shift-invariant?

3.43   A sequence $x(n)$ corresponds to samples of a bandlimited signal using a sampling frequency of 10 kHz. However, the sequence should have been sampled using a sampling frequency $f_s = 12$ kHz. Design a system for digitally changing the sampling rate.

3.44   A signal $x_a(t)$ that is bandlimited to 10 kHz is processed by the following system:

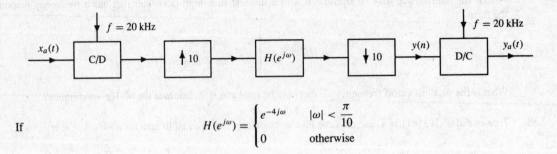

If

$$H(e^{j\omega}) = \begin{cases} e^{-4j\omega} & |\omega| < \dfrac{\pi}{10} \\ 0 & \text{otherwise} \end{cases}$$

express the output $y_a(t)$ in terms of the input $x_a(t)$.

Answers to Supplementary Problems

3.27 $x_1(t) = \cos(1200\pi t)$ and $x_2(t) = \cos(17200\pi t)$.

3.28 (a) $4\Omega_s$. (b) $\Omega_s/3$. (c) Ω_s.

3.29 500 Hz.

3.30 (a) 16 kHz. (b) 18 kHz.

3.31 (a) 720 kHz. (b) $\omega_1 = 65\pi/142$ and $\omega_2 = \pi/2$.

3.32 ω_c decreases.

3.33 $x_a(t) = \dfrac{1}{2000} \dfrac{\sin(2000\pi t)}{\pi t} e^{j2\pi(11000)t}$.

3.34 (a) 56 kHz. (b) 32 kHz.

3.35 10 bits.

3.36 73.51 dB.

3.37 $\omega_c = 0.012\pi$.

3.38 True.

3.39

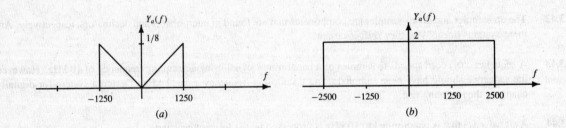

3.40 $H_a(j\Omega) = \begin{cases} \dfrac{1}{1 - \frac{1}{2}e^{-j\Omega T_s}} & |\Omega| < \dfrac{\pi}{T_s} \\ 0 & \text{otherwise} \end{cases}$

3.41 $H(e^{j\omega}) = j\omega/T_s$ for $|\omega| < \pi$.

3.42   Both are linear and shift-varying.

3.43   Up-sample by $L = 6$, filter with a low-pass filter that has a cutoff frequency of $\omega_c = \pi/6$ and a gain of 6, and down-sample by $M = 5$.

3.44 $y_a(t) = x_a(t - 4T_s/10) = x_a(t - 2 \cdot 10^{-5})$.

Chapter 4

The *z*-Transform

4.1 INTRODUCTION

The z-transform is a useful tool in the analysis of discrete-time signals and systems and is the discrete-time counterpart of the Laplace transform for continuous-time signals and systems. The z-transform may be used to solve constant coefficient difference equations, evaluate the response of a linear time-invariant system to a given input, and design linear filters. In this chapter, we will look at the z-transform and examine how it may be used to solve a variety of different problems.

4.2 DEFINITION OF THE *z*-TRANSFORM

In Chap. 2, we saw that the discrete-time Fourier transform (DTFT) of a sequence $x(n)$ is equal to the sum

$$X(e^{j\omega}) = \sum_{n=-\infty}^{\infty} x(n)e^{-jn\omega}$$

However, in order for this series to converge, it is necessary that the signal be absolutely summable. Unfortunately, many of the signals that we would like to consider are not absolutely summable and, therefore, do not have a DTFT. Some examples include

$$x(n) = u(n) \qquad x(n) = (0.5)^n u(-n) \qquad x(n) = \sin n\omega_0$$

The z-transform is a generalization of the DTFT that allows one to deal with such sequences and is defined as follows:

Definition: The z-transform of a discrete-time signal $x(n)$ is defined by[1]

$$X(z) = \sum_{n=-\infty}^{\infty} x(n)z^{-n} \qquad\qquad (4.1)$$

where $z = re^{j\omega}$ is a complex variable. The values of z for which the sum converges define a region in the z-plane referred to as the *region of convergence* (ROC).

Notationally, if $x(n)$ has a z-transform $X(z)$, we write

$$x(n) \overset{z}{\longleftrightarrow} X(z)$$

The z-transform may be viewed as the DTFT of an exponentially weighted sequence. Specifically, note that with $z = re^{j\omega}$,

$$X(z) = \sum_{n=-\infty}^{\infty} x(n)z^{-n} = \sum_{n=-\infty}^{\infty} [r^{-n}x(n)]e^{-jn\omega}$$

and we see that $X(z)$ is the discrete-time Fourier transform of the sequence $r^{-n}x(n)$. Furthermore, the ROC is determined by the range of values of r for which

$$\sum_{n=-\infty}^{\infty} |x(n)r^{-n}| < \infty$$

---

[1]The reader should note that in many mathematics books, and in some engineering books, $X(z)$ is defined in terms of a sum using positive powers of z.

Because the z-transform is a function of a complex variable, it is convenient to describe it using the complex z-plane. With

$$z = \text{Re}(z) + j\text{Im}(z) = re^{j\omega}$$

the axes of the z-plane are the real and imaginary parts of z as illustrated in Fig. 4-1, and the contour corresponding to $|z| = 1$ is a circle of unit radius referred to as the *unit circle*. The z-transform evaluated on the unit circle corresponds to the DTFT,

$$X(e^{j\omega}) = X(z)|_{z=e^{j\omega}} \qquad (4.2)$$

More specifically, evaluating $X(z)$ at points around the unit circle, beginning at $z = 1(\omega = 0)$, through $z = j$ $(\omega = \pi/2)$, to $z = -1(\omega = \pi)$, we obtain the values of $X(e^{j\omega})$ for $0 \leq \omega \leq \pi$. Note that in order for the DTFT of a signal to exist, the unit circle must be within the region of convergence of $X(z)$.

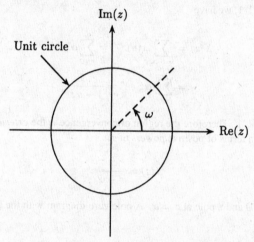

Fig. 4-1.   The unit circle in the complex z-plane.

Many of the signals of interest in digital signal processing have z-transforms that are rational functions of z:

$$X(z) = \frac{B(z)}{A(z)} = \frac{\displaystyle\sum_{k=0}^{q} b(k)z^{-k}}{\displaystyle\sum_{k=0}^{p} a(k)z^{-k}} \qquad (4.3)$$

Factoring the numerator and denominator polynomials, a rational z-transform may be expressed as follows:

$$X(z) = C\frac{\displaystyle\prod_{k=1}^{q}(1 - \beta_k z^{-1})}{\displaystyle\prod_{k=1}^{p}(1 - \alpha_k z^{-1})} \qquad (4.4)$$

The roots of the numerator polynomial, β_k, are referred to as the *zeros* of $X(z)$, and the roots of the denominator polynomial, α_k, are referred to as the *poles*. The poles and zeros uniquely define the functional form of a rational z-transform to within a constant. Therefore, they provide a concise representation for $X(z)$ that is often represented pictorially in terms of a *pole-zero* plot in the z-plane. With a pole-zero plot, the location of each pole is indicated by an "×" and the location of each zero is indicated by an "○", with the region of convergence indicated by shading the appropriate region of the z-plane. The region of convergence is, in general, an *annulus* of the form

$$\alpha < |z| < \beta$$

If $\alpha = 0$, the ROC may also include the point $z = 0$, and if $\beta = \infty$, the ROC may also include infinity. For a rational $X(z)$, the region of convergence will contain no poles. Listed below are three properties of the region of convergence:

1. A finite-length sequence has a z-transform with a region of convergence that includes the entire z-plane except, possibly, $z = 0$ and $z = \infty$. The point $z = \infty$ will be included if $x(n) = 0$ for $n < 0$, and the point $z = 0$ will be included if $x(n) = 0$ for $n > 0$.

2. A right-sided sequence has a z-transform with a region of convergence that is the *exterior* of a circle:

$$\text{ROC} : |z| > \alpha$$

3. A left-sided sequence has a z-transform with a region of convergence that is the *interior* of a circle:

$$\text{ROC} : |z| < \beta$$

EXAMPLE 4.2.1   Let us find the z-transform of the sequence $x(n) = \alpha^n u(n)$. Using the definition of the z-transform and the geometric series given in Table 1-1, we have

$$X(z) = \sum_{n=-\infty}^{\infty} x(n)z^{-n} = \sum_{n=0}^{\infty} \alpha^n z^{-n}$$

$$= \sum_{n=0}^{\infty} (\alpha z^{-1})^n = \frac{1}{1 - \alpha z^{-1}}$$

with the sum converging if $|\alpha z^{-1}| < 1$. Therefore the region of convergence is the *exterior* of a circle defined by the set of points $|z| > |\alpha|$. Expressing $X(z)$ in terms of positive powers of z,

$$X(z) = \frac{z}{z - \alpha}$$

we see that $X(z)$ has a zero at $z = 0$ and a pole at $z = \alpha$. A pole-zero diagram with the region of convergence is shown in the figure below.

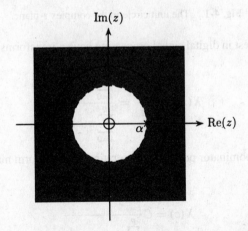

Note that if $|\alpha| < 1$, the unit circle is included within the region of convergence, and the DTFT of $x(n)$ exists.

Example 4.2.1 considered the z-transform of a right-sided sequence, which led to a region of convergence that is the exterior of a circle. The following example considers the z-transform of a left-sided sequence.

EXAMPLE 4.2.2   Let us find the z-transform of the sequence $x(n) = -\alpha^n u(-n - 1)$. Proceeding as in the previous example, we have

$$X(z) = \sum_{n=-\infty}^{\infty} x(n)z^{-n} = -\sum_{n=-\infty}^{-1} \alpha^n z^{-n} = -\sum_{n=0}^{\infty} (\alpha^{-1} z)^{n+1}$$

$$= -\alpha^{-1} z \sum_{n=0}^{\infty} (\alpha^{-1} z)^n = -\frac{\alpha^{-1} z}{1 - \alpha^{-1} z} = \frac{1}{1 - \alpha z^{-1}}$$

with the sum converging if $|\alpha^{-1}z| < 1$ or $|z| < |\alpha|$. A pole-zero diagram with the region of convergence indicated is given in the figure below.

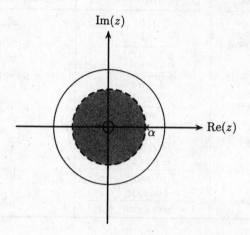

Note that if $|\alpha| \leq 1$, the unit circle is *not* included within the region of convergence, and the DTFT of $x(n)$ does not exist.

Comparing the z-transforms of the signals in Examples 4.2.1 and 4.2.2, we see that they are the same, differing only in their regions of convergence. Thus, the z-transform of a sequence is not uniquely defined until its region of convergence has been specified.

EXAMPLE 4.2.3   Find the z-transform of $x(n) = (\frac{1}{2})^n u(n) - 2^n u(-n-1)$, and find another signal that has the same z-transform but a different region of convergence.

Here we have a sum of two sequences. Therefore, we may find the z-transform of each sequence separately and add them together. From Example 4.2.1, we know that the z-transform of $x_1(n) = (\frac{1}{2})^n u(n)$ is

$$X_1(z) = \frac{1}{1 - \frac{1}{2}z^{-1}} \qquad |z| > \frac{1}{2}$$

and from Example 4.2.2 that the z-transform of $x_2(n) = -2^n u(-n-1)$ is

$$X_2(z) = \frac{1}{1 - 2z^{-1}} \qquad |z| < 2$$

Therefore, the z-transform of $x(n) = x_1(n) + x_2(n)$ is

$$X(z) = \frac{1}{1 - \frac{1}{2}z^{-1}} + \frac{1}{1 - 2z^{-1}} = \frac{2 - \frac{5}{2}z^{-1}}{\left(1 - \frac{1}{2}z^{-1}\right)\left(1 - 2z^{-1}\right)}$$

with a region of convergence $\frac{1}{2} < |z| < 2$, which is the set of all points that are in the ROC of both $X_1(z)$ and $X_2(z)$.

To find another sequence that has the same z-transform, note that because $X(z)$ is a sum of two z-transforms,

$$X(z) = \frac{1}{1 - \frac{1}{2}z^{-1}} + \frac{1}{1 - 2z^{-1}}$$

each term corresponds to the z-transform of either a right-sided or a left-sided sequence, depending upon the region of convergence. Therefore, choosing the right-sided sequences for both terms, it follows that

$$x^1(n) = \left(\frac{1}{2}\right)^n u(n) + 2^n u(n)$$

has the same z-transform as $x(n)$, except that the region of convergence is $|z| > 2$.

Listed in Table 4-1 are a few common z-transform pairs. With these z-transform pairs and the z-transform properties described in the following section, most z-transforms of interest may be easily evaluated.

Table 4-1   Common z-Transform Pairs

Sequence	z-Transform	Region of Convergence
$\delta(n)$	1	all z
$\alpha^n u(n)$	$\dfrac{1}{1 - \alpha z^{-1}}$	$|z| > |\alpha|$
$-\alpha^n u(-n-1)$	$\dfrac{1}{1 - \alpha z^{-1}}$	$|z| < |\alpha|$
$n\alpha^n u(n)$	$\dfrac{\alpha z^{-1}}{(1 - \alpha z^{-1})^2}$	$|z| > |\alpha|$
$-n\alpha^n u(-n-1)$	$\dfrac{\alpha z^{-1}}{(1 - \alpha z^{-1})^2}$	$|z| < |\alpha|$
$\cos(n\omega_0)u(n)$	$\dfrac{1 - (\cos \omega_0)z^{-1}}{1 - 2(\cos \omega_0)z^{-1} + z^{-2}}$	$|z| > 1$
$\sin(n\omega_0)u(n)$	$\dfrac{(\sin \omega_0)z^{-1}}{1 - 2(\cos \omega_0)z^{-1} + z^{-2}}$	$|z| > 1$

4.3   PROPERTIES

Just as with the DTFT, there are a number of important and useful z-transform properties. A few of these properties are described below.

Linearity

As with the DTFT, the z-transform is a *linear* operator. Therefore, if $x(n)$ has a z-transform $X(z)$ with a region of convergence R_x, and if $y(n)$ has a z-transform $Y(z)$ with a region of convergence R_y,

$$w(n) = ax(n) + by(n) \overset{Z}{\longleftrightarrow} W(z) = aX(z) + bY(z)$$

and the ROC of $w(n)$ will *include* the intersection of R_x and R_y, that is,

$$R_w \text{ contains } R_x \cap R_y$$

However, the region of convergence of $W(z)$ *may* be larger. For example, if $x(n) = u(n)$ and $y(n) = u(n-1)$, the ROC of $X(z)$ and $Y(z)$ is $|z| > 1$. However, the z-transform of $w(n) = x(n) - y(n) = \delta(n)$ is the entire z-plane.

Shifting Property

Shifting a sequence (delaying or advancing) multiplies the z-transform by a power of z. That is to say, if $x(n)$ has a z-transform $X(z)$,

$$x(n - n_0) \overset{Z}{\longleftrightarrow} z^{-n_0}X(z)$$

Because shifting a sequence does not affect its absolute summability, shifting does not change the region of convergence. Therefore, the z-transforms of $x(n)$ and $x(n - n_0)$ have the same region of convergence, with the possible exception of adding or deleting the points $z = 0$ and $z = \infty$.

Time Reversal

If $x(n)$ has a z-transform $X(z)$ with a region of convergence R_x that is the annulus $\alpha < |z| < \beta$, the z-transform of the time-reversed sequence $x(-n)$ is

$$x(-n) \overset{Z}{\longleftrightarrow} X(z^{-1})$$

and has a region of convergence $1/\beta < |z| < 1/\alpha$, which is denoted by $1/R_x$.

Multiplication by an Exponential

If a sequence $x(n)$ is multiplied by a complex exponential α^n,

$$\alpha^n x(n) \overset{Z}{\longleftrightarrow} X(\alpha^{-1} z)$$

This corresponds to a scaling of the z-plane. If the region of convergence of $X(z)$ is $r_- < |z| < r_+$, which will be denoted by R_x, the region of convergence of $X(\alpha^{-1} z)$ is $|\alpha| r_- < |z| < |\alpha| r_+$, which is denoted by $|\alpha| R_x$. As a special case, note that if $x(n)$ is multiplied by a complex exponential, $e^{jn\omega_0}$,

$$e^{jn\omega_0} x(n) \overset{Z}{\longleftrightarrow} X(e^{-j\omega_0} z)$$

which corresponds to a rotation of the z-plane.

Convolution Theorem

Perhaps the most important z-transform property is the convolution theorem, which states that convolution in the time domain is mapped into multiplication in the frequency domain, that is,

$$y(n) = x(n) * h(n) \overset{Z}{\longleftrightarrow} Y(z) = X(z)H(z)$$

The region of convergence of $Y(z)$ includes the intersection of R_x and R_y,

$$R_w \text{ contains } R_x \cap R_y$$

However, the region of convergence of $Y(z)$ *may* be larger, if there is a pole-zero cancellation in the product $X(z)H(z)$.

EXAMPLE 4.3.1   Consider the two sequences

$$x(n) = \alpha^n u(n) \qquad h(n) = \delta(n) - \alpha\delta(n-1)$$

The z-transform of $x(n)$ is

$$X(z) = \frac{1}{1 - \alpha z^{-1}} \qquad |z| > |\alpha|$$

and the z-transform of $h(n)$ is

$$H(z) = 1 - \alpha z^{-1} \qquad 0 < |z|$$

However, the z-transform of the convolution of $x(n)$ with $h(n)$ is

$$Y(z) = X(z)H(z) = \frac{1}{1 - \alpha z^{-1}} \cdot (1 - \alpha z^{-1}) = 1$$

which, due to a *pole-zero* cancellation, has a region of convergence that is the entire z-plane.

Conjugation

If $X(z)$ is the z-transform of $x(n)$, the z-transform of the complex conjugate of $x(n)$ is

$$x^*(n) \overset{Z}{\longleftrightarrow} X^*(z^*)$$

As a corollary, note that if $x(n)$ is real-valued, $x(n) = x^*(n)$, then

$$X(z) = X^*(z^*)$$

Derivative

If $X(z)$ is the z-transform of $x(n)$, the z-transform of $nx(n)$ is

$$nx(n) \overset{z}{\longleftrightarrow} -z\frac{dX(z)}{dz}$$

Repeated application of this property allows for the evaluation of the z-transform of $n^k x(n)$ for any integer k.

These properties are summarized in Table 4-2. As illustrated in the following example, these properties are useful in simplifying the evaluation of z-transforms.

Table 4-2   Properties of the z-Transform

Property	Sequence	z-Transform	Region of Convergence
Linearity	$ax(n) + by(n)$	$aX(z) + bY(z)$	Contains $R_x \cap R_y$
Shift	$x(n - n_0)$	$z^{-n_0} X(z)$	R_x
Time reversal	$x(-n)$	$X(z^{-1})$	$1/R_x$
Exponentiation	$\alpha^n x(n)$	$X(\alpha^{-1}z)$	$|\alpha| R_x$
Convolution	$x(n) * y(n)$	$X(z)Y(z)$	Contains $R_x \cap R_y$
Conjugation	$x^*(n)$	$X^*(z^*)$	R_x
Derivative	$nx(n)$	$-z\dfrac{dX(z)}{dz}$	R_x

Note: Given the z-transforms $X(z)$ and $Y(z)$ of $x(n)$ and $y(n)$, with regions of convergence R_x and R_y, respectively, this table lists the z-transforms of sequences that are formed from $x(n)$ and $y(n)$.

EXAMPLE 4.3.2 Let us find the z-transform of $x(n) = n\alpha^n u(-n)$. To find $X(z)$, we will use the time-reversal and derivative properties. First, as we saw in Example 4.2.1,

$$\alpha^n u(n) \overset{z}{\longleftrightarrow} \frac{1}{1 - \alpha z^{-1}} \qquad |z| > \alpha$$

Therefore,

$$\left(\frac{1}{\alpha}\right)^n u(n) \overset{z}{\longleftrightarrow} \frac{1}{1 - \alpha^{-1}z^{-1}} \qquad |z| > \frac{1}{\alpha}$$

and, using the time-reversal property,

$$\alpha^n u(-n) \overset{z}{\longleftrightarrow} \frac{1}{1 - \alpha^{-1}z} \qquad |z| < \alpha$$

Finally, using the derivative property, it follows that the z-transform of $n\alpha^n u(-n)$ is

$$-z\frac{d}{dz}\frac{1}{1 - \alpha^{-1}z} = -\frac{\alpha^{-1}z}{(1 - \alpha^{-1}z)^2} \qquad |z| < \alpha$$

A property that may be used to find the initial value of a causal sequence from its z-transform is the initial value theorem.

Initial Value Theorem

If $x(n)$ is equal to zero for $n < 0$, the initial value, $x(0)$, may be found from $X(z)$ as follows:

$$x(0) = \lim_{z \to \infty} X(z)$$

This property is a consequence of the fact that if $x(n) = 0$ for $n < 0$,

$$X(z) = x(0) + x(1)\,z^{-1} + x(2)\,z^{-2} + \cdots$$

Therefore, if we let $z \to \infty$, each term in $X(z)$ goes to zero except the first.

4.4 THE INVERSE z-TRANSFORM

The z-transform is a useful tool in linear systems analysis. However, just as important as techniques for finding the z-transform of a sequence are methods that may be used to invert the z-transform and recover the sequence $x(n)$ from $X(z)$. Three possible approaches are described below.

4.4.1 Partial Fraction Expansion

For z-transforms that are rational functions of z,

$$X(z) = \frac{\sum\limits_{k=0}^{q} b(k)z^{-k}}{\sum\limits_{k=0}^{p} a(k)z^{-k}} = C\,\frac{\prod\limits_{k=1}^{q}(1-\beta_k z^{-1})}{\prod\limits_{k=1}^{p}(1-\alpha_k z^{-1})}$$

a simple and straightforward approach to find the inverse z-transform is to perform a partial fraction expansion of $X(z)$. Assuming that $p > q$, and that all of the roots in the denominator are simple, $\alpha_i \neq \alpha_k$ for $i \neq k$, $X(z)$ may be expanded as follows:

$$X(z) = \sum_{k=1}^{p} \frac{A_k}{1-\alpha_k z^{-1}} \qquad (4.5)$$

for some constants A_k for $k = 1, 2, \ldots, p$. The coefficients A_k may be found by multiplying both sides of Eq. (4.5) by $(1 - \alpha_k z^{-1})$ and setting $z = \alpha_k$. The result is

$$A_k = \left[(1-\alpha_k z^{-1})X(z)\right]_{z=\alpha_k}$$

If $p \leq q$, the partial fraction expansion must include a polynomial in z^{-1} of order $(p-q)$. The coefficients of this polynomial may be found by long division (i.e., by dividing the numerator polynomial by the denominator). For multiple-order poles, the expansion must be modified. For example, if $X(z)$ has a second-order pole at $z = \alpha_k$, the expansion will include two terms,

$$\frac{B_1}{1-\alpha_k z^{-1}} + \frac{B_2}{(1-\alpha_k z^{-1})^2}$$

where B_1 and B_2 are given by

$$B_1 = \alpha_k \left[\frac{d}{dz}(1-\alpha_k z^{-1})^2 X(z)\right]_{z=\alpha_k}$$

$$B_2 = \left[(1-\alpha_k z^{-1})^2 X(z)\right]_{z=\alpha_k}$$

EXAMPLE 4.4.1 Suppose that a sequence $x(n)$ has a z-transform

$$X(z) = \frac{4 - \frac{7}{4}z^{-1} + \frac{1}{4}z^{-2}}{1 - \frac{3}{4}z^{-1} + \frac{1}{8}z^{-2}} = \frac{4 - \frac{7}{4}z^{-1} + \frac{1}{4}z^{-2}}{\left(1 - \frac{1}{2}z^{-1}\right)\left(1 - \frac{1}{4}z^{-1}\right)}$$

with a region of convergence $|z| > \frac{1}{2}$. Because $p = q = 2$, and the two poles are simple, the partial fraction expansion has the form

$$X(z) = C + \frac{A_1}{1 - \frac{1}{2}z^{-1}} + \frac{A_2}{1 - \frac{1}{4}z^{-1}}$$

The constant C is found by long division:

$$\frac{1}{8}z^{-2} - \frac{3}{4}z^{-1} + 1 \quad \overline{\Big)\ \frac{1}{4}z^{-2} - \frac{7}{4}z^{-1} + 4}$$

$$\frac{\frac{1}{4}z^{-2} - \frac{3}{2}z^{-1} + 2}{-\frac{1}{4}z^{-1} + 2}$$

Therefore, $C = 2$ and we may write $X(z)$ as follows:

$$X(z) = 2 + \frac{2 - \frac{1}{4}z^{-1}}{\left(1 - \frac{1}{2}z^{-1}\right)\left(1 - \frac{1}{4}z^{-1}\right)}$$

Next, for the coefficients A_1 and A_2 we have

$$A_1 = \left[\left(1 - \frac{1}{2}z^{-1}\right)X(z)\right]_{z^{-1}=2} = \left.\frac{4 - \frac{7}{4}z^{-1} + \frac{1}{4}z^{-2}}{1 - \frac{1}{4}z^{-1}}\right|_{z^{-1}=2} = 3$$

and

$$A_2 = \left[\left(1 - \frac{1}{4}z^{-1}\right)X(z)\right]_{z^{-1}=4} = \left.\frac{4 - \frac{7}{4}z^{-1} + \frac{1}{4}z^{-2}}{1 - \frac{1}{2}z^{-1}}\right|_{z^{-1}=4} = -1$$

Thus, the complete partial fraction expansion becomes

$$X(z) = 2 + \frac{3}{1 - \frac{1}{2}z^{-1}} - \frac{1}{1 - \frac{1}{4}z^{-1}}$$

Finally, because the region of convergence is the exterior of the circle $|z| > \frac{1}{2}$, $x(n)$ is the right-sided sequence

$$x(n) = 2\delta(n) + 3\left(\frac{1}{2}\right)^n u(n) - \left(\frac{1}{4}\right)^n u(n)$$

4.4.2 Power Series

The z-transform is a power series expansion,

$$X(z) = \sum_{n=-\infty}^{\infty} x(n)z^{-n} = \cdots + x(-2)z^2 + x(-1)z + x(0) + x(1)z^{-1} + x(2)z^{-2} + \cdots$$

where the sequence values $x(n)$ are the coefficients of z^{-n} in the expansion. Therefore, if we can find the power series expansion for $X(z)$, the sequence values $x(n)$ may be found by simply picking off the coefficients of z^{-n}.

EXAMPLE 4.4.2 Consider the z-transform

$$X(z) = \log(1 + az^{-1}) \qquad |z| > |a|$$

The power series expansion of this function is

$$\log(1 + az^{-1}) = \sum_{n=1}^{\infty} \frac{1}{n}(-1)^{n+1}a^n z^{-n}$$

Therefore, the sequence $x(n)$ having this z-transform is

$$x(n) = \begin{cases} \frac{1}{n}(-1)^{n+1}a^n & n > 0 \\ 0 & n \le 0 \end{cases}$$

4.4.3 Contour Integration

Another approach that may be used to find the inverse z-transform of $X(z)$ is to use contour integration. This procedure relies on Cauchy's integral theorem, which states that if C is a closed contour that encircles the origin in a counterclockwise direction,

$$\frac{1}{2\pi j}\oint_C z^{-k}dz = \begin{cases} 1 & k=1 \\ 0 & k\neq 1 \end{cases}$$

With

$$X(z) = \sum_{n=-\infty}^{\infty} x(n)z^{-n}$$

Cauchy's integral theorem may be used to show that the coefficients $x(n)$ may be found from $X(z)$ as follows:

$$x(n) = \frac{1}{2\pi j}\oint_C X(z)z^{n-1}dz$$

where C is a closed contour within the region of convergence of $X(z)$ that encircles the origin in a counterclockwise direction. Contour integrals of this form may often by evaluated with the help of Cauchy's residue theorem,

$$x(n) = \frac{1}{2\pi j}\oint_C X(z)z^{n-1}dz = \sum\left[\text{residues of } X(z)z^{n-1}\text{ at the poles inside } C\right]$$

If $X(z)$ is a rational function of z with a first-order pole at $z=\alpha_k$,

$$\text{Res}\left[X(z)z^{n-1} \text{ at } z=\alpha_k\right] = \left[(1-\alpha_k z^{-1})X(z)z^{n-1}\right]_{z=\alpha_k}$$

Contour integration is particularly useful if only a few values of $x(n)$ are needed.

4.5 THE ONE-SIDED z-TRANSFORM

The z-transform defined in Sec. 4.2 is the *two-sided*, or *bilateral*, z-transform. The *one-sided*, or *unilateral*, z-transform is defined by

$$X_1(z) = \sum_{n=0}^{\infty} x(n)z^{-n} \tag{4.6}$$

The primary use of the one-sided z-transform is to solve linear constant coefficient difference equations that have initial conditions. Most of the properties of the one-sided z-transform are the same as those for the two-sided z-transform. One that is different, however, is the shift property. Specifically, if $x(n)$ has a one-sided z-transform $X_1(z)$, the one-sided z-transform of $x(n-1)$ is

$$x(n-1) \overset{z}{\longleftrightarrow} z^{-1}X_1(z) + x(-1)$$

It is this property that makes the one-sided z-transform useful for solving difference equations with initial conditions.

EXAMPLE 4.5.1 Consider the linear constant coefficient difference equation

$$y(n) = 0.25y(n-2) + x(n)$$

Let us find the solution to this equation assuming that $x(n) = \delta(n-1)$ with $y(-1)=y(-2)=1$.

We begin by noting that if the one-sided z-transform of $y(n)$ is $Y_1(z)$, the one-sided z-transform of $y(n-2)$ is

$$\sum_{n=0}^{\infty} y(n-2)z^{-n} = y(-2) + y(-1)z^{-1} + \sum_{n=0}^{\infty} y(n)z^{-n-2} = y(-2) + y(-1)z^{-1} + z^{-2}Y_1(z)$$

Therefore, taking the z-transform of both sides of the difference equation, we have

$$Y_1(z) = 0.25[y(-2) + y(-1)z^{-1} + z^{-2}Y_1(z)] + X_1(z)$$

where $X_1(z) = z^{-1}$. Substituting for $y(-1)$ and $y(-2)$, and solving for $Y_1(z)$, we have

$$Y_1(z) = \frac{1}{4} \frac{1 + 5z^{-1}}{1 - \frac{1}{4}z^{-2}}$$

To find $y(n)$, note that $Y_1(z)$ may be expanded as follows:[2]

$$Y_1(z) = \frac{\frac{11}{8}}{1 - \frac{1}{2}z^{-1}} - \frac{\frac{9}{8}}{1 + \frac{1}{2}z^{-1}}$$

Therefore,

$$y(n) = \left[\frac{11}{8}\left(\frac{1}{2}\right)^n - \frac{9}{8}\left(-\frac{1}{2}\right)^n\right]u(n)$$

Solved Problems

Computing z-Transforms

4.1 The z-transform of a sequence $x(n)$ is

$$X(z) = \frac{z + 2z^{-2} + z^{-3}}{1 - 3z^{-4} + z^{-5}}$$

If the region of convergence includes the unit circle, find the DTFT of $x(n)$ at $\omega = \pi$.

If $X(z)$ is the z-transform of $x(n)$, and the unit circle is within the region of convergence, the DTFT of $x(n)$ may be found by evaluating $X(z)$ around the unit circle:

$$X(e^{j\omega}) = X(z)\big|_{z=-1}$$

Therefore, the DTFT at $\omega = \pi$ is

$$X(e^{j\omega})\big|_{\omega=\pi} = X(z)\big|_{z=e^{j\pi}} = X(z)\big|_{z=-1}$$

and we have

$$X(e^{j\omega})\big|_{\omega=\pi} = \frac{z + 2z^{-2} + z^{-3}}{1 - 3z^{-4} + z^{-5}}\bigg|_{z=e^{j\pi}} = \frac{-1 + 2 - 1}{1 - 3 - 1} = 0$$

4.2 Find the z-transform of each of the following sequences:

   (a) $x(n) = 3\delta(n) + \delta(n - 2) + \delta(n + 2)$

   (b) $x(n) = u(n) - u(n - 10)$

   (a) Because this sequence is finite in length, the z-transform is a polynomial,

$$X(z) = 3 + z^{-2} + z^2$$

and the region of convergence is $0 < |z| < \infty$. Note that because $x(n)$ has nonzero values for $n < 0$, the region of convergence does not include $|z| = \infty$, and because $x(n)$ has nonzero values for $n > 0$, the region of convergence does not include the point $z = 0$.

---

[2] See the discussion in Sec. 4.4.1 on partial fraction expansions.

(b) For this sequence,

$$X(z) = \sum_{n=0}^{9} z^{-n} = \frac{1 - z^{-10}}{1 - z^{-1}}$$

which converges for all $|z| > 0$. Note that the roots of the numerator are solutions to the equation

$$z^{10} = 1$$

These roots are

$$z = e^{j2\pi k/10} \qquad k = 0, 1, \ldots, 9$$

which are 10 equally spaced points around the unit circle. Thus, the pole at $z = 1$ in the denominator of $X(z)$ is canceled by the zero at $z = 1$ in the numerator, and the z-transform may also be expressed in the form

$$X(z) = \prod_{k=1}^{9} \left(1 - e^{jk\frac{2\pi}{10}} z^{-1}\right)$$

4.3 Find the z-transform of each of the following sequences:

(a) $x(n) = 2^n u(n) + 3\left(\frac{1}{2}\right)^n u(n)$

(b) $x(n) = \cos(n\omega_0)u(n)$.

(a) Because $x(n)$ is a sum of two sequences of the form $\alpha^n u(n)$, using the linearity property of the z-transform, and the z-transform pair

$$\alpha^n u(n) \overset{z}{\longleftrightarrow} \frac{1}{1 - \alpha z^{-1}} \qquad |z| > |\alpha|$$

we have $$X(z) = \frac{1}{1 - 2z^{-1}} + \frac{3}{1 - \frac{1}{2}z^{-1}} = \frac{4 - \frac{13}{2}z^{-1}}{(1 - 2z^{-1})\left(1 - \frac{1}{2}z^{-1}\right)} \qquad |z| > 2$$

(b) For this sequence we write

$$x(n) = \cos(n\omega_0)u(n) = \tfrac{1}{2}[e^{jn\omega_0} + e^{-jn\omega_0}]u(n)$$

Therefore, the z-transform is

$$X(z) = \tfrac{1}{2}\frac{1}{1 - e^{j\omega_0}z^{-1}} + \tfrac{1}{2}\frac{1}{1 - e^{-j\omega_0}z^{-1}}$$

with a region of convergence $|z| > 1$. Combining the two terms together, we have

$$X(z) = \frac{1 - (\cos\omega_0)z^{-1}}{1 - 2(\cos\omega_0)z^{-1} + z^{-2}} \qquad |z| > 1$$

4.4 Find the z-transform of each of the following sequences. Whenever convenient, use the properties of the z-transform to make the solution easier.

(a) $x(n) = \left(\frac{1}{3}\right)^n u(-n)$

(b) $x(n) = \left(\frac{1}{2}\right)^n u(n+2) + (3)^n u(-n-1)$

(c) $x(n) = \left(\frac{1}{3}\right)^n \cos(n\omega_0)u(n)$

(d) $x(n) = \alpha^{|n|}$

(a) Using the definition of the z-transform we have

$$X(z) = \sum_{n=-\infty}^{\infty} x(n)z^{-n} = \sum_{n=-\infty}^{0} \left(\tfrac{1}{3}\right)^n z^{-n}$$

$$= \sum_{n=0}^{\infty} 3^n z^n = \frac{1}{1 - 3z}$$

where the sum converges for

$$|3z| < 1 \quad \text{or} \quad |z| < \tfrac{1}{3}$$

Alternatively, note that the time-reversed sequence $y(n) = x(-n) = (\tfrac{1}{3})^{-n}u(n)$ has a z-transform given by

$$Y(z) = \frac{1}{1 - 3z^{-1}}$$

with a region of convergence given by $|z| > 3$. Therefore, using the time-reversal property, $Y(z) = X(z^{-1})$, we obtain the same result.

(b) Because $x(n)$ is the sum of two sequences, we will find the z-transform of $x(n)$ by finding the z-transforms of each of these sequences and adding them together. The z-transform of the first sequence may be found easily using the shift property. Specifically, note that because

$$\left(\tfrac{1}{2}\right)^n u(n+2) = 4\left(\tfrac{1}{2}\right)^{n+2} u(n+2)$$

the z-transform of $(\tfrac{1}{2})^n u(n+2)$ is $4z^2$ times the z-transform of $(\tfrac{1}{2})^n u(n)$, that is,

$$\left(\tfrac{1}{2}\right)^n u(n+2) \xleftrightarrow{\;z\;} \frac{4z^2}{1 - \tfrac{1}{2}z^{-1}}$$

which has a region of convergence $|z| > \tfrac{1}{2}$.

The second term is a left-sided exponential and has a z-transform that we have seen before, that is,

$$3^n u(-n-1) \xleftrightarrow{\;z\;} -\frac{1}{1 - 3z^{-1}}$$

with a region of convergence $|z| < 3$.

Finally, for the z-transform of $x(n)$, we have

$$X(z) = \frac{4z^2}{1 - \tfrac{1}{2}z^{-1}} - \frac{1}{1 - 3z^{-1}}$$

with a region of convergence $\tfrac{1}{2} < |z| < 3$.

(c) As we saw in Problem 4.3(b), the z-transform of $\cos(n\omega_0)u(n)$ is

$$\cos(n\omega_0)u(n) \xleftrightarrow{\;z\;} \frac{1 - (\cos\omega_0)z^{-1}}{1 - 2(\cos\omega_0)z^{-1} + z^{-2}}; \qquad |z| > 1$$

Therefore, using the exponentiation property,

$$\alpha^n x(n) \xleftrightarrow{\;z\;} X(\alpha^{-1}z)$$

we have

$$\left(\tfrac{1}{3}\right)^n \cos(n\omega_0)u(n) \xleftrightarrow{\;z\;} \frac{1 - \tfrac{1}{3}(\cos\omega_0)z^{-1}}{1 - \tfrac{2}{3}(\cos\omega_0)z^{-1} + \tfrac{1}{9}z^{-2}}$$

with a region of convergence $|z| > \tfrac{1}{3}$.

(d) Writing $x(n)$ as

$$x(n) = \alpha^n u(n) + \alpha^{-n} u(-n) - \delta(n)$$

we may use the linearity and time-reversal properties to write

$$X(z) = \frac{1}{1 - \alpha z^{-1}} + \frac{1}{1 - \alpha z} - 1 \qquad \tfrac{1}{2} < |z| < 2$$

which may be simplified to

$$X(z) = \frac{1 - \alpha^2}{(1 - \alpha z^{-1})(1 - \alpha z)} \qquad \tfrac{1}{2} < |z| < 2$$

4.5   Without explicitly solving for $X(z)$, find the region of convergence of the z-transform of each of the following sequences:

(a) $x(n) = \left[\left(\frac{1}{2} \right)^n + \left(\frac{3}{4} \right)^n \right] u(n - 10)$

(b) $x(n) = \begin{cases} 1 & -10 \le n \le 10 \\ 0 & \text{otherwise} \end{cases}$

(c) $x(n) = 2^n u(-n)$

(a)   Because the first sequence is right-sided, the region of convergence is the exterior of a circle. With a pole at $z = \frac{1}{2}$ coming from the term $\left(\frac{1}{2} \right)^n$, and a pole at $z = \frac{3}{4}$ coming from the term $\left(\frac{3}{4} \right)^n$, it follows that the region of convergence must be $|z| > \frac{3}{4}$.

(b)   This sequence is finite in length. Therefore, the region of convergence is at least $0 < |z| < \infty$. Because $x(n)$ has nonzero values for $n < 0$ and for $n > 0$, $z = 0$ and $z = \infty$ are not included within the ROC.

(c)   Because this sequence is left-sided, the region of convergence is the interior of a circle. With a pole at $z = 2$, it follows that the region of convergence is $|z| < 2$.

4.6   Find the z-transform of the sequence $y(n) = \sum_{k=-\infty}^{n} x(k)$ in terms of the z-transform of $x(n)$.

There are two ways to approach this problem. The first is to note that $x(n)$ may be written in terms of $y(n)$ as follows:

$$x(n) = y(n) - y(n - 1)$$

Therefore, if we transform both sides of this equation, and use the shift property of the z-transform, we find

$$X(z) = Y(z) - z^{-1} Y(z)$$

Solving for $Y(z)$, we find

$$Y(z) = \frac{1}{1 - z^{-1}} X(z)$$

Thus,

$$y(n) = \sum_{k=-\infty}^{n} x(k) \overset{z}{\longleftrightarrow} \frac{1}{1 - z^{-1}} X(z)$$

which is referred to as the *summation property*.

The second approach is to note that $y(n)$ is the convolution of $x(n)$ with a unit step,

$$y(n) = x(n) * u(n)$$

Therefore, using the convolution theorem, we have

$$Y(z) = X(z) U(z)$$

and, with $U(z) = 1/(1 - z^{-1})$, we obtain the same result as before. For the region of convergence, note that because the ROC of $U(z)$ is $|z| > 1$, the ROC of $Y(z)$ will be at least

$$R_y = R_x \cap \{ |z| > 1 \}$$

where R_x is the ROC of $X(z)$.

4.7   Find the z-transform of the sequence $y(n)$ where

$$y(n) = \sum_{k=-n}^{n} \alpha^{|k|}; \qquad n \ge 0$$

and $y(n) = 0$ for $n < 0$. Assume that $|\alpha| < 1$.

For this sequence, we may use a variation of the summation property derived in Prob. 4.6. Specifically, recall that if

$$y(n) = x(n) * u(n) = \sum_{k=-\infty}^{n} x(k)$$

then

$$Y(z) = \frac{X(z)}{1 - z^{-1}}$$

Now consider the two-sided summation,

$$y(n) = \sum_{k=-n}^{n} x(k)$$

which may be written as

$$y(n) = \sum_{k=0}^{n} x(k) + \sum_{k=-n}^{0} x(k) - x(0) = \sum_{k=0}^{n} x(k) + \sum_{k=0}^{n} x(-k) - x(0)$$

Therefore, if we let

$$x_1(n) = \begin{cases} x(n) & n \geq 0 \\ 0 & n < 0 \end{cases}$$

and

$$x_2(n) = \begin{cases} x(-n) & n \geq 0 \\ 0 & n < 0 \end{cases}$$

then

$$\sum_{k=0}^{n} x(k) = x_1(n) * u(n) \qquad \sum_{k=0}^{n} x(-k) = x_2(n) * u(n)$$

Therefore, we have

$$Y(z) = \frac{X_1(z) + X_2(z)}{1 - z^{-1}} - x(0)$$

Finally, with $x(n) = \alpha^{|n|}$, it follows that $x_1(n) = x_2(n) = \alpha^n u(n)$, and $x(0) = 1$. Thus,

$$Y(z) = \frac{2X_1(z)}{1 - z^{-1}} - 1 = \frac{2}{(1 - \alpha z^{-1})(1 - z^{-1})} - 1$$

$$= \frac{1 + (1 + \alpha)z^{-1} - \alpha z^{-2}}{(1 - \alpha z^{-1})(1 - z^{-1})}$$

with a region of convergence $|z| > 1$.

4.8 Let $x(n)$ be a finite-length sequence that is nonzero only for $0 \leq n \leq N - 1$, and consider the one-sided periodic sequence, $y(n)$, that is formed by periodically extending $x(n)$ as follows:

$$y(n) = \sum_{k=0}^{\infty} x(n - kN)$$

Express the z-transform of $y(n)$ in terms of $X(z)$ and find the region of convergence of $Y(z)$.

The one-sided periodic sequence $y(n)$ may be written as the convolution of $x(n)$ with the pulse train

$$p_N(n) = \sum_{k=0}^{\infty} \delta(n - kN)$$

In other words,

$$y(n) = x(n) * p_N(n) = \sum_{k=0}^{\infty} x(n - kN)$$

Therefore, the z-transform of $y(n)$ is the product of the z-transforms of $x(n)$ and $p_N(n)$. Because $p_N(n)$ is a sum of shifted unit samples, and because the z-transform of $\delta(n - kN)$ is equal to z^{-kN}, the z-transform of $p_N(n)$ is

$$P_N(z) = \sum_{k=0}^{\infty} z^{-kN} = \frac{1}{1 - z^{-N}} \qquad |z| > 1$$

Thus, the z-transform of the one-sided periodic sequence $y(n)$ is

$$Y(z) = X(z)P_N(z) = \frac{X(z)}{1 - z^{-N}}$$

Because $x(n)$ is finite in length and zero for $n < 0$, the region of convergence for $X(z)$ is $|z| > 0$. Therefore, the region of convergence of $Y(z)$ is $|z| > 1$.

4.9   Consider the sequence shown in the figure below.

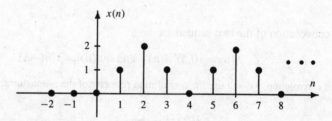

The sequence repeats periodically with a period $N = 4$ for $n \geq 0$ and is zero for $n < 0$. Find the z-transform of this sequence along with its region of convergence.

This is a problem that may be solved easily using the property derived in Prob. 4.8. Because

$$x(n) = \sum_{k=0}^{\infty} w(n - kN)$$

where $N = 4$ and

$$w(n) = \delta(n - 1) + 2\delta(n - 2) + \delta(n - 3)$$

then

$$W(z) = z^{-1}[1 + 2z^{-1} + z^{-2}]$$

and we have

$$X(z) = \frac{z^{-1}[1 + 2z^{-1} + z^{-2}]}{1 - z^{-4}}$$

Because $x(n)$ is right-sided and $X(z)$ has four poles at $|z| = 1$, the region of convergence is $|z| > 1$.

Properties

4.10   Use the z-transform to perform the convolution of the following two sequences:

$$h(n) = \begin{cases} \left(\frac{1}{2}\right)^n & 0 \leq n \leq 2 \\ 0 & \text{else} \end{cases}$$

$$x(n) = \delta(n) + \delta(n - 1) + 4\delta(n - 2)$$

The convolution theorem for z-transforms states that if $y(n) = h(n) * x(n)$, the z-transform of $y(n)$ is $Y(z) = H(z)X(z)$. With

$$H(z) = 1 + \frac{1}{2}z^{-1} + \frac{1}{4}z^{-2}$$

$$X(z) = 1 + z^{-1} + 4z^{-2}$$

it follows that

$$Y(z) = H(z)X(z) = \left(1 + \frac{1}{2}z^{-1} + \frac{1}{4}z^{-2}\right)(1 + z^{-1} + 4z^{-2})$$

Multiplying these two polynomials, we have

$$
\begin{aligned}
Y(z) = 1 &+ z^{-1} && + 4z^{-2} \\
&+ \tfrac{1}{2}z^{-1} && + \tfrac{1}{2}z^{-2} && + 2z^{-3} \\
& && + \tfrac{1}{4}z^{-2} && + \tfrac{1}{4}z^{-3} && + z^{-4} \\
= 1 &+ \tfrac{3}{2}z^{-1} && + \tfrac{19}{4}z^{-2} && + \tfrac{9}{4}z^{-3} && + z^{-4}
\end{aligned}
$$

By inspection, we then have for the sequence $y(n)$,

$$y(n) = \delta(n) + \tfrac{3}{2}\delta(n-1) + \tfrac{19}{4}\delta(n-2) + \tfrac{9}{4}\delta(n-3) + \delta(n-4)$$

4.11 Evaluate the convolution of the two sequences

$$h(n) = (0.5)^n u(n) \quad \text{and} \quad x(n) = 3^n u(-n)$$

To evaluate this convolution, we will use the convolution property of the z-transform. The z-transform of $h(n)$ is

$$H(z) = \frac{1}{1 - \frac{1}{2}z^{-1}} \qquad |z| > \tfrac{1}{2}$$

and the z-transform of $x(n)$ may be found from the time-reversal and shift properties, or directly as follows:

$$
\begin{aligned}
X(z) &= \sum_{n=-\infty}^{\infty} x(n)z^{-n} = \sum_{n=-\infty}^{0} 3^n z^{-n} \\
&= \sum_{n=0}^{\infty} \left(\tfrac{1}{3}z\right)^n = \frac{1}{1 - \frac{1}{3}z} = -\frac{3z^{-1}}{1 - 3z^{-1}} \qquad |z| < 3
\end{aligned}
$$

Therefore, the z-transform of the convolution, $y(n) = x(n) * h(n)$, is

$$Y(z) = -\frac{1}{1 - \frac{1}{2}z^{-1}} \cdot \frac{3z^{-1}}{1 - 3z^{-1}}$$

The region of convergence is the intersection of the regions $|z| > \frac{1}{2}$ and $|z| < 3$, which is $\frac{1}{2} < |z| < 3$. To find the inverse z-transform, we perform a partial fraction expansion of $Y(z)$,

$$Y(z) = \frac{A}{1 - \frac{1}{2}z^{-1}} + \frac{B}{1 - 3z^{-1}}$$

where

$$A = \left[\left(1 - \tfrac{1}{2}z^{-1}\right)Y(z)\right]_{z=\frac{1}{2}} = \tfrac{6}{5}$$

and

$$B = [(1 - 3z^{-1})Y(z)]_{z=3} = -\tfrac{6}{5}$$

Therefore, it follows that

$$y(n) = \left(\tfrac{6}{5}\right)\left(\tfrac{1}{2}\right)^n u(n) + \left(\tfrac{6}{5}\right)3^n u(-n-1)$$

4.12 Let $x(n)$ be an absolutely summable sequence,

$$\sum_{n=-\infty}^{\infty} |x(n)| < \infty$$

with a rational z-transform. If $X(z)$ has a pole at $z = \frac{1}{3}$ and $\lim_{|z|\to\infty} X(z) = 1$, what can be said about the extent of $x(n)$ (i.e., finite-in-length, right-sided, etc.)?

Because $x(n)$ is an absolutely summable sequence, the ROC of $X(z)$ includes the unit circle, $|z| = 1$. With a pole at $z = \frac{1}{3}$, the region of convergence will either be an annulus of the form $r_- < |z| < r_+$, or it will be the exterior of a circle, $r_- < |z|$. However, because $X(z)$ converges as $|z| \to \infty$, the region of convergence will be the exterior of a circle, and it follows that $x(n)$ is right-sided (infinite in length) with $x(n) = 0$ for $n < 0$.

4.13 Find the z-transform of $x(n) = |n|(\frac{1}{2})^{|n|}$.

Using the derivative property and the z-transform pair

$$\left(\tfrac{1}{2}\right)^n u(n) \overset{z}{\longleftrightarrow} \frac{1}{1 - \frac{1}{2}z^{-1}}$$

it follows that the z-transform of $w(n) = n(\frac{1}{2})^n u(n)$ is

$$W(z) = -z\frac{d}{dz}\frac{1}{1 - \frac{1}{2}z^{-1}} = \frac{\frac{1}{2}z^{-1}}{\left(1 - \frac{1}{2}z^{-1}\right)^2}$$

Because $x(n)$ may be written as

$$x(n) = |n|\left(\tfrac{1}{2}\right)^{|n|} = n\left(\tfrac{1}{2}\right)^n u(n) - n\left(\tfrac{1}{2}\right)^{-n} u(-n)$$

using linearity and the time-reversal property, we have

$$X(z) = \frac{\frac{1}{2}z^{-1}}{\left(1 - \frac{1}{2}z^{-1}\right)^2} + \frac{\frac{1}{2}z}{\left(1 - \frac{1}{2}z\right)^2} = \frac{\frac{5}{8}z + \frac{5}{8}z^{-1} - 1}{\left(1 - \frac{1}{2}z^{-1}\right)^2\left(1 - \frac{1}{2}z\right)^2}$$

which has a region of convergence $\frac{1}{2} < |z| < 2$.

4.14 Let $y(n)$ be a sequence that is generated from a sequence $x(n)$ as follows:

$$y(n) = \sum_{k=-\infty}^{n} kx(k)$$

(a) Show that $y(n)$ satisfies the time-varying difference equation

$$y(n) - y(n-1) = nx(n)$$

and show that

$$Y(z) = \frac{-z^2}{z-1}\frac{dX(z)}{dz}$$

where $X(z)$ and $Y(z)$ are the z-transforms of $x(n)$ and $y(n)$, respectively.

(b) Use this property to find the z-transform of

$$y(n) = \sum_{k=0}^{n} k\left(\tfrac{1}{3}\right)^k \qquad n \geq 0$$

(a) From the definition of $y(n)$, we see that

$$y(n-1) = \sum_{k=-\infty}^{n-1} kx(k)$$

and it follows immediately that

$$y(n) - y(n-1) = nx(n)$$

From this difference equation, we may take the z-transform of both sides. Because

$$nx(n) \overset{z}{\longleftrightarrow} -z\frac{dX(z)}{dz}$$

then

$$Y(z) - z^{-1}Y(z) = -z\frac{dX(z)}{dz}$$

or

$$Y(z) = \frac{-z}{1-z^{-1}}\frac{dX(z)}{dz} = \frac{-z^2}{z-1}\frac{dX(z)}{dz}$$

(b) To find the z-transform of the given sequence, note that

$$y(n) = \sum_{k=-\infty}^{n} kx(k)$$

where

$$x(n) = \left(\tfrac{1}{3}\right)^n u(n)$$

Because the z-transform of $x(n)$ is

$$X(z) = \frac{1}{1-\tfrac{1}{3}z^{-1}} \qquad |z| > \tfrac{1}{3}$$

then

$$Y(z) = \frac{-z^2}{z-1}\frac{dX(z)}{dz} = \frac{-z^2}{z-1}\frac{-\tfrac{1}{3}z^{-2}}{\left(1-\tfrac{1}{3}z^{-1}\right)^2} = \frac{\tfrac{1}{3}z^{-1}}{\left(1-\tfrac{1}{3}z^{-1}\right)^2(1-z^{-1})}$$

Because $x(n)$ is right-sided, then the region of convergence is the exterior of a circle. Having poles at $z = 1$ and $z = \tfrac{1}{2}$, it follows that the region of convergence is $|z| > 1$.

4.15 Find the value of $x(0)$ for the sequence that has a z-transform

$$X(z) = \frac{1}{1-az^{-1}} \qquad |z| > a$$

Taking the limit of $X(z)$ as $z \to \infty$, we see that $X(z) \to 1$. Because the limit exists, $x(n)$ is causal, and $x(0) = 1$.

4.16 Find the value of $x(0)$ for the sequence that has a z-transform

$$X(z) = \frac{z}{\left(1-\tfrac{1}{2}z^{-1}\right)\left(1-\tfrac{1}{3}z^{-2}\right)} \qquad |z| > \tfrac{1}{2}$$

Because the region of convergence of $X(z)$ is the exterior of a circle, $x(n)$ is right-sided. However, if we write $X(z)$ in terms of positive powers of z,

$$X(z) = \frac{z^4}{\left(z-\tfrac{1}{2}\right)\left(z^2-\tfrac{1}{3}\right)}$$

we see that $X(z) \to \infty$ as $|z| \to \infty$. Therefore, $x(n)$ is not causal. However, because $x(n)$ is right-sided, it may be delayed so that it is causal. Specifically, if we delay $x(n)$ by 1 to form the sequence $y(n) = x(n-1)$,

$$Y(z) = \frac{z^3}{\left(z - \frac{1}{2}\right)\left(z^2 - \frac{1}{3}\right)}$$

which approaches 1 as $|z| \to \infty$. Thus, $y(n)$ is causal, and we conclude that $y(0) = x(-1) = 1$. Because

$$X(z) = x(-1)z + \sum_{n=0}^{\infty} x(n)z^{-n}$$

$X(z) - x(-1)z$ is the z-transform of a causal sequence, and it follows from the initial value theorem that

$$x(0) = \lim_{|z| \to \infty} [X(z) - x(-1)z]$$

With

$$X(z) - x(-1)z = X(z) - z = \frac{z^4}{\left(z - \frac{1}{2}\right)\left(z^2 - \frac{1}{3}\right)} - z$$

$$= \frac{z^4 - z\left(z^3 - \frac{1}{2}z^2 - \frac{1}{3}z + \frac{1}{6}\right)}{\left(z - \frac{1}{2}\right)\left(z^2 - \frac{1}{3}\right)}$$

we have

$$x(0) = \lim_{|z| \to \infty} [X(z) - x(-1)z] = \frac{1}{2}$$

4.17 Generalize the initial value theorem to find the value of a causal sequence $x(n)$ at $n = 1$, and find $x(1)$ when

$$X(z) = \frac{2 + 6z^{-1}}{4 - 2z^{-2} + 13z^{-3}}$$

If $x(n)$ is causal,

$$X(z) = x(0) + x(1)z^{-1} + x(2)z^{-2} + \cdots$$

Therefore, note that if we subtract $x(0)$ from $X(z)$,

$$X(z) - x(0) = x(1)z^{-1} + x(2)z^{-2} + \cdots$$

Multiplying both sides of this equation by z, we have

$$z[X(z) - x(0)] = x(1) + x(2)z^{-1} + \cdots$$

If we let $z \to \infty$, we obtain the value for $x(1)$,

$$x(1) = \lim_{|z| \to \infty} \{z[X(z) - x(0)]\}$$

For the given z-transform we see that

$$x(0) = \lim_{|z| \to \infty} X(z) = \frac{1}{2}$$

Therefore,

$$X(z) - \frac{1}{2} = \frac{2 + 6z^{-1}}{4 - 2z^{-2} + 13z^{-3}} - \frac{1}{2} = \frac{6z^{-1} + z^{-2} - \frac{13}{2}z^{-3}}{4 - 2z^{-2} + 13z^{-3}}$$

and

$$x(1) = \lim_{|z| \to \infty} \{z[X(z) - x(0)]\} = \frac{3}{2}$$

4.18 Let $x(n)$ be a left-sided sequence that is equal to zero for $n > 0$. If

$$X(z) = \frac{3z^{-1} + 2z^{-2}}{3 - z^{-1} + z^{-2}}$$

find $x(0)$.

For a left-sided sequence that is zero for $n > 0$, the z-transform is

$$X(z) = x(0) + x(-1)z + x(-2)z^2 + \cdots$$

Therefore, it follows that

$$x(0) = \lim_{z \to 0} X(z)$$

For the given z-transform, we see that

$$x(0) = \lim_{z \to 0} X(z) = \lim_{z \to 0} \frac{3z^{-1} + 2z^{-2}}{3 - z^{-1} + z^{-2}} = \lim_{z \to 0} \frac{3z + 2}{3z^2 - z + 1} = 2$$

4.19 If $x(n)$ is real and even with a rational z-transform, show that

$$X(z) = X(z^{-1})$$

and describe what constraints this places on the poles and zeros of $X(z)$.

If $x(n)$ is even,

$$x(n) = x(-n)$$

Therefore, it follows immediately from the time-reversal property that

$$X(z) = X(z^{-1})$$

If $X(z)$ has a zero at $z = z_0$,

$$X(z_0) = 0$$

then

$$X\left(z_0^{-1}\right) = 0$$

which implies that $X(z)$ will also have a zero at $z = 1/z_0$. The same holds true for poles. That is, if there is a pole at z_0, there must also be a pole at $z = 1/z_0$.

4.20 Use the derivative property to find the z-transform of the following sequences:

(a) $x(n) = n\left(\frac{1}{2}\right)^n u(n - 2)$

(b) $x(n) = \frac{1}{n}(-2)^{-n} u(-n - 1)$

(a) The derivative property states that if $X(z)$ is the z-transform of $x(n)$,

$$nx(n) \overset{z}{\longleftrightarrow} -z\frac{d}{dz}X(z)$$

If we let $x(n) = nw(n)$, where

$$w(n) = \left(\tfrac{1}{2}\right)^n u(n - 2) = \tfrac{1}{4}\left(\tfrac{1}{2}\right)^{n-2} u(n - 2)$$

from the delay property and the z-transform pair

$$\alpha^n u(n) \overset{z}{\longleftrightarrow} \frac{1}{1 - \alpha z^{-1}} \qquad |z| > |\alpha|$$

it follows that

$$W(z) = \frac{\frac{1}{4}z^{-2}}{1 - \frac{1}{2}z^{-1}} \qquad |z| > \frac{1}{2}$$

Therefore, using the derivative property, we have the z-transform of $x(n)$,

$$X(z) = -z\frac{d}{dz}W(z) = \frac{1}{2}z^{-2}\frac{1 - \frac{1}{4}z^{-1}}{\left(1 - \frac{1}{2}z^{-1}\right)^2}$$

(b) Evaluating the z-transform of this sequence directly is difficult due to the factor of n^{-1}. However, if we define a new sequence, $y(n)$, as follows,

$$y(n) = nx(n) = (-2)^{-n}u(-n-1)$$

the z-transform of $y(n)$ is easily determined to be

$$Y(z) = \frac{-1}{1 + \frac{1}{2}z^{-1}} \qquad |z| < \frac{1}{2}$$

Noting the relationship between $x(n)$ and $y(n)$, we can apply the derivative property to set up a differential equation for $X(z)$,

$$-z\frac{d}{dz}X(z) = \frac{-1}{1 + \frac{1}{2}z^{-1}}$$

or

$$\frac{d}{dz}X(z) = \frac{1}{z + \frac{1}{2}}$$

The solution to this differential equation is

$$X(z) = \log\left(z + \tfrac{1}{2}\right)$$

and the region of convergence is $|z| < \frac{1}{2}$.

4.21 Up-sampling is an operation that stretches a sequence in time by inserting zeros between the sequence values. For example, up-sampling a sequence $x(n)$ by a factor of L results in the sequence

$$y(n) = \begin{cases} x\left(\dfrac{n}{L}\right) & n = 0, \pm L, \pm 2L, \ldots \\ 0 & \text{otherwise} \end{cases}$$

Express the z-transform of $y(n)$ in terms of the z-transform of $x(n)$.

Because $y(n)$ is equal to zero for all $n \neq kL$, with $y(n)$ equal to $x(n/L)$ for $n = kL$, the z-transform of the up-sampled signal is

$$Y(z) = \sum_{n=-\infty}^{\infty} y(n)z^{-n} = \sum_{n=-\infty}^{\infty} x(n)z^{-nL} = \sum_{n=-\infty}^{\infty} x(n)(z^L)^{-n} = X(z^L)$$

If $X(z)$ converges for $\alpha < |z| < \beta$, $Y(z)$ will converge for $\alpha < |z|^L < \beta$, or

$$\alpha^{1/L} < |z| < \beta^{1/L}$$

4.22 Find the z-transform of the sequence

$$x(n) = \begin{cases} \alpha^{n/10} & n = 0, 10, 20, \ldots \\ 0 & \text{else} \end{cases}$$

where $|\alpha| < 1$.

We recognize $x(n)$ as an exponential sequence that has been up-sampled by a factor of 10 (see Prob. 4.21). Therefore, because

$$\alpha^n u(n) \overset{z}{\longleftrightarrow} \frac{1}{1 - \alpha z^{-1}} \qquad |z| > \alpha$$

the z-transform of $x(n)$ is

$$X(z) = \frac{1}{1 - \alpha z^{-10}} \qquad |z| > \alpha^{1/10}$$

Inverse z-Transforms

4.23 Find the inverse of each of the following z-transforms:

  (a) $X(z) = 4 + 3(z^2 + z^{-2}) \qquad 0 < |z| < \infty$

  (b) $X(z) = \dfrac{1}{1 - \frac{1}{2}z^{-1}} + \dfrac{3}{1 - \frac{1}{3}z^{-1}} \qquad |z| > \frac{1}{2}$

  (c) $X(z) = \dfrac{1}{1 + 3z^{-1} + 2z^{-2}} \qquad |z| > 2$

  (d) $X(z) = \dfrac{1}{(1 - z^{-1})(1 - z^{-2})} \qquad |z| > 1$

  (a) Because $X(z)$ is a finite-order polynomial, $x(n)$ is a finite-length sequence. Therefore, $x(n)$ is the coefficient that multiplies z^{-n} in $X(z)$. Thus, $x(0) = 4$ and $x(2) = x(-2) = 3$.

  (b) This z-transform is a sum of two first-order rational functions of z. Because the region of convergence of $X(z)$ is the exterior of a circle, $x(n)$ is a right-sided sequence. Using the z-transform pair for a right-sided exponential, we may invert $X(z)$ easily as follows:

$$x(n) = \left(\tfrac{1}{2}\right)^n u(n) + 3\left(\tfrac{1}{3}\right)^n u(n)$$

  (c) Here we have a rational function of z with a denominator that is a quadratic in z. Before we can find the inverse z-transform, we need to factor the denominator and perform a partial fraction expansion:

$$X(z) = \frac{1}{1 + 3z^{-1} + 2z^{-2}} = \frac{1}{(1 + 2z^{-1})(1 + z^{-1})}$$
$$= \frac{2}{1 + 2z^{-1}} - \frac{1}{1 + z^{-1}}$$

Because $x(n)$ is right-sided, the inverse z-transform is

$$x(n) = 2(-2)^n u(n) - (-1)^n u(n)$$

  (d) One way to invert this z-transform is to perform a partial fraction expansion. With

$$X(z) = \frac{1}{(1 - z^{-1})(1 - z^{-2})} = \frac{1}{(1 - z^{-1})^2(1 + z^{-1})}$$
$$= \frac{A}{1 + z^{-1}} + \frac{B_1}{1 - z^{-1}} + \frac{B_2}{(1 - z^{-1})^2}$$

the constants A, B_1, and B_2 are as follows:

$$A = [(1 + z^{-1})X(z)]_{z=-1} = \tfrac{1}{4}$$

$$B_1 = \left[\frac{d}{dz}(1 - z^{-1})^2 X(z)\right]_{z=1} = \left[\frac{z^{-2}}{(1 + z^{-1})^2}\right]_{z=1} = \tfrac{1}{4}$$

$$B_2 = [(1 - z^{-1})^2 X(z)]_{z=1} = \tfrac{1}{2}$$

Inverse transforming each term, we have

$$x(n) = \tfrac{1}{4}[(-1)^n + 1 + 2(n+1)]u(n)$$

Another way to invert this z-transform is to note that $x(n)$ is the convolution of the two sequences,

$$x(n) = x_1(n) * x_2(n)$$

where $x_1(n) = u(n)$ and $x_2(n)$ is a step function that is up-sampled by a factor of 2. Because

$$x_1(n) * x_2(n) = \{1, 1, 2, 2, 3, 3, 4, 4, \ldots\}$$

we have the same result as before.

4.24   Find the inverse z-transform of the second-order system

$$X(z) = \frac{1 + \tfrac{1}{4}z^{-1}}{\left(1 - \tfrac{1}{2}z^{-1}\right)^2} \qquad |z| > \tfrac{1}{2}$$

Here we have a second-order pole at $z = \tfrac{1}{2}$. The partial fraction expansion for $X(z)$ is

$$X(z) = \frac{A_1}{1 - \tfrac{1}{2}z^{-1}} + \frac{A_2}{\left(1 - \tfrac{1}{2}z^{-1}\right)^2}$$

The constant A_1 is

$$A_1 = \tfrac{1}{2}\left[\frac{d}{dz}\left(1 - \tfrac{1}{2}z^{-1}\right)^2 X(z)\right]_{z=1/2} = \tfrac{1}{2}\left[-\tfrac{1}{4}z^{-2}\right]_{z=1/2} = -\tfrac{1}{2}$$

and the constant A_2 is

$$A_2 = \left[\left(1 - \tfrac{1}{2}z^{-1}\right)^2 X(z)\right]_{z=1/2} = \tfrac{3}{2}$$

Therefore,

$$X(z) = -\frac{\tfrac{1}{2}}{1 - \tfrac{1}{2}z^{-1}} + \frac{\tfrac{3}{2}}{\left(1 - \tfrac{1}{2}z^{-1}\right)^2}$$

and

$$x(n) = -\left(\tfrac{1}{2}\right)^{n+1} u(n) + 3(n+1)\left(\tfrac{1}{2}\right)^{n+1} u(n)$$

4.25   Find the inverse of each of the following z-transforms:

(a) $X(z) = \log\left(1 - \tfrac{1}{2}z^{-1}\right) \qquad |z| > \tfrac{1}{2}$

(b) $X(z) = e^{1/z}$, with $x(n)$ a right-sided sequence

(a)   There are several ways to solve this problem. One is to look up or compute the power series expansion of the log function. Another way is to differentiate $X(z)$. Specifically, because

$$\frac{d}{dz}X(z) = \frac{\tfrac{1}{2}z^{-2}}{1 - \tfrac{1}{2}z^{-1}}$$

if we multiply both sides of this equation by $(-z)$, we have

$$Y(z) = -z\frac{d}{dz}X(z) = -\frac{\tfrac{1}{2}z^{-1}}{1 - \tfrac{1}{2}z^{-1}}$$

Note that the region of convergence for $X(z)$ is $|z| > \frac{1}{2}$. Because the region of convergence for $Y(z)$ is the same as it is for $X(z)$, the inverse z-transform of $Y(z)$ is

$$y(n) = -\left(\tfrac{1}{2}\right)^n u(n-1)$$

Now, from the derivative property, $y(n) = nx(n)$, and it follows that

$$x(n) = -\tfrac{1}{n}\left(\tfrac{1}{2}\right)^n u(n-1)$$

(b) For this z-transform, we could determine the inverse by finding the power series expansion of $X(z)$. However, another approach is to do what we did in part (a) and take the derivative. Differentiating $X(z)$, we find

$$\frac{d}{dz}X(z) = -z^{-2}X(z)$$

Multiplying both sides by $(-z)$, we have

$$-z\frac{d}{dz}X(z) = z^{-1}X(z)$$

and taking the inverse z-transform gives

$$nx(n) = x(n-1)$$

which is a recursion for $x(n)$. To solve this recursion, we need an initial condition. Because $x(n)$ is a right-sided sequence, we may use the initial value theorem to find $x(0)$. Specifically,

$$x(0) = \lim_{|z| \to \infty} X(z) = 1$$

Thus, the recursion that we want to solve is

$$x(n) = \frac{1}{n}x(n-1) \qquad n > 0$$

with $x(0) = 1$. The solution for $n > 0$ is

$$x(n) = \frac{1}{n!}$$

and we have

$$x(n) = \delta(n) + \frac{1}{n!}u(n-1)$$

4.26 Find the inverse z-transform of $X(z) = \sin z$.

To find the inverse z-transform of $X(z) = \sin z$, we expand $X(z)$ in a Taylor series about $z = 0$ as follows:

$$X(z) = X(z)\Big|_{z=0} + z\frac{dX(z)}{dz}\Big|_{z=0} + \frac{z^2}{2!}\frac{d^2X(z)}{dz^2}\Big|_{z=0} + \cdots + \frac{z^n}{n!}\frac{d^nX(z)}{dz^n}\Big|_{z=0} + \cdots$$

$$= z - \frac{z^3}{3!} + \frac{z^5}{5!} - \cdots = \sum_{n=0}^{\infty}(-1)^n\frac{z^{2n+1}}{(2n+1)!}$$

Because

$$X(z) = \sum_{n=-\infty}^{\infty} x(n)z^{-n}$$

we may associate the coefficients in the Taylor series expansion with the sequence values $x(n)$. Thus, we have

$$x(n) = (-1)^n\frac{1}{(2|n|+1)!} \qquad n = -1, -3, -5, \ldots$$

4.27   Evaluate the following integral:

$$\frac{1}{2\pi j} \oint_C \frac{1 + 2z^{-1} - z^{-2}}{\left(1 - \frac{1}{2}z^{-1}\right)\left(1 - \frac{2}{3}z^{-1}\right)} z^3 dz$$

where the contour of integration C is the unit circle.

Recall that for a sequence $x(n)$ that has a z-transform $X(z)$, the sequence may be recovered using contour integration as follows:

$$x(n) = \frac{1}{2\pi j} \oint_c X(z)z^{n-1}dz \qquad (4.7)$$

Therefore, the integral that is to be evaluated corresponds to the value of the sequence $x(n)$ at $n = 4$ that has a z-transform

$$X(z) = \frac{1 + 2z^{-1} - z^{-2}}{\left(1 - \frac{1}{2}z^{-1}\right)\left(1 - \frac{2}{3}z^{-1}\right)}$$

Thus, we may find $x(n)$ using a partial fraction expansion of $X(z)$ and then evaluate the sequence at $n = 4$. With this approach, however, we are finding the values of $x(n)$ for all n. Alternatively, we could perform long division and divide the numerator of $X(z)$ by the denominator. The coefficient multiplying z^{-4} would then be the value of $x(n)$ at $n = 4$, and the value of the integral. However, because we are only interested in the value of the sequence at $n = 4$, the easiest approach is to evaluate the integral directly using the Cauchy integral theorem. The value of the integral is equal to the sum of the residues of the poles of $X(z)z^3$ inside the unit circle. Because

$$X(z)z^3 = z^3 \frac{z^2 + 2z - 1}{\left(z - \frac{1}{2}\right)\left(z - \frac{2}{3}\right)}$$

has poles at $z = \frac{1}{2}$ and $z = \frac{2}{3}$,

$$\text{Res}[X(z)z^3]_{z=\frac{1}{2}} = \left[z^3 \frac{z^2 + 2z - 1}{z - \frac{2}{3}}\right]_{z=\frac{1}{2}} = -\frac{3}{16}$$

and

$$\text{Res}[X(z)z^3]_{z=\frac{2}{3}} = \left[z^3 \frac{z^2 + 2z - 1}{z - \frac{1}{2}}\right]_{z=\frac{2}{3}} = \frac{112}{81}$$

Therefore, we have

$$\frac{1}{2\pi j} \oint_c X(z)z^3 dz = \frac{112}{81} - \frac{3}{16} = 1.1952$$

4.28   Find the inverse z-transform of

$$X(z) = \frac{1}{1 - \alpha^{10}z^{-10}} \qquad |z| > |\alpha|$$

Note that the denominator of $X(z)$ is a tenth-order polynomial. Although the roots may be found easily, performing a partial fraction expansion would be time consuming. For this problem, it is much better to exploit the properties of the z-transform. Note, for example, that

$$X(z) = Y(z^{10}) \qquad \text{where} \qquad Y(z) = \frac{1}{1 - \alpha^{10}z^{-1}}$$

Because

$$y(n) = \alpha^{10n}u(n)$$

we may use the up-sampling property (Prob. 4.21) to obtain

$$x(n) = \begin{cases} y\left(\dfrac{n}{10}\right) & n = 0, \pm 10, \pm 20, \ldots \\ 0 & \text{otherwise} \end{cases}$$

Therefore, we have

$$x(n) = \begin{cases} \alpha^n & n = 0, 10, 20, \ldots \\ 0 & \text{otherwise} \end{cases}$$

4.29   In many cases one is interested in computing the inverse z-transform of a rational function

$$X(z) = \frac{B(z)}{A(z)} = \frac{\displaystyle\sum_{k=0}^{q} b(k)z^{-k}}{\displaystyle\sum_{k=0}^{p} a(k)z^{-k}}$$

Because a partial fraction expansion requires knowledge of the roots of $A(z)$, if the order of the denominator is large, finding the roots may be difficult. Although a partial fraction expansion would give a closed-form solution for $x(n)$ for all n, if one only wants to plot $x(n)$ for a limited range of values for n, a closed-form expression is not required. Given that $x(n) = 0$ for $n < 0$, find a recursion that generates $x(n)$ for $n \geq 0$.

If we consider $x(n)$ to be the unit sample response of a linear shift-invariant system, we may straightforwardly specify the filter in terms of a linear constant coefficient difference equation. This leads to a recursively computable difference equation for $x(n)$. Specifically, note that because

$$X(z)A(z) = B(z)$$

we may express this in the time domain as follows:

$$x(n) * a(n) = b(n)$$

Writing out this convolution explicitly, we have

$$\sum_{k=0}^{p} a(k)x(n-k) = b(n)$$

Bringing the first term out of the summation and dividing by $a(0)$ gives

$$x(n) = \frac{b(n)}{a(0)} - \sum_{k=1}^{p} \frac{a(k)}{a(0)} x(n-k)$$

Therefore, given that $x(n) = 0$ for $n < 0$, this recursion allows us to compute $x(n)$ for all $n \geq 0$. For example,

$$x(0) = \frac{b(0)}{a(0)}$$

$$x(1) = \frac{b(1)}{a(0)} - \frac{a(1)}{a(0)} x(0)$$

$$\vdots$$

Note that $b(n) = 0$ for $n > q$. Thus, for $n > q$, the recursion simplifies to

$$x(n) = -\sum_{k=1}^{p} \frac{a(k)}{a(0)} x(n-k) \qquad n > q$$

One-Sided z-Transforms

4.30   Find the one-sided z-transform of the following sequences:

(a) $x(n) = \left(\frac{1}{3}\right)^n u(n+3)$

(b) $x(n) = \delta(n-5) + \delta(n) + 2^{n-1}u(-n)$

In the following, let $x_+(n)$ denote the sequence that is formed from $x(n)$ by setting $x(n)$ equal to zero for $n < 0$, that is,

$$x_+(n) = \begin{cases} x(n) & n \geq 0 \\ 0 & n < 0 \end{cases}$$

(a)   Because $x_+(n) = \left(\frac{1}{3}\right)^n u(n)$, the one-sided z-transform of $x(n)$ is

$$X_1(z) = \frac{1}{1 - \frac{1}{3}z^{-1}} \qquad |z| > \frac{1}{3}$$

(b)   For this sequence, because

$$x_+(n) = \delta(n-5) + \delta(n) + 2^{-1}\delta(n)$$

then

$$X_1(z) = z^{-5} + 1 + \frac{1}{2} = 1.5 + z^{-5}$$

4.31   Let $X_1(z)$ be the one-sided z-transform of $x(n)$. Find the one-sided z-transform of $y(n) = x(n+1)$.

The one-sided z-transform of $x(n)$ is

$$X_1(z) = x(0) + x(1)z^{-1} + x(2)z^{-2} + \cdots = \sum_{n=0}^{\infty} x(n)z^{-n}$$

If $x(n)$ is advanced in time by one, $y(n) = x(n+1)$, the one-sided z-transform of $y(n)$ is

$$Y_1(z) = \sum_{n=0}^{\infty} y(n)z^{-n} = \sum_{n=0}^{\infty} x(n+1)z^{-n}$$

Therefore, $$Y_1(z) = x(1) + x(2)z^{-1} + x(3)z^{-2} + \cdots$$

Comparing this to $X_1(z)$, we see that

$$Y_1(z) = z[X_1(z) - x(0)]$$

4.32   Consider the LCCDE

$$y(n) - \frac{1}{4}y(n-2) = \delta(n) \qquad n \geq 0$$

Find a set of initial conditions on $y(n)$ for $n < 0$ so that $y(n) = 0$ for $n \geq 0$.

The one-sided z-transform of the LCCDE is

$$Y_1(z) - \frac{1}{4}[z^{-2}Y_1(z) + y(-1)z^{-1} + y(-2)] = 1$$

Solving for $Y_1(z)$, we have

$$Y_1(z) = \frac{1 + \frac{1}{4}[y(-2) + y(-1)z^{-1}]}{1 - \frac{1}{4}z^{-2}}$$

In order for $y(n)$ to be equal to zero for $n \geq 0$, $Y_1(z)$ must be equal to zero. This will be the case when

$$1 + \frac{1}{4}y(-2) = 0$$
$$\frac{1}{4}y(-1) = 0$$

or $$y(-2) = -4 \qquad y(-1) = 0$$

4.33 Consider a system described by the difference equation

$$y(n) = y(n-1) - y(n-2) + 0.5x(n) + 0.5x(n-1)$$

Find the response of this system to the input

$$x(n) = (0.5)^n u(n)$$

with initial conditions $y(-1) = 0.75$ and $y(-2) = 0.25$.

This is the same problem as Prob. 1.37. Whereas this difference equation was solved in Chap. 1 by finding the particular and homogeneous solutions, here we will use the one-sided z-transform.

First, we take the one-sided z-transform of each term in the difference equation

$$Y(z) = z^{-1}Y(z) + y(-1) - [z^{-2}Y(z) + z^{-1}y(-1) + y(-2)] + \tfrac{1}{2}X(z) + \tfrac{1}{2}z^{-1}X(z)$$

Substituting the given values for the initial conditions, we have

$$Y(z) = z^{-1}Y(z) + \tfrac{3}{4} - z^{-2}Y(z) - \tfrac{3}{4}z^{-1} - \tfrac{1}{4} + \tfrac{1}{2}X(z) + \tfrac{1}{2}z^{-1}X(z)$$

Collecting all of the terms that contain $Y(z)$ onto the left side of the equation gives

$$Y(z)[1 - z^{-1} + z^{-2}] = \tfrac{1}{2} - \tfrac{3}{4}z^{-1} + \tfrac{1}{2}X(z) + \tfrac{1}{2}z^{-1}X(z)$$

Because $x(n) = (\tfrac{1}{2})^n u(n)$,

$$X(z) = \frac{1}{1 - \tfrac{1}{2}z^{-1}}$$

which gives

$$Y(z) = \frac{\tfrac{1}{2} - \tfrac{3}{4}z^{-1}}{1 - z^{-1} + z^{-2}} + \frac{\tfrac{1}{2} + \tfrac{1}{2}z^{-1}}{\left(1 - \tfrac{1}{2}z^{-1}\right)(1 - z^{-1} + z^{-2})}$$

Expanding the second term using a partial fraction expansion, we have

$$Y(z) = \frac{\tfrac{1}{2} - \tfrac{3}{4}z^{-1}}{1 - z^{-1} + z^{-2}} + \frac{\tfrac{1}{2}}{1 - \tfrac{1}{2}z^{-1}} + \frac{z^{-1}}{1 - z^{-1} + z^{-2}}$$

or

$$Y(z) = \frac{\tfrac{1}{2}}{1 - \tfrac{1}{2}z^{-1}} + \frac{\tfrac{1}{2} + \tfrac{1}{4}z^{-1}}{1 - z^{-1} + z^{-2}}$$

Therefore, the solution is

$$y(n) = \left(\tfrac{1}{2}\right)^{n+1} u(n) + \left[\tfrac{\sqrt{3}}{6}\sin\left(\frac{n\pi}{3}\right) + \tfrac{\sqrt{3}}{3}\sin(n-1)\frac{\pi}{3}\right]u(n)$$

4.34 A digital filter that is implemented on a DSP chip is described by the linear constant coefficient difference equation

$$y(n) = \tfrac{3}{4}y(n-1) - \tfrac{1}{8}y(n-2) + x(n)$$

In evaluating the performance of the filter, the unit sample response is measured (i.e., the response $y(n)$ to the input $x(n) = \delta(n)$ is determined). The internal storage registers on the chip, however, are not set to zero prior to applying the input. Therefore, the output of the filter contains the effect of the initial conditions, which are

$$y(-1) = -1 \qquad \text{and} \qquad y(-2) = 1$$

Determine the response of the filter for all $n \geq 0$ and compare it with the zero state response (i.e., the output with $y(-1) = y(-2) = 0$).

Here we want to solve a difference equation that has initial conditions. Using the one-sided z-transform, we have

$$Y(z) = \tfrac{3}{4}\{z^{-1}Y(z) + y(-1)\} - \tfrac{1}{8}\{z^{-2}Y(z) + y(-1)z^{-1} + y(-2)\} + X(z)$$

With $X(z) = 1$ and the given initial conditions, this becomes

$$Y(z) = \tfrac{3}{4}\{z^{-1}Y(z) - 1\} - \tfrac{1}{8}\{z^{-2}Y(z) - z^{-1} + 1\} + 1$$

Solving for $Y(z)$, we find

$$Y(z) = \tfrac{1}{8}\frac{1 + z^{-1}}{1 - \tfrac{3}{4}z^{-1} + \tfrac{1}{8}z^{-2}} = \tfrac{1}{8}\frac{1 + z^{-1}}{\left(1 - \tfrac{1}{4}z^{-1}\right)\left(1 - \tfrac{1}{2}z^{-1}\right)}$$

Performing a partial fraction expansion gives

$$Y(z) = \frac{-\tfrac{5}{8}}{1 - \tfrac{1}{4}z^{-1}} + \frac{\tfrac{3}{4}}{1 - \tfrac{1}{2}z^{-1}}$$

Thus, with an inverse z-transform we have

$$y(n) = \left[-\tfrac{5}{8}\left(\tfrac{1}{4}\right)^{n} + \tfrac{3}{4}\left(\tfrac{1}{2}\right)^{n}\right]u(n)$$

The zero state response, on the other hand, is simply the unit sample response of the filter. With

$$H(z) = \frac{1}{1 - \tfrac{3}{4}z^{-1} + \tfrac{1}{8}z^{-2}} = \frac{2}{1 - \tfrac{1}{2}z^{-1}} - \frac{1}{1 - \tfrac{1}{4}z^{-1}}$$

it follows that

$$h(n) = \left[2\left(\tfrac{1}{2}\right)^{n} - \left(\tfrac{1}{4}\right)^{n}\right]u(n)$$

Applications

4.35   There are two kinds of particles inside a nuclear reactor. Every second, an α particle will split into eight β particles and a β particle will split into an α particle and two β particles. If there is a single α particle in the reactor at time $t = 0$, how may particles are there altogether at time $t = 100$?

In this problem we need to begin by writing down, in mathematical terms, what is happening within the reactor. Let $\alpha(n)$ be the number of α particles in the reactor at time n, and let $\beta(n)$ be the number of β particles. Because there are eight β particles created from each α particle and two from each β particle, we have

$$\beta(n) = 8\alpha(n-1) + 2\beta(n-1)$$

Also, because one α particle is created from each β particle,

$$\alpha(n) = \beta(n-1)$$

Substituting the second equation into the first, we have

$$\beta(n) = 8\beta(n-2) + 2\beta(n-1)$$

which is an equation that defines how many β particles there are in the reactor at time n. Because there is one α particle in the reactor at time $n = 0$, it follows that there are eight β particles at time $n = 1$. Therefore, the initial condition associated with $\beta(n)$ is $\beta(1) = 8$, and this may be incorporated into the equation as follows:

$$\beta(n) = 8\beta(n-2) + 2\beta(n-1) + 8\delta(n-1) \qquad n \geq 1$$

with $\beta(n) = 0$ for $n < 1$. Using z-transforms, we may solve this equation for $\beta(n)$ as follows:

$$B(z) = \frac{8z^{-1}}{1 - 2z^{-1} - 8z^{-2}} = \frac{\frac{4}{3}}{1 - 4z^{-1}} - \frac{\frac{4}{3}}{1 + 2z^{-1}}$$

Taking the inverse z-transform, we have

$$\beta(n) = \tfrac{4}{3}(4)^n u(n) - \tfrac{4}{3}(-2)^n u(n)$$

Finally, because the number of α particles at time n is equal to the number of β particles at time $(n-1)$, the total number of particles at time $n = 100$ is

$$N = \beta(100) + \beta(99) = \tfrac{4}{3}\left[(4)^{100} - (-2)^{100} + (4)^{99} - (-2)^{99}\right] = \tfrac{4}{3}\left[5(4)^{99} + (-2)^{99}\right]$$

4.36 A \$100,000 mortgage is to be paid off in 360 *equal* monthly payments of d dollars. Interest, compounded monthly, is charged at the rate of 10 percent per annum on the unpaid balance (e.g., after the first month the total debt equals \$100,000 $+ \frac{0.10}{12}$\$100,000). Find the payment d so that the mortgage is paid in full after 30 years, leaving a net balance of zero.

This is the same problem that was solved in Prob. 1.39. Here, however, we will use the z-transform to find the solution.

The total unpaid balance at the end of the nth month, in the absence of any additional loans or payments, is equal to the unpaid balance in the previous month plus the interest charged on the unpaid balance for the previous month. Therefore, if $y(n)$ is the balance at the end of the nth month,

$$y(n) = y(n-1) + \beta y(n-1)$$

where β is the interest charged on the unpaid balance. In addition, the balance must be adjusted by the amount of money leaving the bank into your pocket, which is simply the amount borrowed in the nth month and the amount paid to the bank in the nth month. Thus

$$y(n) = y(n-1) + \beta y(n-1) + x_b(n) - x_p(n)$$

where $x_b(n)$ is the amount borrowed in the nth month, and $x_p(n)$ is the amount paid in the nth month. Combining terms, we have

$$y(n) - \nu y(n-1) = x_b(n) - x_p(n) = x(n)$$

where $\nu = 1 + \beta = 1 + 0.10/12$, and $x(n)$ is the net amount of money in the nth month that leaves the bank. Because a principal of p dollars is borrowed during month zero, and payments of d dollars begin with month 1, the *input* $x(n)$ is

$$x(n) = x_b(n) - x_p(n) = p\delta(n) - du(n-1)$$

and the difference equation for $y(n)$ becomes

$$y(n) - \nu y(n-1) = p\delta(n) - du(n-1)$$

Expressing this difference equation in terms of z-transforms, we have

$$Y(z) - \nu z^{-1} Y(z) = p - d\frac{z^{-1}}{1 - z^{-1}}$$

Solving for $Y(z)$, we find

$$Y(z) = \frac{p - (p+d)z^{-1}}{(1 - \nu z^{-1})(1 - z^{-1})} = \frac{1}{1 - \nu}\left[\frac{p + d - p\nu}{1 - \nu z^{-1}} - \frac{d}{1 - z^{-1}}\right]$$

Taking the inverse z-transforms yields

$$y(n) = \frac{1}{1 - \nu}[(p + d - p\nu)\nu^n - d]u(n)$$

We now want to find the value of d so that the mortgage is retired after 360 equal monthly payments. That is, we want to find d so that

$$y(360) = \frac{1}{1-v}[(p+d-pv)v^{360} - d] = 0$$

Solving for d, we have

$$d = \frac{p(1-v)}{1-v^{360}}v^{360}$$

With $v = \frac{12.1}{12}$ and $p = 100,000$ we have

$$d = 877.57$$

which is the same as we had previously calculated.

4.37   A generalized Fibonacci sequence is a sequence of numbers, $x(n)$, that satisfies the difference equation

$$x(n+2) = x(n) + x(n+1) \qquad \text{for} \qquad n \geq 0$$

That is, $x(n)$ is the sum of the two previous values. The classical Fibonacci sequence results when the initial conditions are $x(0) = 0$ and $x(1) = 1$. The Fibonacci numbers occur in such unsuspecting places as the number of ancestors in succeeding generations of the male bee, the input impedance of a resistor ladder network, and the spacing of buds on the branch of a tree.

(a) Find a closed-form expression for $x(n)$.

(b) Show that the ratio $x(n)/x(n+1)$ approaches the limit $2/(1+\sqrt{5})$ as $n \to \infty$. This ratio is known as the *golden mean* and was said by the ancient Greeks to be the ratio of the sides of the rectangle that has the most pleasing proportions.

(c) Show that the Fibonacci sequence has the following properties:

1. $x^2(n) + x^2(n+1) = x(2n+1)$
2. $x^2(n+2) - x^2(n+1) = x(n)x(n+3)$

(a) Here we have a second-order linear constant coefficient difference equation that we want to solve. Let us begin by rewriting it in a slightly different form. Specifically, consider the following

$$x(n) - x(n-1) - x(n-2) = \delta(n-1)$$

where we assume that $x(n) = 0$ for $n < 0$ (i.e, initial rest). Written in this form with the delayed unit sample on the right-hand side, we note that $x(0) = 0$ and $x(1) = 1$ as desired and $x(n+2) = x(n) + x(n+1)$ for $n > 0$. The solution to this difference equation may be found using z-transforms as follows:

$$X(z) - z^{-1}X(z) - z^{-2}X(z) = z^{-1}$$

Solving for $X(z)$, we have

$$X(z) = \frac{z^{-1}}{1 - z^{-1} - z^{-2}}$$

The poles of $X(z)$ are located at $z = (1 \pm \sqrt{5})/2$, and the partial fraction expansion of $X(z)$ is

$$X(z) = \frac{1}{\sqrt{5}}\left[\frac{1}{1 - \left(\frac{1+\sqrt{5}}{2}\right)z^{-1}} - \frac{1}{1 - \left(\frac{1-\sqrt{5}}{2}\right)z^{-1}}\right]$$

Taking the inverse z-transform of $X(z)$, we find

$$x(n) = \frac{1}{\sqrt{5}}\left[\left(\frac{1+\sqrt{5}}{2}\right)^n - \left(\frac{1-\sqrt{5}}{2}\right)^n\right]u(n)$$

(b) Starting with the difference equation that defines the Fibonacci sequence, divide both sides by $x(n+1)$:

$$\frac{x(n+2)}{x(n+1)} = \frac{x(n)}{x(n+1)} + 1$$

If we define $r(n)$ to be the ratio of two successive Fibonacci numbers

$$r(n) = \frac{x(n+1)}{x(n)}$$

we have

$$r(n+1) = \frac{1}{r(n)} + 1$$

or

$$r(n+1)r(n) = 1 + r(n)$$

If we let $n \rightarrow \infty$, and define $r(\infty) = \lim_{n \to \infty} r(n)$, we have

$$r^2(\infty) = 1 + r(\infty)$$

Solving this quadratic equation for $r(\infty)$, we find

$$r(\infty) = \frac{1 \pm \sqrt{5}}{2}$$

However, because $r(n) > 0$, it follows that $r(\infty)$ is the positive root, which is

$$r(\infty) = \frac{1 + \sqrt{5}}{2}$$

Finally, because

$$\frac{x(n)}{x(n+1)} = \frac{1}{r(n)}$$

then

$$\lim_{n \to \infty} \frac{x(n)}{x(n+1)} = \frac{2}{1 + \sqrt{5}}$$

(c) For the first property, we may simply substitute the closed-form expression for the Fibonacci sequence into the equation, and verify that it is true. For the second property, from the definition of the Fibonacci sequence we have

$$x^2(n+2) = x^2(n) + x^2(n+1) + 2x(n)x(n+1)$$

which we may rewrite as

$$x^2(n+2) - x^2(n+1) = x(n)[x(n) + 2x(n+1)]$$

However, note that

$$x(n+3) = x(n+1) + x(n+2) = x(n) + 2x(n+1)$$

Substituting this into the previous equation, we have the desired property.

4.38 A savings account pays interest at the rate of 5 percent per year with interest compounded monthly.

(a) If $50 is deposited into the account every month for 60 months, find the balance in the account at the end of the 60 months. Assume that the money is deposited on the first day of the month so that, at the end of the month, an entire month's interest has been earned.

(b) If no deposits are made for the next 60 months, find the account balance at the end of the next 60-month period.

(c) Instead of being compounded monthly, suppose that the bank offers to compound the interest daily. Compute the account balance at the end of 60 months and 120 months and compare your balances with those obtained when the interest is compounded monthly.

(a) The savings account balance at the beginning of the nth month is equal to the balance in the previous month plus the amount deposited in the nth month plus the interest earned on the balance from the previous month. Therefore, if $y(n)$ is the balance at the beginning of the nth month,

$$y(n) = y(n-1) + \beta y(n-1) + x(n)$$

where β is the interest earned on the account, and $x(n)$ is the amount deposited into the savings account in the nth month. Taking z-transforms, and solving for $Y(z)$, we have

$$Y(z) = \frac{X(z)}{1 - \nu z^{-1}}$$

where $\nu = 1 + \beta$. With \$50 deposits beginning with month number zero, $x(n) = 50u(n)$, and

$$Y(z) = 50 \frac{1}{(1 - \nu z^{-1})(1 - z^{-1})}$$

Performing a partial fraction expansion of $Y(z)$, we have

$$Y(z) = \frac{50}{1 - \nu}\left[\frac{1}{1 - z^{-1}} - \frac{\nu}{1 - \nu z^{-1}}\right]$$

Taking the inverse z-transform, we have

$$y(n) = \frac{50}{1 - \nu}[1 - \nu^{n+1}]u(n) \tag{4.8}$$

With $\nu = 1 + \beta$, and $\beta = \frac{0.05}{12}$, at the end of 60 months, after earning 1 month's interest, but prior to making the next deposit, the balance is

$$y(60) = \frac{50}{1 - \nu}[1 - \nu^{61}] - 50 = 3{,}412.47$$

(b) With no deposits for the next 60 months, the balance at the end of the first 60 months simply grows as

$$y(n) = y(60) \cdot \nu^{n-60} \qquad n \geq 60$$

Therefore, $$y(120) = 4{,}379.42$$

(c) With the interest compounded daily, let us compute the effective monthly interest rate. Assuming a balance of \$1 at the beginning of the month, the difference equation that describes the daily balance, $w(n)$, is

$$w(n) = w(n-1) + \beta w(n-1) + \delta(n)$$

where $\beta = \frac{0.05}{365}$. Using z-transforms as we did in part (a), the solution to this difference equation is

$$w(n) = \nu^n u(n)$$

where $\nu = 1 + \beta$. Assuming that a month is 30 days long, for 1 month's interest we have

$$w(30) = \nu^{30} = 1.004175$$

Using $\nu = 1.004175$ in Eq. (4.8), we have

$$y(60) = 3{,}465.37$$
$$y(120) = 4{,}449.56$$

4.39 The *deterministic autocorrelation sequence* corresponding to a sequence $x(n)$ is defined as

$$r_x(n) = \sum_{k=-\infty}^{\infty} x(k)x(n+k)$$

(a) Express $r_x(n)$ as the convolution of two sequences, and find the z-transform of $r_x(n)$ in terms of the z-transform of $x(n)$.

(b) If $x(n) = a^n u(n)$, where $|a| < 1$, find the autocorrelation sequence, $r_x(n)$, and its z-transform.

(a) From the definition of the deterministic autocorrelation, we see that $r_x(n)$ is the convolution of $x(n)$ with $x(-n)$,

$$r_x(n) = x(n) * x(-n)$$

Therefore, using the time-reversal property of the z-transform, it follows that

$$R_x(z) = X(z)X(z^{-1})$$

If the region of convergence of $X(z)$ is R_x, the region of convergence of $R_x(z)$ will be the intersection of the regions R_x and $1/R_x$. Therefore, if this intersection is to be nonempty, R_x must include the unit circle.

(b) With $x(n) = a^n u(n)$, the z-transform is

$$X(z) = \frac{1}{1 - az^{-1}} \qquad |z| > |a|$$

and the z-transform of the autocorrelation sequence is

$$R_x(z) = \frac{1}{(1 - az^{-1})(1 - az)} \qquad |a| < |z| < \frac{1}{|a|}$$

The autocorrelation sequence may be found by computing the inverse z-transform of $R_x(z)$. Performing a partial fraction expansion of $R_x(z)$, we have

$$R_x(z) = \frac{1}{(1 - az^{-1})(1 - az)} = \frac{-a^{-1}z^{-1}}{(1 - az^{-1})(1 - a^{-1}z^{-1})} = \frac{1}{1 - a^2}\left[\frac{1}{1 - az^{-1}} - \frac{1}{1 - a^{-1}z^{-1}}\right]$$

Thus, because the region of convergence is $|a| < z < 1/|a|$, the inverse z-transform is

$$r_x(n) = \frac{1}{1 - a^2}[a^n u(n) + a^{-n}u(-n - 1)] = \frac{1}{1 - a^2}a^{|n|}$$

4.40 In many disciplines, differential equations play a major role in characterizing the behavior of various phenomena. Obtaining an approximate solution to a differential equation with the use of a digital computer requires that the differential equation be put into a form that is suitable for digital computation. This problem presents a transformation procedure that will convert a differential equation into a *difference equation*, which may then be solved by a digital computer. Consider a first-order differential equation of the form

$$\frac{d}{dt}y_a(t) + \alpha y_a(t) = x_a(t)$$

where $y_a(0) = y_0$. Because numerical techniques are to be used, we will restrict our attention to investigating $y_a(t)$ at sampling instants nT where T is the sampling period. Evaluating the differential equation at $t = nT$, we have

$$\frac{d}{dt}y_a(nT) + \alpha y_a(nT) = x_a(nT)$$

From calculus we know that the derivative of a function $y_a(t)$ at $t = nT$ is simply the slope of the function at $t = nT$. This slope may be approximated by the relationship

$$\frac{d}{dt}y_a(nT) \approx \frac{1}{T}[y_a(nT) - y_a(nT - T)]$$

(a) Insert this approximation into the sampled differential equation above and find a difference equation that relates $y(n) = y_a(nT)$ and $x(n) = x_a(nT)$, and specify the appropriate initial conditions.

(b) With $x_a(t) = u(t)$ and $y_a(0^-) = 1$, numerically solve the differential equation using the difference equation approximation obtained above.

(c) Compare your approximation to the exact solution.

(a) With

$$\frac{d}{dt}y_a(t) + \alpha y_a(t) = x_a(t) \qquad y_a(0^-) = y_0$$

using the approximation

$$\frac{d}{dt}y_a(nT) \approx \frac{1}{T}\left[y_a(nT) - y_a(nT - T)\right]$$

we have

$$\frac{1}{T}[y_a(nT) - y_a(nT - T)] + \alpha y_a(nT) = x_a(nT) \qquad y_a(0^-) = y_0$$

If we let $y(n) = y_a(nT)$ and $x(n) = x_a(nT)$,

$$\frac{1}{T}[y(n) - y(n - 1)] + \alpha y(n) = x(n) \qquad y(0) = y_0$$

With

$$a = \frac{1}{1 + \alpha T}$$

this becomes

$$y(n) - ay(n - 1) = aT x(n) \qquad y(0) = y_0$$

(b) Using the one-sided z-transform to solve this difference equation, we have

$$Y_1(z) - a[z^{-1}Y_1(z) + y(-1)] = aT X(z)$$

We must now derive the initial condition on $y(n)$ at time $n = -1$ from the initial condition at $n = 0$. From the difference equation, we have

$$y(0) - ay(-1) = aT x(0)$$

With $y(0) = 1$ and $x(0) = 1$, the initial condition becomes

$$y(-1) = \frac{1 - aT}{a}$$

With $x_a(t) = u(t)$ or $x(n) = u(n)$,

$$X(z) = \frac{1}{1 - z^{-1}}$$

Therefore, using the given initial condition, we have

$$Y_1(z) = \frac{aT}{(1 - az^{-1})(1 - z^{-1})} + \frac{1 - aT}{1 - az^{-1}}$$

Performing a partial fraction expansion gives

$$Y_1(z) = \frac{\frac{aT}{1-a}}{1-z^{-1}} - \frac{\frac{a^2T}{1-a}}{1-az^{-1}} + \frac{1-aT}{1-az^{-1}}$$

and we may find $y(n)$ by taking the inverse z-transform:

$$y(n) = \frac{aT}{1-a} - \frac{a^2T}{1-a}a^n + (1-aT)a^n \qquad n \geq 0$$

Because

$$\frac{aT}{1-a} = \frac{T}{1+\alpha T} \cdot \frac{1+\alpha T}{\alpha T} = \frac{1}{\alpha}$$

this may be written as

$$y(n) = \frac{1}{\alpha} + \left(1 - \frac{1}{\alpha}\right)\left(\frac{1}{1+\alpha T}\right)^n \qquad n \geq 0$$

(c) The solution to the differential equation is a sum of two terms. The first is the homogeneous solution, which is

$$y_h(t) = Ae^{-\alpha t}$$

where A is a constant that is selected in order to satisfy the initial condition $y(0^-) = 1$. The second is the particular solution, which is

$$y_p(t) = \frac{1}{\alpha}$$

Thus, the total solution is

$$y_a(t) = \frac{1}{\alpha} + Ae^{-\alpha t}$$

Evaluating this at time $t = 0^-$,

$$y_a(0^-) = \frac{1}{\alpha} + A$$

we see that in order to match the initial conditions, we must have

$$A = 1 - \frac{1}{\alpha}$$

Therefore,

$$y_a(t) = \frac{1}{\alpha} + \left(1 - \frac{1}{\alpha}\right)e^{-\alpha t} \qquad t \geq 0$$

If we compare this to the approximation in part (b), note that if $T \ll 1$,

$$e^{-\alpha t}|_{t=nT} = (e^{\alpha T})^{-n} \approx (1+\alpha T)^{-n}$$

and

$$y_a(nT) = \frac{1}{\alpha} + \left(1 - \frac{1}{\alpha}\right)\left(\frac{1}{1+\alpha T}\right)^n$$

Supplementary Problems

z-Transforms

4.41   Find the z-transform of

$$x(n) = \begin{cases} \left(\frac{1}{2}\right)^{|n|} & |n| < 10 \\ 0 & \text{else} \end{cases}$$

4.42 The z-transform of a sequence $x(n)$ is

$$X(z) = \frac{1 - 4z^{-1} + 2z^{-2}}{1 - 3z^{-1} + 0.5z^{-2}}$$

If the region of convergence includes the unit circle, find the DTFT of $x(n)$ at $\omega = \pi/2$.

4.43 Find the z-transform of each of the following sequences:

(a) $x(n) = (-1)^n u(n)$

(b) $x(n) = \frac{1}{n} u(n-1)$

(c) $x(n) = z \cosh(\alpha n) u(n)$

4.44 Find the z-transform of the sequence

$$y(n) = \begin{cases} \sum_{k=0}^{n}(1+k) & n \ge 0 \\ 0 & n < 0 \end{cases}$$

4.45 Find the z-transform of the sequence

$$x(n) = \binom{N}{n} = \frac{N!}{n!(N-n)!} \qquad 0 \le n \le N$$

4.46 How many different sequences have a z-transform given by

$$H(z) = \frac{1 - 2z^{-1} + 3z^{-2}}{\left(1 - \frac{1}{8}z^{-1} + \frac{1}{4}z^{-2}\right)\left(1 + \frac{1}{3}z^{-1}\right)}$$

4.47 The sequence $y(n)$ is formed from $x(n)$ by

$$y(n) = \sum_{k=0}^{n} kx(k)$$

where $X(z) = \sin z^{-1}$. Find $Y(z)$.

4.48 If $x(n)$ is an absolutely summable sequence with a rational z-transform that has poles at $z = \frac{1}{2}$ and $z = 2$, what can be said about the extent of $x(n)$ (i.e., finite in length, right-sided, etc.)?

Properties

4.49 A right-sided sequence $x(n)$ has a z-transform $X(z)$ given by

$$X(z) = \frac{2z^{-8} + z^{-6} - 2z^{-3} + 4}{z^{-8} + 2z^{-5} - 4z^{-3}}$$

Find the values of $x(n)$ for all $n < 0$.

4.50 Use the z-transform to perform the convolution of the following two sequences:

$$x(n) = \delta(n) - 2\delta(n-2)$$
$$h(n) = 2\delta(n) - 2\delta(n-1) + 3\delta(n-2) + \delta(n-3)$$

4.51 Evaluate the following summation:

$$\sum_{n=0}^{\infty} \left(\tfrac{1}{4}\right)^n \cos\left(\frac{2n\pi}{8}\right)$$

4.52   Find the value of $x(0)$ for the sequence that has a z-transform

$$X(z) = \frac{z + 3 + z^{-1}}{\left(1 - \frac{1}{4}z^{-1}\right)\left(1 - \frac{1}{3}z^{-2}\right)} \qquad |z| > \frac{1}{3}$$

4.53   A right-sided sequence has a z-transform

$$X(z) = \frac{z^{-7} + 3z^{-4} - 2z^{-3} + 1}{3z^{-9} + 8z^{-7} - 4z^{-2}}$$

Find the index and the value of the first nonzero value of $x(n)$.

Inverse z-Transforms

4.54   Find the inverse z-transform of

(a) $X(z) = \dfrac{1}{(1 - 0.4z^{-1})^2} \qquad |z| > 0.4$

(b) $X(z) = \dfrac{1}{1 - 0.2z^{-2}} \qquad |z| > 0.2$

4.55   Find the inverse z-transform of $X(z) = \cos z^{-1}$. Assume that the ROC includes the unit circle, $|z| = 1$.

4.56   Find the inverse z-transform of $X(z) = e^z$. Assume that the ROC includes the unit circle, $|z| = 1$.

4.57   Find the inverse z-transform of

$$X(z) = \frac{z^5 - 3}{1 - z^{-5}} \qquad |z| > 1$$

4.58   Find the inverse z-transform of

$$X(z) = \frac{1 + \frac{1}{4}z^{-1}}{\left(1 - \frac{1}{2}z^{-1}\right)^2} \qquad |z| < \frac{1}{2}$$

4.59   If

$$X(z) = \frac{1}{z - 2} \qquad |z| < 2$$

find the values of $x(n)$ at $n = -2$ and $n = -1$ using contour integration.

4.60   Use the residue theorem to find the value of $x(n)$ at $n = 10$ when

$$X(z) = \frac{1 - 3z^{-4}}{(1 - 0.2z^{-1})(1 + 0.6z^{-1})} \qquad |z| > 0.6$$

4.61   Find the inverse z-transform of

$$X(z) = \log(1 - 2z) \qquad |z| < \frac{1}{2}$$

One-Sided z-Transforms

4.62   Find the one-sided z-transform of the sequences $x(n) = \left(\frac{1}{4}\right)^{|n|}$.

4.63   Let $X_1(z)$ be the one-sided z-transform of $x(n)$.

(a) Find the one-sided z-transform of $y(n) = x(n - 1)$.

(b) Find the one-sided z-transform of $y(n) = x(n + 3)$.

4.64   Find the solution to the following linear constant coefficient difference equations:

   (a) $y(n) = \frac{1}{2}y(n-1) + x(n)$ with $x(n) = u(n)$ and $y(-1) = \frac{1}{4}$.

   (b) $y(n) = y(n-1) - y(n-2) + 2u(n)$ with $y(-1) = 2$ and $y(-2) = 1$.

   (c) $y(n) + y(n-2) = \delta(n)$ with $y(-1) = 1$ and $y(-2) = 0$.

4.65   The sequence $y(n)$ is the solution to the LCCDE

$$y(n) - \frac{3}{2}y(n-1) + \frac{1}{2}y(n-2) = x(n) \qquad n \geq 0$$

with $x(n) = \delta(n)$. Find a set of initial conditions on $y(n)$ for $n < 0$ so that $y(n) = 1$ for $n \geq 0$.

4.66   Consider the following difference equation:

$$y(n) + y(n-2) = x(n) + x(n-1)$$

If $x(n) = 10u(n)$ and $y(-2) = -10$ and $y(-1) = 0$, find the output sequence $y(n)$ for $n \geq 0$.

Applications

4.67   Determine the number of years that are required for an investment of money in a savings account to double if the money is compounded monthly at an annual rate of (a) 5 percent and (b) 10 percent.

4.68   Suppose that $x(n)$ has a z-transform

$$X(z) = \frac{1}{1 - az^{-1}} - \frac{1}{1 - bz^{-1}}$$

with $|a| < 1$ and $|b| < 1$ and a region of convergence that includes the unit circle. (a) Find the deterministic autocorrelation sequence $r_x(n)$. (b) Find another sequence that has the same autocorrelation.

Answers to Supplimentary Problems

4.41 $\dfrac{\frac{3}{4} - \left(\frac{1}{2}z^{-1}\right)^{10}\left(1 - \frac{1}{2}z\right) - \left(\frac{1}{2}z\right)^{10}\left(1 - \frac{1}{2}z^{-1}\right)}{\left(1 - \frac{1}{2}z^{-1}\right)\left(1 - \frac{1}{2}z\right)}$.

4.42 $\dfrac{-1 + 4j}{\frac{1}{2} + 3j}$.

4.43   (a) $\dfrac{1}{1 + z^{-1}}$, $|z| > 1$.

   (b) $-\log(1 - z^{-1})$, $|z| > 1$.

   (c) $\dfrac{1}{1 - e^{\alpha}z^{-1}} + \dfrac{1}{1 - e^{-\alpha}z^{-1}}$ $|z| > \max(e^{\alpha}, e^{-\alpha})$.

4.44 $Y(z) = \dfrac{1}{(1 - z^{-1})^3}$.

4.45 $X(z) = (1 + z^{-1})^N$.

4.46   Three.

4.47 $\dfrac{\cos z^{-1}}{z - 1}$

4.48   Two-sided.

4.49 $x(-3) = -1$ is the only nonzero value for $n < 0$.

4.50 $x(n) * h(n) = 2\delta(n) - 2\delta(n-1) - \delta(n-2) + 5\delta(n-3) - 6\delta(n-4) - 2\delta(n-5)$.

4.51 $\dfrac{16 - 2\sqrt{2}}{17 - 4\sqrt{2}}$

4.52 $x(0) = \frac{13}{4}$.

4.53 $x(-2) = -\frac{1}{4}$.

4.54   (a) $(n+1)(0.4)^n u(n)$.

      (b) $x(n) = \begin{cases} (0.2)^{n/2} & n \text{ even} \\ 0 & \text{else} \end{cases}$

4.55 $\displaystyle\sum_{k=0}^{\infty} \dfrac{1}{(2k)!} (-1)^k \delta(n - 2k)$.

4.56 $x(n) = \dfrac{1}{(-n)!} u(-n)$.

4.57 $x(n) = \delta(n+5) - 2 \displaystyle\sum_{k=0}^{\infty} \delta(n - 5k)$.

4.58 $-n\left(\frac{1}{2}\right)^{n+1} u(-n-1) - (n+1)\left(\frac{1}{2}\right)^n u(-n-2) = \delta(n+1) - (3n+2)\left(\frac{1}{2}\right)^{n+1} u(-n-2)$.

4.59 $x(-1) = -\frac{1}{4}$ and $x(-2) = -\frac{1}{8}$.

4.60 $\frac{3}{4}(-0.6)^6[(-0.6)^4 - 3] + \frac{1}{4}(0.2)^6[(0.2)^4 - 3]$.

4.61 $n^{-1}2^{-n} u(-n-1)$.

4.62 $X_1(z) = \dfrac{1}{1 - \frac{1}{4}z^{-1}}$.

4.63   (a) $z^{-1} X_1(z) + x(-1)$.

      (b) $[X_1(z) - x(0) - x(1)z^{-1} - x(2)z^{-2}]z^3$.

4.64   (a) $\left[2 - \frac{7}{8}\left(\frac{1}{2}\right)^n\right] u(n)$.

      (b) $\left[2 + \frac{2}{\sqrt{3}} \sin(n+1)\pi/3\right] u(n)$.

      (c) $[\cos(n\pi/2) - \sin(n\pi/2)] u(n)$.

4.65 $y(-1) = 1$ and $y(-2) = 3$.

4.66 $y(n) = 10 + 10\sqrt{2}\cos(n\pi/2 - \pi/4), \qquad n \geq 0$.

4.67   (a) 167 months. (b) 84 months.

4.68   (a) $r_x(n) = \dfrac{1}{1 - a^2} a^{|n|} + \dfrac{1}{1 - b^2} b^{|n|} - \dfrac{1}{1 - ab}[(a^n + b^n)u(n) + (a^{-n} + b^{-n})u(-n-1)]$.

      (b) $x'(n) = a^n u(n) - b^{-n} u(-n-1)$.

Chapter 5

Transform Analysis of Systems

5.1 INTRODUCTION

Given a linear shift-invariant system with a unit sample response $h(n)$, the input and output are related by a convolution sum

$$y(n) = h(n) * x(n) = \sum_{k=-\infty}^{\infty} h(k)x(n-k)$$

As discussed in Chap. 2, this relationship implies that $Y(e^{j\omega}) = X(e^{j\omega})H(e^{j\omega})$ where $H(e^{j\omega})$, the frequency response of the system, is the discrete-time Fourier transform of $h(n)$. This relationship between $x(n)$ and $y(n)$ may also be expressed in the z-transform domain as

$$Y(z) = X(z)H(z)$$

where $H(z)$, the z-transform of $h(n)$, is the *system function* of the LSI system. The system function is very useful in the description and analysis of LSI systems. In this chapter, we look at the characterization of a linear shift-invariant system in terms of its system function and discuss special types of LSI systems such as linear phase systems, allpass systems, minimum phase systems, and feedback networks.

5.2 SYSTEM FUNCTION

The frequency response of a linear shift-invariant system is the discrete-time Fourier transform of the unit sample response, and the *system function* is the z-transform of the unit sample response:

$$H(z) = \sum_{n=-\infty}^{\infty} h(n)z^{-n} \tag{5.1}$$

The frequency response may be derived from the system function by evaluating $H(z)$ around the unit circle:

$$H(e^{j\omega}) = H(z)\big|_{z=e^{j\omega}}$$

If the z-transform of the input to a linear shift-invariant system with a system function $H(z)$ is $X(z)$, the z-transform of the output is

$$Y(z) = H(z)X(z)$$

For linear shift-invariant systems that are described by a linear constant coefficient difference equation,

$$y(n) + \sum_{k=1}^{p} a(k)y(n-k) = \sum_{k=0}^{q} b(k)x(n-k)$$

the system function is a rational function of z:

$$H(z) = \frac{\displaystyle\sum_{k=0}^{q} b(k)z^{-k}}{1 + \displaystyle\sum_{k=1}^{p} a(k)z^{-k}} = A\frac{\displaystyle\prod_{k=1}^{q}(1 - \beta_k z^{-1})}{\displaystyle\prod_{k=1}^{p}(1 - \alpha_k z^{-1})} \tag{5.2}$$

183

Therefore, the system function is defined, to within a scale factor, by the location of its poles, α_k, and zeros, β_k. Note that each term in the numerator

$$1 - \beta_k z^{-1} = \frac{z - \beta_k}{z}$$

contributes a zero to the system function at $z = \beta_k$ and a pole to the system function at $z = 0$. Similarly, each term in the denominator contributes a pole at $z = \alpha_k$ and a zero at $z = 0$. Therefore, including the poles and zeros that may lie at $z = 0$ or $z = \infty$, the number of zeros in $H(z)$ is equal to the number of poles.

If the unit sample response is real-valued, $H(z)$ is a conjugate symmetric function of z,

$$H(z) = H^*(z^*)$$

and the complex poles and zeros occur in conjugate symmetric pairs (i.e., if there is a complex pole (zero) at $z = z_0$, there is also a complex pole (zero) at $z = z_0^*$).

5.2.1 Stability and Causality

Stability and causality impose some constraints on the system function of a linear shift-invariant system.

Stability

The unit sample response of a stable system must be absolutely summable:

$$\sum_{n=-\infty}^{\infty} |h(n)| < \infty$$

Note that because this is equivalent to the condition that

$$\sum_{n=-\infty}^{\infty} |h(n)| z^{-n} < \infty$$

for $|z| = 1$, the region of convergence of the system function must include the unit circle if the system is stable.

Causality

Because the unit sample response of a causal system is right-sided, $h(n) = 0$ for $n < 0$, the region of convergence of $H(z)$ will be the exterior of a circle, $|z| > \alpha$. Because no poles may lie within the region of convergence, all of the poles of $H(z)$ must lie on or inside the circle $|z| \leq \alpha$.

Causality imposes some tight constraints on a linear shift-invariant system. The first of these is the Paley-Wiener theorem.

> **Paley-Wiener Theorem:** If $h(n)$ has finite energy and $h(n) = 0$ for $n < 0$,
>
> $$\int_{-\pi}^{\pi} \left| \ln|H(e^{j\omega})| \right| d\omega < \infty$$

One of the consequences of this theorem is that the frequency response of a stable and causal system cannot be zero over any finite band of frequencies. Therefore, any stable ideal frequency selective filter will be noncausal.

Causality also places restrictions on the real and imaginary parts of the frequency response. For example, if $h(n)$ is real, $h(n)$ may be decomposed into its even and odd parts as follows:

$$h(n) = h_e(n) + h_o(n)$$

where

$$h_e(n) = \tfrac{1}{2}[h(n) + h(-n)]$$

and
$$h_o(n) = \tfrac{1}{2}[h(n) - h(-n)]$$

If $h(n)$ is causal, it is uniquely defined by its even part:

$$h(n) = 2h_e(n)u(n) - h_e(n)\delta(n)$$

If $h(n)$ is absolutely summable, the DTFT of $h(n)$ exists, and $H(e^{j\omega})$ may be written in terms of its real and imaginary parts as follows:

$$H(e^{j\omega}) = H_R(e^{j\omega}) + jH_I(e^{j\omega})$$

Therefore, because $H_R(e^{j\omega})$ is the DTFT of the even part of $h(n)$, it follows that if $h(n)$ is real, stable, and causal, $H(e^{j\omega})$ is uniquely defined by its real part. This implies a relationship between the real and imaginary parts of $H(e^{j\omega})$, which is given by

$$H_I(e^{j\omega}) = -\frac{1}{2\pi} \int_{-\pi}^{\pi} H_R(e^{j\theta}) \cot\left(\frac{\omega - \theta}{2}\right) d\theta \qquad (5.3)$$

This integral is called a *discrete Hilbert transform*. Specifically, $H_I(e^{j\omega})$ is the discrete Hilbert transform of $H_R(e^{j\omega})$.

Realizable Systems

A realizable system is one that is both stable and causal. A realizable system will have a system function with a region of convergence of the form $|z| > \alpha$ where $0 \le \alpha < 1$. Therefore, any poles of $H(z)$ must lie *inside* the unit circle. For example, the first-order system

$$H(z) = \frac{b(0)}{1 + a(1)z^{-1}} \qquad |z| > |a(1)|$$

will be realizable (stable and causal) if and only if

$$|a(1)| < 1$$

For the second-order system,

$$H(z) = \frac{b(0)}{1 + a(1)z^{-1} + a(2)z^{-2}}$$

$H(z)$ has two zeros at the origin and poles at

$$\alpha_1, \alpha_2 = -\frac{a(1)}{2} \pm \sqrt{\frac{a^2(1) - 4a(2)}{4}}$$

These roots satisfy the following two equations:

$$a(1) = -(\alpha_1 + \alpha_2)$$
$$a(2) = \alpha_1 \cdot \alpha_2$$

From these equations, it follows that the roots of $H(z)$ will be inside the unit circle if and only if (see Prob. 8.29)

$$|a(2)| < 1$$
$$|a(1)| < 1 + a(2)$$

These constraints define a *stability triangle* in the coefficient plane as shown in Fig. 5-1. Thus, a causal second-order system will be stable if and only if the coefficients $a(1)$ and $a(2)$ lie inside this triangle. This result is of special interest, because second-order systems are the basic building blocks for higher-order systems. If the coefficients lie in the shaded region above the parabola defined by the equation

$$a^2(1) - 4a(2) = 0$$

the roots are complex; otherwise they are real.

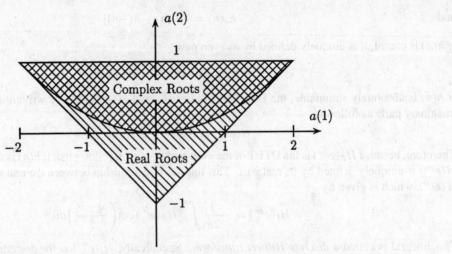

Fig. 5-1. The stability triangle, which is defined by the lines $|a(2)| < 1$ and $|a(1)| < 1 + a(2)$. The shaded region above the parabola $a^2(1) - 4a(2) = 0$ contains the values of $a(1)$ and $a(2)$ that correspond to complex roots.

5.2.2 Inverse Systems

For a linear shift-invariant system with a system function $H(z)$, the *inverse system* is defined to be the system that has a system function $G(z)$ such that

$$H(z) \cdot G(z) = 1$$

In other words, the cascade of $H(z)$ with $G(z)$ produces the identity system. In terms of $H(z)$, the inverse is simply

$$G(z) = \frac{1}{H(z)}$$

For example, if $H(z)$ is a rational function of z as given in Eq. (5.2), the inverse system is

$$G(z) = A^{-1} \frac{\prod_{k=1}^{p}(1 - \alpha_k z^{-1})}{\prod_{k=1}^{q}(1 - \beta_k z^{-1})}$$

Thus, the poles of $H(z)$ become the zeros of $G(z)$, and the zeros of $H(z)$ become the poles of $G(z)$. The region of convergence that is associated with the inverse system is determined by the requirement that $H(z)$ and $G(z)$ have overlapping regions of convergence.[1]

EXAMPLE 5.2.1 If

$$H(z) = \frac{1 - 0.5z^{-1}}{1 - 0.8z^{-1}} \qquad |z| > 0.8$$

the inverse system is

$$G(z) = \frac{1 - 0.8z^{-1}}{1 - 0.5z^{-1}}$$

There are two possible regions of convergence for $g(n)$. The first is $|z| > \frac{1}{2}$, and the second is $|z| < \frac{1}{2}$. Because $|z| < \frac{1}{2}$ does not overlap the region of convergence for $H(z)$, the only possibility for the inverse system is $|z| > \frac{1}{2}$. In this case, the unit sample response is

$$g(n) = \left(\tfrac{1}{2}\right)^n u(n) - 0.8\left(\tfrac{1}{2}\right)^{n-1} u(n-1)$$

---

[1]If this were not the case, $H(z)G(z)$ would not be the identity system, because the region of convergence would be empty.

which is stable and causal. However, suppose that

$$H(z) = \frac{0.5 - z^{-1}}{1 - 0.8z^{-1}} \qquad |z| > 0.8$$

In this case, the inverse system is

$$G(z) = \frac{1 - 0.8z^{-1}}{0.5 - z^{-1}} = 2\frac{1 - 0.8z^{-1}}{1 - 2z^{-1}}$$

where the region of convergence may be either $|z| > 2$ or $|z| < 2$. Because both regions of convergence overlap the region of convergence of $H(z)$, both are valid inverse systems. The first, which has a region of convergence $|z| > 2$, has a unit sample response

$$g(n) = 2(2)^n u(n) - 1.6(2)^{n-1} u(n-1)$$

and is causal but unstable. The second, with a region of convergence $|z| < 2$, has a unit sample response

$$g(n) = -2(2)^n u(-n-1) + 1.6(2)^{n-1} u(-n)$$

and is stable but noncausal.

5.2.3   Unit Sample Response for Rational System Functions

A linear shift-invariant system with a rational system function may be written in factored form as follows:

$$H(z) = A\frac{\displaystyle\prod_{k=1}^{q}(1 - \beta_k z^{-1})}{\displaystyle\prod_{k=1}^{p}(1 - \alpha_k z^{-1})} \tag{5.4}$$

Assuming only first-order poles, with $\alpha_k \neq \beta_l$ for all k and l, if $p > q$, $H(z)$ may be expanded using a partial fraction expansion as follows:

$$H(z) = \sum_{k=1}^{p} \frac{A_k}{1 - \alpha_k z^{-1}}$$

If the system is causal, the unit sample response is

$$h(n) = \sum_{k=1}^{p} A_k \alpha_k^n u(n) \tag{5.5}$$

When $p \leq q$, the partial fraction expansion has the form

$$H(z) = \sum_{k=0}^{q-p} B_k z^{-k} + \sum_{k=1}^{p} \frac{A_k}{1 - \alpha_k z^{-1}}$$

and, if the system is causal, the unit sample response becomes

$$h(n) = \sum_{k=0}^{q-p} B_k \delta(n-k) + \sum_{k=1}^{p} A_k \alpha_k^n u(n)$$

If $p = 0$, $H(z)$ has only zeros,

$$H(z) = \sum_{k=0}^{q} b(k) z^{-k}$$

and $h(n)$ is finite in length with

$$h(n) = \sum_{k=0}^{q} b(k)\delta(n-k) \tag{5.6}$$

These systems are called finite-length impulse response (FIR) filters. If $p > 0$, $H(z)$ is infinite in length, and these systems are called infinite-length impulse response (IIR) filters.

If $h(n)$ is real, $H(z) = H^*(z^*)$, and the complex poles and zeros of $H(z)$ occur in complex-conjugate pairs. For example, if $\alpha_k = r_k e^{j\omega_k}$ is a complex-valued pole, $\alpha_k^* = r_k e^{-j\omega_k}$ will also be a pole. This symmetry implies that the complex terms in Eq. (5.5) may be combined to form terms of the form

$$C_k r_k^n \cos(n\omega_k + \phi_k)$$

5.2.4 Frequency Response for Rational System Functions

The frequency response of a linear shift-invariant system may be found from the system function by evaluating $H(z)$ on the unit circle. For a rational function of z, the frequency response may be found geometrically from the poles and zeros of $H(z)$. With $H(z)$ written in factored form as in Eq. (5.4), the frequency response is

$$H(e^{j\omega}) = A \frac{\displaystyle\prod_{k=1}^{q}(1 - \beta_k e^{-j\omega})}{\displaystyle\prod_{k=1}^{p}(1 - \alpha_k e^{-j\omega})} \tag{5.7}$$

Because the magnitude of the frequency response is

$$|H(e^{j\omega})| = |A| \frac{\displaystyle\prod_{k=1}^{q}|1 - \beta_k e^{-j\omega}|}{\displaystyle\prod_{k=1}^{p}|1 - \alpha_k e^{-j\omega}|}$$

$|H(e^{j\omega})|$ is $|A|$ times the product of the terms $|1 - \beta_k e^{-j\omega}|$ divided by the product of the terms $|1 - \alpha_k e^{-j\omega}|$. Each term in the numerator

$$|1 - \beta_k e^{-j\omega}| = |e^{j\omega} - \beta_k|$$

is the length of the vector from the zero at $z = \beta_k$ to the unit circle at $z = e^{j\omega}$ (labeled $\mathbf{v}_1$ in Fig. 5-2). Similarly, each term in the denominator

$$|1 - \alpha_k e^{-j\omega}| = |e^{j\omega} - \alpha_k|$$

is the length of the vector from the pole at $z = \alpha_k$ to the unit circle at $z = e^{j\omega}$ (labeled $\mathbf{v}_2$ in Fig. 5-2). When a pole is close to the unit circle, $\alpha_k = r_k e^{j\omega_k}$ with $r_k \approx 1$, the magnitude of the frequency response becomes large for $\omega \approx \omega_k$ because the length of the vector from the pole to the unit circle becomes small. Similarly, if there is a zero close to the unit circle, $\beta_k = r_k e^{j\omega_k}$ with $r_k \approx 1$, the magnitude of the frequency response becomes small for $\omega \approx \omega_k$ (if the zero is *on* the unit circle at $z = e^{j\omega_k}$, $H(e^{j\omega_k}) = 0$).

The analysis for the phase is similar. Assuming that A is a positive real number, the phase corresponding to the frequency response $H(e^{j\omega})$ given by Eq. (5.7) is

$$\phi_h(\omega) = \sum_{k=1}^{q} \arg(1 - \beta_k e^{-j\omega}) - \sum_{k=1}^{p} \arg(1 - \alpha_k e^{-j\omega})$$

Thus, $\phi_h(\omega)$ is the sum of the phases associated with the terms $(1 - \beta_k e^{-j\omega})$, minus the sum of the phases of the terms $(1 - \alpha_k e^{-j\omega})$. Because

$$1 - \beta_k e^{-j\omega} = e^{-j\omega}(e^{j\omega} - \beta_k)$$

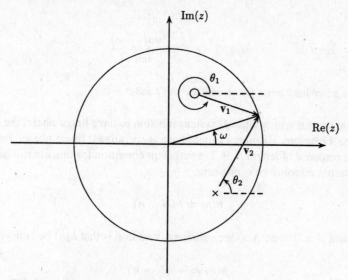

Fig. 5-2. Evaluating the frequency response geometrically from the
poles and zeros of the system function.

then
$$\arg(1 - \beta_k e^{-j\omega}) = \arg(e^{j\omega} - \beta_k) - \omega = \theta_1 - \omega$$

where θ_1 is the angle subtended by the vector from the zero at $z = \beta_k$ to the unit circle at $z = e^{j\omega}$ (see Fig. 5-2).
Similarly, for each term in the denominator

$$\arg(1 - \alpha_k e^{-j\omega}) = \theta_2 - \omega$$

where θ_2 is the angle of the vector from the pole at $z = \alpha_k$ to the unit circle at $z = e^{j\omega}$. When a pole (zero)
is close to the unit circle, the phase decreases (increases) rapidly as we move past the pole (zero). Because the
group delay is the negative of the derivative of the phase, this implies that the group delay is large and positive
close to a pole and large and negative when close to a zero.

5.3 SYSTEMS WITH LINEAR PHASE

A linear shift-invariant system is said to have *linear phase* if the frequency response has the form

$$H(e^{j\omega}) = |H(e^{j\omega})|e^{-j\alpha\omega}$$

where α is a real number. Thus, linear phase systems have a constant group delay,

$$\tau_h(\omega) = \alpha$$

In some applications, one is interested in designing systems that have what is referred to as *generalized linear
phase*. A system is said to have generalized linear phase if the frequency response has the form

$$H(e^{j\omega}) = A(e^{j\omega})e^{-j(\alpha\omega - \beta)} \tag{5.8}$$

where $A(e^{j\omega})$ is a real-valued (possibly bipolar) function of ω, and β is a constant. Often, the term *linear phase*
is used to denote a system that has either linear or generalized linear phase.

EXAMPLE 5.3.1   Consider the FIR system with a unit sample response

$$h(n) = \begin{cases} 1 & n = 0, 1, \ldots, N \\ 0 & \text{else} \end{cases}$$

The frequency response is

$$H(e^{j\omega}) = e^{-jN\omega/2}\, \frac{\sin\!\left(\frac{N+1}{2}\omega\right)}{\sin\!\left(\frac{\omega}{2}\right)}$$

Therefore, this system has generalized linear phase, with $\alpha = N/2$ and $\beta = 0$.

In order for a causal system with a rational system function to have linear phase, the unit sample response must be finite in length. Therefore, IIR filters cannot have (generalized) linear phase. For an FIR filter with a real-valued unit sample response of length $N + 1$, a sufficient condition for this filter to have generalized linear phase is that the unit sample response be symmetric,

$$h(n) = h(N - n)$$

In this case, $\alpha = N/2$ and $\beta = 0$ or π. Another sufficient condition is that $h(n)$ be antisymmetric,

$$h(n) = -h(N - n)$$

which corresponds to the case in which $\alpha = N/2$ and $\beta = \pi/2$ or $3\pi/2$.

Linear phase filters may be classified into four types, depending upon whether $h(n)$ is symmetric or antisymmetric and whether N is even or odd. Each of these filters has specific constraints on the locations of the zeros in $H(z)$ which, in turn, place constraints on the frequency response magnitude. For each of these types, which are described below, it is assumed that $h(n)$ is real-valued, and that $h(0)$ is the first nonzero value of $h(n)$.

Type I Linear Phase Filters

A type I linear phase filter has a symmetric unit sample response,

$$h(n) = h(N - n) \qquad 0 \le n \le N$$

and N is even. The center of symmetry is about the point $\alpha = N/2$, which is an integer, as illustrated in Fig. 5-3(a).

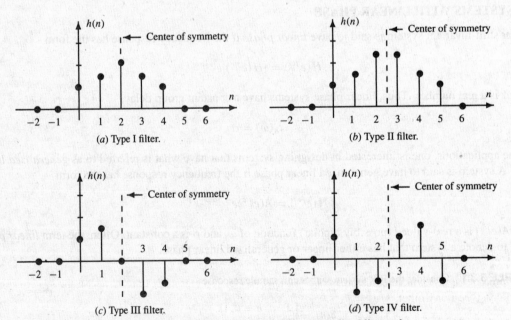

Fig. 5-3. Symmetries in the unit sample response for generalized linear phase systems.

The frequency response of a type I linear phase filter may be expressed in the form

$$H(e^{j\omega}) = e^{-jN\omega/2} \sum_{k=0}^{N/2} a(k) \cos(k\omega)$$ (5.9)

where

$$a(k) = 2h\left(\frac{N}{2} - k\right) \qquad k = 1, 2, \ldots, \frac{N}{2}$$

$$a(0) = h\left(\frac{N}{2}\right)$$

Type II Linear Phase Filters

A type II linear phase filter has a symmetric unit sample response, and N is odd. Therefore, the center of symmetry of $h(n)$ occurs at the half-integer value $\alpha = N/2$, as illustrated in Fig. 5-3(b). The frequency response of a type II linear phase filter may be written as

$$H(e^{j\omega}) = e^{-jN\omega/2} \sum_{k=1}^{(N+1)/2} b(k) \cos\left[\left(k - \tfrac{1}{2}\right)\omega\right]$$ (5.10)

where

$$b(k) = 2h\left(\frac{N+1}{2} - k\right) \qquad k = 1, 2, \ldots, \frac{(N+1)}{2}$$

Type III Linear Phase Filters

A type III linear phase filter has a unit sample response that is antisymmetric,

$$h(n) = -h(N - n)$$

and N is even. Therefore, $h(n)$ is antisymmetric about $\alpha = N/2$, which is an integer, as illustrated in Fig. 5-3(c). The frequency response of a type III linear phase filter may be written as

$$H(e^{j\omega}) = je^{-jN\omega/2} \sum_{k=1}^{N/2} c(k) \sin(k\omega)$$ (5.11)

where

$$c(k) = 2h\left(\frac{N}{2} - k\right) \qquad k = 1, 2, \ldots, \frac{N}{2}$$

Type IV Linear Phase Filters

A type IV linear phase filter has a unit sample response that is antisymmetric, and N is odd. Therefore, $h(n)$ is antisymmetric about the half-integer value $\alpha = N/2$, and the frequency response has the form

$$H(e^{j\omega}) = je^{-jN\omega/2} \sum_{k=1}^{(N+1)/2} d(k) \sin\left[\left(k - \tfrac{1}{2}\right)\omega\right]$$ (5.12)

where

$$d(k) = 2h\left(\frac{N+1}{2} - k\right) \qquad k = 1, 2, \ldots, \frac{N}{2}$$

The z-Transform of Linear Phase Systems

The symmetries in the unit sample response of a linear phase system impose constraints on the system function $H(z)$. For a type I or II filter, $h(n) = h(N - n)$, which implies that

$$H(z) = z^{-N} H(z^{-1})$$

Similarly, for a type III or IV linear phase filter, $h(n) = -h(N - n)$, which implies that

$$H(z) = -z^{-N} H(z^{-1})$$

In both cases, if $H(z)$ is equal to zero at $z = z_0$, $H(z)$ must also be zero at $z = 1/z_0$. Therefore, the zeros of $H(z)$ occur in reciprocal pairs. In addition, with $h(n)$ being real-valued, complex zeros occur in conjugate reciprocal pairs. Thus, the constraints on the zeros of a linear phase filter are as follows. First, $H(z)$ may have one or more zeros at $z = \pm 1$. Second, $H(z)$ may have complex-conjugate zeros on the unit circle at $z = e^{\pm j\omega_k}$ or reciprocal zeros on the real axis at $z = \alpha$ and $z = 1/\alpha$. Finally, $H(z)$ may have groups of four zeros in conjugate reciprocal pairs at $z = r_k e^{\pm j\omega_k}$ and $z = \frac{1}{r_k} e^{\pm j\omega_k}$. These constraints are illustrated in Fig. 5-4.

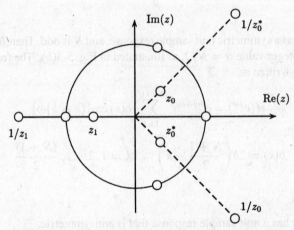

Fig. 5-4. Constraints on the zeros of the system function of an FIR system with generalized linear phase and a real unit sample response. Types III and IV filters must have a zero at $z = 1$, whereas types II and III filters must have a zero at $z = -1$.

The cases of $z = 1$ and $z = -1$ deserve special attention. Evaluating the system function at $z = -1$ for a type II filter, we have

$$H(-1) = (-1)^N H(-1)$$

Because N is odd, this implies that

$$H(-1) = -H(-1)$$

which will be true only if $H(-1) = 0$. Therefore, a type II linear phase filter *must* have a zero at $z = -1$. Similarly, evaluating $H(z)$ at $z = -1$ for a type III filter, we have

$$H(-1) = -(-1)^N H(-1)$$

which, because N is even, requires that there be a zero at $z = -1$. Because the system function evaluated at $z = -1$ is equal to the frequency response at $\omega = \pi$,

$$H(e^{j\omega})|_{\omega=\pi} = 0 \qquad \text{Types II and III filters} \qquad (5.13)$$

For types III and IV filters, evaluating the system function at $z = 1$, we find

$$H(1) = -H(1)$$

which will be true only if $H(z)$ is zero at $z = 1$. Therefore, types III and IV linear phase filters must have a zero at $z = 1$, which implies that

$$H(e^{j\omega})|_{\omega=0} = 0 \qquad \text{Types III and IV filters} \qquad (5.14)$$

5.4 ALLPASS FILTERS

An allpass filter has a frequency response with a constant magnitude,

$$|H(e^{j\omega})| = 1$$

This unit magnitude constraint constrains the poles and zeros of a rational system function to occur in conjugate reciprocal pairs:

$$H(z) = \prod_{k=1}^{N} \frac{z^{-1} - a_k^*}{1 - a_k z^{-1}} \tag{5.15}$$

Thus, if $H(z)$ has a pole at $z = a_k$, $H(z)$ must have a zero at the conjugate reciprocal location $z = 1/a_k^*$. If $h(n)$ is real-valued, the complex roots in Eq. (5.15) occur in conjugate pairs, and if these conjugate pairs are combined to form second-order factors, the system function may be written as

$$H(z) = \prod_{k=1}^{N_1} \frac{z^{-1} - b_k}{1 - b_k z^{-1}} \prod_{k=1}^{N_2} \frac{d_k - c_k z^{-1} + z^{-2}}{1 - c_k z^{-1} + d_k z^{-2}}$$

where the coefficients b_k, c_k, and d_k are real. If an allpass filter $H(z)$ is stable and causal, the poles of $H(z)$ lie inside the unit circle, $|a_k| < 1$. Figure 5-5 shows a typical pole-zero plot for an allpass filter. Allpass filters are useful for *group delay equalization* to compensate for phase nonlinearities.

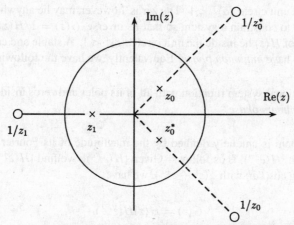

Fig. 5-5.   Illustration of the conjugate reciprocal symmetry constraint that is placed on the poles and zeros of an allpass system.

A stable allpass filter has a group delay that is nonnegative for all ω. This follows from the fact that, for a first-order allpass factor of the form

$$H(z) = \frac{z^{-1} - \alpha^*}{1 - \alpha z^{-1}} \tag{5.16}$$

where $\alpha = re^{j\theta}$, the group delay is

$$\tau(\omega) = \frac{1 - r^2}{|1 - re^{j\theta}e^{-j\omega}|^2}$$

Therefore, with $0 \le r < 1$, it follows that $\tau(\omega) > 0$. Because a general allpass filter has a group delay that is a sum of terms of this form, the group delay of a rational, stable, and causal allpass filter is nonnegative.

A filter may be cascaded with an allpass filter without changing the magnitude of the frequency response. If the pole of the allpass filter cancels a zero, the zero is replaced with one at the conjugate reciprocal location.

Thus, *flipping* one or more zeros of the system function about the unit circle does not change the magnitude of the frequency response.

EXAMPLE 5.4.1 For a filter that has a system function

$$H(z) = \frac{1 - 0.2z^{-1}}{1 - 0.5z^{-1}}$$

the magnitude of the frequency response will not be changed if it is cascaded with the allpass filter

$$H_{ap}(z) = \frac{z^{-1} - 0.2}{1 - 0.2z^{-1}}$$

This allpass filter flips the zero at $z = 0.2$ in $H(z)$ to its reciprocal location, $z = 5$, and the new filter has a system function

$$G(z) = \frac{z^{-1} - 0.2}{1 - 0.5z^{-1}}$$

5.5   MINIMUM PHASE SYSTEMS

A stable and causal linear shift-invariant system with a rational system function of the form given in Eq. (5.2) has all of its poles inside the unit circle, $|\alpha_k| < 1$. The zeros, however, may lie anywhere in the z-plane. In some applications, it is necessary to constrain a system so that its inverse, $G(z) = 1/H(z)$, is also stable and causal. This requires that the zeros of $H(z)$ lie inside the unit circle, $|\beta_k| < 1$. A stable and causal filter that has a stable and causal inverse is said to have *minimum phase*. Equivalently, we have the following definition:

> **Definition:**   A rational system function with all of its poles and zeros inside the unit circle is said to be have *minimum phase*.

A minimum phase system is uniquely defined by the magnitude of its Fourier transform, $|H(e^{j\omega})|$. The procedure to find $H(z)$ from $|H(e^{j\omega})|$ is as follows. Given $|H(e^{j\omega})|$, we find $|H(e^{j\omega})|^2$, which is a function of $\cos(k\omega)$. Then, by replacing $\cos(k\omega)$ with $\frac{1}{2}(z^k + z^{-k})$, we have

$$G(z) = H(z)H(z^{-1})$$

Finally, the minimum phase system is then formed from the poles and zeros of $G(z)$ that are *inside* the unit circle.

EXAMPLE 5.5.1 Let $H(z)$ be a minimum phase system with a Fourier transform magnitude

$$|H(e^{j\omega})|^2 = \tfrac{17}{16} - \tfrac{1}{2}\cos\omega$$

Expressing $\cos\omega$ in terms of complex exponentials,

$$|H(e^{j\omega})|^2 = \tfrac{17}{16} - \tfrac{1}{4}e^{j\omega} - \tfrac{1}{4}e^{-j\omega}$$

and replacing $e^{j\omega}$ with z and $e^{-j\omega}$ with z^{-1}, we have

$$G(z) = H(z)H(z^{-1}) = \tfrac{17}{16} - \tfrac{1}{4}z - \tfrac{1}{4}z^{-1} = \left(1 - \tfrac{1}{4}z^{-1}\right)\left(1 - \tfrac{1}{4}z\right)$$

Thus, the minimum phase system is

$$H(z) = 1 - \tfrac{1}{4}z^{-1}$$

A stable and causal system may always be factored into a product of a minimum phase system with an allpass system:

$$H(z) = H_{\min}(z) \cdot H_{ap}(z) \tag{5.17}$$

The procedure for performing this factorization is as follows. First, all of the zeros of $H(z)$ that are outside the unit circle are reflected inside the unit circle to their conjugate reciprocal location. The resulting system function is minimum phase, $H_{\min}(z)$. Then, the allpass filter is selected so that it reflects the appropriate set of zeros of $H_{\min}(z)$ back outside the unit circle.

EXAMPLE 5.5.2   For the system function

$$H(z) = \frac{1 - 2z^{-1}}{(1 - 0.2z^{-1})(1 - 0.7z^{-1})}$$

the minimum phase factor is

$$H(z) = \frac{z^{-1} - 2}{(1 - 0.2z^{-1})(1 - 0.7z^{-1})}$$

Then, to reflect the zero at $z = 0.5$ back outside the unit circle to $z = 2$, we use the allpass factor

$$H_{ap}(z) = \frac{1 - 2z^{-1}}{z^{-1} - 2}$$

Two properties of minimum phase systems are as follows. First, of all systems that have the same Fourier transform magnitude, the minimum phase system has the minimum group delay. This follows from the factorization given in Eq. (5.17). Specifically, let $H_{\min}(z)$ be a minimum phase system, and let $H(z)$ be another system with the same magnitude. The group delay for $H(z)$ may be written as

$$\tau_h(\omega) = \tau_{\min}(\omega) + \tau_{ap}(\omega)$$

where $\tau_{ap}(\omega)$ is the group delay of a stable and causal allpass system. Because $\tau_{ap}(\omega) > 0$, the group delay of $H(z)$ will be larger than the group delay of the minimum phase system $H_{\min}(z)$. Furthermore, because the phase is the negative of the integral of the group delay, the minimum phase system is also said to have the *minimum phase-lag*.

The second property of minimum phase systems is that they have the minimum energy delay. Specifically, if $h_{\min}(n)$ is the unit sample response of a minimum phase system, and $h(n)$ is the unit sample response of another causal system that has the same magnitude response,

$$\sum_{k=0}^{n} |h(k)|^2 < \sum_{k=0}^{n} |h_{\min}(k)|^2$$

for any $n \geq 0$.

5.6 FEEDBACK SYSTEMS

Feedback systems are used in many applications such as stabilization of unstable systems, compensation of nonideal elements, tracking systems, and inverse system design. The general configuration of a discrete-time feedback system is shown in Fig. 5-6. The system $H(z)$ is referred to as the system function of the *forward* path, and $G(z)$ is referred to as the system function of the *feedback* path. The system function relating the input $x(n)$ to the output $y(n)$ is called the *closed-loop* system function and is denoted by $Q(z)$. Because

$$Y(z) = H(z)[X(z) - G(z)Y(z)]$$

the closed-loop system function is

$$Q(z) = \frac{Y(z)}{X(z)} = \frac{H(z)}{1 + G(z)H(z)} \tag{5.18}$$

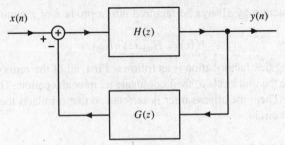

Fig. 5-6. A feedback network.

If $H(z)$ and $G(z)$ are rational functions of z,

$$H(z) = \frac{N_h(z)}{D_h(z)} \qquad G(z) = \frac{N_g(z)}{D_g(z)}$$

the closed-loop system function may be written as

$$Q(z) = \frac{N_h(z)D_g(z)}{D_g(z)D_h(z) + N_g(z)N_h(z)}$$

Therefore, the poles of the *closed-loop* system $Q(z)$ are the roots of the equation

$$D_g(z)D_h(z) + N_g(z)N_h(z) = 0 \qquad\qquad (5.19)$$

With the appropriate order and coefficients for $G(z)$, the poles may be placed anywhere in the z-plane.

EXAMPLE 5.6.1   Suppose that we have an unstable system with system function

$$H(z) = \frac{1}{1 - 1.2z^{-1}}$$

Placed in a feedback network with

$$G(z) = K$$

the system function of the closed-loop system is

$$Q(z) = \frac{Y(z)}{X(z)} = \frac{H(z)}{1 + KH(z)} = \frac{1}{(1 - 1.2z^{-1}) + K} = \frac{1}{(1+K) - 1.2z^{-1}}$$

which has a pole at $z = 1.2/(1 + K)$. Therefore, this system will be stable for all $K > 0.2$.

Solved Problems

System Function

5.1   If the input to a linear shift-invariant system is

$$x(n) = \left(\tfrac{1}{2}\right)^n u(n) + 2^n u(-n - 1)$$

the output is

$$y(n) = 6\left(\tfrac{1}{2}\right)^n u(n) - 6\left(\tfrac{3}{4}\right)^n u(n)$$

Find the system function, $H(z)$, and determine whether or not the system is stable and/or causal.

In order to find the system function, recall that $H(z) = Y(z)/X(z)$. Because we are given both $x(n)$ and $y(n)$, all that is necessary to find $H(z)$ is to evaluate the z-transform of $x(n)$ and $y(n)$ and divide. With

$$X(z) = \frac{1}{1 - \frac{1}{2}z^{-1}} - \frac{1}{1 - 2z^{-1}} = \frac{-\frac{3}{2}z^{-1}}{\left(1 - \frac{1}{2}z^{-1}\right)\left(1 - 2z^{-1}\right)} \qquad \frac{1}{2} < |z| < 2$$

and
$$Y(z) = \frac{6}{1 - \frac{1}{2}z^{-1}} - \frac{6}{1 - \frac{3}{4}z^{-1}} = \frac{-\frac{3}{2}z^{-1}}{\left(1 - \frac{1}{2}z^{-1}\right)\left(1 - \frac{3}{4}z^{-1}\right)} \qquad |z| > \frac{3}{4}$$

Then,
$$H(z) = \frac{Y(z)}{X(z)} = \frac{(1 - 2z^{-1})}{\left(1 - \frac{3}{4}z^{-1}\right)}$$

For the region of convergence of $H(z)$, we have two possibilities. Either $|z| > \frac{3}{4}$ or $|z| < \frac{3}{4}$. Because the region of convergence of $Y(z)$ is $|z| > \frac{3}{4}$ and includes the intersection of the regions of convergence of $X(z)$ and $H(z)$, the region of convergence of $H(z)$ must be $|z| > \frac{3}{4}$.

Because the region of convergence of $H(z)$ includes the unit circle, $h(n)$ is stable, and because the region of convergence is the exterior of a circle and includes $z = \infty$, $h(n)$ is causal.

5.2   When the input to a linear shift-invariant system is

$$x(n) = 2u(n)$$

the output is

$$y(n) = \left[4\left(\tfrac{1}{2}\right)^n - 3\left(-\tfrac{3}{4}\right)^n\right]u(n)$$

Find the unit sample response of the system.

One approach that we may use to solve this problem is to evaluate $H(z) = Y(z)/X(z)$ and then compute the inverse z-transform. Note, however, that we are given the response of the system to a step with an amplitude of 2, and we are asked to find the unit sample response. Because

$$\delta(n) = u(n) - u(n-1)$$

if we let $s(n)$ be the step response, it follows from linearity that

$$h(n) = s(n) - s(n-1)$$

Therefore, from the response given above, we have

$$h(n) = \tfrac{1}{2}[y(n) - y(n-1)]$$

or
$$h(n) = \tfrac{1}{2}\left[4\left(\tfrac{1}{2}\right)^n - 3\left(-\tfrac{3}{4}\right)^n\right]u(n) - \tfrac{1}{2}\left[4\left(\tfrac{1}{2}\right)^{n-1} - 3\left(-\tfrac{3}{4}\right)^{n-1}\right]u(n-1)$$

$$= \tfrac{1}{2}\delta(n) + \left[-2\left(\tfrac{1}{2}\right)^n - \tfrac{7}{2}\left(-\tfrac{3}{4}\right)^n\right]u(n-1)$$

5.3   A causal linear shift-invariant system is characterized by the difference equation

$$y(n) = \tfrac{1}{4}y(n-1) + \tfrac{1}{8}y(n-2) + x(n) - x(n-1)$$

Find the system function, $H(z)$, and the unit sample response, $h(n)$.

To find the system function, we take the z-transform of the difference equation,

$$Y(z) = \tfrac{1}{4}z^{-1}Y(z) + \tfrac{1}{8}z^{-2}Y(z) + X(z) - z^{-1}X(z)$$

or

$$Y(z)\left[1 - \tfrac{1}{4}z^{-1} - \tfrac{1}{8}z^{-2}\right] = X(z)[1 - z^{-1}]$$

Therefore, the system function is

$$H(z) = \frac{Y(z)}{X(z)} = \frac{1 - z^{-1}}{1 - \tfrac{1}{4}z^{-1} - \tfrac{1}{8}z^{-2}} = \frac{1 - z^{-1}}{\left(1 - \tfrac{1}{2}z^{-1}\right)\left(1 + \tfrac{1}{4}z^{-1}\right)}$$

Because the system is causal, the region of convergence is $|z| > \tfrac{1}{2}$.

To find the unit sample response, we perform a partial fraction expansion of $H(z)$,

$$H(z) = \frac{A}{1 - \tfrac{1}{2}z^{-1}} + \frac{B}{1 + \tfrac{1}{4}z^{-1}}$$

where

$$A = \left(1 - \tfrac{1}{2}z^{-1}\right)H(z)\Big|_{z^{-1}=2} = \frac{1 - z^{-1}}{1 + \tfrac{1}{4}z^{-1}}\bigg|_{z^{-1}=2} = -\tfrac{2}{3}$$

$$B = \left(1 + \tfrac{1}{4}z^{-1}\right)H(z)\Big|_{z^{-1}=-4} = \frac{1 - z^{-1}}{1 - \tfrac{1}{2}z^{-1}}\bigg|_{z^{-1}=-4} = \tfrac{5}{3}$$

Therefore,

$$H(z) = \frac{-\tfrac{2}{3}}{1 - \tfrac{1}{2}z^{-1}} + \frac{\tfrac{5}{3}}{1 + \tfrac{1}{4}z^{-1}}$$

and the unit sample response is

$$h(n) = -\tfrac{2}{3}\left(\tfrac{1}{2}\right)^n u(n) + \tfrac{5}{3}\left(-\tfrac{1}{4}\right)^n u(n)$$

5.4   A causal linear shift-invariant system has a system function

$$H(z) = \frac{1 + z^{-1}}{1 - \tfrac{1}{2}z^{-1}}$$

Find the z-transform of the input, $x(n)$, that will produce the output

$$y(n) = -\tfrac{1}{3}\left(\tfrac{1}{4}\right)^n u(n) - \tfrac{4}{3}(2)^n u(-n - 1)$$

To find the input to a linear shift-invariant filter that will produce a given output $y(n)$, we use the relationship $Y(z) = H(z)X(z)$ to solve for $X(z)$:

$$X(z) = \frac{Y(z)}{H(z)}$$

Computing the z-transform of $y(n)$, we have

$$Y(z) = \frac{-\tfrac{1}{3}}{1 - \tfrac{1}{4}z^{-1}} + \frac{\tfrac{4}{3}}{1 - 2z^{-1}} = \frac{1 + \tfrac{1}{3}z^{-1}}{\left(1 - \tfrac{1}{4}z^{-1}\right)(1 - 2z^{-1})}$$

Therefore,

$$X(z) = \frac{\left(1 + \tfrac{1}{3}z^{-1}\right)\left(1 - \tfrac{1}{2}z^{-1}\right)}{\left(1 - \tfrac{1}{4}z^{-1}\right)(1 - 2z^{-1})(1 + z^{-1})} = \frac{A}{1 - \tfrac{1}{4}z^{-1}} + \frac{B}{1 - 2z^{-1}} + \frac{C}{1 + z^{-1}}$$

where
$$A = \left(1 - \tfrac{1}{4}z^{-1}\right)X(z)\Big|_{z^{-1}=4} = \frac{1 - \tfrac{1}{6}z^{-1} - \tfrac{1}{6}z^{-2}}{(1 - 2z^{-1})(1 + z^{-1})}\Big|_{z^{-1}=4} = \tfrac{1}{15}$$

$$B = (1 - 2z^{-1})X(z)\Big|_{z^{-1}=\frac{1}{2}} = \frac{1 - \tfrac{1}{6}z^{-1} - \tfrac{1}{6}z^{-2}}{\left(1 - \tfrac{1}{4}z^{-1}\right)(1 + z^{-1})}\Big|_{z^{-1}=\frac{1}{2}} = \tfrac{2}{3}$$

$$C = (1 + z^{-1})X(z)\Big|_{z=-1} = \frac{1 - \tfrac{1}{6}z^{-1} - \tfrac{1}{6}z^{-2}}{\left(1 - \tfrac{1}{4}z^{-1}\right)(1 - 2z^{-1})}\Big|_{z=-1} = \tfrac{4}{15}$$

Because $h(n)$ is causal, the region of convergence for $H(z)$ is $|z| > \tfrac{1}{2}$. With the region of convergence of $Y(z)$ the annulus $\tfrac{1}{4} < |z| < 2$, the region of convergence of $X(z)$ is $\tfrac{1}{4} < |z| < 1$. Therefore,

$$x(n) = \tfrac{1}{15}\left(\tfrac{1}{4}\right)^n u(n) - \tfrac{2}{3}2^n u(-n-1) - \tfrac{4}{15}(-1)^n u(-n-1)$$

5.5 Show that if $h(n)$ is real, and $H(z)$ is rational,

$$H(z) = A\frac{\displaystyle\prod_{k=1}^{q}(1 - \beta_k z^{-1})}{\displaystyle\prod_{k=1}^{p}(1 - \alpha_k z^{-1})}$$

the poles and zeros of $H(z)$ occur in conjugate pairs.

It follows from the symmetry property of the z-transform that if $h(n)$ is real, $H(z) = H^*(z^*)$. Therefore,

$$H(z) = \left[A\frac{\displaystyle\prod_{k=1}^{q}(1 - \beta_k(z^*)^{-1})}{\displaystyle\prod_{k=1}^{p}(1 - \alpha_k(z^*)^{-1})}\right]^* = A^*\frac{\displaystyle\prod_{k=1}^{q}(1 - \beta_k^* z^{-1})}{\displaystyle\prod_{k=1}^{p}(1 - \alpha_k^* z^{-1})}$$

and the result follows.

5.6 Without evaluating the inverse z-transform, determine which of the following z-transforms *could* be the system function of a *causal but not necessarily stable* discrete-time linear shift-invariant system:

(a) $X(z) = \dfrac{\left(1 - \tfrac{1}{2}z^{-1}\right)^2}{\left(1 - \tfrac{1}{3}z^{-1}\right)}$

(b) $X(z) = \dfrac{(z - 1)^3}{\left(z - \tfrac{1}{4}\right)^2}$

(c) $X(z) = \dfrac{\left(z - \tfrac{1}{2}\right)^3}{\left(z - \tfrac{1}{3}\right)^4}$

(d) $X(z) = \dfrac{\left(z - \tfrac{1}{3}\right)^4}{\left(z - \tfrac{1}{2}\right)^3}$

A causal sequence is one that is equal to zero for $n < 0$. Therefore, the z-transform of a causal sequence may be written as a one-sided summation:

$$X(z) = \sum_{n=0}^{\infty} x(n)z^{-n}$$

What distinguishes the z-transform of a causal signal from one that is not is the fact that $X(z)$ does not contain any positive powers of z. Consequently, if we let $|z| \to \infty$, $X(z) \to x(0)$, which is a statement of the initial value theorem. It follows, therefore, that if $x(n)$ is causal, this limit must be finite. For noncausal signals, on the other hand, this limit will tend to infinity, because the z-transform will contain positive powers of z. For example, the sequence $x(n) = u(n+1)$ has a z-transform

$$X(z) = z + \sum_{n=0}^{\infty} z^{-n} = \frac{z}{1-z^{-1}}$$

and
$$\lim_{|z|\to\infty} X(z) = \infty$$

Thus, a z-transform may be the system function of a causal system only if

$$\lim_{|z|\to\infty} X(z) < \infty$$

Of the transforms listed, only (a) and (c) have a finite limit as $|z| \to \infty$ and, therefore, are the only ones that could be the z-transform of a causal signal.

5.7   The result of a particular computer-aided filter design is the following causal second-order filter:

$$H(z) = \frac{1 + 2z^{-1} + z^{-2}}{1 - 2z^{-1} + 1.33z^{-2}}$$

Show that this filter is unstable, and find a causal and stable filter that has the same magnitude response as $H(z)$.

This filter is clearly not stable, because the coefficient for z^{-2} in the denominator, which is the product of the roots of $H(z)$, is greater than 1. Specifically, if the poles of $H(z)$ are α_1 and α_2, then $\alpha_1 \cdot \alpha_2 = 1.33$, and this implies that at least one of the roots is outside the unit circle. Because the discriminant of the polynomial is negative,

$$[(2)^2 - 4(1.33)] < 0$$

the roots are complex with $\alpha_1 = re^{j\theta}$ and $\alpha_2 = re^{-j\theta}$ where $r = \sqrt{1.33}$ and $\theta = \cos^{-1}(1/\sqrt{1.33})$.

Recall that if we form a new system function given by $H'(z) = H(z)G_{ap}(z)$, where $G_{ap}(z)$ is an allpass filter of the form

$$G_{ap}(z) = \frac{1 + \alpha z^{-1} + \beta z^{-2}}{\beta + \alpha z^{-1} + z^{-2}}$$

$|H'(e^{j\omega})| = |H(e^{j\omega})|$. Therefore, if

$$G_{ap}(z) = \frac{1 - 2z^{-1} + 1.33z^{-2}}{1.33 - 2z^{-1} + z^{-2}}$$

the effect of $G_{ap}(z)$ is to replace the pair of complex poles in $H(z)$ that are outside the unit circle with a complex pole pair inside the unit circle at the reciprocal locations while preserving the magnitude response. Thus, a stable filter that has a frequency response with the same magnitude as $H(e^{j\omega})$ is the following:

$$H'(z) = \frac{1 + 2z^{-1} + z^{-2}}{1.33 - 2z^{-1} + z^{-2}}$$

5.8   The system function of a discrete-time linear shift-invariant system is $H(z)$. Assume that $H(z)$ is a rational function of z and that $H(z)$ is causal and stable. Determine which of the following systems are stable and which are causal:

(a) $G(z) = H(z)H^*(z^*)$

(b) $G(z) = H'(z)$ where $H'(z) = \dfrac{d}{dz}[H(z)]$

(c) $G(z) = H(z^{-1})$

(d) $G(z) = H(-z)$

With $H(z)$ a rational function of z, if $h(n)$ is stable and causal, the poles of $H(z)$ (if any) are inside the unit circle, and the region of convergence is the exterior of a circle and includes the unit circle.

(a) If $H(z)$ is the z-transform of $h(n)$, then $H^*(z^*)$ is the z-transform of $h^*(n)$, and the region of convergence is the same as that for $H(z)$. Because the region of convergence of $G(z) = H(z)H^*(z^*)$ includes the regions of convergence of $H(z)$ and $H^*(z^*)$, the region of convergence of $G(z)$ will be the exterior of a circle and include the unit circle. Therefore, $g(n)$ is stable and causal.

(b) Recall that if $H(z)$ is the z-transform of $h(n)$,

$$nh(n) \xleftrightarrow{\;z\;} -zH'(z)$$

Therefore, delaying the sequence $nh(n)$ by 1 yields the following z-transform pair:

$$(n-1)h(n-1) \xleftrightarrow{\;z\;} -H'(z)$$

and, clearly, $(n-1)h(n-1)$ will be causal if $h(n)$ is causal. Finally, because $H(z)$ is a rational function of z,

$$H(z) = \frac{N(z)}{D(z)}$$

and we have

$$H'(z) = \frac{D(z)N'(z) - N(z)D'(z)}{D^2(z)}$$

Therefore, if the poles of $H(z)$ are inside the unit circle, the poles of $G(z)$ are inside the unit circle, and $g(n)$ is stable.

(c) With $G(z) = H(z^{-1})$, note that if $H(z)$ has a pole at $z = z_0$, $G(z)$ will have a pole at $z = 1/z_0$. Therefore, all of the poles of $G(z)$ will be outside the unit circle, and $g(n)$ cannot be both stable and causal. Because the replacement of z with z^{-1} corresponds to a time reversal,

$$h(-n) \xleftrightarrow{\;z\;} H(z^{-1})$$

$g(n)$ is noncausal. Furthermore, because time-reversing a sequence does not affect its absolute summability,

$$\sum_{n=-\infty}^{\infty} |h(n)| = \sum_{n=-\infty}^{\infty} |h(-n)|$$

the region of convergence for $G(z)$ will include the unit circle. Thus, $g(n)$ is stable.

(d) With $G(z) = H(-z)$, note that replacing z with $-z$ corresponds to modulating $h(n)$ by $(-1)^n$:

$$G(z) = H(-z) = \sum_{n=-\infty}^{\infty} h(n)(-z)^{-n} = \sum_{n=-\infty}^{\infty} (-1)^n h(n)z^{-n}$$

Therefore, if $h(n)$ is causal and stable, so is $g(n)$.

5.9 Find the inverse system of

$$H(z) = \frac{1 - 2z^{-1}}{1 - 0.6z^{-1}} \qquad |z| > 0.6$$

The inverse system is

$$G(z) = \frac{1 - 0.6z^{-1}}{1 - 2z^{-1}}$$

and there are two possible regions of convergence: $|z| > 2$ and $|z| < 2$. Because both of these overlap with the region of convergence for $H(z)$, both are valid inverse systems. For $|z| > 2$, the unit sample response is

$$g(n) = 2^n u(n) - 0.6(2)^{n-1} u(n-1)$$

which is a causal but unstable system. For $|z| < 2$, the unit sample response is

$$g(n) = -2^n u(-n-1) + 0.6(2)^{n-1} u(-n)$$

which is stable but noncausal. Note, however, that the system

$$\tilde{G}(z) = \frac{1 - 0.6z^{-1}}{z^{-1} - 2} \qquad |z| > 0.5$$

is both stable and causal, and the magnitude of the frequency response is the same as that of the inverse system. Therefore, this system is realizable, and the system that is the cascade of $H(z)$ with $\tilde{G}(z)$ has a frequency response with a magnitude of 1.

5.10   Let $H(z)$ be a stable and causal filter with a system function

$$H(z) = A \frac{(1 - e^{j\omega_0} z^{-1})(1 - e^{-j\omega_0} z^{-1})}{(1 - re^{j\omega_0} z^{-1})(1 - re^{-j\omega_0} z^{-1})}$$

(a)   Make a pole-zero plot of the system function, and use geometric arguments to show that if $r \approx 1$, the system is a notch filter.

(b)   At what frequency does $|H(e^{j\omega})|$ reach its maximum value?

(a)   This system has a pair of complex zeros on the unit circle at $z = e^{\pm j\omega_0}$ and a pair of complex poles just inside the unit circle at $z = re^{\pm j\omega_0}$. A pole-zero diagram for $H(z)$ is shown in the figure below.

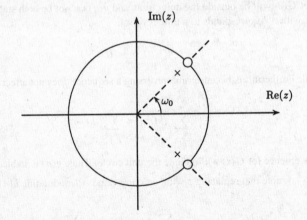

The first thing to note is that, due to the zeros that are on the unit circle, the frequency response goes to zero at $\omega = \pm \omega_0$. The second thing to observe is that, as we move away from the unit circle zeros, the lengths of the vectors from the poles to the unit circle approach the lengths of the vectors from the zeros to the unit circle. Furthermore, the closer r is to 1, the more rapidly the lengths of these vectors become the same. Therefore, if $r \approx 1$, $H(e^{j\omega})$ is a *notch filter*, with a frequency response that is approximately constant except within a narrow band of frequencies around $\omega = \pm \omega_0$, where the frequency response goes to zero.

(b)   The magnitude of the frequency response increases monotonically as we more away from the unit circle zeros. Therefore, $|H(e^{j\omega})|$ will reach its maximum value either at $\omega = 0$ or $\omega = \pi$. Because the frequency response at $\omega = 0$ is

$$A \frac{(1 - e^{j\omega_0})(1 - e^{-j\omega_0})}{(1 - re^{j\omega_0})(1 - re^{-j\omega_0})} = A \frac{2 - 2\cos\omega_0}{1 + r^2 - 2r\cos\omega_0}$$

and the frequency response at $\omega = \pi$ is

$$A\frac{(1+e^{j\omega_0})(1+e^{-j\omega_0})}{(1+re^{j\omega_0})(1+re^{-j\omega_0})} = A\frac{2+2\cos\omega_0}{1+r^2+2r\cos\omega_0}$$

$|H(e^{j\omega})|$ will reach its maximum value at $\omega = 0$ if $\pi/2 < \omega_0 < \pi$, and it will reach its maximum value at $\omega = \pi$ if $0 < \omega_0 < \pi/2$.

5.11 A signal $y(n)$ contains a primary signal, $x(n)$, plus two echos:

$$y(n) = x(n) + \tfrac{1}{2}x(n-n_d) + \tfrac{1}{4}x(n-2n_d)$$

Find a realizable filter that will recover $x(n)$ from $y(n)$.

Because $Y(z)$ is related to $X(z)$ as follows:

$$Y(z) = \left(1 + \tfrac{1}{2}z^{-n_d} + \tfrac{1}{4}z^{-2n_d}\right)X(z)$$

the inverse filter is

$$G(z) = \frac{1}{1+\tfrac{1}{2}z^{-n_d}+\tfrac{1}{4}z^{-2n_d}}$$

We must check, however, to see whether or not this filter is realizable. First, note that we may write $G(z)$ as

$$G(z) = F(z^{n_d})$$

where

$$F(z) = \frac{1}{1+\tfrac{1}{2}z^{-1}+\tfrac{1}{4}z^{-2}}$$

The poles of $F(z)$ are at

$$z = -\tfrac{1}{4}(1 \pm j\sqrt{3})$$

which are *inside* the unit circle at a radius of $r = 0.5$. Therefore, the poles of $G(z)$ are inside the unit circle, at a radius of $r' = (0.5)^{-n_d}$, and $G(z)$ is realizable.

5.12 A causal system with a real-valued unit sample response has a frequency response with a real part given by

$$H_R(e^{j\omega}) = 1 + 0.2\cos 2\omega$$

Find $h(n)$ and $H(e^{j\omega})$.

We are given $H_R(e^{j\omega})$ and are asked to find

$$H(e^{j\omega}) = H_R(e^{j\omega}) + jH_I(e^{j\omega})$$

Although we could find $H_I(e^{j\omega})$ using the discrete Hilbert transform, an easier approach is as follows. Because $h(n)$ is causal, the even and odd parts of $h(n)$ are

$$h_e(n) = \begin{cases} \tfrac{1}{2}h(n) & n \neq 0 \\ h(0) & n = 0 \end{cases}$$

and

$$h_o(n) = \begin{cases} \tfrac{1}{2}h(n) & n > 0 \\ 0 & n = 0 \\ -\tfrac{1}{2}h(n) & n < 0 \end{cases}$$

Therefore, the relationship between $h_e(n)$ and $h_o(n)$ is as follows:

$$h_o(n) = \text{sgn}(n) \cdot h_e(n)$$

where

$$\text{sgn}(n) = \begin{cases} 1 & n > 0 \\ 0 & n = 0 \\ -1 & n < 0 \end{cases}$$

The inverse DTFT of $H_R(e^{j\omega})$, which is the even part of $h(n)$, is

$$h_e(n) = \delta(n) + 0.1\delta(n-2) + 0.1\delta(n+2)$$

Thus,

$$h_o(n) = \text{sgn}(n)h_e(n) = 0.1\delta(n-2) - 0.1\delta(n+2)$$

and $H_I(e^{j\omega})$, the discrete-time Fourier transform of $h_o(n)$, is

$$H_I(e^{j\omega}) = -0.2\sin(2\omega)$$

Therefore,

$$H(e^{j\omega}) = H_R(e^{j\omega}) + jH_I(e^{j\omega}) = 1 + 0.2\cos 2\omega - j0.2\sin(2\omega) = 1 + 0.2e^{-2j\omega}$$

and

$$h(n) = \delta(n) + 0.2\delta(n-2)$$

5.13 A second-order system has two poles at $z = 0.5$ and a pair of complex zeros at $z = e^{\pm j\pi/2}$. Geometrically find the gain, A, of the filter so that $|H(e^{j\omega})|$ is equal to unity at $\omega = 0$.

Because the length of the vectors from the two zeros at $z = e^{\pm j\pi/2}$ to the point $z = 1$ on the unit circle is equal to $\sqrt{2}$, and because the distance from the two poles at $z = 0.5$ to $z = 1$ is equal to 0.5, the magnitude of the frequency response at $\omega = 0$ is

$$|H(e^{j\omega})|_{\omega=0} = A\frac{\sqrt{2} \cdot \sqrt{2}}{0.5 \cdot 0.5} = 8A$$

Therefore, the desired gain is

$$A = \tfrac{1}{8}$$

Systems with Linear Phase

5.14 Derive Eq. (5.9) for the frequency response of a type I linear phase filter.

A type I linear phase filter satisfies the symmetry condition

$$h(n) = h(N - n)$$

and N is even. The symmetry condition is equivalent to

$$h\left(\frac{N}{2} - k\right) = h\left(\frac{N}{2} + k\right) \qquad k = 1, \ldots, \frac{N}{2}$$

Therefore, the frequency response may be written as follows:

$$H(e^{j\omega}) = \sum_{n=0}^{N} h(n)e^{-jn\omega} = h\left(\frac{N}{2}\right)e^{-jN\omega/2} + \sum_{k=1}^{N/2} h\left(\frac{N}{2} - k\right)e^{-j(\frac{N}{2}-k)\omega} + \sum_{k=1}^{N/2} h\left(\frac{N}{2} + k\right)e^{-j(\frac{N}{2}+k)\omega}$$

Factoring out a linear phase term, $e^{-jN\omega/2}$, from each sum, and using the symmetry of $h(n)$, we have

$$H(e^{j\omega}) = h\left(\frac{N}{2}\right)e^{-jN\omega/2} + e^{-jN\omega/2}\sum_{k=1}^{N/2} h\left(\frac{N}{2} - k\right)[e^{jk\omega} + e^{-jk\omega}]$$

$$= e^{-jN\omega/2}\left[h\left(\frac{N}{2}\right) + \sum_{k=1}^{N/2} 2h\left(\frac{N}{2} - k\right)\cos(k\omega)\right] \qquad (5.20)$$

Therefore, we may write the frequency response as follows:

$$H(e^{j\omega}) = e^{-jN\omega/2} \sum_{k=0}^{N/2} a(k)\cos(k\omega)$$

where

$$a(k) = 2h\left(\frac{N}{2} - k\right) \qquad k = 1, 2, \ldots, \frac{N}{2}$$

$$a(0) = h\left(\frac{N}{2}\right)$$

which is the desired result.

5.15 Derive Eq. (5.10) for the frequency response of a type II linear phase filter.

For a type II linear phase filter,

$$h(n) = h(N - n)$$

where N is odd. Therefore, $h(n)$ is symmetric about the half-integer, $N/2$, and the symmetry condition is equivalent to

$$h\left(\frac{N+1}{2} - k\right) = h\left(\frac{N-1}{2} + k\right) \qquad k = 1, \ldots, \frac{N+1}{2}$$

Thus, the frequency response may be written as

$$H(e^{j\omega}) = \sum_{n=0}^{N} h(n)e^{-jn\omega} = \sum_{k=1}^{(N+1)/2} h\left(\frac{N+1}{2} - k\right)e^{-j(\frac{N+1}{2}-k)\omega} + \sum_{k=1}^{(N+1)/2} h\left(\frac{N-1}{2} + k\right)e^{-j(\frac{N-1}{2}+k)\omega}$$

Factoring out a linear phase term $e^{-jN\omega/2}$ and using the symmetry of $h(n)$, we have

$$H(e^{j\omega}) = e^{-jN\omega/2} \sum_{k=1}^{(N+1)/2} h\left(\frac{N+1}{2} - k\right)\left[e^{j(k-\frac{1}{2})\omega} + e^{-j(k-\frac{1}{2})\omega}\right]$$

$$= e^{-jN\omega/2} \sum_{k=1}^{(N+1)/2} 2h\left(\frac{N+1}{2} - k\right)\cos\left[\left(k - \tfrac{1}{2}\right)\omega\right] \qquad (5.21)$$

Therefore,

$$H(e^{j\omega}) = e^{-jN\omega/2} \sum_{k=1}^{(N+1)/2} b(k)\cos\left[\left(k - \tfrac{1}{2}\right)\omega\right]$$

where

$$b(k) = 2h\left(\frac{N+1}{2} - k\right) \qquad k = 1, 2, \ldots, \frac{N+1}{2}$$

which is the desired result.

5.16 How would the derivations in the previous two problems be modified to find the form of the frequency response for types III and IV linear phase filters?

The only difference between a type I and a type III linear phase filter is that

$$h(n) = h(N - n)$$

for a type I filter, and

$$h(n) = -h(N - n)$$

for a type III filter. Therefore, $h(N/2) = 0$, and Eq. (5.20) is modified as follows:

$$H(e^{j\omega}) = e^{-jN\omega/2} \sum_{k=1}^{N/2} h\left(\frac{N}{2} - k\right)[e^{jk\omega} - e^{-jk\omega}]$$

$$= e^{-jN\omega/2} \sum_{k=1}^{N/2} 2jh\left(\frac{N}{2} - k\right)\sin(k\omega)$$

Thus, it follows that the frequency response may be written as

$$H(e^{j\omega}) = je^{-jN\omega/2} \sum_{k=1}^{N/2} c(k) \sin(k\omega)$$

where

$$c(k) = 2h\left(\frac{N}{2} - k\right) \qquad k = 1, 2, \ldots, \frac{N}{2}$$

The only modification required in Prob. 5.15 to find the form of the frequency response for a type IV linear phase filter is to use the fact that $h(n)$ is odd to rewrite Eq. (5.21) as follows:

$$H(e^{j\omega}) = e^{-jN\omega/2} \sum_{k=1}^{(N+1)/2} h\left(\frac{N+1}{2} - k\right)\left[e^{j(k-\frac{1}{2})\omega} - e^{-j(k-\frac{1}{2})\omega}\right]$$

$$= e^{-jN\omega/2} \sum_{k=1}^{(N+1)/2} 2jh\left(\frac{N+1}{2} - k\right) \sin\left[\left(k - \tfrac{1}{2}\right)\omega\right]$$

Therefore, the frequency response is

$$H(e^{j\omega}) = je^{-jN\omega/2} \sum_{k=1}^{(N+1)/2} d(k) \sin\left[\left(k - \tfrac{1}{2}\right)\omega\right]$$

where

$$d(k) = 2h\left(\frac{N+1}{2} - k\right) \qquad k = 1, 2, \ldots, \frac{N}{2}$$

5.17 Show that a system with a complex unit sample response has generalized linear phase if

$$h(n) = \pm h^*(N - n)$$

If $H(e^{j\omega})$ is the DTFT of $h(n)$, it follows from the delay property and the time-reversal property that the DTFT of $h(N - n)$ is

$$h(N - n) \overset{DTFT}{\Longleftrightarrow} e^{-jN\omega} H(e^{-j\omega})$$

Applying the conjugation property, we then have

$$h^*(N - n) \overset{DTFT}{\Longleftrightarrow} e^{-jN\omega} H^*(e^{j\omega})$$

Now, let us consider the case in which $h(n)$ is conjugate symmetric, $h(n) = h^*(N - n)$. Then

$$H(e^{j\omega}) = e^{-jN\omega} H^*(e^{j\omega})$$

and, expressing $H(e^{j\omega})$ and $H^*(e^{j\omega})$ in terms of their magnitude and phase, we have

$$H(e^{j\omega}) = |H(e^{j\omega})|e^{j\phi_h(\omega)}$$

and

$$H^*(e^{j\omega}) = |H(e^{j\omega})|e^{-j\phi_h(\omega)}$$

Therefore, it follows that

$$e^{j\phi_h(\omega)} = e^{-jN\omega}e^{-j\phi_h(\omega)} \qquad\qquad (5.22)$$

or

$$2\phi_h(\omega) = -N\omega + 2\pi k(\omega)$$

where $k(\omega)$ is an integer for each ω. Solving for the phase, we have

$$\phi_h(\omega) = -\frac{N\omega}{2} + \pi k(\omega)$$

Therefore, $$H(e^{j\omega}) = |H(e^{j\omega})|e^{-jN\omega/2}e^{j\pi k(\omega)} = A(e^{j\omega})e^{-jN\omega/2}$$

where $A(e^{j\omega})$ is a real-valued (in general bipolar) function of ω. Thus, $h(n)$ has linear phase.
    For the case in which $h(n)$ is conjugate antisymmetric,

$$h(n) = -h^*(N - n)$$

Eq. (5.22) becomes

$$e^{j\phi_h(\omega)} = -e^{-jN\omega}e^{-j\phi_h(\omega)}$$

or $$e^{j\phi_h(\omega)} = e^{j\pi}e^{-jN\omega}e^{-j\phi_h(\omega)}$$

Therefore, $$2\phi_h(\omega) = -N\omega + \pi + 2\pi k(\omega)$$

where again $k(\omega)$ is an integer for each ω. Solving for the phase, we have

$$\phi_h(\omega) = -\frac{N\omega}{2} + \frac{\pi}{2} + \pi k(\omega)$$

Therefore, $$H(e^{j\omega}) = |H(e^{j\omega})|e^{-jN\omega/2}e^{j\pi/2}e^{j\pi k(\omega)} = A(e^{j\omega})e^{-jN\omega/2}e^{j\pi/2}$$

where $A(e^{j\omega})$ is a real-valued function of ω, and $h(n)$ has generalized linear phase.

5.18   The relationship between the input and the output of an FIR system is as follows:

$$y(n) = \sum_{k=0}^{N} b(k)x(n - k)$$

Find the coefficients $b(k)$ of the smallest-order filter that satisfies the following conditions:

1.   The filter has (generalized) linear phase.
2.   It completely rejects a sinusoid of frequency $\omega_0 = \pi/3$.
3.   The magnitude of the frequency response is equal to 1 at $\omega = 0$ and $\omega = \pi$.

To reject a sinusoid of frequency $\omega_0 = \pi/3$, the system function must have a pair of zeros on the unit circle at $z = e^{\pm j\pi/3}$. Therefore, $H(z)$ must contain a (linear phase) factor of the form

$$H_1(z) = (1 - e^{j\pi/3}z^{-1})(1 - e^{-j\pi/3}z^{-1}) = 1 - 2\cos\left(\frac{\pi}{3}\right)z^{-1} + z^{-2} = 1 - z^{-1} + z^{-2}$$

Note that if

$$H(z) = A(1 - z^{-1} + z^{-2})$$

the magnitude of the frequency response at $\omega = 0$ is

$$H(e^{j\omega})|_{\omega=0} = A$$

and the magnitude of the frequency response at $\omega = \pi$ is

$$H(e^{j\omega})|_{\omega=\pi} = 3A$$

Thus, no value for A will allow us to simultaneously satisfy both unit magnitude constraints, and it is necessary to add another linear phase term to $H(z)$. To minimize the order of the filter, we will pick a factor of the form

$$H_2(z) = 1 + Bz^{-1} + z^{-2}$$

In this case, the filter becomes

$$H(z) = A(1 - z^{-1} + z^{-2}) \cdot (1 + Bz^{-1} + z^{-2})$$

Now, with two free parameters, A and B, we should be able to satisfy the magnitude constraints. With

$$H(e^{j\omega})|_{\omega=0} = A(2 + B) = 1$$

and

$$H(e^{j\omega})|_{\omega=\pi} = 3A(2 - B) = 1$$

solving for A and B, we find

$$A = \tfrac{1}{3} \qquad B = 1$$

Therefore, the filter is

$$H(z) = \tfrac{1}{3}(1 - z^{-1} + z^{-2}) \cdot (1 + z^{-1} + z^{-2}) = \tfrac{1}{3}(1 + z^{-2} + z^{-4})$$

5.19 Show that if $h(n)$ is real, and 2α is an integer, the constraint

$$h(n) = h(2\alpha - n)$$

is sufficient, but not necessary, for $h(n)$ to be the unit sample response of a system with generalized linear phase.

To show that this symmetry condition is sufficient for a system to have generalized linear phase, if we let $H(e^{j\omega})$ be the frequency response, the symmetry condition implies that

$$H(e^{j\omega}) = H(e^{-j\omega})e^{-j2\alpha\omega}$$

Expressing $H(e^{j\omega})$ in polar form,

$$H(e^{j\omega}) = |H(e^{j\omega})|e^{j\phi_h(\omega)}$$

this becomes

$$|H(e^{j\omega})|e^{j\phi_h(\omega)} = |H(e^{-j\omega})|e^{j\phi_h(-\omega)}e^{-j2\alpha\omega}$$

Because the magnitude is an even function, $|H(e^{j\omega})| = |H(e^{-j\omega})|$, and the phase is odd, $\phi_h(\omega) = -\phi_h(-\omega)$,

$$e^{j\phi_h(\omega)} = e^{-j\phi_h(\omega)-j2\alpha\omega}$$

Therefore, the terms in the exponentials must be equal to within an integer multiple of 2π,

$$\phi_h(\omega) = -\phi_h(\omega) - 2\alpha\omega + 2\pi k$$

where k is an integer. Solving for $\phi_h(\omega)$, we have

$$\phi_h(\omega) = -\alpha\omega + \pi k$$

and it follows that the system has generalized linear phase.
  To show that this condition is not necessary, note that if

$$H(e^{j\omega}) = e^{-j\omega\alpha}$$

then $H(e^{j\omega})$ has linear phase. However, the unit sample response is

$$h(n) = \frac{1}{2\pi}\int_{-\pi}^{\pi} e^{-j\omega\alpha}e^{jn\omega}d\omega = \frac{1}{2\pi}\int_{-\pi}^{\pi} e^{j\omega(n-\alpha)}d\omega$$

$$= \frac{1}{2\pi}\frac{1}{j(n-\alpha)}e^{j\omega(n-\alpha)}\Big|_{-\pi}^{\pi} = \frac{\sin \pi(n-\alpha)}{\pi(n-\alpha)}$$

which is not symmetric about an integer index unless $2\alpha = n_d$ is an integer.

5.20 An FIR linear phase filter has a unit sample response that is real with $h(n) = 0$ for $n < 0$ and $n > 7$. If $h(0) = 1$ and the system function has a zero at $z = 0.4e^{j\pi/3}$ and a zero at $z = 3$, what is $H(z)$?

Because $h(n) = 0$ for $n < 0$ and for $n > 7$, $H(z)$ has seven zeros. With a complex zero at $z = 0.4e^{j\pi/3}$, because $h(n)$ is real, there must be another zero at the conjugate location, $z = 0.4e^{-j\pi/3}$. This conjugate pair of zeros produces the second-order factor

$$H_1(z) = (1 - 0.4e^{j\pi/3}z^{-1})(1 - 0.4e^{-j\pi/3}z^{-1}) = 1 - 0.4z^{-1} + 0.16z^{-2}$$

The linear phase constraint requires that there be a pair of zeros at the reciprocal locations. Therefore, $H(z)$ must also contain the factor

$$H_2(z) = 0.16 - 0.4z^{-1} + z^{-2}$$

The system function also contains a zero at $z = 3$. Again, the linear phase constraint requires that there also be a zero at $z = \frac{1}{3}$. Thus, $H(z)$ also has the factor

$$H_3(z) = (1 - 3z^{-1})\left(1 - \tfrac{1}{3}z^{-1}\right)$$

and we have

$$H(z) = A(1 - 0.4z^{-1} + 0.16z^{-2}) \cdot (0.16 - 0.4z^{-1} + z^{-2}) \cdot (1 - 3z^{-1})\left(1 - \tfrac{1}{3}z^{-1}\right)$$

Finally, because the coefficient of the zero-order term in this polynomial is $0.16A$, A must be equal to

$$A = \tfrac{1}{0.16}$$

in order to make $h(0) = 1$.

5.21 Let $x(n)$ be a finite-length sequence that has a z-transform

$$X(z) = A \prod_{k=1}^{N-1} (1 - \alpha_k z^{-1})$$

with no conjugate reciprocal zeros, i.e., $\alpha_k \neq 1/\alpha_l^*$ for any k and l. Show that $x(n)$ is uniquely defined to within a constant by the phase of its discrete-time Fourier transform.

Let $x_1(n)$ be a finite-length sequence that has a z-transform with no conjugate reciprocal zeros and a DTFT with the same phase as $x(n)$. Then

$$X_1(z) = G(z)X(z)$$

where $G(z)$ is the system function of a filter that has zero phase. Thus, $G(z)$ must have zeros in conjugate reciprocal pairs

$$(1 - \alpha_k z^{-1})(z^{-1} - \alpha_k^*)$$

and/or poles in conjugate reciprocal pairs

$$\frac{1}{(1 - \alpha_k z^{-1})(z^{-1} - \alpha_k^*)}$$

However, conjugate reciprocal zeros in $G(z)$ are not allowed, because this would imply that $X_1(z)$ has conjugate reciprocal zeros. Similarly, because $X(n)$ is finite in length, $G(z)$ cannot have poles in conjugate reciprocal pairs because this would imply that $x_1(n)$ is infinite in length. Therefore, $G(z)$ must be a constant, and the result follows.

Allpass Filters

5.22 Consider a linear shift-invariant system with system function

$$H(z) = \frac{z^{-1} - a^*}{1 - az^{-1}}$$

where $|a| < 1$.

(a) Find a difference equation to implement this system.

(b) Show that this system is an allpass system (i.e., one for which the magnitude of the frequency response is constant).

(c) $H(z)$ is to be cascaded with a system $G(z)$ so that the overall system function is unity. If $G(z)$ is to be a stable system, find the unit sample response, $g(n)$.

(a) Because
$$H(z) = \frac{Y(z)}{X(z)} = \frac{z^{-1} - a^*}{1 - az^{-1}}$$

cross-multiplying, we have
$$Y(z)[1 - az^{-1}] = X(z)[z^{-1} - a^*]$$

Taking the inverse z-transform of both sides of the equation gives
$$y(n) - ay(n-1) = x(n-1) - a^* x(n)$$

or
$$y(n) = ay(n-1) + x(n-1) - a^* x(n)$$

which is the desired difference equation.

(b) To show that this system is an allpass filter, note that the frequency response is
$$H(e^{j\omega}) = H(z)|_{z=e^{j\omega}} = \frac{e^{-j\omega} - a^*}{1 - ae^{-j\omega}}$$

Therefore, the squared magnitude is
$$|H(e^{j\omega})|^2 = H(e^{j\omega})H^*(e^{j\omega}) = \frac{e^{-j\omega} - a^*}{1 - ae^{-j\omega}} \cdot \frac{e^{j\omega} - a}{1 - a^* e^{j\omega}} = \frac{1 + |a|^2 - 2\text{Re}(a^* e^{j\omega})}{1 + |a|^2 - 2\text{Re}(a^* e^{j\omega})} = 1$$

and $H(e^{j\omega})$ is an allpass filter.

(c) The inverse system is
$$G(z) = \frac{1}{H(z)} = \frac{1 - az^{-1}}{z^{-1} - a^*} = -\frac{1}{a^*} \frac{1 - az^{-1}}{1 - (a^* z)^{-1}}$$

which has a pole at $z = 1/a^*$ and a zero at $z = a$. Because $|a| < 1$, the pole is *outside* the unit circle. Therefore, if $g(n)$ is to be stable, the region of convergence must be $|z| < 1/|a|$. Thus, $g(n)$ is the left-sided sequence
$$g(n) = (a^*)^{-n-1}u(-n-1) - a(a^*)^{-(n-1)}u(-n)$$

5.23 The system function of a causal FIR filter is
$$H(z) = (1 - 0.3z^{-1} + 0.9z^{-2})(1 - 0.5z^{-1})$$

Find three other causal FIR filters with $h(0) > 1$ that have a frequency response with the same magnitude.

This filter has a pair of complex zeros and one real zero. The magnitude of the frequency response of this filter will not be changed if it is cascaded with an allpass filter that flips the zeros to their reciprocal location. Therefore, three other FIR filters that have the same magnitude response are
$$H_1(z) = (1 - 0.3z^{-1} + 0.9z^{-2})(0.5 - z^{-1})$$
$$H_2(z) = (0.9 - 0.3z^{-1} + z^{-2})(1 - 0.5z^{-1})$$
$$H_3(z) = (0.9 - 0.3z^{-1} + z^{-2})(0.5 - z^{-1})$$

Note that each of these filters is causal with $h(0) > 1$. The causality constraint along with the condition that $h(0) > 1$ prevents $h(n)$ from being shifted or scaled by (-1), two operations that do not change the Fourier transform magnitude.

5.24 Let $x(n)$ be a finite-length sequence that is zero for $n < 0$ and $n > N$. If $x(n)$ is allowed to be complex, what is the maximum number of distinct finite-length sequences that have the same Fourier transform magnitude as $x(n)$?

Let $X(z)$ be the z-transform of $x(n)$, which is of the form

$$X(z) = x(0) \prod_{k=1}^{N} (1 - \beta_k z^{-1})$$

Each zero may be reflected about the unit circle by multiplying by an allpass filter

$$H_k(z) = \frac{\beta_k^* - z^{-1}}{1 - \beta_k z^{-1}}$$

without changing the magnitude of $X(e^{j\omega})$. Because there are two possible locations for each of the N zeros, the number of distinct finite-length sequences (ignoring delays and multiplication by a unit magnitude complex number) is 2^N.

5.25 Show that the group delay of an allpass filter is nonnegative for all ω.

If α is real and $|\alpha| < 1$, the group delay of a filter that has a system function

$$H_1(z) = \frac{1}{1 - \alpha z^{-1}}$$

which has a single pole at $z = \alpha$, is (see Prob. 2.19)

$$\tau_1(\omega) = -\frac{\alpha^2 - \alpha \cos\omega}{1 + \alpha^2 - 2\alpha \cos\omega}$$

Similarly, the group delay for a filter with the system function

$$H_2(z) = 1 - \alpha z^{-1}$$

which has a single zero at $z = \alpha$, is

$$\tau_2(\omega) = -\tau_1(\omega)$$

Furthermore, if

$$H_3(z) = z^{-1} - \alpha = -\alpha \left[1 - \frac{1}{\alpha} z^{-1} \right]$$

the group delay is

$$\tau_3(\omega) = \frac{\alpha^{-2} - \alpha^{-1} \cos\omega}{1 + \alpha^{-2} - 2\alpha^{-1} \cos\omega} = \frac{1 - \alpha \cos\omega}{1 + \alpha^2 - 2\alpha \cos\omega}$$

Therefore, the group delay of a single allpass factor of the form

$$H_0(z) = \frac{z^{-1} - \alpha}{1 - \alpha z^{-1}}$$

is

$$\tau_0(\omega) = \tau_1(\omega) + \tau_3(\omega) = -\frac{\alpha^2 - \alpha \cos\omega}{1 + \alpha^2 - 2\alpha \cos\omega} + \frac{1 - \alpha \cos\omega}{1 + \alpha^2 - 2\alpha \cos\omega}$$

$$= \frac{1 - \alpha^2}{1 + \alpha^2 - 2\alpha \cos\omega} = \frac{1 - \alpha^2}{|1 - \alpha e^{-j\omega}|^2}$$

which, because $|\alpha| < 1$, is positive for all ω.

For complex roots, the allpass factors have the form

$$H(z) = \frac{z^{-1} - \alpha^*}{1 - \alpha z^{-1}}$$

Therefore, with $\alpha = |\alpha|e^{j\theta}$, the frequency response is

$$H(e^{j\omega}) = \frac{e^{-j\omega} - |\alpha|e^{-j\theta}}{1 - |\alpha|e^{j\theta}e^{-j\omega}}$$

$$= e^{-j\theta}\frac{e^{-j(\omega-\theta)} - |\alpha|}{1 - |\alpha|e^{-j(\omega-\theta)}}$$

and the group delay is

$$\tau(\omega) = \tau_0(\omega - \theta) = \frac{1 - \alpha^2}{|1 - \alpha e^{-j(\omega-\theta)}|^2}$$

which is nonnegative for all ω.

5.26 Show that the phase of an allpass filter with $h(n)$ real, if plotted as a continuous function of ω, is nonpositive for all ω.

The group delay is minus the derivative of the phase. Therefore, the phase is related to the group delay as follows:

$$\phi_h(\omega) = -\int_0^\omega \tau_h(\theta)d\theta + \phi_h(0) \qquad 0 \le \omega < \pi$$

Because the general form for the frequency response of an allpass filter is

$$H(e^{j\omega}) = \prod_{k=1}^{N_1} \frac{e^{-j\omega} - b_k}{1 - b_k e^{-j\omega}} \prod_{k=1}^{N_2} \frac{d_k - c_k e^{-j\omega} + e^{-2j\omega}}{1 - c_k e^{-j\omega} + d_k e^{-2j\omega}}$$

then

$$H(e^{j\omega})|_{\omega=0} = \prod_{k=1}^{N_1} \frac{1 - b_k}{1 - b_k} \prod_{k=1}^{N_2} \frac{d_k - c_k + 1}{1 - c_k + d_k} = 1$$

Thus, $\phi_h(0) = 0$, and the positivity of $\tau_h(\omega)$ makes the phase nonpositive.

Minimum Phase

5.27 Suppose that $H(z)$ and $G(z)$ are rational and have minimum phase. Which of the following filters have minimum phase?

(a) $H(z)G(z)$

(b) $H(z) + G(z)$

(a) If $H(z)$ and $G(z)$ have minimum phase, neither $H(z)$ nor $G(z)$ have any poles or zeros outside the unit circle. Because the poles and zeros of $H(z)G(z)$ are the union of the poles and zeros of $H(z)$ and $G(z)$, $H(z)G(z)$ will not have any poles or zeros outside the unit circle and, therefore, has minimum phase.

(b) If $H(z)$ and $G(z)$ have minimum phase, it is not necessarily true that $H(z) + G(z)$ will have minimum phase. We may show this by a simple counter example. If

$$H(z) = \frac{A}{1 - 0.5z^{-1}}$$

and

$$G(z) = \frac{B}{1 - 0.75z^{-1}}$$

both $H(z)$ and $G(z)$ have minimum phase. However, the sum

$$H(z) + G(z) = \frac{(A + B) - (0.5B + 0.75A)z^{-1}}{(1 - 0.5z^{-1})(1 - 0.75z^{-1})}$$

may have a zero anywhere in the z-plane by choosing the appropriate values for A and B. For example, because $H(z) + G(z)$ has a zero at

$$z = \frac{0.5B + 0.75A}{A + B}$$

to place a zero at $z = 2$, we may set $A = 1$ and solve the following equation for B:

$$2 = \frac{0.5B + 0.75}{1 + B}$$

which gives $B = -\frac{5}{6}$.

5.28   A nonminimum phase causal sequence $x(n)$ has a z-transform

$$X(z) = \frac{\left(1 - \frac{3}{2}z^{-1}\right)\left(1 + \frac{1}{3}z^{-1}\right)\left(1 + \frac{5}{3}z^{-1}\right)}{(1 - z^{-1})^2\left(1 - \frac{1}{4}z^{-1}\right)}$$

For what values of the constant α will the sequence $y(n) = \alpha^n x(n)$ be minimum phase?

Multiplying a sequence by α^n moves the poles and zeros radially by a factor of α:

$$Y(z) = X\left(\frac{z}{\alpha}\right) = \frac{\left(1 - \frac{3}{2}\alpha z^{-1}\right)\left(1 + \frac{1}{3}\alpha z^{-1}\right)\left(1 + \frac{5}{3}\alpha z^{-1}\right)}{(1 - \alpha z^{-1})^2\left(1 - \frac{1}{4}\alpha z^{-1}\right)}$$

In order for $Y(z)$ to be minimum phase, all of the poles and zeros must be inside the unit circle. Because the singularity (pole or zero) of $X(z)$ that is the furthest from the unit circle is the zero at $z = -\frac{5}{3}$, $y(n)$ will be minimum phase if $|\alpha| < \frac{3}{5}$.

5.29   A causal linear shift-invariant system has a system function

$$H(z) = \frac{(1 - 2z^{-2})(1 + 0.4z^{-1})}{1 - 0.85z^{-1}}$$

Find a factorization for $H(z)$ of the form

$$H(z) = H_{\min}(z)H_{ap}(z)$$

where $H_{\min}(z)$ has minimum phase, and $H_{ap}(z)$ is an allpass filter.

The system function $H(z)$ has a nonminimum phase factor, $(1 - 2z^{-2})$, which may be written as the product of a minimum phase term and an allpass factor as follows:

$$1 - 2z^{-2} = (z^{-2} - 2) \cdot \frac{1 - 2z^{-2}}{z^{-2} - 2}$$

Therefore, $H(z)$ may be written as the product of a minimum phase system with an allpass system as follows:

$$H(z) = \frac{(1 + 0.4z^{-1})(z^{-2} - 2)}{1 - 0.85z^{-1}} \cdot \frac{1 - 2z^{-2}}{z^{-2} - 2}$$

5.30   A causal linear shift-invariant system has a system function

$$H(z) = \frac{(3 + z^{-1})(2 - 3z^{-1})}{1 - \frac{1}{2}z^{-1}}$$

Find a factorization for $H(z)$ of the form

$$H(z) = H_{\min}(z)H_{lp}(z)$$

where $H_{\min}(z)$ has minimum phase, and $H_{lp}(z)$ is a linear phase system.

This system is not minimum phase because the factor $(2 - 3z^{-1})$ corresponds to a zero outside the unit circle at $z = \frac{3}{2}$. However, we may express this factor as the product of a minimum phase term with a linear phase term as follows. First, we reflect the zero about the unit circle and replace it with a pole:

$$\frac{1}{2z^{-1} - 3}$$

Then, we multiply this term with a linear phase factor that has a zero at $z = \frac{2}{3}$ and a zero at $z = \frac{3}{2}$:

$$\frac{1}{2z^{-1} - 3} \cdot (2z^{-1} - 3)(2 - 3z^{-1})$$

Thus, the factorization for $H(z)$ is

$$H(z) = \frac{3 + z^{-1}}{\left(1 - \frac{1}{2}z^{-1}\right)(2z^{-1} - 3)} \cdot (2z^{-1} - 3)(2 - 3z^{-1})$$

5.31 Find a real-valued causal sequence with $x(0) > 0$ and

$$|X(e^{j\omega})|^2 = (1 + a^2) - 2a\cos\omega$$

We begin by expressing $|X(e^{j\omega})|^2$ in terms of complex exponentials:

$$|X(e^{j\omega})|^2 = (1 + a^2) - ae^{j\omega} - ae^{-j\omega}$$

Replacing $e^{j\omega}$ with z, and $e^{-j\omega}$ with z^{-1}, we have

$$G(z) = X(z)X(z^{-1}) = (1 + a^2) - az - az^{-1} = (1 - az^{-1})(1 - az)$$

Therefore, a real-valued causal sequence with the given magnitude with $x(0) > 1$ is

$$x(n) = \delta(n) - a\delta(n - 1)$$

5.32 Find the minimum phase system that has a magnitude response given by

$$|H(e^{j\omega})|^2 = \frac{\frac{5}{4} - \cos\omega}{\frac{10}{9} - \frac{2}{3}\cos\omega}$$

To solve this problem, we begin by expressing $|H(e^{j\omega})|^2$ in terms of complex exponentials as follows:

$$|H(e^{j\omega})|^2 = \frac{\frac{5}{4} - \frac{1}{2}e^{j\omega} - \frac{1}{2}e^{-j\omega}}{\frac{10}{9} - \frac{1}{3}e^{j\omega} - \frac{1}{3}e^{-j\omega}}$$

Replacing $e^{j\omega}$ by z, and $e^{-j\omega}$ by z^{-1}, this becomes

$$H(z)H(z^{-1}) = \frac{\frac{5}{4} - \frac{1}{2}z - \frac{1}{2}z^{-1}}{\frac{10}{9} - \frac{1}{3}z - \frac{1}{3}z^{-1}} = \frac{\left(1 - \frac{1}{2}z^{-1}\right)\left(1 - \frac{1}{2}z\right)}{\left(1 - \frac{1}{3}z^{-1}\right)\left(1 - \frac{1}{3}z\right)}$$

The minimum phase system is then formed by extracting the poles and zeros that are inside the unit circle:

$$H_{\min}(z) = \frac{1 - \frac{1}{2}z^{-1}}{1 - \frac{1}{3}z^{-1}}$$

5.33 Use the initial value theorem to show that if $h_{\min}(n)$ is a minimum phase sequence, and if $h(n)$ is a causal sequence with the same Fourier transform magnitude, then

$$|h(0)| < |h_{\min}(0)|$$

The initial value theorem states that for a causal sequence, the initial value may be found from the z-transform as follows:

$$h(0) = \lim_{z \to \infty} H(z)$$

Let $h_{\min}(n)$ be a minimum phase sequence, and let $h(n)$ be the nonminimum phase sequence that is formed by reflecting a zero from inside the unit circle at $z = a$ to its conjugate reciprocal location at $z = 1/a^*$:

$$H(z) = H_{\min}(z) \cdot \frac{z^{-1} - a^*}{1 - az^{-1}} \qquad |a| < 1$$

Because $(z^{-1} - a^*)/(1 - az^{-1})$ is an allpass filter, $h(n)$ and $h_{\min}(n)$ have the same Fourier transform magnitude. Using the initial value theorem, we may compare the value of $h(0)$ to $h_{\min}(0)$:

$$h(0) = \lim_{z \to \infty} \left[H_{\min}(z) \cdot \frac{z^{-1} - a^*}{1 - az^{-1}} \right] = -a^* h_{\min}(0)$$

Therefore,

$$|h(0)| = |a| \cdot |h_{\min}(0)|$$

and because $|a| < 1$, $|h(0)| < |h_{\min}(0)|$. Because the magnitude of $|h(0)|$ is reduced each time that a zero of $H_{\min}(z)$ is flipped outside the unit circle,

$$|h(0)| < |h_{\min}(0)|$$

for any sequence $h(n)$ that has a Fourier transform with the same magnitude as that of $h_{\min}(n)$.

5.34 Prove the minimum energy delay property for minimum phase sequences.

Let $h_{\min}(n)$ be a minimum phase sequence, and let α_k be a zero of $H_{\min}(z)$. Then $H_{\min}(z)$ may be written as

$$H_{\min}(z) = (1 - \alpha_k z^{-1}) G_{\min}(z) \tag{5.23}$$

where $G_{\min}(z)$ is another minimum phase sequence. Because $H_{\min}(z)$ is minimum phase, $|\alpha_k| < 1$. Let $H(z)$ be the causal nonminimum phase sequence that is formed by replacing the zero at $z = \alpha_k$ with a zero at $z = 1/\alpha_k^*$:

$$H(z) = (z^{-1} - \alpha_k^*) G_{\min}(z) \tag{5.24}$$

Because

$$H(z) = \frac{z^{-1} - \alpha_k^*}{1 - \alpha_k z^{-1}} H_{\min}(z)$$

then

$$|H(e^{j\omega})| = |H_{\min}(e^{j\omega})|$$

Expressing Eqs. (5.23) and (5.24) in the time domain, we have

$$h_{\min}(n) = g_{\min}(n) - \alpha_k g_{\min}(n-1)$$
$$h(n) = g_{\min}(n-1) - \alpha_k^* g_{\min}(n)$$

Now, let us evaluate the difference between the partial sums of $|h_{\min}(n)|^2$ and $|h(n)|^2$:

$$\sum_{l=0}^{n} |h_{\min}(l)|^2 - \sum_{l=0}^{n} |h(l)|^2 = \sum_{l=0}^{n} |g_{\min}(l) - \alpha_k g_{\min}(l-1)|^2 - \sum_{l=0}^{n} \left| g_{\min}(l-1) - \alpha_k^* g_{\min}(l) \right|^2$$

Expanding the square and canceling the common terms, this becomes

$$\sum_{l=0}^{n} |h_{\min}(l)|^2 - \sum_{l=0}^{n} |h(l)|^2 = \sum_{l=0}^{n} (1 - |\alpha_k|^2) |g_{\min}(l)|^2 - \sum_{l=0}^{n} (1 - |\alpha_k|^2) |g_{\min}(l-1)|^2$$

$$= (1 - |\alpha_k|^2) |g_{\min}(n)|^2$$

which is greater than zero because $|\alpha_k| < 1$. Therefore,

$$\sum_{l=0}^{n} |h_{\min}(l)|^2 > \sum_{l=0}^{n} |h(l)|^2$$

Because $g_{min}(n)$ is minimum phase, this procedure may be repeated for any remaining zeros in $G_{min}(z)$. Therefore, it follows that any causal nonminimum phase sequence that has the same Fourier transform magnitude as $H_{min}(z)$ will have a partial sum that is smaller than that for $h_{min}(n)$.

Feedback

5.35 Suppose that we have an unstable second-order system

$$H(z) = \frac{1}{1 - z^{-1} + 1.44z^{-2}}$$

that we would like to stabilize with the feedback system shown below.

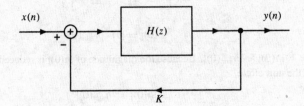

Find the system function of the closed-loop system, $Q(z)$, and determine the values for the feedback gain K that result in a stable system.

The system function of the feedback network is

$$Q(z) = \frac{H(z)}{1 + KH(z)} = \frac{1}{1 + K - z^{-1} + 1.44z^{-2}} = \frac{\dfrac{1}{1+K}}{1 - \dfrac{1}{1+K}z^{-1} + \dfrac{1.44}{1+K}z^{-2}}$$

Therefore, this system will be stable if

$$\left| \frac{1.44}{1+K} \right| < 1$$

which implies that

$$K > 0.44$$

and if

$$\left| \frac{1}{1+K} \right| < 1 + \frac{1.44}{1+K}$$

which is automatically satisfied by the first condition, $K > 0.44$.

5.36 Let $H(z)$ be an unstable system with

$$H(z) = \frac{1}{1 - 1.5z^{-1} - 3z^{-2}}$$

(a) Using a feedback system of the form

$$G(z) = Kz^{-1}$$

determine the values for the gain K, if any, that will stabilize this system.

(b) Repeat part (a) using a feedback system of the form

$$G(z) = Kz^{-2}$$

(a) The system function of the feedback network is

$$Q(z) = \frac{H(z)}{1 + H(z) \cdot K z^{-1}} = \frac{1}{(1 - 1.5z^{-1} - 3z^{-2}) + K z^{-1}} = \frac{1}{1 - (1.5 - K)z^{-1} - 3z^{-2}}$$

Because the coefficient multiplying the term z^{-2} is larger than 1, this system will be unstable for all K.

(b) With $G(z) = K z^{-2}$, the closed-loop system function becomes

$$Q(z) = \frac{H(z)}{1 + H(z) \cdot K z^{-2}} = \frac{1}{1 - 1.5z^{-1} + (K - 3)z^{-2}}$$

This system will not be stable unless

$$|a(2)| = |K - 3| < 1$$

which requires that

$$2 < K < 4$$

In addition, however, we must have

$$|a(1)| = 1.5 < 1 + a(2) = 1 + (K - 3) = K - 2$$

which requires that $K > 3.5$. Therefore, in order for this system to be stable, we must have

$$3.5 < K < 4$$

5.37 Let $H(z)$ be a plant with a system function

$$H(z) = \frac{1 + 0.2z^{-1}}{1 + 1.1z^{-1}}$$

Find a feedback system $G(z)$ of the form

$$G(z) = \frac{1 + cz^{-1}}{1 + dz^{-1}}$$

that will place a second-order pole in the closed-loop system at $z = 1$.

The system function of the closed-loop system is

$$Q(z) = \frac{H(z)}{1 + G(z)H(z)} = \frac{(1 + 0.2z^{-1})(1 + dz^{-1})}{(1 + 1.1z^{-1})(1 + dz^{-1}) + (1 + 0.2z^{-1})(1 + cz^{-1})}$$

or

$$Q(z) = \frac{(1 + 0.2z^{-1})(1 + dz^{-1})}{2 + (1.3 + c + d)z^{-1} + (1.1d + 0.2c)z^{-2}}$$

To place a second-order pole at $z = 1$, we want to find c and d so that

$$\frac{1.3 + c + d}{2} = -2 \qquad \frac{1.1d + 0.2c}{2} = 1$$

The solution is $c = -8.7$ and $d = 3.4$.

Supplementary Problems

The System Function

5.38   The input to a causal linear shift-invariant system is

$$x(n) = u(-n-1) + \left(\tfrac{1}{2}\right)^n u(n)$$

The z-transform of the output of this system is

$$Y(z) = \frac{-\tfrac{1}{2}z^{-1}}{\left(1 - \tfrac{1}{2}z^{-1}\right)(1 + z^{-1})}$$

Find the system function $H(z)$ of the filter.

5.39   A causal linear shift-invariant digital filter has a system function given by

$$H(z) = \frac{1}{(1 - z^{-1})^2\left(1 - \tfrac{3}{4}z^{-1} + \tfrac{1}{8}z^{-2}\right)}$$

Determine whether or not the filter is stable.

5.40   The system function of a linear shift-invariant system is

$$H(z) = e^{1/z}$$

If $h(n)$ is a right-sided sequence, is this system stable?

5.41   Let $x(n)$ be a real-valued, causal sequence with a discrete-time Fourier transform

$$X(e^{j\omega}) = X_R(e^{j\omega}) + jX_I(e^{j\omega})$$

If

$$X_R(e^{j\omega}) = 1 + a\cos\omega$$

find $x(n)$.

5.42   The system function of a linear shift-invariant system is

$$H(z) = \log\left(1 - \tfrac{1}{3}z^{-1}\right) \qquad |z| > \tfrac{1}{3}$$

Is this system causal?

5.43   Which of the following z-transforms *could* be the system function of a causal system?

(a) $X(z) = \dfrac{d}{dz}\dfrac{(1 - 2z^{-1})^2}{(3 - 2z^{-1})^2}$

(b) $X(z) = \dfrac{(z-1)^3}{\left(z^{-1} - \tfrac{1}{4}\right)^2}$

5.44   The system function of a causal filter is

$$H(z) = \frac{1}{1 + az^{-1} + 0.3z^{-2}}$$

For what values of a will this filter be stable?

5.45   A stable filter has a system function

$$H(z) = \frac{(1 - 3z^{-1})\left(1 - \tfrac{1}{3}z^{-1}\right)}{1 - 0.2z^{-1} + 0.4z^{-2}}$$

Find a stable and causal system $G(e^{j\omega})$ such that

$$|G(e^{j\omega})H(e^{j\omega})| = 1$$

5.46   If the frequency response of a stable linear shift-invariant system is real and even, will the inverse system be stable?

Systems with Linear Phase

5.47   The system function of an FIR filter is

$$H(z) = (1 + 0.2z^{-1} + 0.8z^{-2})^2$$

Find a linear phase system that has a frequency response with the same magnitude.

5.48   An FIR filter with generalized linear phase has the following properties:

1. $h(n)$ is real, and $h(n) = 0$ for $n < 0$ and for $n > 5$.
2. $\sum_{n=0}^{5}(-1)^n h(n) = 0$.
3. $H(z)$ is equal to zero at $z = 0.7e^{j\pi/4}$.
4. $\int_{-\pi}^{\pi} H(e^{j\omega})d\omega = 4\pi$.

Find $H(z)$.

5.49   An FIR filter with a real-valued unit sample response has a group delay

$$\tau(\omega) = 2$$

If the system function has a zero at $z = \frac{1}{2}j$, and $H(z)|_{z=1} = 1$, find $h(n)$.

5.50   Let $x(n)$ be a sequence that is equal to zero for $n < 0$ and $n > 5$. If the z-transform of $x(n)$ is

$$X(z) = 3(1 - 0.2z^{-1})(1 + 0.5z^{-1} + 0.8z^{-2})(1 + 0.4z^{-1} - 0.5z^{-2})$$

how many other sequences are equal to zero for $n < 0$ and $n > 5$, have the same initial value as $x(n)$, and have the same phase?

Allpass Filters

5.51   The system function of an FIR filter is

$$H(z) = 1 + 0.2z^{-1} - 0.5z^{-2} + 0.8z^{-3} + 0.4z^{-4}$$

Find another causal FIR filter with $h(n) = 0$ for $n > 4$ that has the same frequency response magnitude.

5.52   A causal and stable allpass filter has a unit sample response that is real. The system function contains three poles, one of which is at $z = 0.8$. If $H(z)$ has a zero at $z = 2e^{j\pi/4}$, what is $H(z)$?

5.53   A linear shift-invariant system has a system function

$$H(z) = \frac{\sum_{k=0}^{P} b(k)z^{-k}}{1 + \sum_{k=1}^{P} a(k)z^{-k}}$$

If $H(z)$ is an allpass filter, what is the relationship between the numerator coefficients $b(k)$ and the denominator coefficients $a(k)$?

Minimum Phase

5.54   What can you say about the poles and zeros of a minimum phase system $H_{min}(z)$ if there is an allpass system $H_{ap}(z)$ such that

$$H_{min}(z)H_{ap}(z) = H_{lp}(z)$$

where $H_{lp}(z)$ is a causal (generalized) linear phase system?

5.55   Find the minimum phase system that has a magnitude response given by

$$|H(e^{j\omega})|^2 = \frac{4}{(1+a^2) - 2a\cos\omega}, \qquad |a| < 1$$

5.56   Suppose that $H(z)$ and $G(z)$ are rational and have minimum phase. Which of the following also have minimum phase?

(a) $H^{-1}(z)$,

(b) $H(z)/G(z)$,

(c) $z^{-1}H(z)$?

5.57   A causal linear shift-invariant discrete-time system has a system function

$$H(z) = \frac{(1 - 0.7z^{-1})(1 - j2z^{-1})(1 + j2z^{-1})}{(1 - 0.8z^{-1})(1 + 0.8z^{-1})}$$

(a) Find a minimum phase system function $H_{\text{min}}(z)$ and an allpass system function $H_{\text{ap}}(z)$ such that

$$H(z) = H_{\text{min}}(z)H_{\text{ap}}(z)$$

(b) Find a minimum phase system function $H_{\text{min}}(z)$ and a linear phase system function $H_{\text{lp}}(z)$ such that

$$H(z) = H_{\text{min}}(z)H_{\text{lp}}(z)$$

5.58   Let $x(n)$ be a real-valued minimum phase sequence. Find another real-valued minimum phase sequence $y(n)$ such that $x(0) = y(0)$ and $y(n) = |x(n)|$.

5.59   Find two different real-valued sequences that satisfy the following constraints:

1. $x(0) = 0$ and $x(1) > 0$.
2. $|X(e^{j\omega})|^2 = \frac{17}{16} - \frac{1}{2}\cos\omega$.

Feedback

5.60   If a feedback system of the form $G(z) = K$ is used to compensate the system

$$H(z) = \frac{1}{1 + 0.8z^{-1} + 1.6z^{-2}}$$

for what values of K will the closed-loop system be stable?

5.61   For the system

$$H(z) = \frac{1}{1 + 1.2z^{-1} + 1.5z^{-2}}$$

find a feedback system of the form

$$G(z) = 1 + g(1)z^{-1} + g(2)z^{-2}$$

that will move the poles of $H(z)$ to $z = 0.5$ and $z = -0.5$.

5.62   Find the closed-loop system function of a feedback network with

(a) $H(z) = \dfrac{z^{-1}}{1 - \frac{1}{2}z^{-1}}$ and $G(z) = \frac{2}{3} - \frac{1}{6}z^{-1}$.

(b) $H(z) = \frac{2}{3} - \frac{1}{6}z^{-1}$ and $G(z) = \dfrac{z^{-1}}{1 - \frac{1}{2}z^{-1}}$.

Answers to Supplementary Problems

5.38 $H(z) = \dfrac{1 - z^{-1}}{1 + z^{-1}}.$

5.39   Unstable.

5.40   Yes.

5.41 $x(n) = \delta(n) + a\delta(n - 1).$

5.42   Yes.

5.43   (a) Yes. (b) No.

5.44 $|a| < 1.3.$

5.45 $G(z) = \dfrac{1 - 0.2z^{-1} + 0.4z^{-2}}{(z^{-1} - 3)\left(1 - \frac{1}{3}z^{-1}\right)}.$

5.46   Yes.

5.47 $G(z) = (1 + 0.2z^{-1} + 0.8z^{-2})(0.8 + 0.2z^{-1} + z^{-2}).$

5.48 $H(z) = \dfrac{2}{0.49}(1 - 0.7\sqrt{2}z^{-1} + 0.49z^{-2})(0.49 - 0.7\sqrt{2}z^{-1} + z^{-2})(1 + z^{-1}).$

5.49 $h(n) = \frac{4}{25}\left[\delta(n) + \frac{17}{4}\delta(n - 2) + \delta(n - 4)\right].$

5.50   None.

5.51 $H(z) = 0.4 + 0.8z^{-1} - 0.5z^{-2} + 0.2z^{-3} + z^{-4}.$

5.52 $H(z) = \dfrac{z^{-1} - 0.8}{1 - 0.8z^{-1}} \cdot \dfrac{z^{-2} - \frac{\sqrt{2}}{2}z^{-1} + \frac{1}{4}}{1 - \frac{\sqrt{2}}{2}z^{-1} + \frac{1}{4}z^{-2}}.$

5.53 $b(k) = a(p - k)$ for $k = 0, 1, \ldots, p - 1$, and $b(p) = 1.$

5.54 $H_{\min}(z)$ is FIR with each zero having even order (i.e., $H_{\min}(z) = G^2(z)$ where $G(z)$ is a minimum phase system).

5.55 $H(z) = \dfrac{2}{1 - az^{-1}}.$

5.56   (a) and (b) have minimum phase but (c) does not.

5.57   (a) $H_{\min}(z) = \dfrac{(1 - 0.7z^{-1})}{(1 - 0.8z^{-1})(1 + 0.8z^{-1})}\left(1 - \frac{1}{2}jz^{-1}\right)\left(1 + \frac{1}{2}jz^{-1}\right)$

$\qquad\qquad H_{\mathrm{ap}}(z) = \dfrac{(1 - j2z^{-1})(1 + j2z^{-1})}{\left(1 - \frac{1}{2}jz^{-1}\right)\left(1 + \frac{1}{2}jz^{-1}\right)}.$

$\qquad$ (b) $H_{\min}(z) = \dfrac{(1 - 0.7z^{-1})}{(1 - 0.8z^{-1})(1 + 0.8z^{-1})}\dfrac{1}{\left(1 - \frac{1}{2}jz^{-1}\right)\left(1 + \frac{1}{2}jz^{-1}\right)}$

$\qquad\qquad H_{\mathrm{lp}}(z) = \left(1 - \frac{1}{2}jz^{-1}\right)\left(1 + \frac{1}{2}jz^{-1}\right)(1 - 2jz^{-1})(1 + 2jz^{-1}).$

5.58 $y(n) = (-1)^n x(n)$.

5.59 $x_1(n) = \delta(n-1) - \frac{1}{4}\delta(n-2)$ and $x_2(n) = \frac{1}{4}\delta(n-1) - \delta(n-2)$.

5.60 $K > 0.6$.

5.61 $g(1) = -1.2$ and $g(2) = -2$.

5.62   (a) $Q(z) = \dfrac{z^{-1}}{1 + \frac{1}{6}z^{-1} - \frac{1}{6}z^{-2}}$.

      (b) $Q(z) = \dfrac{\frac{2}{3} - \frac{2}{3}z^{-1} + \frac{1}{12}z^{-2}}{1 + \frac{1}{6}z^{-1} - \frac{1}{6}z^{-2}}$.

Chapter 6

The DFT

6.1 INTRODUCTION

In previous chapters, we have seen how to represent a sequence in terms of a linear combination of complex exponentials using the discrete-time Fourier transform (DTFT) and how the sequence values may be used as the coefficients in a power series expansion of a complex-valued function of z. For finite-length sequences there is another representation, called the discrete Fourier transform (DFT). Unlike the DTFT, which is a continuous function of a continuous variable, ω, the DFT is a sequence that corresponds to samples of the DTFT. Such a representation is very useful for digital computations and for digital hardware implementations. In this chapter, we look at the DFT, explore its properties, and see how it may be used to perform such tasks as digital filtering and evaluating the frequency response of a linear shift-invariant system.

6.2 DISCRETE FOURIER SERIES

Let $\tilde{x}(n)$ be a periodic sequence with a period N:

$$\tilde{x}(n) = \tilde{x}(n + N)$$

Although, strictly speaking, $\tilde{x}(n)$ does not have a Fourier transform because it is not absolutely summable, it can be expressed in terms of a discrete Fourier series (DFS) as follows:

$$\tilde{x}(n) = \frac{1}{N} \sum_{k=0}^{N-1} \tilde{X}(k) e^{j2\pi nk/N} \tag{6.1}$$

which is a decomposition of $\tilde{x}(n)$ into a sum of N harmonically related complex exponentials. The values of the discrete Fourier series coefficients, $\tilde{X}(k)$, may be derived by multiplying both sides of this expansion by $e^{-j2\pi nl/N}$, summing over one period, and using the fact that the complex exponentials are orthogonal:

$$\sum_{k=0}^{N-1} e^{j2\pi n(k-l)/N} = \begin{cases} N & k = l \\ 0 & k \neq l \end{cases}$$

The result is

$$\tilde{X}(k) = \sum_{n=0}^{N-1} \tilde{x}(n) e^{-j2\pi nk/N} \tag{6.2}$$

Note that the DFS coefficients are periodic with a period N:

$$\tilde{X}(k + N) = \tilde{X}(k)$$

Equations (6.1) and (6.2) form a DFS pair, and we write

$$\tilde{x}(n) \overset{DFS}{\Longleftrightarrow} \tilde{X}(k)$$

EXAMPLE 6.2.1   Let us find the discrete Fourier series representation for the sequence

$$\tilde{x}(n) = \sum_{k=-\infty}^{\infty} x(n - 10k)$$

where
$$x(n) = \begin{cases} 1 & 0 \le n < 5 \\ 0 & \text{else} \end{cases}$$

Note that $\tilde{x}(n)$ is a periodic sequence with a period $N = 10$. Therefore, the DFS coefficients are

$$\tilde{X}(k) = \sum_{n=0}^{9} \tilde{x}(n) e^{-j2\pi nk/10} = \sum_{n=0}^{4} e^{-j2\pi nk/10} = \frac{1 - e^{-j\pi k}}{1 - e^{-j\pi k/5}}$$

which, for $0 \le k \le 9$, may be simplified to

$$\tilde{X}(k) = \begin{cases} 5 & k = 0 \\ \dfrac{2}{1 - e^{-j\pi k/5}} & k \text{ odd} \\ 0 & k \text{ even} \end{cases}$$

The DFS coefficients for all other values of k may be found from the periodicity of $\tilde{X}(k)$:

$$\tilde{X}(k + N) = \tilde{X}(k)$$

A notational simplification that is often used for the DFS is to define

$$W_N \equiv e^{-j2\pi/N}$$

for the complex exponentials and write the DFS pair as follows:

$$\tilde{x}(n) = \frac{1}{N} \sum_{k=0}^{N-1} \tilde{X}(k) W_N^{-nk}$$

$$\tilde{X}(k) = \sum_{n=0}^{N-1} \tilde{x}(n) W_N^{nk}$$

The discrete Fourier series has a number of useful and interesting properties. A few of these properties are described below.

Linearity

The DFS pair satisfies the property of linearity. Specifically, if $\tilde{x}_1(n)$ and $\tilde{x}_2(n)$ are periodic with period N, the DFS coefficients of the sum are equal to the sum of the coefficients for $\tilde{x}_1(n)$ and $\tilde{x}_2(n)$ individually,

$$\tilde{x}_1(n) + \tilde{x}_2(n) \overset{DFS}{\Longleftrightarrow} \tilde{X}_1(k) + \tilde{X}_2(k)$$

Shift

If a periodic sequence $\tilde{x}(n)$ is shifted, the DFS coefficients are multiplied by a complex exponential. In other words, if $\tilde{X}(k)$ are the DFS coefficients for $\tilde{x}(n)$, the DFS coefficients for $\tilde{y}(n) = \tilde{x}(n - n_0)$ are

$$\tilde{Y}(k) = W_N^{kn_0} \tilde{X}(k)$$

Similarly, if $\tilde{x}(n)$ is multiplied by a complex exponential,

$$\tilde{y}(n) = W_N^{nk_0} \tilde{x}(n)$$

the DFS coefficients of $\tilde{x}(n)$ are shifted:

$$\tilde{Y}(k) = \tilde{X}(k + k_0)$$

Periodic Convolution

If $\tilde{h}(n)$ and $\tilde{x}(n)$ are periodic with a period N with DFS coefficients $\tilde{H}(k)$ and $\tilde{X}(k)$, respectively, the sequence with DFS coefficients

$$\tilde{Y}(k) = \tilde{H}(k)\tilde{X}(k)$$

is formed by *periodically convolving* $\tilde{h}(n)$ with $\tilde{x}(n)$ as follows:

$$\tilde{y}(n) = \sum_{k=0}^{N-1} \tilde{h}(k)\tilde{x}(n-k) \tag{6.3}$$

Notationally, the periodic convolution of two sequences is written as

$$\tilde{y}(n) = \tilde{h}(n) \circledast \tilde{x}(n)$$

The only difference between periodic and linear convolution is that, with periodic convolution, the sum is only evaluated over a single period, whereas with linear convolution the sum is taken over all values of k.

EXAMPLE 6.2.2   Let us periodically convolve the two sequences pictured below that have a period $N = 6$.

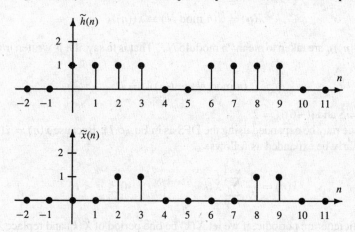

The periodic convolution of two sequences may be performed graphically, analytically, or using the DFS. In this problem, we will use the graphical approach. We begin by plotting $\tilde{x}(n-k)$ versus k. This sequence, for $n = 0$, is illustrated below.

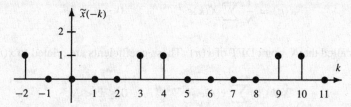

The value of $\tilde{y}(0)$ is then found by summing the product $\tilde{h}(k)\tilde{x}(-k)$ from $k = 0$ to $k = 5$. The result is $\tilde{y}(0) = 1$. Next, $\tilde{x}(-k)$ is shifted to the right by one and multiplied by $\tilde{h}(k)$. Because the only two nonzero values of $\tilde{x}(1-k)$ are at $k = 4, 5$, the product $\tilde{h}(k)\tilde{x}(1-k)$ is equal to zero, and $\tilde{y}(1) = 0$. This process is continued until we have one period of $\tilde{y}(n)$. The result is illustrated below.

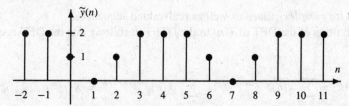

6.3 DISCRETE FOURIER TRANSFORM

The DFT is an important decomposition for sequences that are finite in length. Whereas the DTFT is a mapping from a sequence to a function of a continuous variable, ω,

$$x(n) \stackrel{DTFT}{\Longleftrightarrow} X(e^{j\omega})$$

the DFT is a mapping from a sequence, $x(n)$, to another sequence, $X(k)$,

$$x(n) \stackrel{DFT}{\Longleftrightarrow} X(k)$$

The DFT may be easily developed from the discrete Fourier series representation for periodic sequences. Let $x(n)$ be a finite-length sequence of length N that is equal to zero outside the interval $[0, N-1]$. A periodic sequence $\tilde{x}(n)$ may be formed from $x(n)$ as follows:

$$\tilde{x}(n) = \sum_{k=-\infty}^{\infty} x(n+kN)$$

This periodic extension may be expressed as follows:

$$\tilde{x}(n) = x(n \bmod N) \equiv x((n))_N$$

where $(n \bmod N)$ and $((n))_N$ are taken to mean "n modulo N." That is to say, if n is written in the form $n = kN + l$ where $0 \leq l < N$,

$$(n \bmod N) = ((n))_N = l$$

For example, $((13))_8 = 5$ and $((-6))_8 = 2$.

A periodic sequence may be expanded using the DFS as in Eq. (6.1). Because $x(n) = \tilde{x}(n)$ for $n = 0, 1, \ldots, N-1$, $x(n)$ may similarly be expanded as follows:

$$x(n) = \frac{1}{N} \sum_{k=0}^{N-1} \tilde{X}(k) e^{j2\pi nk/N} \qquad 0 \leq n < N$$

Because the DFS coefficients are periodic, if we let $X(k)$ be one period of $\tilde{X}(k)$ and replace $\tilde{X}(k)$ in the sum with $X(k)$, then we have

$$x(n) = \frac{1}{N} \sum_{k=0}^{N-1} X(k) e^{j2\pi nk/N} \qquad 0 \leq n < N \tag{6.4}$$

The sequence $X(k)$ is called the N-point DFT of $x(n)$. These coefficients are related to $x(n)$ as follows:

$$X(k) = \sum_{n=0}^{N-1} x(n) e^{-j2\pi nk/N} \qquad 0 \leq k < N \tag{6.5}$$

Equations (6.4) and (6.5) form a DFT pair, and we write

$$x(n) \stackrel{DFT}{\Longleftrightarrow} X(k)$$

This expansion is valid for *complex-valued* as well as real-valued sequences.

Comparing the definition of the DFT of $x(n)$ to the DTFT, it follows that the DFT coefficients are *samples* of the DTFT:

$$X(k) = \sum_{n=0}^{N-1} x(n) e^{-j2\pi nk/N} = \sum_{n=-\infty}^{\infty} x(n) e^{-j2\pi nk/N} = X(e^{j\omega})|_{\omega=2\pi k/N}$$

Alternatively, the DFT coefficients correspond to N samples of $X(z)$ that are taken at N equally spaced points around the unit circle:

$$X(k) = X(z)\big|_{z=\exp\{j2\pi k/N\}}$$

6.4 DFT PROPERTIES

In this section, we list some of the properties of the DFT. Because each sequence is assumed to be finite in length, some care must be exercised in manipulating DFTs.

Linearity

If $x_1(n)$ and $x_2(n)$ have N-point DFTs $X_1(k)$ and $X_2(k)$, respectively,

$$ax_1(n) + bx_2(n) \overset{DFT}{\Longleftrightarrow} aX_1(k) + bX_2(k)$$

In using this property, it is important to ensure that the DFTs are the same length. If $x_1(n)$ and $x_2(n)$ have different lengths, the shorter sequence must be *padded* with zeros in order to make it the same length as the longer sequence. For example, if $x_1(n)$ is of length N_1 and $x_2(n)$ is of length N_2 with $N_2 > N_1$, $x_1(n)$ may be considered to be a sequence of length N_2 with the last $N_2 - N_1$ values equal to zero, and DFTs of length N_2 may be taken for both sequences.

Symmetry

If $x(n)$ is real-valued, $X(k)$ is *conjugate symmetric*,

$$X(k) = X^*((-k)) = X^*((N-k))_N$$

and if $x(n)$ is imaginary, $X(k)$ is *conjugate antisymmetric*,

$$X(k) = -X^*((-k)) = -X^*((N-k))_N$$

Circular Shift

The circular shift of a sequence $x(n)$ is defined as follows:

$$x((n-n_0))_N \mathcal{R}_N(n) = \tilde{x}(n-n_0)\mathcal{R}_N(n)$$

where n_0 is the amount of the shift and $\mathcal{R}_N(n)$ is a rectangular window:

$$\mathcal{R}_N(n) = \begin{cases} 1 & 0 \leq n < N \\ 0 & \text{else} \end{cases}$$

A circular shift may be visualized as follows. Suppose that the values of a sequence $x(n)$, from $n = 0$ to $n = N - 1$, are marked around a circle as illustrated in Fig. 6-1 or in an eight-point sequence. A circular shift to the right by n_0 corresponds to a rotation of the circle n_0 positions in a clockwise direction. An example illustrating the circular shift of a four-point sequence is shown in Fig. 6-2. Another way to circularly shift a sequence is to form the periodic sequence $\tilde{x}(n)$, perform a linear shift, $\tilde{x}(n - n_0)$, and then extract one period of $\tilde{x}(n - n_0)$ by multiplying by a rectangular window.

If a sequence is circularly shifted, the DFT is multiplied by a complex exponential,

$$x((n-n_0))_N \mathcal{R}_N(n) \overset{DFT}{\Longleftrightarrow} W_N^{n_0 k} X(k) \qquad (6.6)$$

Similarly, with a circular shift of the DFT, $X((k - k_0))_N$, the sequence is multiplied by a complex exponential,

$$W_N^{nk_0} x(n) \overset{DFT}{\Longleftrightarrow} X((k+k_0))_N \qquad (6.7)$$

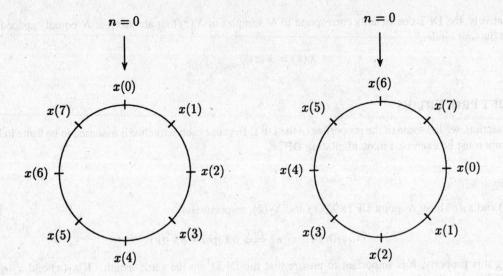

(a) An eight-point sequence.   (b) Circular shift by two.

Fig. 6-1. Visualizing a circular shift by rotating a circle that has the sequence values written around the circle.

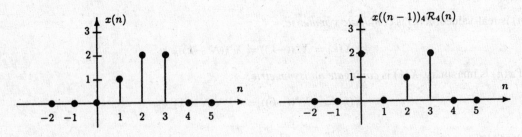

(a) A discrete-time signal of length $N = 4$.   (b) Circular shift by one.

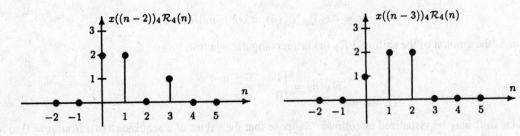

(c) Circular shift by two.   (d) Circular shift by three.

Fig. 6-2. The circular shift of a four-point sequence.

Circular Convolution

Let $h(n)$ and $x(n)$ be finite-length sequences of length N with N-point DFTs $H(k)$ and $X(k)$, respectively. The sequence that has a DFT equal to the product $Y(k) = H(k)X(k)$ is

$$y(n) = \left[\sum_{k=0}^{N-1} \tilde{h}(k)\tilde{x}(n-k) \right] \mathcal{R}_N(n) = \left[\sum_{k=0}^{N-1} \tilde{h}(n-k)\tilde{x}(k) \right] \mathcal{R}_N(n) \tag{6.8}$$

where $\tilde{x}(n)$ and $\tilde{h}(n)$ are the periodic extensions of the sequences $x(n)$ and $h(n)$, respectively. Because $\tilde{h}(n) = h(n)$ for $0 \leq n < N$, the sum in Eq. (6.8) may also be written as

$$y(n) = \left[\sum_{k=0}^{N-1} h(k)\tilde{x}(n-k) \right] \mathcal{R}_N(n) \qquad (6.9)$$

The sequence $y(n)$ in Eq. (6.9) is the *N-point circular convolution* of $h(n)$ with $x(n)$, and it is written as

$$y(n) = h(n) \textcircled{N} x(n) = x(n) \textcircled{N} h(n)$$

The circular convolution of two finite-length sequences $h(n)$ and $x(n)$ is equivalent to one period of the periodic convolution of the periodic sequences $\tilde{h}(n)$ and $\tilde{x}(n)$,

$$y(n) = h(n) \textcircled{N} x(n) = [\tilde{h}(n) \circledast \tilde{x}(n)]\mathcal{R}_N(n)$$

In general, circular convolution is not the same as *linear* convolution, and N-point circular convolution is different, in general, from M-point circular convolution when $M \neq N$.

EXAMPLE 6.4.1   Let us perform the four-point circular convolution of the two sequences $h(n)$ and $x(n)$ shown below.

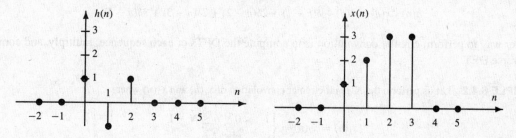

The four-point circular convolution is

$$y(n) = \left[\sum_{k=0}^{3} \tilde{h}(n-k)\tilde{x}(k) \right] \mathcal{R}_4(n)$$

which may be performed graphically, as follows. The value of $y(n)$ at $n = 0$ is

$$y(0) = \sum_{k=0}^{3} \tilde{h}(-k)\tilde{x}(k)$$

Shown in the figure below is a plot of the sequence $\tilde{h}(-k)\mathcal{R}_4(k)$.

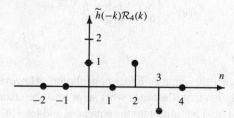

To evaluate $y(0)$, we multiply this sequence by $x(k)$ and sum the product from $k = 0$ to $k = 3$. The result is $y(0) = 1$. Next, to find the value of $y(1)$, we evaluate the sum

$$y(1) = \sum_{k=0}^{3} \tilde{h}(1-k)\tilde{x}(k)$$

Shown in the figure below is a plot of $\tilde{h}(1-k)\mathcal{R}_4(n)$.

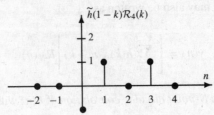

Multiplying by $\tilde{x}(k)$ and summing from $k = 0$ to $k = 3$, we find that $y(1) = 4$. Repeating for $n = 2$ and $n = 3$, we have

$$y(2) = \sum_{k=0}^{3} \tilde{h}(2-k)\tilde{x}(k) = 2$$

$$y(3) = \sum_{k=0}^{3} \tilde{h}(3-k)\tilde{x}(k) = 2$$

Therefore, $y(n) = h(n) \,④\, x(n) = \delta(n) + 4\delta(n-1) + 2\delta(n-2) + 2\delta(n-3)$

By comparison, the linear convolution of $h(n)$ with $x(n)$ is the following six-length sequence:

$$h(n) * x(n) = \delta(n) + \delta(n-1) + 2\delta(n-2) + 2\delta(n-3) + 3\delta(n-5)$$

Another way to perform circular convolution is to compute the DFTs of each sequence, multiply, and compute the inverse DFT.

EXAMPLE 6.4.2   Let us perform the N-point circular convolution of $x_1(n)$ and $x_2(n)$ where

$$x_1(n) = x_2(n) = \begin{cases} 1 & 0 \le n \le N-1 \\ 0 & \text{else} \end{cases}$$

Because the N-point DFTs of $x_1(n)$ and $x_2(n)$ are

$$X_1(k) = X_2(k) = \sum_{n=0}^{N-1} W_N^{nk} = \begin{cases} N & k = 0 \\ 0 & \text{else} \end{cases}$$

then $X(k) = X_1(k)X_2(k) = \begin{cases} N^2 & k = 0 \\ 0 & \text{else} \end{cases}$

Therefore, the N-point circular convolution of $x_1(n)$ with $x_2(n)$ is the inverse DFT of $X(k)$, which is

$$x(n) = \begin{cases} N & 0 \le n \le N-1 \\ 0 & \text{else} \end{cases}$$

Circular Versus Linear Convolution

In general, circular convolution is not the same as linear convolution. However, there is a simple relationship between circular and linear convolution that illustrates what steps must be taken in order to ensure that they are the same. Specifically, let $x(n)$ and $h(n)$ be finite-length sequences and let $y(n)$ be the linear convolution

$$y(n) = x(n) * h(n)$$

The N-point circular convolution of $x(n)$ with $h(n)$ is related to $y(n)$ as follows:

$$h(n) \,⊗\, x(n) = \left[\sum_{k=-\infty}^{\infty} y(n + kN) \right] \mathcal{R}_N(n)   \qquad (6.10)$$

In other words, the circular convolution of two sequences is found by performing the linear convolution and *aliasing* the result.

An important property that follows from Eq. (6.10) is that if $y(n)$ is of length N or less, $y(n - kN)$ $\mathcal{R}_N(n) = 0$ for $k \neq 0$ and

$$h(n) \circledN x(n) = h(n) * x(n)$$

that is, circular convolution is equivalent to linear convolution. Thus, if $h(n)$ and $x(n)$ are finite-length sequences of length N_1 and N_2, respectively, $y(n) = h(n) * x(n)$ is of length $N_1 + N_2 - 1$, and the N-point circular convolution is equivalent to linear convolution provided $N \geq N_1 + N_2 - 1$.

EXAMPLE 6.4.3　Let us find the four-point circular convolution of the sequences $h(n)$ and $x(n)$ in Example 6.4.1. Because the linear convolution is

$$y(n) = \delta(n) + \delta(n - 1) + 2\delta(n - 2) + 2\delta(n - 3) + 3\delta(n - 5)$$

we may set up a table to evaluate the sum

$$h(n) \circledN x(n) = \left[\sum_{k=-\infty}^{\infty} y(n + kN) \right] \mathcal{R}_N(n)$$

This is done by listing the values of the sequence $y(n + kN)$ in a table and summing these values for $n = 0, 1, 2, 3$. Note that the only sequences that have nonzero values in the interval $0 \leq n \leq 3$ are $y(n)$ and $y(n + 4)$, and these are the only sequences that need be listed. Thus, we have

n	0	1	2	3	4	5	6	7
$y(n)$	1	1	2	2	0	3	0	0
$y(n + 4)$	0	3	0	0	0	0	0	0
$h(n) \circled{4} x(n)$	1	4	2	2	–	–	–	–

Summing the columns for $0 \leq n \leq 3$, we have

$$h(n) \circled{4} x(n) = \delta(n) + 4\delta(n - 1) + 2\delta(n - 2) + 3\delta(n - 3)$$

which is the same as computed in Example 6.4.1.

6.5 SAMPLING THE DTFT

Let $x(n)$ be a sequence with a DTFT $X(e^{j\omega})$, and consider the finite-length sequence $y(n)$ of length N whose DFT coefficients are obtained by sampling $X(e^{j\omega})$ at $\omega_k = 2\pi k/N$:

$$Y(k) = X(e^{j\omega})|_{\omega = 2\pi k/N} \qquad k = 0, 1, \ldots, N - 1 \qquad (6.11)$$

Because the DTFT is equal to the z-transform evaluated around the unit circle, these DFT coefficients may also be obtained by sampling $X(z)$ at N equally spaced points around the unit circle at $z_k = \exp\{j2\pi k/N\}$:

$$Y(k) = X(z)|_{z = \exp\{j2\pi k/N\}} \qquad k = 0, 1, \ldots, N - 1$$

These sampling points are illustrated in Fig. 6-3 for $N = 8$. To express the sequence values $y(n)$ in terms of $x(n)$, we begin by finding the inverse DFT of $Y(k)$:

$$y(n) = \frac{1}{N} \sum_{k=0}^{N-1} Y(k) e^{j2\pi nk/N} \qquad (6.12)$$

Because the DFT coefficients $Y(k)$ are samples of the DTFT of $x(n)$,

$$Y(k) = X(e^{j\omega})|_{\omega = 2\pi k/N} = \sum_{l=-\infty}^{\infty} x(l) e^{-j2\pi lk/N}$$

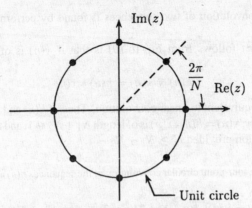

Fig. 6-3. Sampling the z-transform at eight equally
spaced points around the unit circle.

Substituting this expression for $Y(k)$ into Eq. (6.12) gives

$$y(n) = \frac{1}{N} \sum_{k=0}^{N-1} \left\{ \sum_{l=-\infty}^{\infty} x(l) e^{-j2\pi lk/N} \right\} e^{j2\pi nk/N} = \frac{1}{N} \sum_{l=-\infty}^{\infty} x(l) \left\{ \sum_{k=0}^{N-1} e^{j2\pi(n-l)k/N} \right\}$$

The term in brackets is equal to N when $l = n + mN$ where m is an integer, and it is equal to zero otherwise. Therefore,

$$y(n) = \left[\sum_{m=-\infty}^{\infty} x(n - mN) \right] \mathcal{R}_N(n) \tag{6.13}$$

and it follows that $y(n)$ is formed by *aliasing* $x(n)$ in time.

6.6 LINEAR CONVOLUTION USING THE DFT

The DFT provides a convenient way to perform convolutions without having to evaluate the convolution sum. Specifically, if $h(n)$ is N_1 points long and $x(n)$ is N_2 points long, $h(n)$ may be linearly convolved with $x(n)$ as follows:

1. Pad the sequences $h(n)$ and $x(n)$ with zeros so that they are of length $N \geq N_1 + N_2 - 1$.
2. Find the N-point DFTs of $h(n)$ and $x(n)$.
3. Multiply the DFTs to form the product $Y(k) = H(k)X(k)$.
4. Find the inverse DFT of $Y(k)$.

It would appear that there is considerably more effort involved in performing convolutions using DFTs. However, significant computational savings may be realized with this approach if the DFTs are computed efficiently. As we will see in Chap. 7, the fast Fourier transform (FFT) provides such an algorithm.

In spite of its computational advantages, there are some difficulties with the DFT approach. For example, if $x(n)$ is *very long*, we must commit a significant amount of time computing very long DFTs and in the process accept very long processing delays. In some cases, it may even be possible that $x(n)$ is *too long* to compute the DFT. The solution to these problems is to use *block convolution*, which involves segmenting the signal to be filtered, $x(n)$, into sections. Each section is then filtered with the FIR filter $h(n)$, and the filtered sections are pieced together to form the sequence $y(n)$. There are two block convolution techniques. The first is overlap-add, and the second is overlap-save.

Overlap-Add

Let $x(n)$ be a sequence that is to be convolved with a causal FIR filter $h(n)$ of length L:

$$y(n) = h(n) * x(n) = \sum_{k=0}^{L-1} h(k)x(n-k)$$

Assume that $x(n) = 0$ for $n < 0$ and that the length of $x(n)$ is much greater than L. In the overlap-add method, $x(n)$ is partitioned into nonoverlapping subsequences of length M as illustrated in Fig. 6-4. Thus, $x(n)$ may be written as a sum of shifted finite-length sequences of length M,

$$x(n) = \sum_{i=0}^{\infty} x_i(n - Mi)$$

where

$$x_i(n) = \begin{cases} x(n+Mi) & n = 0, 1, \ldots, M-1 \\ 0 & \text{else} \end{cases}$$

Therefore, the linear convolution of $x(n)$ with $h(n)$ is

$$y(n) = x(n) * h(n) = \sum_{i=0}^{\infty} x_i(n - Mi) * h(n) = \sum_{i=0}^{\infty} y_i(n - Mi) \qquad (6.14)$$

where $y_i(n)$ is the linear convolution of $x_i(n)$ with $h(n)$,

$$y_i(n) = x_i(n) * h(n)$$

Because each sequence $y_i(n)$ is of length $N = L + M - 1$, it may be found by multiplying the N-point DFTs of $x_i(n)$ and $h(n)$. The reason for the name *overlap-add* is that, for each i, the sequences $y_i(n)$ and $y_{i+1}(n)$ overlap at $(N - M)$ points and, in performing the sum in Eq. (6.14), these overlapping points are *added*.

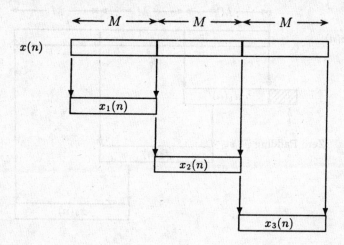

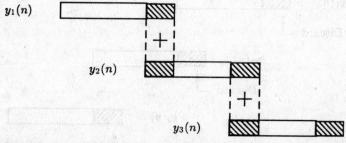

Fig. 6-4. Partitioning a sequence into subsequences of length M for the ovelap-add method of block convolution.

Overlap-Save

The second way that the DFT may be used to perform linear convolution is to use the *overlap-save method*. This method takes advantage of the fact that the aliasing that occurs in circular convolution only affects a portion of the sequence. For example, if $x(n)$ and $h(n)$ are finite-length sequences of lengths L and N, respectively, the linear convolution $y(n)$ is a finite-length sequence of lengths $N + L - 1$. Therefore, assuming that $N > L$, if we perform an N-point circular convolution of $x(n)$ with $h(n)$,

$$h(n) \otimes x(n) = \left[\sum_{k=-\infty}^{\infty} y(n + kN) \right] \mathcal{R}_N(n)$$

Because $y(n + N)$ is the only term that is aliased into the interval $0 \leq n \leq N - 1$, and because $y(n + N)$ only overlaps the first $L - 1$ values of $y(n)$, the remaining values in the circular convolution will not be aliased. In other words, the first $L - 1$ values of the circular convolution are not equal to the linear convolution, whereas the last $M = N - L + 1$ values are the same (see Fig. 6-5). Thus, with the appropriate *partitioning* of the input sequence $x(n)$, linear convolution may be performed by *piecing together* circular convolutions. The procedure is as follows:

1. Let $x_1(n)$ be the sequence

$$x_1(n) = \begin{cases} 0 & 0 \leq n < L - 1 \\ x(n - L + 1) & L - 1 \leq n \leq N - 1 \end{cases}$$

2. Perform the N-point circular convolution of $x_1(n)$ with $h(n)$ by forming the product $H(k)X_1(k)$ and then finding the inverse DFT, $y_1(n)$. The first $L - 1$ values of the circular convolution are aliased, and the last

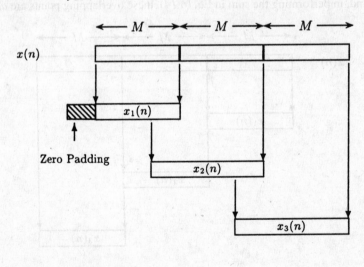

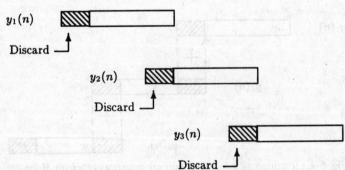

Fig. 6-5. Illustration of the overlap-save method of block convolution.

$N - L + 1$ values correspond to the *linear* convolution of $x(n)$ with $h(n)$. Due to the zero padding at the start of $x_1(n)$, these last $N - L + 1$ values are the first $N - L + 1$ values of $y(n)$:

$$y(n) = y_1(n + L - 1) \qquad 0 \le n \le N - L$$

3. Let $x_2(n)$ be the N-point sequence that is extracted from $x(n)$ with the first $L - 1$ values overlapping with those of $x_1(n)$.

4. Perform an N-point circular convolution of $x_2(n)$ with $h(n)$ by forming the product $H(k)X_2(k)$ and taking the inverse DFT. The first $L - 1$ values of $y_2(n)$ are discarded and the final $N - L + 1$ values are saved and concatenated with the saved values of $y_1(n)$:

$$y(n + N - L + 1) = y_2(n + L - 1) \qquad 0 \le n \le N - L$$

5. Steps 3 and 4 are repeated until all of the values in the linear convolution have been evaluated.

The reason for the name *overlap-save* is that $x(n)$ is partitioned into overlapping sequences of length N and, after performing the N-point circular convolution, only the last $N - L + 1$ values are *saved*.

Solved Problems

Discrete Fourier Series

6.1 Find the DFS expansion of the sequence

$$\tilde{x}(n) = A \cos\left(\frac{n\pi}{2}\right)$$

Because $\tilde{x}(n)$ is periodic with a period $N = 4$, the DFS coefficients may be found by evaluating the sum

$$\tilde{X}(k) = \sum_{n=0}^{3} \tilde{x}(n)e^{-j2\pi nk/4}$$

Alternatively, if we express $\tilde{x}(n)$ in terms of complex exponentials,

$$\tilde{x}(n) = \frac{A}{2}e^{j2\pi n/4} + \frac{A}{2}e^{-j2\pi n/4}$$

and use the fact that

$$\tilde{x}(n) = \frac{1}{4}\sum_{n=0}^{3} \tilde{X}(k)e^{j2\pi nk/4} \tag{6.15}$$

we may *pick off* the DFS coefficients directly as follows. Using the periodicity of the complex exponentials,

$$e^{-j\frac{2\pi}{4}kn} = e^{j\frac{2\pi}{4}(4-k)n}$$

we may express $\tilde{x}(n)$ as

$$\tilde{x}(n) = \frac{A}{2}e^{j\frac{2\pi}{4}n} + \frac{A}{2}e^{j\frac{2\pi}{4}3n} \tag{6.16}$$

Comparing Eqs. (6.15) and (6.16), we see that

$$\tilde{X}(1) = 2A \qquad \tilde{X}(3) = 2A$$

with $\tilde{X}(0) = \tilde{X}(2) = 0$.

6.2   If $\tilde{x}(n)$ is a periodic sequence with a period N,

$$\tilde{x}(n) = \tilde{x}(n+N)$$

$\tilde{x}(n)$ is also periodic with period $2N$. Let $\tilde{X}(k)$ denote the DFS coefficients when $\tilde{x}(n)$ is considered to be periodic with a period N, and let $\tilde{X}_2(k)$ be the DFS coefficients when the period is assumed to be $2N$. Express the DFS coefficients $\tilde{X}_2(k)$ in terms of $\tilde{X}(k)$.

If we consider $\tilde{x}(n)$ to be periodic with a period $2N$, the DFS coefficients are

$$\tilde{X}_2(k) = \sum_{n=0}^{2N-1} \tilde{x}(n) e^{-j\frac{2\pi}{2N}nk}$$

Because $\tilde{x}(n) = \tilde{x}(n+N)$, this sum may be written as

$$\tilde{X}_2(k) = \sum_{n=0}^{N-1} \tilde{x}(n)\left[e^{-j\frac{2\pi}{2N}nk} + e^{-j\frac{2\pi}{2N}(n+N)k} \right] = \sum_{n=0}^{N-1} \tilde{x}(n) e^{-j\frac{2\pi}{2N}nk}[1 + e^{-j\pi k}]$$

Note that the term in brackets is equal to 2 when k is even, and it is equal to zero when k is odd. When k is even,

$$\tilde{X}_2(k) = 2\sum_{n=0}^{N-1} \tilde{x}(n) e^{-j\frac{2\pi}{N}n(k/2)} = 2\tilde{X}\left(\frac{k}{2}\right)$$

Therefore, the DFS coefficients $\tilde{X}_2(k)$ are

$$\tilde{X}_2(k) = \begin{cases} 2\tilde{X}\left(\dfrac{k}{2}\right) & k = 0, 2, \ldots, 2N-2 \\ 0 & k = 1, 3, \ldots, 2N-1 \end{cases}$$

6.3   If $\tilde{x}_1(n)$ and $\tilde{x}_2(n)$ are periodic with period N with DFS coefficients $\tilde{X}_1(k)$ and $\tilde{X}_2(k)$, respectively, show that the sequence with DFS coefficients $\tilde{X}(k) = \tilde{X}_1(k)\tilde{X}_2(k)$ is equal to the periodic convolution of $\tilde{x}_1(n)$ and $\tilde{x}_2(n)$:

$$\tilde{x}(n) = \sum_{k=0}^{N-1} \tilde{x}_1(k)\tilde{x}_2(n-k)$$

Given that $\tilde{X}(k) = \tilde{X}_1(k)\tilde{X}_2(k)$, the sequence $\tilde{x}(n)$ is

$$\tilde{x}(n) = \frac{1}{N}\sum_{k=0}^{N-1} \tilde{X}_1(k)\tilde{X}_2(k) e^{j2\pi nk/N} \tag{6.17}$$

Because we would like to express $\tilde{x}(n)$ in terms of $\tilde{x}_1(n)$ and $\tilde{x}_2(n)$, we begin by substituting

$$\tilde{X}_1(k) = \sum_{l=0}^{N-1} \tilde{x}_1(l) e^{-j2\pi kl/N}$$

for $\tilde{X}_1(k)$ into Eq. (6.17). With this substitution, we have

$$\tilde{x}(n) = \frac{1}{N}\sum_{k=0}^{N-1} \tilde{X}_2(k)\left[\sum_{l=0}^{N-1} \tilde{x}_1(l) e^{-j2\pi kl/N}\right] e^{j2\pi nk/N}$$

Rearranging the sums and combining the exponentials yields

$$\tilde{x}(n) = \frac{1}{N}\sum_{l=0}^{N-1} \tilde{x}_1(l) \sum_{k=0}^{N-1} \tilde{X}_2(k) e^{j2\pi(n-l)k/N}$$

Note that

$$\frac{1}{N}\sum_{k=0}^{N-1}\tilde{X}_2(k)e^{j2\pi(n-l)k/N} = \tilde{x}_2(n-l)$$

Therefore, we have

$$\tilde{x}(n) = \sum_{l=0}^{N-1}\tilde{x}_1(l)\tilde{x}_2(n-l)$$

as was to be shown.

6.4 Let $x_a(t)$ be a periodic continuous-time signal

$$x_a(t) = A\cos(200\pi t) + B\cos(500\pi t)$$

that is sampled with a sampling frequency $f_s = 1$ kHz. Find the DFS coefficients of the sampled signal $\tilde{x}(n) = x_a(nT_s)$.

With a sampling frequency $f_s = 1$ kHz, the sampling period is $T_s = 1/f_s = 10^{-3}$, and the sampled signal is

$$\tilde{x}(n) = x_a(nT_s) = A\cos\left(\frac{\pi}{5}n\right) + B\cos\left(\frac{\pi}{2}n\right)$$

The first term is periodic with a period $N_1 = 10$, and the second is periodic with a period $N_2 = 4$. Therefore, the sum is periodic with period $N = 20$, and we may write

$$\tilde{x}(n) = A\cos\left(\frac{2\pi}{20}2n\right) + B\cos\left(\frac{2\pi}{20}5n\right)$$

Expressing $\tilde{x}(n)$ in terms of complex exponentials, we have

$$\tilde{x}(n) = \frac{A}{2}e^{j\frac{2\pi}{20}2n} + \frac{A}{2}e^{-j\frac{2\pi}{20}2n} + \frac{B}{2}e^{j\frac{2\pi}{20}5n} + \frac{B}{2}e^{-j\frac{2\pi}{20}5n}$$

Using the periodicity of the complex exponentials, it follows that

$$e^{-j\frac{2\pi}{20}2n} = e^{j\frac{2\pi}{20}18n} \qquad e^{-j\frac{2\pi}{20}5n} = e^{j\frac{2\pi}{20}15n}$$

As a result, $\tilde{x}(n)$ may be written as

$$\tilde{x}(n) = \frac{A}{2}e^{j\frac{2\pi}{20}2n} + \frac{A}{2}e^{j\frac{2\pi}{20}18n} + \frac{B}{2}e^{j\frac{2\pi}{20}5n} + \frac{B}{2}e^{j\frac{2\pi}{20}15n}$$

which is in the form of a DFS decomposition,

$$\tilde{x}(n) = \frac{1}{20}\sum_{k=0}^{19}\tilde{X}(k)e^{j\frac{2\pi}{20}nk}$$

Thus, by inspection, we see that

$$\tilde{X}(2) = \tilde{X}(18) = 10A \qquad \tilde{X}(5) = \tilde{X}(15) = 10B$$

with the other DFS coefficients from $k = 0$ to $k = 19$ equal to zero.

The Discrete Fourier Transform

6.5 Compute the N-point DFT of each of the following sequences:

(a) $x_1(n) = \delta(n)$

(b) $x_2(n) = \delta(n - n_0)$, where $0 < n_0 < N$

(c) $x_3(n) = \alpha^n \qquad 0 \le n < N$

(d) $x_4(n) = u(n) - u(n - n_0)$, where $0 < n_0 < N$

(a)   The DFT of the unit sample may be easily evaluated from the definition of the DFT:

$$X_1(k) = \sum_{n=0}^{N-1} \delta(n)W_N^{nk} = 1 \qquad k = 0, 1, \ldots, N-1$$

Another approach, however, is to recall that the DFT corresponds to samples of the z-transform $X_1(z)$ at N equally spaced points around the unit circle. Because $X_1(z) = 1$, it follows that $X_1(k) = 1$.

(b)   For the second sequence, we may again evaluate the DFT directly from the definition of the DFT. Let us instead, however, sample the z-transform. We know that $X_2(z) = z^{-n_0}$. Therefore, sampling $X_2(z)$ at the points $z = W_N^{-k}$ for $k = 0, 1, \ldots, N-1$, we find

$$X_2(k) = W_N^{n_0 k} \qquad k = 0, 1, \ldots, N-1$$

(c)   For $x_3(n)$, the DFT may be found directly as follows:

$$X_3(k) = \sum_{n=0}^{N-1} x_3(n)W_N^{nk} = \sum_{n=0}^{N-1} \alpha^n W_N^{nk}$$

$$= \sum_{n=0}^{N-1} \left(\alpha W_N^k\right)^n = \frac{1 - \left(\alpha W_N^k\right)^N}{1 - \alpha W_N^k} \qquad k = 0, 1, \ldots, N-1$$

(d)   The DFT of the pulse, $x_4(n) = u(n) - u(n - n_0)$, may be evaluated directly as follows:

$$X_4(k) = \sum_{n=0}^{n_0-1} W_N^{nk} = \frac{1 - W_N^{kn_0}}{1 - W_N^k}$$

Factoring out a complex exponential $W_N^{kn_0/2}$ from the numerator and a complex exponential $W_N^{k/2}$ from the denominator, the DFT may be written as

$$X_4(k) = W_N^{k(n_0-1)/2} \frac{W_N^{-kn_0/2} - W_N^{kn_0/2}}{W_N^{-k/2} - W_N^{k/2}} = e^{-j\frac{2\pi k}{N}\left(\frac{n_0-1}{2}\right)} \frac{\sin(n_0 \pi k/N)}{\sin(\pi k/N)} \qquad k = 0, 1, \ldots, N-1$$

6.6   Find the 10-point inverse DFT of

$$X(k) = \begin{cases} 3 & k = 0 \\ 1 & 1 \le k \le 9 \end{cases}$$

To find the inverse DFT, note that $X(k)$ may be expressed as follows:

$$X(k) = 1 + 2\delta(k) \qquad 0 \le k \le 9$$

Written in this way, the inverse DFT may be easily determined. Specifically, note that the inverse DFT of a constant is a unit sample:

$$x_1(n) = \delta(n) \overset{DFT}{\Longleftrightarrow} X_1(k) = 1$$

Similarly, the DFT of a constant is a unit sample:

$$x_2(n) = 1 \overset{DFT}{\Longleftrightarrow} X_2(k) = N\delta(k)$$

Therefore, it follows that

$$x(n) = \tfrac{1}{5} + \delta(n)$$

6.7 Find the N-point DFT of the sequence

$$x(n) = \cos(n\omega_0) \qquad 0 \le n \le N - 1$$

Compare the values of the DFT coefficients $X(k)$ when $\omega_0 = 2\pi k_0/N$ to those when $\omega_0 \ne 2\pi k_0/N$. Explain the difference.

To find the N-point DFT of this sequence, it is easier if we write the cosine in terms of complex exponentials:

$$x(n) = \tfrac{1}{2}e^{jn\omega_0} + \tfrac{1}{2}e^{-jn\omega_0}$$

Evaluating the DFT of each of these terms, we find

$$X(k) = \sum_{n=0}^{N-1} x(n)e^{-j\frac{2\pi}{N}nk} = \frac{1}{2}\sum_{n=0}^{N-1} e^{-jn(\frac{2\pi}{N}k-\omega_0)} + \frac{1}{2}\sum_{n=0}^{N-1} e^{-jn(\frac{2\pi}{N}k+\omega_0)} \qquad (6.18)$$

At this point, note that if $\omega_0 = 2\pi k_0/N$,

$$X(k) = \frac{1}{2}\sum_{n=0}^{N-1} e^{-jn\frac{2\pi}{N}(k-k_0)} + \frac{1}{2}\sum_{n=0}^{N-1} e^{-jn\frac{2\pi}{N}(k+k_0)}$$

Because the first term is a sum of a complex exponential of frequency $\omega_0 = 2\pi(k - k_0)/N$, the sum will be equal to zero unless $k = k_0$, in which case the sum is equal to N. Similarly, the second sum is equal to zero unless $k = N - k_0$, in which case the sum is equal to N. Therefore, if $\omega_0 = 2\pi k_0/N$, the DFT coefficients are

$$X(k) = \begin{cases} \dfrac{N}{2} & k = k_0 \text{ and } k = N - k_0 \\ 0 & \text{otherwise} \end{cases}$$

In the general case, when $\omega_0 \ne 2\pi k_0/N$, we must use the geometric series to evaluate Eq. (6.18):

$$X(k) = \frac{1}{2}\sum_{n=0}^{N-1} e^{-jn(\frac{2\pi}{N}k-\omega_0)} + \frac{1}{2}\sum_{n=0}^{N-1} e^{-jn(\frac{2\pi}{N}k+\omega_0)}$$

$$= \frac{1}{2}\frac{1 - e^{-jN(\frac{2\pi}{N}k-\omega_0)}}{1 - e^{-j(\frac{2\pi}{N}k-\omega_0)}} + \frac{1}{2}\frac{1 - e^{-jN(\frac{2\pi}{N}k+\omega_0)}}{1 - e^{-j(\frac{2\pi}{N}k+\omega_0)}}$$

Factoring out a complex exponential from the numerator and one from the denominator, we have

$$X(k) = \frac{1}{2}e^{-j(\frac{N-1}{2})(\frac{2\pi}{N}k-\omega_0)}\frac{\sin\left(\pi k - \frac{N\omega_0}{2}\right)}{\sin\left(\frac{\pi k}{N} - \frac{\omega_0}{2}\right)} + \frac{1}{2}e^{-j(\frac{N-1}{2})(\frac{2\pi}{N}k+\omega_0)}\frac{\sin\left(\pi k + \frac{N\omega_0}{2}\right)}{\sin\left(\frac{\pi k}{N} + \frac{\omega_0}{2}\right)}$$

Note that, unless ω_0 is an integer multiple of $2\pi/N$, $X(k)$ is, in general, nonzero for each k. The reason for this difference between these two cases comes from the fact that $X(k)$ corresponds to samples of the DTFT of $x(n)$, which is

$$X(e^{j\omega}) = \sum_{n=0}^{N-1} \cos(n\omega_0)e^{-jn\omega} = \frac{1}{2}e^{-j(\frac{N-1}{2})(\omega-\omega_0)}\frac{\sin N(\omega - \omega_0)/2}{\sin(\omega - \omega_0)/2}$$

$$+ \frac{1}{2}e^{-j(\frac{N-1}{2})(\omega+\omega_0)}\frac{\sin N(\omega + \omega_0)/2}{\sin(\omega + \omega_0)/2}$$

When sampled at N equally spaced points over the interval $[0, 2\pi]$, the sample values will, in general, be nonzero. However, if $\omega_0 = 2\pi k_0/N$, all of the samples except those at $k = k_0$ and $k = N - k_0$ occur at the zeros of the sine function.

6.8 Find the N-point DFT of the sequence

$$x(n) = 4 + \cos^2\left(\frac{2\pi n}{N}\right) \qquad n = 0, 1, \ldots, N - 1$$

The DFT of this sequence may be evaluated by expanding the cosine as a sum of complex exponentials:

$$x(n) = 4 + \tfrac{1}{4}\left[e^{j2\pi n/N} + e^{-j2\pi n/N}\right]^2 = 4 + \tfrac{1}{2} + \tfrac{1}{4}e^{j4\pi n/N} + \tfrac{1}{4}e^{-j4\pi n/N}$$

Using the periodicity of the complex exponentials, we may write $x(n)$ as follows:

$$x(n) = \tfrac{9}{2} + \tfrac{1}{4}e^{j\frac{2\pi}{N}(2n)} + \tfrac{1}{4}e^{j\frac{2\pi}{N}(N-2)n}$$

Therefore, the DFT coefficients are

$$X(k) = \begin{cases} \tfrac{9}{2}N & k = 0 \\ \tfrac{1}{4}N & k = 2 \text{ and } k = N-2 \\ 0 & \text{else} \end{cases}$$

6.9 Suppose that we are given a program to find the DFT of a complex-valued sequence $x(n)$. How can this program be used to find the inverse DFT of $X(k)$?

A program to find the DFT of a sequence $x(n)$ evaluates the sum

$$X(k) = \sum_{n=0}^{N-1} x(n) W_N^{nk} \tag{6.19}$$

and produces the sequence of DFT coefficients $X(k)$. What we would like to do is to use this program to find the inverse DFT of $X(k)$, which is

$$x(n) = \frac{1}{N} \sum_{k=0}^{N-1} X(k) W_N^{-nk} \tag{6.20}$$

Note that the only difference between the forward and the inverse DFT is the factor of $1/N$ in the inverse DFT and the sign of the complex exponentials. Therefore, if we conjugate both sides of Eq. (6.20) and multiply by N, we have

$$Nx^*(n) = \sum_{k=0}^{N-1} X^*(k) W_N^{nk}$$

Comparing this to Eq. (6.19), we see that the sum on the right is the DFT of the sequence $X^*(k)$. Thus, if $X^*(k)$ is used as the input in the DFT program, the output will be $Nx^*(n)$. Conjugating this output and dividing by N produces the sequence $x(n)$. Therefore, the procedure is as follows:

1. Conjugate the DFT coefficients $X(k)$ to produce the sequence $X^*(k)$.
2. Use the program to find the DFT of the sequence $X^*(k)$.
3. Conjugate the result obtained in step 2, and divide by N.

DFT Properties

6.10 Consider the finite-length sequence

$$x(n) = \delta(n) + 2\delta(n-5)$$

(a) Find the 10-point discrete Fourier transform of $x(n)$.

(b) Find the sequence that has a discrete Fourier transform

$$Y(k) = e^{j2k\frac{2\pi}{10}} X(k)$$

where $X(k)$ is the 10-point DFT of $x(n)$.

(c) Find the 10-point sequence $y(n)$ that has a discrete Fourier transform

$$Y(k) = X(k)W(k)$$

where $X(k)$ is the 10-point DFT of $x(n)$, and $W(k)$ is the 10-point DFT of the sequence

$$w(n) = \begin{cases} 1 & 0 \le n \le 6 \\ 0 & \text{otherwise} \end{cases}$$

(a) The DFT of $x(n)$ is easily seen to be

$$X(k) = 1 + 2W_N^{5k} = 1 + 2e^{-j\frac{2\pi}{10}5k} = 1 + 2(-1)^k$$

(b) Multiplying $X(k)$ by a complex exponential of the form $W_N^{kn_0}$ corresponds to a circular shift of $x(n)$ by n_0. In this case, because $n_0 = -2$, $x(n)$ is circularly shifted to the left by 2, and we have

$$y(n) = x((n+2))_{10} = 2\delta(n-3) + \delta(n-8)$$

(c) Multiplying $X(k)$ by $W(k)$ corresponds to the circular convolution of $x(n)$ with $w(n)$. To perform the circular convolution, we may find the linear convolution and alias the result. The linear convolution of $x(n)$ with $w(n)$ is

$$z(n) = x(n) * w(n) = [1, 1, 1, 1, 1, 3, 3, 2, 2, 2, 2, 2]$$

and the circular convolution is

$$y(n) = \left[\sum_{k=-\infty}^{\infty} z(n-10k) \right] \mathcal{R}_{10}(n)$$

Because $z(n)$ and $z(n+10)$ are the only two sequences in the sum that have nonzero values for $0 \le n < 10$, using a table to list the values of $z(n)$ and $z(n+10)$, and summing for $n = 0, 1, 2, \ldots, 9$, we have

n	0	1	2	3	4	5	6	7	8	9	10	11
$z(n)$	1	1	1	1	1	3	3	2	2	2	2	2
$z(n+10)$	2	2	0	0	0	0	0	0	0	0	0	0
$y(n)$	3	3	1	1	1	3	3	2	2	2	–	–

Thus, the 10-point circular convolution is

$$y(n) = [3, 3, 1, 1, 1, 3, 3, 2, 2, 2]$$

6.11 Consider the sequence

$$x(n) = 4\delta(n) + 3\delta(n-1) + 2\delta(n-2) + \delta(n-3)$$

Let $X(k)$ be the six-point DFT of $x(n)$.

(a) Find the finite-length sequence $y(n)$ that has a six-point DFT

$$Y(k) = W_6^{4k} X(k)$$

(b) Find the finite-length sequence $w(n)$ that has a six-point DFT that is equal to the real part of $X(k)$,

$$W(k) = \text{Re}\{X(k)\}$$

(c) Find the finite-length sequence $q(n)$ that has a three-point DFT

$$Q(k) = X(2k) \qquad k = 0, 1, 2$$

(a) The sequence $y(n)$ is formed by multiplying the DFT of $x(n)$ by the complex exponential W_6^{4k}. Because this corresponds to a circular shift of $x(n)$ by 4,

$$y(n) = x((n-4))_6$$

it follows that

$$y(n) = 4\delta(n-4) + 3\delta(n-5) + 2\delta(n) + \delta(n-1)$$

(b) The real part of $X(k)$ is

$$\text{Re}\{X(k)\} = \tfrac{1}{2}[X(k) + X^*(k)]$$

To find the inverse DFT of $\text{Re}\{X(k)\}$, we need to evaluate the inverse DFT of $X^*(k)$. Because

$$X^*(k) = \left[\sum_{n=0}^{N-1} x(n)W_N^{nk}\right]^* = \sum_{n=0}^{N-1} x^*(n)W_N^{-nk}$$

$$= \sum_{n=0}^{N-1} x^*(n)W_N^{(N-n)k} = \sum_{n=0}^{N-1} x^*((N-n))_N W_N^{nk}$$

$X^*(k)$ is the DFT of $x^*((N-n))_N$. Therefore, the inverse DFT of $\text{Re}\{X(k)\}$ is

$$w(n) = \tfrac{1}{2}[x(n) + x^*((N-n))_N]$$

With $N = 6$, this becomes

$$w(n) = \left[4, \tfrac{3}{2}, 1, 1, 1, \tfrac{3}{2}\right]$$

(c) The sequence $q(n)$ is of length three with a DFT $Q(k) = X(2k)$ for $k = 0, 1, 2$ where $X(k)$ is the six-point DFT of $x(n)$. Because the coefficients $X(k)$ are samples of $X(z)$ at six equally spaced points around the unit circle, $X(2k)$ for $k = 0, 1, 2$ corresponds to three equally spaced samples of $X(z)$ around the unit circle. Therefore,

$$q(n) = \left[\sum_{r=-\infty}^{\infty} x(n-3r)\right]\mathcal{R}_3(n)$$

With $x(n) = 0$ outside the interval $0 \le n \le 3$, it follows that

$$q(0) = x(0) + x(3) = 5$$
$$q(1) = x(1) = 3$$
$$q(2) = x(2) = 2$$

and we have

$$q(n) = 5\delta(n) + 3\delta(n-1) + 2\delta(n-2)$$

6.12 Consider the sequence

$$x(n) = \delta(n) + 2\delta(n-2) + \delta(n-3)$$

(a) Find the four-point DFT of $x(n)$.

(b) If $y(n)$ is the four-point circular convolution of $x(n)$ with itself, find $y(n)$ and the four-point DFT $Y(k)$.

(c) With $h(n) = \delta(n) + \delta(n-1) + 2\delta(n-3)$, find the four-point circular convolution of $x(n)$ with $h(n)$.

(a) The four-point DFT of $x(n)$ is

$$X(k) = \sum_{n=0}^{3} x(n)W_4^{nk} = 1 + 2W_4^{2k} + W_4^{3k}$$

(b) With $y(n) = x(n) \textcircled{N} x(n)$, it follows that $Y(k) = X^2(k)$:

$$Y(k) = \left(1 + 2W_4^{2k} + W_4^{3k}\right)\left(1 + 2W_4^{2k} + W_4^{3k}\right)$$
$$= 1 + 4W_4^{2k} + 2W_4^{3k} + 4W_4^{4k} + 4W_4^{5k} + W_4^{6k}$$

Because

$$W_4^{4k} = 1 \qquad W_4^{5k} = W_4^k \qquad W_4^{6k} = W_4^{2k}$$

the expression for $Y(k)$ may be simplified to

$$Y(k) = 5 + 4W_4^k + 5W_4^{2k} + 2W_4^{3k}$$

Therefore, $y(n) = 5\delta(n) + 4\delta(n-1) + 5\delta(n-2) + 2\delta(n-3)$

(c) With $h(n) = \delta(n) + \delta(n-1) + 2\delta(n-3)$, the four-point circular convolution of $x(n)$ with $h(n)$ may be found using the tabular method. Because, the linear convolution of $x(n)$ with $h(n)$ is

$$y(n) = x(n) * h(n) = [1, \ 1, \ 2, \ 5, \ 1, \ 4, \ 2]$$

then

n	0	1	2	3	4	5	6	7	8
$y(n)$	1	1	2	5	1	4	2	0	0
$y(n+4)$	1	4	2	0	0	0	0	0	0
$z(n)$	2	5	4	5	–	–	–	–	–

or

$$z(n) = 2\delta(n) + 5\delta(n-1) + 4\delta(n-2) + 5\delta(n-3)$$

6.13 Let $x(n)$ be the sequence

$$x(n) = 2\delta(n) + \delta(n-1) + \delta(n-3)$$

The five-point DFT of $x(n)$ is computed and the resulting sequence is squared:

$$Y(k) = X^2(k)$$

A five-point inverse DFT is then computed to produce the sequence $y(n)$. Find the sequence $y(n)$.

The sequence $y(n)$ has a five-point DFT that is equal to the product $Y(k) = X(k)X(k)$. Therefore, $y(n)$ is the five-point circular convolution of $x(n)$ with itself:

$$y(n) = \left[\sum_{k=0}^{4} x(k)x((n-k))_5 \right] \mathcal{R}_5(n)$$

A simple way to evaluate this circular convolution is to perform the linear convolution $y'(n) = x(n) * x(n)$ and alias the result:

$$y(n) = \left[\sum_{k=-\infty}^{\infty} y'(n-5k) \right] \mathcal{R}_5(n)$$

The linear convolution of $x(n)$ with itself is easily seen to be

$$y'(n) = [4, 4, 1, 4, 2, 0, 1]$$

Using the tabular method for computing the circular convolution, we have

n	0	1	2	3	4	5	6	7
$y'(n)$	4	4	1	4	2	0	1	0
$y'(n+5)$	0	1	0	0	0	0	0	0
$y(n)$	4	5	1	4	2	–	–	–

Therefore, $y(n) = 4\delta(n) + 5\delta(n-1) + \delta(n-2) + 4\delta(n-3) + 2\delta(n-4)$

6.14 Consider the two sequences

$$x(n) = \delta(n) + 3\delta(n-1) + 3\delta(n-2) + 2\delta(n-3)$$
$$h(n) = \delta(n) + \delta(n-1) + \delta(n-2) + \delta(n-3)$$

If we form the product

$$Y(k) = X(k)H(k)$$

where $X(k)$ and $H(k)$ are the five-point DFTs of $x(n)$ and $h(n)$, respectively, and take the inverse DFT to form the sequence $y(n)$, find the sequence $y(n)$.

Because $Y(k)$ is the product of two 5-point DFTs, $H(k)$ and $X(k)$, $y(n)$ is the five-point circular convolution of $h(n)$ with $x(n)$. We may find $y(n)$ by performing the circular convolution analytically (or graphically) or by finding the linear convolution and aliasing the result or by multiplying DFTs and finding the inverse DFT. In this problem, because $h(n)$ is a simple sequence, we will use the analytic approach.

The five-point circular convolution of $x(n)$ with $h(n)$ is

$$y(n) = x(n) \circledS h(n) = \sum_{k=0}^{4} h(k)x((n-k))_5 \qquad n = 0, 1, 2, 3, 4$$

Because $h(n) = 1$ for $n = 0, 1, 2, 3$, and $h(4) = 0$, the five-point convolution is

$$y(n) = x(n) \circledS h(n) = \sum_{k=0}^{3} x((n-k))_5 \qquad n = 0, 1, 2, 3, 4$$

Therefore, the circular convolution is equal to the sum of the values of the circularly shifted sequence $x((n-k))_5$ from $k = 0$ to $k = 3$. Because $x(n)$ is

$$x(n) = [1, \ 3, \ 3, \ 2, \ 0]$$

(recall that $x(n)$ is considered to be a sequence of length five), $x((-n))_5$ is formed by reading the sequence values *backward*, beginning with $n = 0$:

$$x((-n))_5 = [1, \ 0, \ 2, \ 3, \ 3]$$

Thus, $y(0)$ is the sum of the first four values of $x((-n))_5$, which gives $y(0) = 6$. Circularly shifting this sequence to the right by 1, we have

$$x((1-n))_5 = [3, \ 1, \ 0, \ 2, \ 3]$$

and summing the first four values gives $y(1) = 6$. Continuing with this process, we find $y(2) = 7$, $y(3) = 9$, and $y(4) = 8$.

6.15   Let $x(n)$ and $h(n)$ be finite-length sequences that are six points long, and let $X(k)$ and $H(k)$ be the eight-point DFTs of $x(n)$ and $h(n)$, respectively. If we form the product

$$Y(k) = X(k)H(k)$$

and take the inverse DFT to form the sequence $y(n)$, find the values of n for which $y(n)$ is equal to the linear convolution

$$z(n) = \sum_{k=-\infty}^{\infty} x(k)h(n-k)$$

If the linear convolution of two sequences is M points long, for an N-point circular convolution with $N < M$, the first $M - N$ points will be aliased. With $x(n)$ and $h(n)$ both of length six, $z(n) = x(n) * h(n)$ will be 11 points long. Therefore, with an eight-point circular convolution, the first three points will be aliased, and the last five will be equal to the linear convolution.

6.16   If $Y(k) = H(k)X(k)$ where $H(k)$ and $X(k)$ are the N-point DFTs of the finite-length sequences $h(n)$ and $x(n)$, respectively, show that

$$y(n) = \left[\sum_{k=0}^{N-1} \tilde{h}(k)\tilde{x}(n-k) \right] \mathcal{R}_N(n)$$

The sequence that has an N-point DFT equal to $Y(k) = H(k)X(k)$ is

$$y(n) = \frac{1}{N} \sum_{k=0}^{N-1} H(k)X(k)W_N^{-nk} \qquad n = 0, 1, \ldots, N - 1$$

Because we would like to express $y(n)$ in terms of $x(n)$ and $h(n)$, let us substitute

$$H(k) = \sum_{l=0}^{N-1} h(l) W_N^{lk}$$

into the expression for $y(n)$ as follows:

$$y(n) = \frac{1}{N} \sum_{k=0}^{N-1} X(k) \sum_{l=0}^{N-1} h(l) W_N^{lk} W_N^{-nk} \qquad n = 0, 1, \ldots, N-1$$

Interchanging the order of the summations gives

$$y(n) = \sum_{l=0}^{N-1} h(l) \left[\frac{1}{N} \sum_{k=0}^{N-1} X(k) W_N^{-k(n-l)} \right] \qquad n = 0, 1, \ldots, N-1$$

However, note that the term in brackets is equal to $x((n-l))_N$. Therefore, it follows that

$$y(n) = \sum_{l=0}^{N-1} h(l) x((n-l))_N \qquad n = 0, 1, \ldots, N-1$$

which is equivalent to

$$y(n) = \left[\sum_{k=0}^{N-1} \tilde{h}(k) \tilde{x}(n-k) \right] \mathcal{R}_N(n)$$

as was to be shown.

6.17   Let $y(n)$ be the linear convolution of the two finite-length sequences, $h(n)$ and $x(n)$, of length N,

$$y(n) = h(n) * x(n)$$

and let $y_N(n)$ be the N-point circular convolution

$$y_N(n) = h(n) \circledN x(n) = \left[\sum_{k=0}^{N-1} h(k) \tilde{x}(n-k) \right] \mathcal{R}_N(n)$$

Derive the following relationship between $y(n)$ and $y_N(n)$:

$$y_N(n) = \left[\sum_{k=-\infty}^{\infty} y(n+kN) \right] \mathcal{R}_N(n)$$

There are several ways to derive this relationship. One is to examine what happens when the DTFT of $y(n)$ is sampled. Alternatively, this result may be derived from a systems point of view as follows. First, note that $y_N(n)$ is equal to one period of the *linear* convolution of the finite-length sequence $h(n)$ with the periodic sequence $\tilde{x}(n)$:

$$y_N(n) = [h(n) * \tilde{x}(n)] \mathcal{R}_N(n)$$

If we let

$$p_N(n) = \sum_{k=-\infty}^{\infty} \delta(n-kN)$$

then the periodic sequence $\tilde{x}(n)$ is formed by linearly convolving $x(n)$ with $p_N(n)$:

$$\tilde{x}(n) = x(n) * p_N(n)$$

Therefore, the N-point circular convolution may be written as

$$y_N(n) = \{h(n) * [x(n) * p_N(n)]\} \mathcal{R}_N(n)$$

which is illustrated in the figure below.

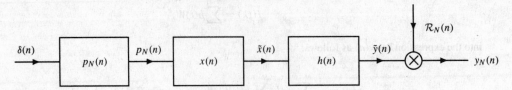

Because the first three systems are linear and shift-invariant, the order of these systems may be interchanged as illustrated in the following figure:

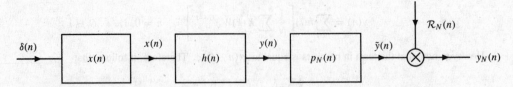

However, note that the output of the second filter, $y(n)$, is the linear convolution of $h(n)$ with $x(n)$. This sequence is then convolved with $p_N(n)$, which gives the periodic sequence

$$\tilde{y}(n) = y(n) * p_N(n) = \sum_{k=-\infty}^{\infty} y(n + kN)$$

This sequence is then multiplied by the rectangular window $\mathcal{R}_N(n)$,

$$y_N(n) = \left[\sum_{k=-\infty}^{\infty} y(n + kN) \right] \mathcal{R}_N(n)$$

which is the relationship that was to be established.

6.18 How may we compute the N-point DFT of two real-valued sequences, $x_1(n)$ and $x_2(n)$, using one N-point DFT?

The DFTs of two real-valued sequences may be found from one N-point DFT as follows. First, we form the N-point complex sequence

$$x(n) = x_1(n) + jx_2(n)$$

After finding the N-point DFT of $x(n)$, we extract $X_1(k)$ and $X_2(k)$ from $X(k)$ by exploiting the symmetry properties of the DFT. Specifically, recall that the DFT of a real-valued sequence is conjugate symmetric,

$$X(k) = X^*((N - k))_N$$

and the DFT of an imaginary sequence is conjugate antisymmetric,

$$X(k) = -X^*((N - k))_N$$

Therefore, because

$$X(k) = X_1(k) + X_2(k)$$

with $X_1(k)$ the DFT of a real-valued sequence, then

$$X_1(k) = \tfrac{1}{2}[X(k) + X^*((N - k))_N]$$

which is the conjugate symmetric part of $X(k)$. Similarly, because $X_2(k)$ is the DFT of an imaginary sequence,

$$X_2(k) = \tfrac{1}{2}[X(k) - X^*((N - k))_N]$$

which is the conjugate antisymmetric part of $X(k)$.

6.19   Let $x_1(n)$ and $x_2(n)$ be N-point sequences with N-point DFTs $X_1(k)$ and $X_2(k)$, respectively. Find an expression for the N-point DFT of the product $x(n) = x_1(n)x_2(n)$ in terms of $X_1(k)$ and $X_2(k)$.

Just as we have seen with the DTFT, there is a duality in the DFT properties. We have seen, for example, that multiplying a sequence by a complex exponential results in a circular shift of the DFT coefficients. Similarly, multiplying the DFT coefficients by a complex exponential results in a circular shift of the sequence. Therefore, because multiplying DFT coefficients corresponds to a circular convolution of the sequences, we expect that the multiplication of two sequences would result in the circular convolution of their DFTs. To establish this property, we begin by noting that

$$X(k) = \sum_{n=0}^{N-1} x_1(n)x_2(n)W_N^{nk} \tag{6.21}$$

Because we would like to express $X(k)$ in terms of $X_1(k)$ and $X_2(k)$, we substitute the following expression for $x_2(n)$ into Eq. (6.21):

$$x_2(n) = \frac{1}{N} \sum_{l=0}^{N-1} X_2(l)W_N^{-ln}$$

The result is

$$X(k) = \frac{1}{N} \sum_{n=0}^{N-1} x_1(n)W_N^{nk} \sum_{l=0}^{N-1} X_2(l)W_N^{-ln}$$

Interchanging the order of the summations, this becomes

$$X(k) = \frac{1}{N} \sum_{l=0}^{N-1} X_2(l) \sum_{n=0}^{N-1} x_1(n)W_N^{n(k-l)}$$

Recognizing that the second sum is $X_1((k-l))_N$, we have

$$X(k) = \frac{1}{N} \sum_{l=0}^{N-1} X_2(l)X_1((k-l))_N$$

Therefore, $X(k)$ is $1/N$ times the circular convolution of $X_1(k)$ with $X_2(k)$:

$$X(k) = \frac{1}{N} X_1(k) \, \circledN \, X_2(k)$$

6.20   If $x_1(n)$ and $x_2(n)$ are N-point sequences with N-point DFTs $X_1(k)$ and $X_2(k)$, respectively, show that

$$\sum_{n=0}^{N-1} x_1(n)x_2^*(n) = \frac{1}{N} \sum_{k=0}^{N-1} X_1(k)X_2^*(k) \tag{6.22}$$

This result is easy to derive if we use the properties of the DFT that we already have. First, note that if $X(k)$ is the N-point DFT of $x(n) = x_1(n)x_2^*(n)$, then

$$X(0) = \sum_{n=0}^{N-1} x(n) = \sum_{n=0}^{N-1} x_1(n)x_2^*(n)$$

Second, note that the DFT of $x_2^*(n)$ is $X_2^*((-k))_N$,

$$x_2^*(n) \overset{DFT}{\Longleftrightarrow} X_2^*((-k))_N = X_2^*((N-k))_N \qquad k = 0, 1, \ldots, N-1$$

Finally, recall that if $x(n) = x_1(n)x_2^*(n)$, the N-point DFT of $x(n)$ is $1/N$ times the circular convolution of $X_1(k)$ and $X_2^*((-k))_N$ (see Prob. 6.19):

$$X(k) = \frac{1}{N} \sum_{l=0}^{N-1} X_1(l)X_2^*((k+l))_N$$

Evaluating $X(k)$ at $k = 0$, we have the desired result

$$\sum_{n=0}^{N-1} x_1(n)x_2^*(n) = \frac{1}{N}\sum_{l=0}^{N-1} X_1(l)X_2^*(l)$$

6.21   Evaluate the sum

$$S = \sum_{n=0}^{N-1} x_1(n)x_2^*(n)$$

when

$$x_1(n) = \cos\left(\frac{2\pi n k_1}{N}\right) \qquad x_2(n) = \cos\left(\frac{2\pi n k_2}{N}\right)$$

The N-point DFT of $x_1(n)$ is zero, except for $k = k_1$ and $k = N - k_1$, when it is equal to $N/2$. Similarly, the N-point DFT of $x_2(n)$ is zero, except for $k = k_2$ and $k = N - k_2$, when it is equal to $N/2$. Using the results of Prob. 6.20,

$$\sum_{n=0}^{N-1} x_1(n)x_2^*(n) = \frac{1}{N}\sum_{k=0}^{N-1} X_1(k)X_2^*(k)$$

we see that if $k_1 = k_2$ (or $k_1 = N - k_2$),

$$\sum_{n=0}^{N-1} x_1(n)x_2^*(n) = \frac{1}{N}\left[\frac{N^2}{4} + \frac{N^2}{4}\right] = \frac{N}{2}$$

and the sum is equal to zero otherwise.

6.22   Let $x(n)$ be an N-point sequence with an N-point DFT $X(k)$. Derive Parseval's theorem,

$$\sum_{n=0}^{N-1} |x(n)|^2 = \frac{1}{N}\sum_{k=0}^{N-1} |X(k)|^2$$

This property is an immediate consequence of the property derived in Prob. 6.20. Specifically, Parseval's theorem follows from Eq. (6.22) by setting $x_1(n) = x_2(n) = x(n)$.

6.23   Let $x(n)$ be a sequence that is zero outside the interval $[0, N - 1]$ with a z-transform $X(z)$. Listed below are four sequences of length $2N$ that are derived from $x(n)$. Find an expression for the DFT of each sequence in terms of samples of $X(z)$.

(a) $y_1(n) = \begin{cases} x(n) & 0 \le n < N \\ 0 & N \le n \le 2N \end{cases}$

(b) $y_2(n) = x(n) + x(n - N)$

(c) $y_3(n) = \begin{cases} x\left(\dfrac{n}{2}\right) & n \text{ even} \\ 0 & n \text{ odd} \end{cases}$

(a)   The $2N$-point sequence $y_1(n)$ is formed from $x(n)$ by padding with zeros. Therefore, $Y_1(k)$ corresponds to $2N$ equally spaced samples of $X(z)$ around the unit circle:

$$Y_1(k) = X\left(e^{j\frac{2\pi}{2N}k}\right)$$

(b)   The sequence $y_2(n)$ is formed by adding $x(n - N)$ to $x(n)$ (i.e., a delayed version of $x(n)$). If $X(z)$ is the z-transform of $x(n)$, the z-transform of $x(n - N)$ is $z^{-N}X(z)$. Therefore,

$$Y_2(z) = [1 + z^{-N}]X(z)$$

and $Y_2(k)$ is

$$Y_2(k) = [(1 + z^{-N})X(z)]_{z=\exp\{j2\pi k/2N\}} = \left[1 + e^{-j\frac{2\pi}{2N}Nk}\right]X\left(e^{j\frac{2\pi}{2N}k}\right)$$

$$= [1 + (-1)^k]X\left(e^{j\frac{2\pi}{2N}k}\right) = \begin{cases} 2X\left(e^{j\frac{2\pi}{2N}k}\right) & k \text{ even} \\ 0 & k \text{ odd} \end{cases}$$

(c) The third sequence is formed by *up-sampling* by a factor of 2 (i.e., by stretching $x(n)$ in time by a factor of 2 and inserting a zero between each sample). The z-transform of $y_3(n)$ is

$$Y_3(z) = \sum_{\substack{n=0 \\ n \text{ even}}}^{2N-1} x\left(\frac{n}{2}\right)z^{-n} = \sum_{n=0}^{N-1} x(n)z^{-2n} = X(z^2)$$

Therefore, the $2N$-point DFT is

$$Y_3(k) = Y_3(z)|_{z=e^{j2\pi k/2N}} = X\left(e^{j2\pi k/N}\right) \qquad k = 0, 1, \dots, 2N-1$$

Thus, the DFT coefficients $Y_3(k)$ correspond to two periods of the coefficients $X(k)$.

Sampling the DTFT

6.24 Let $h(n)$ be a finite-length sequence of length N with $h(n) = 0$ for $n < 0$ and $n \geq N$. The discrete-time Fourier transform of $h(n)$ is sampled at $3N$ equally spaced points:

$$\omega_k = \frac{2\pi k}{3N} \qquad k = 0, 1, \dots, 3N-1$$

Find the sequence $g(n)$ that is the inverse DFT of the $3N$ samples $H(k) = H(e^{j\omega_k})$.

Because $h(n)$ is a finite-length sequence of length N, it may be recovered from its N-point DFT, which corresponds to N equally spaced samples of $H(e^{j\omega})$. A sequence of length N may also be considered to be a sequence of length $3N$, with the last $2N$ samples having a value of zero. The inverse DFT of the $3N$ equally spaced samples of $H(e^{j\omega})$ corresponds to this $3N$-point sequence. Thus,

$$g(n) = \begin{cases} h(n) & n = 0, 1, \dots, N-1 \\ 0 & \text{else} \end{cases}$$

6.25 Consider the finite-length sequence

$$x(n) = \begin{bmatrix} 1, 1, 1, 1, 1, 1 \end{bmatrix}$$

and let $X(z)$ be its z-transform. If we sample $X(z)$ at $z_k = \exp(j\frac{2\pi}{4}k)$ for $k = 0, 1, 2, 3$, we obtain a set of DFT coefficients $X(k)$. Find the sequence, $y(n)$, that has a four-point DFT equal to these samples.

Sampling $X(z)$ at four equally spaced points around the unit circle produces an aliased version of $x(n)$:

$$y(n) = \left[\sum_{k=-\infty}^{\infty} x(n - 4k)\right]\mathcal{R}_4(n)$$

Using the tabular method to evaluate this sum, noting that $x(n)$ and $x(n+4)$ are the only sequences that have nonzero values for $0 \leq n \leq 3$, we have

n	0	1	2	3	4	5	6	7
$x(n)$	1	1	1	1	1	1	0	0
$x(n+4)$	1	1	0	0	0	0	0	0
$y(n)$	2	2	1	1	–	–	–	–

Therefore, $y(n) = 2\delta(n) + 2\delta(n-1) + \delta(n-2) + \delta(n-3)$

6.26 Consider a finite-length sequence $x(n)$ that is zero outside the interval $[0, N-1]$. Suppose that we form a new sequence $\tilde{x}(n)$ as follows:

$$\tilde{x}(n) = \sum_{k=-\infty}^{\infty} x(n - kM)$$

where $M < N$. Find the M-point DFT of the sequence

$$y(n) = \begin{cases} \tilde{x}(n) & 0 \le n < M \\ 0 & \text{otherwise} \end{cases}$$

expressing the answer in terms of the DTFT of $x(n)$.

This problem is most easily solved if we take advantage of what we know about the DFT. Recall that the M-point DFT corresponds to M samples of the DTFT, $X(e^{j\omega})$, at $\omega_k = 2\pi k/M$ for $k = 0, 1, \ldots, M-1$. In addition, if these samples are used for the DFT coefficients of a sequence $y(n)$ of length M, $y(n)$ is related to $x(n)$ as follows:

$$y(n) = \left\{ \sum_{k=-\infty}^{\infty} x(n - kM) \right\} \mathcal{R}_N(n)$$

which is the sequence defined above. Therefore, the M-point DFT of $y(n)$ is

$$Y(k) = X(e^{j\omega})|_{\omega = 2\pi k/M}$$

6.27 The unit sample response of a single pole filter is

$$h(n) = \left(\tfrac{1}{3}\right)^n u(n)$$

The frequency response of this filter is sampled at $\omega_k = 2\pi k/16$ for $k = 0, 1, \ldots, 15$. The resulting samples are

$$G(k) = H(e^{j\omega})|_{\omega = 2\pi k/16} \qquad k = 0, 1, 2, \ldots, 15$$

Find $g(n)$, the 16-point inverse DFT of $G(k)$.

The straightforward but tedious way to solve this problem would be to find the DTFT of $h(n)$,

$$H(e^{j\omega}) = \frac{1}{1 - \tfrac{1}{3} e^{-j\omega}}$$

sample it at the given frequencies,

$$G(k) = H(e^{j\omega})|_{\omega = 2\pi k/16} = \frac{1}{1 - \tfrac{1}{3} e^{-j2\pi k/16}} \qquad k = 0, 1, \ldots, 15$$

and then find the inverse DFT. Another approach is to use the *frequency sampling theorem* given in Eq. (*6.13*), which states that if the DTFT of a sequence $h(n)$ is sampled at N equally spaced frequencies between zero and 2π, the sequence $g(n)$ that has these samples as its DFT coefficients is the time-aliased sequence

$$g(n) = \left\{ \sum_{k=-\infty}^{\infty} h(n - kN) \right\} \mathcal{R}_N(n)$$

Therefore, with $h(n) = \left(\tfrac{1}{3}\right)^n u(n)$, it follows that

$$g(n) = \left\{ \sum_{k=-\infty}^{\infty} \left(\tfrac{1}{3}\right)^{n-kN} u(n - kN) \right\} \mathcal{R}_N(n)$$

Because $u(n - kN)$ is equal to zero for $0 \le n \le N-1$ when $k \ge 0$, this sum may be simplified to

$$g(n) = \left\{ \sum_{k=-\infty}^{0} \left(\tfrac{1}{3}\right)^{n-kN} \right\} \mathcal{R}_N(n)$$

Evaluating the sum, we find

$$\sum_{k=-\infty}^{0} \left(\tfrac{1}{3}\right)^{n-kN} = \left(\tfrac{1}{3}\right)^{n} \sum_{k=0}^{\infty} \left(\tfrac{1}{3}\right)^{kN} = \left(\tfrac{1}{3}\right)^{n} \frac{1}{1-\left(\tfrac{1}{3}\right)^{N}}$$

Therefore,

$$g(n) = \left(\tfrac{1}{3}\right)^{n} \frac{1}{1-\left(\tfrac{1}{3}\right)^{N}} \mathcal{R}_N(n)$$

6.28 The DFT of a sequence $x(n)$ corresponds to N equally spaced samples of its z-transform, $X(z)$, around the unit circle starting at $z=1$.

(a) If we want to sample the z-transform on a circle of radius r, how should $x(n)$ be modified so that the DFT will correspond to samples of $X(z)$ at the desired radius?

(b) Suppose that we would like to shift the samples around the unit circle. In particular, consider the N samples that are equally spaced around the unit circle with the first sample at $z = \exp\{j\pi/N\}$. How should the sequence $x(n)$ be modified so that the DFT will correspond to samples of $X(z)$ at these points?

(a) The z-transform of $x(n)$ is

$$X(z) = \sum_{n=-\infty}^{\infty} x(n)z^{-n}$$

If we sample $X(z)$ at N equally spaced points around a circle of radius r, we have

$$X\left(re^{j\frac{2\pi}{N}k}\right) = \sum_{n=-\infty}^{\infty} x(n)\left(re^{j\frac{2\pi}{N}k}\right)^{-n} = \sum_{n=-\infty}^{\infty} x(n)r^{-n}e^{-j\frac{2\pi}{N}nk} \qquad k=0,1,\dots,N-1$$

which is the z-transform of $r^{-n}x(n)$ sampled at N equally spaced points around the unit circle. Therefore, N equally spaced samples of $X(z)$ around a circle of radius r may be found by computing the N-point DFT of $r^{-n}x(n)$.

(b) Here, we want to rotate the DFT samples by an amount equal to π/N. In other words, we would like to find

$$X\left(e^{j\frac{2\pi}{N}k+j\frac{\pi}{N}}\right) = \sum_{n=-\infty}^{\infty} x(n)\left(e^{j\frac{2\pi}{N}k+j\frac{\pi}{N}}\right)^{-n} = \sum_{n=-\infty}^{\infty} \left[x(n)e^{-j\frac{\pi}{N}n}\right]e^{-j\frac{2\pi}{N}nk}$$

Therefore, to find N samples of $X(z)$ that are equally spaced around the unit circle, with the first sample at $z = \exp\{j\pi/N\}$, we multiply $x(n)$ by $e^{-j\pi n/N}$ and find the N-point DFT of the resulting sequence.

Linear Convolution Using the DFT

6.29 Two finite-length sequences, $x_1(n)$ and $x_2(n)$, that are zero outside the interval $[0, 99]$ are circularly convolved to form a new sequence $y(n)$,

$$y(n) = x_1(n) \circledN x_2(n)$$

where $N = 100$. If $x_1(n)$ is nonzero only for $10 \le n \le 39$, determine the values of n for which $y(n)$ is guaranteed to be equal to the *linear* convolution of $x_1(n)$ and $x_2(n)$.

Because

$$y(n) = \sum_{k=0}^{99} x_2(k)x_1((n-k))_{100}$$

the values of n for which $y(n)$ is equal to the linear convolution of $x_1(n)$ with $x_2(n)$ are those values of n in the interval $[0, 99]$ for which the circular shift $x_1((n-k))_{100}$ is equal to the linear shift $x_1(n-k)$. With $x_1(n)$ nonzero only over the interval $[10, 39]$ we see that $x_1((n-k))_{100} = x_1(n-k)$ for n in the interval $[39, 99]$. Therefore, the circular convolution and the linear convolution are equal for $39 \le n \le 99$.

6.30   We would like to linearly convolve a 3000-point sequence with a linear shift-invariant filter whose unit sample response is 60 points long. To utilize the computational efficiency of the fast Fourier transform algorithm, the filter is to be implemented using 128-point discrete Fourier transforms and inverse discrete Fourier transforms. If the overlap-add method is used, how many DFTs are needed to complete the filtering operation?

With overlap-add, $x(n)$ is partitioned into nonoverlapping sequences of length M. If $h(n)$ is of length L, $x_i(n) * h(n)$ is of length $L + M - 1$. Therefore, we must use a DFT of length $N \geq L + M - 1$. Here, we have set $N = 128$, and $h(n)$ is of length $L = 60$. Therefore, $x(n)$ must be partitioned into sequences of length

$$M = N - L + 1 = 69$$

Because $x(n)$ is 3000 points long, we will have 44 sequences (with the last sequence containing only 33 nonzero values). Thus, to perform the convolution we need:

1. One DFT to compute $H(k)$
2. 44 DFTs for $X_i(k)$
3. 44 inverse DFTs for $Y_i(k) = H(k)X_i(k)$

for a total of 45 DFTs and 44 inverse DFTs.

6.31   Suppose that we are given 10 s of speech that has been sampled at a rate of 8 kHz and that we would like to filter it with an FIR filter $h(n)$ of length $L = 64$. Using the overlap-save method with 1024-point DFTs, how many DFTs and inverse DFTs are necessary to perform the convolution?

Sampling 10 s of speech with a sampling frequency of 8 kHz generates $N = 10 \cdot 8000 = 8 \cdot 10^4$ samples $x(n)$. If we segment the speech into records of length 1024 and perform the circular convolution of these segments with $h(n)$, the first 63 values will be aliased, and the last $1024 - 63 = 961$ values will be equivalent to the linear convolution. Therefore, each circular convolution generates 961 valid data points. Because the filtered signal $y(n) = x(n) * h(n)$ is of length $8 \cdot 10^4 + 63 = 80,063$, $x(n)$ will be segmented into

$$L = \frac{80,063}{961} = 83.3$$

or 84 overlapping sections. Therefore, to perform the convolution, we need 85 DFTs and 84 inverse DFTs.

6.32   If an 8-point sequence $x(n)$ is convolved (linearly) with a 3-point sequence $h(n)$, the result is a 10-point sequence $y(n) = x(n) * h(n)$. Suppose that we would like to construct the entire output $y(n)$ from two 6-point circular convolutions:

$$y_1(n) = x_1(n) \;⑥\; g(n) \qquad y_2(n) = x_2(n) \;⑥\; g(n)$$

where

$$g(n) = \begin{cases} h(n) & n = 0, 1, 2 \\ 0 & n = 3, 4, 5 \end{cases}$$

$$x_1(n) = \begin{cases} x(n) & n = 0, 1, 2, 3 \\ 0 & n = 4, 5 \end{cases}$$

$$x_2(n) = \begin{cases} x(n+4) & n = 0, 1, 2, 3 \\ 0 & n = 4, 5 \end{cases}$$

If the values of $y_1(n)$ and $y_2(n)$ are as tabulated below,

n	0	1	2	3	4	5
$y_1(n)$	1	−2	−3	2	1	3
$y_2(n)$	2	−3	−4	3	−2	−2

find the sequence $y(n)$.

In this problem, note that $x(n)$ has been partitioned into two 4-point sequences, $x_1(n)$ and $x_2(n)$. Because $h(n)$ is of length three, the linear convolution of $h(n)$ with $x_1(n)$ and $x_2(n)$ are both of length six. Therefore, the six-point circular convolutions are equal to the linear convolution, and $y(n) = x(n) * h(n)$ is given by

$$y(n) = y_1(n) + y_2(n - 4)$$

which is tabulated below.

n	0	1	2	3	4	5	6	7	8	9
$y_1(n)$	1	-2	-3	2	1	3	0	0	0	0
$y_2(n - 4)$	–	–	–	–	2	-3	-4	3	-2	-2
$y(n)$	1	-2	-3	2	3	0	-4	3	-2	-2

Applications

6.33 Consider a linear shift-invariant system characterized by the linear constant coefficient difference equation

$$y(n) = \sum_{k=1}^{p} a(k)y(n - k) + \sum_{k=0}^{q} b(k)x(n - k)$$

Describe a method that may be used to plot N samples of the frequency response $H(e^{j\omega})$ using N-point DFTs.

The unit sample response of the linear shift-invariant system that is described by this difference equation is infinite in duration. Finding N samples of the frequency response can be accomplished by computing the N-point DFT of the time-aliased signal

$$g(n) = \left[\sum_{k=-\infty}^{\infty} h(n - kN) \right] \mathcal{R}_N(n)$$

However, an easier method is as follows. The system function is

$$H(z) = \frac{B(z)}{A(z)} = \frac{\sum_{k=0}^{q} b(k)z^{-k}}{1 - \sum_{k=1}^{p} a(k)z^{-k}}$$

Therefore, samples of $H(z)$ at N equally spaced points around the unit circle may be found as follows:

$$H(k) = \frac{B(k)}{A(k)}$$

where $A(k)$ and $B(k)$ are the N-point DFTs of the denominator and numerator sequences, respectively, that is,

$$A(k) = 1 - \sum_{k=1}^{p} a(k)e^{-j2\pi nk/N} \qquad k = 0, 1, \ldots, N - 1$$

$$B(k) = \sum_{k=0}^{q} b(k)e^{-j2\pi nk/N} \qquad k = 0, 1, \ldots, N - 1$$

6.34 In many applications, it is necessary to multiply a sequence by a window $w(n)$. Let $x(n)$ be an N-point sequence, and let $w(n)$ be a Hamming window:

$$w(n) = \tfrac{1}{2} + \tfrac{1}{2} \cos\left[\tfrac{2\pi}{N} \left(n - \tfrac{N}{2} \right) \right]$$

How would you find the DFT of the windowed sequence, $x(n)w(n)$, from the DFT of the unwindowed sequence?

Let us express the Hamming window in terms of complex exponentials:

$$w(n) = \tfrac{1}{2} + \tfrac{1}{4}e^{j\frac{2\pi}{N}(n-\frac{N}{2})} + \tfrac{1}{4}e^{-j\frac{2\pi}{N}(n-\frac{N}{2})}$$

$$= \tfrac{1}{2} + \tfrac{1}{4}e^{-j\pi}e^{j\frac{2\pi}{N}n} + \tfrac{1}{4}e^{j\pi}e^{-j\frac{2\pi}{N}n}$$

$$= \tfrac{1}{2} - \tfrac{1}{4}e^{j\frac{2\pi}{N}n} - \tfrac{1}{4}e^{-j\frac{2\pi}{N}n}$$

Therefore, $$x(n)w(n) = \tfrac{1}{2}x(n) - \tfrac{1}{4}e^{j\frac{2\pi}{N}n}x(n) - \tfrac{1}{4}e^{-j\frac{2\pi}{N}n}x(n)$$

Because the DFT of $e^{j2\pi n/N}x(n)$ is $X((k-1))_N$, and the DFT of $e^{-j2\pi n/N}x(n)$ is $X((k+1))_N$,

$$x(n)w(n) \overset{DFT}{\Longleftrightarrow} \tfrac{1}{2}X(k) - \tfrac{1}{4}X((k-1))_N - \tfrac{1}{4}X((k+1))_N$$

6.35 A signal $x_a(t)$ that is bandlimited to 10 kHz is sampled with a sampling frequency of 20 kHz. The DFT of $N = 1000$ samples of $x(n)$ is then computed, that is,

$$X(k) = \sum_{n=0}^{N-1} x(n)e^{-j\frac{2\pi}{N}nk}$$

with $N = 1000$.

(a) To what analog frequency does the index $k = 150$ correspond? What about $k = 800$?

(b) What is the spacing between the spectral samples?

(a) With a sampling frequency $\Omega_s = 2\pi/T_s = 2\pi(20 \cdot 10^3)$, the discrete frequency ω is related to the analog frequency Ω by

$$\omega = \Omega T_s$$

or

$$\omega = \frac{\Omega}{20,000}$$

With an N-point DFT, the DTFT is sampled at the N frequencies

$$\omega_k = \frac{2\pi}{N}k \qquad k = 0, 1, \ldots, N - 1$$

Therefore, $X(k)$ corresponds to an analog frequency of

$$\Omega_k = 20,000 \cdot \omega_k = \frac{2\pi}{N}20,000k$$

or

$$f_k = 20,000\frac{k}{N}$$

Thus, with $N = 1000$, the index $k = 150$ corresponds to $f = 3$ kHz.

For $k = 800$, we need to be careful. Because $X(e^{j\omega})$ is periodic,

$$X(e^{j\omega}) = X\left(e^{j(\omega+2\pi)}\right)$$

$k = 800$ corresponds to the frequency

$$\omega_k = \frac{2\pi}{N}k = \frac{2\pi}{N}(k - N) = -200\frac{2\pi}{N}$$

with $N = 1000$, this is $w_k = -0.4\pi$. In analog frequency, this corresponds to

$$\Omega_k = -0.4\pi \cdot 20,000 = -8000\pi$$

or

$$f_k = -4000 \text{ Hz}$$

(b) The spacing between spectral samples is

$$\Delta f = \frac{20{,}000}{N} = 20 \text{ Hz}$$

6.36 An important operation in digital signal processing is *correlation*. Correlations are used in applications such as target detection and frequency (spectrum) estimation. The *correlation* of two signals, $x_1(n)$ and $x_2(n)$, is defined by

$$r_{x_1,x_2}(m) = \sum_{n=-\infty}^{\infty} x_1(n)\, x_2^*(m+n)$$

which is the convolution of $x_1(m)$ with $x_2^*(-m)$,

$$r_{x_1,x_2}(m) = x_1(m) * x_2^*(-m)$$

Given two finite-length sequences $x_1(n)$ and $x_2(n)$ of length N, the N-point *circular correlation* is

$$r'_{x_1,x_2}(m) = \left\{ \sum_{n=0}^{N-1} \tilde{x}_1(n)\tilde{x}_2^*((m+n))_N \right\} R_N(n)$$

where $\tilde{x}_1(n)$ and $\tilde{x}_2^*(n)$ are the periodic extensions of the finite-length sequences $\tilde{x}_1(n)$ and $x_2^*(n)$, respectively. What is the DFT of the circular correlation $r'_{x_1,x_2}(m)$?

The circular correlation of $x_1(n)$ and $x_2^*(n)$ is one period of the periodic convolution of $\tilde{x}_1(n)$ with $\tilde{x}_2^*(-n)$. This, in turn, is equal to one period of the periodic convolution of $\tilde{x}_1(n)$ with $\tilde{x}^*(N-n)$, which is the same as the circular convolution of $x_1(n)$ with $x_2^*(N-n)$:

$$r'_{x_1,x_2}(m) = x_1(m) \, \textcircled{N} \, x_2^*(N-m)$$

Consequently, the DFT of $r'_{x_1,x_2}(m)$ is equal to the product of the DFTs of $x_1(m)$ and $x_2^*(N-m)$:

$$R'_{x_1,x_2}(k) = \text{DFT}\,[x_1(m)]\, \text{DFT}\left[x_2^*(N-m)\right]$$

Because the DFT of $x_2^*(N-m)$ is

$$\sum_{m=0}^{N-1} x_2^*(N-m)W_N^{mk} = \sum_{m=0}^{N-1} x_2^*(m)W_N^{(N-m)k} = \sum_{m=0}^{N-1} x_2^*(m)W_N^{-mk} = X_2^*(k)$$

then

$$r'_{x_1,x_2}(m) \overset{DFT}{\Longleftrightarrow} X_1(k)X_2^*(k)$$

6.37 Consider the two sequences

$$x_1(n) = \cos\left(\frac{2\pi n}{N}\right) \qquad x_2(n) = \sin\left(\frac{2\pi n}{N}\right)$$

(a) Find the N-point circular convolution of $x_1(n)$ with $x_2(n)$.

(b) Find the N-point circular *correlation* of $x_1(n)$ with itself (this is referred to as the *autocorrelation*).

(c) Find the N-point circular *correlation* of $x_1(n)$ with $x_2(n)$.

(a) We may perform the circular convolution of $x_1(n)$ with $x_2(n)$ in any one of several different ways. However, because there are only two coefficients in the DFT of $x_1(n)$ and $x_2(n)$ that are nonzero, the easiest approach is to find the inverse DFT of the product $X_1(k)X_2(k)$. With

$$x_1(n) = \tfrac{1}{2}\left[e^{j2\pi n/N} + e^{-j2\pi n/N}\right] = \tfrac{1}{2}\left[e^{j2\pi n/N} + e^{j2\pi(N-1)n/N}\right]$$

and

$$x_2(n) = \tfrac{1}{2j}\left[e^{j2\pi n/N} - e^{j2\pi(N-1)n/N}\right]$$

the N-point DFTs are

$$X_1(k) = \begin{cases} \dfrac{N}{2} & k = 1 \text{ and } k = N-1 \\ 0 & \text{else} \end{cases}$$

$$X_2(k) = \begin{cases} \dfrac{N}{2j} & k = 1 \\ -\dfrac{N}{2j} & k = N-1 \\ 0 & \text{else} \end{cases}$$

Therefore,

$$X_1(k)X_2(k) = \begin{cases} \dfrac{N^2}{4j} & k = 1 \\ -\dfrac{N^2}{4j} & k = N-1 \\ 0 & \text{else} \end{cases}$$

and the inverse DFT is a sinusoid with an amplitude of $N/2$:

$$x(n) = x_1(n) \text{Ⓝ} x_2(n) = \tfrac{1}{2}N \sin\left(\frac{2\pi n}{N}\right)$$

(b) For the DFT of the circular autocorrelation of $x_1(n)$, we have

$$R_x(k) = |X_1(k)|^2 = \begin{cases} \dfrac{N^2}{4} & k = 1 \text{ and } k = N-1 \\ 0 & \text{else} \end{cases}$$

and it follows that

$$r_x(n) = \tfrac{1}{2}N \cos\left(\frac{2\pi n}{N}\right)$$

(c) Finally, for the DFT of the circular correlation of $x_1(n)$ with $x_2(n)$, we have

$$R_{x_1x_2}(k) = X_1(k)X_2^*(k) = \begin{cases} -\dfrac{N^2}{4j} & k = 1 \\ \dfrac{N^2}{4j} & k = N-1 \\ 0 & \text{else} \end{cases}$$

Therefore,

$$r_{x_1x_2}(n) = -\tfrac{1}{2}N \sin\left(\frac{2\pi n}{N}\right)$$

6.38 It is known that $y(n)$ is the output of a stable LSI system that has a system function

$$H(z) = \frac{\beta}{1 - \alpha z^{-1}}$$

The input $x(n)$ is completely unknown, and we would like to recover $x(n)$ from $y(n)$. The following procedure is proposed for recovering part of $x(n)$ from $y(n)$.

1. Using N values of $y(n)$ for $0 \le n < N$, calculate the N-point DFT

$$Y(k) = \text{DFT}\{y(n)\}$$

2. Form the sequence

$$V(k) = \frac{1}{\beta}\left[1 - \alpha W_N^k\right]Y(k)$$

3.   Invert $V(k)$ to obtain $v(n)$

$$v(n) = \text{IDFT}\{V(k)\}$$

For what values of n in the range $n = 0, 1, \ldots, N - 1$ is it *true* that $x(n) = v(n)$?

To recover $x(n)$, ideally we would take $y(n)$ and convolve it with the *inverse system*

$$G(z) = \frac{1}{\beta}[1 - \alpha z^{-1}]$$

Therefore, $x(n)$ may be recovered exactly by convolving $y(n)$ with the FIR filter

$$g(n) = \frac{1}{\beta}[\delta(n) - \alpha\delta(n - 1)]$$

which is of length $L = 2$. Therefore, because multiplying the DFT of $g(n)$ with the DFT of $y(n)$ is equivalent to performing the circular convolution,

$$v(n) = g(n) \,\,\textcircled{N}\,\, y(n)$$

$v(n)$ will be equal to $x(n)$ only for $1 \le n \le N - 1$.

6.39   The N-point circular convolution of two sequences $x(n)$ and $h(n)$ of length N may be written in matrix form as follows:

$$\mathbf{y} = \mathbf{Hx}$$

where $\mathbf{H}$ is an $N \times N$ *circulant* matrix, and $\mathbf{x}$ and $\mathbf{y}$ are vectors that contain the signal values $x(0), x(1), \ldots, x(N - 1)$ and $y(0), y(1), \ldots, y(N - 1)$, respectively. Determine the form of the matrix $\mathbf{H}$.

The circular convolution of $x(n)$ with $h(n)$ is

$$y(n) \,\,\textcircled{N}\,\, h(n) = \left\{ \sum_{k=0}^{N-1} x(k)h((n - k))_N \right\} \mathcal{R}_N(n)$$

For example, $y(0)$ is the sum of the products of $x(k)$ with the circularly time-reversed sequence $h((-k))_N$:

$$y(0) = h((0))_N x(0) + h((-1))_N x(1) + h((-2))_N x(2) + \cdots + h((-N + 1))_N x(N - 1)$$
$$= h(0)x(0) + h(N - 1)x(1) + h(N - 2)x(2) + \cdots + h(1)x(N - 1)$$

Next, for $y(1)$, we circularly shift $h((-k))_N$ to the right by 1 and multiply by the sequence values $x(k)$:

$$y(1) = h(1)x(0) + h(0)x(1) + h(N - 1)x(2) + \cdots + h(2)x(N - 1)$$

This process continues until we get to the last value, $y(N - 1)$, which is

$$y(N - 1) = h(N - 1)x(0) + h(N - 2)x(1) + h(N - 3)x(2) + \cdots + h(0)x(N - 1)$$

If we arrange these equations in matrix form, we have

$$\begin{bmatrix} y(0) \\ y(1) \\ y(2) \\ \vdots \\ y(N - 1) \end{bmatrix} = \begin{bmatrix} h(0) & h(N - 1) & h(N - 2) & \cdots & h(1) \\ h(1) & h(0) & h(N - 1) & \cdots & h(2) \\ h(2) & h(1) & h(0) & \cdots & h(3) \\ \vdots & \vdots & \vdots & & \vdots \\ h(N - 1) & h(N - 2) & h(N - 3) & \cdots & h(0) \end{bmatrix} \begin{bmatrix} x(0) \\ x(1) \\ x(2) \\ \vdots \\ x(N - 1) \end{bmatrix}$$

Note that the second row of $\mathbf{H}$ is formed by circularly shifting the first row to the right by 1. This shift corresponds to a circular shift of the sequence $h(n)$. Similarly, the third row is formed by shifting the second row by 1, and so on. Due to this circular property, $\mathbf{H}$ is said to be a *circulant* matrix.

Supplementary Problems

Discrete Fourier Series

6.40   Find the DFS coefficients for the sequence

$$\tilde{x}(n) = \cos\left(\frac{2\pi n}{10}\right) + \sin\left(\frac{2\pi n}{10}\right)$$

6.41   Find the DFS coefficients for the sequence of period $N = 8$ whose first four values are equal to 1 and the last four are equal to 0.

6.42   If $\tilde{x}(n)$ is a periodic sequence with a period N,

$$x(n) = x(n + N)$$

$\tilde{x}(n)$ is also periodic with period $3N$. Let $\tilde{X}(k)$ denote the DFS coefficient of $\tilde{x}(n)$ when considered to be periodic with a period N, and let $\tilde{X}_3(k)$ be the DFS coefficients of $\tilde{x}(n)$ when considered to be periodic with a period $3N$. Express the DFS coefficients $\tilde{X}_3(k)$ in terms of $\tilde{X}(k)$.

6.43   If the DFS coefficients of a periodic sequence $\tilde{x}(n)$ are real, $\tilde{X}(k) = \tilde{X}^*(k)$, what does this imply about $\tilde{x}(n)$?

The Discrete Fourier Transform

6.44   Find the 10-point DFT of each of the following sequences:
(a) $x(n) = \delta(n) + \delta(n - 5)$
(b) $x(n) = u(n) - u(n - 6)$

6.45   Find the 10-point DFT of the sequence

$$x(n) = \cos\left(\frac{3\pi}{5}n\right) \cdot \sin\left(\frac{4\pi}{5}n\right)$$

6.46   Find the 10-point inverse DFT of

$$X(k) = \begin{cases} 3 & k = 0 \\ 2 & k = 3, 7 \\ 1 & \text{else} \end{cases}$$

6.47   Find the N-point DFT of the sequence

$$x(n) = (-1)^n \qquad 0 \le n \le N - 1$$

where N is an even number.

6.48   Find the 16-point inverse DFT of

$$X(k) = \cos\left(\frac{2\pi}{16}3k\right) + 3j \sin\left(\frac{2\pi}{16}5k\right)$$

DFT Properties

6.49   If $x(n)$ is a finite-length sequence of length four with a four-point DFT $X(k)$, find the four-point DFT of each of the following sequences in terms of $X(k)$:
(a) $x(n) + \delta(n)$

(b) $x((3 - n))_4$

(c) $\frac{1}{2}[x(n) + x^*((-n))_4]$

6.50   If $X(k)$ is the 10-point DFT of the sequence

$$x(n) = \delta(n - 1) + 2\delta(n - 4) - \delta(n - 7)$$

what sequence, $y(n)$, has a 10-point DFT

$$Y(k) = 2X(k) \cos\left(\frac{6\pi k}{N}\right)$$

6.51   If the 10-point DFTs of $x(n) = \delta(n) - \delta(n - 1)$ and $h(n) = u(n) - u(n - 10)$ are $X(k)$ and $H(k)$, respectively, find the sequence $w(n)$ that corresponds to the 10-point inverse DFT of the product $H(k)X(k)$.

6.52   Let $x(n)$ be a sequence that is zero outside the interval $[0, N - 1]$ with a z-transform $X(z)$. If

$$y(n) = x(n) + x(N - n)$$

find the $2N$-point DFT of $y(n)$, and express it in terms of $X(z)$.

6.53   If $x(n)$ is real and $x(n) = x(N - n)$, what can you say about the N-point DFT of $x(n)$?

6.54   If $x(n) = \delta(n) + 2\delta(n - 2) - \delta(n - 5)$ has a 10-point DFT $X(k)$, find the inverse DFT of (a) $\text{Re}[X(k)]$ and (b) $\text{Im}[X(k)]$.

6.55   If $x(n)$ has an N-point DFT $X(k)$, find the N-point DFT of $y(n) = \cos(2\pi n/N)x(n)$.

6.56   Find the inverse DFT of $Y(k) = |X(k)|^2$ where $X(k)$ is the 10-point DFT of the sequence $x(n) = u(n) - u(n - 6)$.

6.57   If $X(k)$ is the N-point DFT of $x(n)$, what is the N-point DFT of the sequence $y(n) = X(n)$?

6.58   Evaluate the sum

$$S = \sum_{n=0}^{15} x_1(n)x_2^*(n)$$

when

$$x_1(n) = \cos\left(\frac{3\pi n}{8}\right)$$

and

$$X_2(k) = 3 \qquad 0 \le k \le 15$$

Sampling the DTFT

6.59   The z-transform of the sequence

$$x(n) = u(n) - u(n - 7)$$

is sampled at five points around the unit circle,

$$X(k) = X(z)\big|_{z=e^{j2\pi k/5}} \qquad k = 0, 1, 2, 3, 4$$

Find the inverse DFT of $X(k)$.

Linear Convolution Using the DFT

6.60   How many DFTs and inverse DFTs of length $N = 128$ are necessary to linearly convolve a sequence $x(n)$ of length 1000 with a sequence $h(n)$ of length 64 using the overlap-add method? Repeat for the overlap-save method.

6.61   A sequence $x(n)$ of length $N_1 = 100$ is circularly convolved with a sequence $h(n)$ of length $N_2 = 64$ using DFTs of length $N = 128$. For what values of n will the circular convolution be equal to the linear convolution?

Applications

6.62   A continuous-time signal $x_a(t)$ is sampled with a sampling frequency of $f_s = 2$ kHz. If a 1000-point DFT of 1000 samples is computed, what is the spacing between the frequency samples $X(k)$ in terms of the analog frequency?

6.63   Given $X(k)$, the N-point DFT of $x(n)$, how would you compute the N-point DFT of the windowed sequence $y(n) = w(n) x(n)$ where $w(n)$ is a Blackman window,

$$w(n) = 0.42 - 0.5 \cos\frac{2\pi n}{N-1} + 0.08 \cos\frac{4\pi n}{N-1}; \qquad 0 \le n \le N-1$$

Answers to Supplementary Problems

6.40 $\tilde{X}(1) = 5 - 5j$ and $\tilde{X}(9) = 5 + 5j$, with $X(k) = 0$ for $k = 0$ and $k = 2, \dots, 8$.

6.41 $\tilde{X}(0) = 4, \tilde{X}(k) = 2/(1 - e^{-j\pi k/4})$ for $k = 1, 3, 5, 7$, and $\tilde{X}(k) = 0$ for $k = 2, 4, 6$.

6.42 $\tilde{X}_3(k) = \begin{cases} 3\tilde{X}\left(\dfrac{k}{3}\right) & k = 0, 3, \dots, 3N - 3 \\ 0 & \text{else} \end{cases}$

6.43 $\tilde{x}(n)$ is conjugate symmetric, $\tilde{x}(n) = \tilde{x}^*(N - n)$.

6.44   (a) $X(k) = 1 + e^{-j\pi k}$. (b) $X(k) = \dfrac{1 - e^{-j\frac{2\pi}{10}6k}}{1 - e^{-j\frac{2\pi}{10}k}}$.

6.45 $X(1) = X(7) = \frac{5}{2j}$ and $X(3) = X(9) = -\frac{5}{2j}$.

6.46 $x(n) = \delta(n) + \frac{1}{5} + \frac{1}{5} \cos(3\pi n/5)$.

6.47 $X(k) = N\delta\left(k - \frac{N}{2}\right)$.

6.48 $x(n) = \frac{1}{2}[\delta(n - 3) + \delta(n - 13)] + \frac{3}{2}[-\delta(n - 5) + \delta(n - 11)]$.

6.49   (a) $1 + X(k)$. (b) $W_4^{3k} X((4 - k))_4$. (c) $\mathrm{Re}[X(k)]$.

6.50 $y(n) = -\delta(n) + 2\delta(n - 1) + 2\delta(n - 7) + \delta(n - 8)$.

6.51 $w(n) = 0$.

6.52 $X(e^{j\pi k/N}) + (-1)^k X(e^{j\pi(2N-k)/N})$.

6.53 $X(k) = X(N - k)$.

6.54   (a) $x(n) = \delta(n) + \delta(n - 2) - \delta(n - 5) + \delta(n - 8)$. (b) $x(n) = \delta(n - 2) - \delta(n - 8)$.

6.55 $Y(k) = \frac{1}{2}[X((k - 1))_N + X((k + 1))_N]$ for $0 \le k \le N - 1$.

6.56 $y(n) = [6, 5, 4, 3, 2, 2, 2, 3, 4, 5]$.

6.57 $Nx((N - n))_N$.

6.58 3.

6.59 [2, 1, 1, 1, 1].

6.60 17 DFTs and 16 IDFTs for overlap-add and the same for overlap save.

6.61 $35 \leq n \leq 127$.

6.62 $\Delta f = 2$ kHz.

6.63 $Y(k) = 0.42X(k) - 0.25[X((k-1))_N + X((k+1))_N] + 0.04[X((k-2))_N + X((k+2))_N]$.

Chapter 7

The Fast Fourier Transform

7.1 INTRODUCTION

In Chap. 6 we saw that the discrete Fourier transform (DFT) could be used to perform convolutions. In this chapter we look at the computational requirements of the DFT and derive some *fast algorithms* for computing the DFT. These algorithms are known, generically, as *fast Fourier transforms* (FFTs). We begin with the radix-2 decimation-in-time FFT, an algorithm published in 1965 by Cooley and Tukey. We then look at mixed-radix FFT algorithms and the prime factor FFT.

7.2 RADIX-2 FFT ALGORITHMS

The N-point DFT of an N-point sequence $x(n)$ is

$$X(k) = \sum_{n=0}^{N-1} x(n) W_N^{nk} \tag{7.1}$$

Because $x(n)$ may be either real or complex, evaluating $X(k)$ requires on the order of N complex multiplications and N complex additions for each value of k. Therefore, because there are N values of $X(k)$, computing an N-point DFT requires N^2 complex multiplications and additions.

The basic strategy that is used in the FFT algorithm is one of "divide and conquer," which involves decomposing an N-point DFT into successively smaller DFTs. To see how this works, suppose that the length of $x(n)$ is even (i.e., N is divisible by 2). If $x(n)$ is *decimated* into two sequences of length $N/2$, computing the $N/2$-point DFT of each of these sequences requires approximately $(N/2)^2$ multiplications and the same number of additions. Thus, the two DFTs require $2(N/2)^2 = \frac{1}{2}N^2$ multiplies and adds. Therefore, if it is possible to find the N-point DFT of $x(n)$ from these two $N/2$-point DFTs in fewer than $N^2/2$ operations, a savings has been realized.

7.2.1 Decimation-in-Time FFT

The decimation-in-time FFT algorithm is based on splitting (decimating) $x(n)$ into smaller sequences and finding $X(k)$ from the DFTs of these decimated sequences. This section describes how this decimation leads to an efficient algorithm when the sequence length is a power of 2.

Let $x(n)$ be a sequence of length $N = 2^\nu$, and suppose that $x(n)$ is split (decimated) into two subsequences, each of length $N/2$. As illustrated in Fig. 7-1, the first sequence, $g(n)$, is formed from the even-index terms,

$$g(n) = x(2n) \qquad n = 0, 1, \ldots, \frac{N}{2} - 1$$

and the second, $h(n)$, is formed from the odd-index terms,

$$h(n) = x(2n + 1) \qquad n = 0, 1, \ldots, \frac{N}{2} - 1$$

In terms of these sequences, the N-point DFT of $x(n)$ is

$$X(k) = \sum_{n=0}^{N-1} x(n) W_N^{nk} = \sum_{n \text{ even}} x(n) W_N^{nk} + \sum_{n \text{ odd}} x(n) W_N^{nk}$$

$$= \sum_{l=0}^{\frac{N}{2}-1} g(l) W_N^{2lk} + \sum_{l=0}^{\frac{N}{2}-1} h(l) W_N^{(2l+1)k} \tag{7.2}$$

262

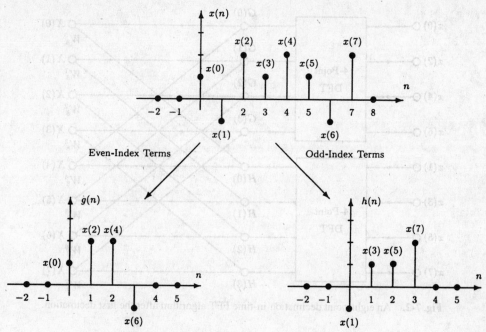

Fig. 7-1. Decimating a sequence of length $N = 8$ by a factor of 2.

Because $W_N^{2lk} = W_{N/2}^{lk}$, Eq. (7.2) may be written as

$$X(k) = \sum_{l=0}^{\frac{N}{2}-1} g(l)W_{N/2}^{lk} + W_N^k \sum_{l=0}^{\frac{N}{2}-1} h(l)W_{N/2}^{lk}$$

Note that the first term is the $N/2$-point DFT of $g(n)$, and the second is the $N/2$-point DFT of $h(n)$:

$$X(k) = G(k) + W_N^k H(k) \qquad k = 0, 1, \ldots, N - 1 \tag{7.3}$$

Although the $N/2$-point DFTs of $g(n)$ and $h(n)$ are sequences of length $N/2$, the periodicity of the complex exponentials allows us to write

$$G(k) = G\left(k + \frac{N}{2}\right) \qquad H(k) = H\left(k + \frac{N}{2}\right)$$

Therefore, $X(k)$ may be computed from the $N/2$-point DFTs $G(k)$ and $H(k)$. Note that because

$$W_N^{k+N/2} = W_N^k W_N^{N/2} = -W_N^k$$

then

$$W_N^{k+\frac{N}{2}} H\left(k + \frac{N}{2}\right) = -W_N^k H(k)$$

and it is only necessary to form the products $W_N^k H(k)$ for $k = 0, 1, \ldots, N/2 - 1$. The complex exponentials multiplying $H(k)$ in Eq. (7.3) are called *twiddle factors*. A block diagram showing the computations that are necessary for the first *stage* of an eight-point decimation-in-time FFT is shown in Fig. 7-2.

If $N/2$ is even, $g(n)$ and $h(n)$ may again be decimated. For example, $G(k)$ may be evaluated as follows:

$$G(k) = \sum_{n=0}^{\frac{N}{2}-1} g(n)W_{N/2}^{nk} = \sum_{n \text{ even}}^{\frac{N}{2}-1} g(n)W_{N/2}^{nk} + \sum_{n \text{ odd}}^{\frac{N}{2}-1} g(n)W_{N/2}^{nk}$$

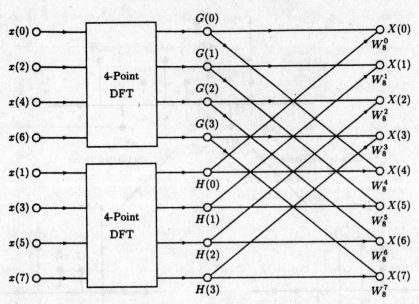

Fig. 7-2. An eight-point decimation-in-time FFT algorithm after the first decimation.

As before, this leads to

$$G(k) = \sum_{n=0}^{\frac{N}{4}-1} g(2n)W_{N/4}^{nk} + W_{N/2}^{k} \sum_{n=0}^{\frac{N}{4}-1} g(2n+1)W_{N/4}^{nk}$$

where the first term is the $N/4$-point DFT of the even samples of $g(n)$, and the second is the $N/4$-point DFT of the odd samples. A block diagram illustrating this decomposition is shown in Fig. 7-3. If N is a power of 2, the decimation may be continued until there are only two-point DFTs of the form shown in Fig. 7-4.

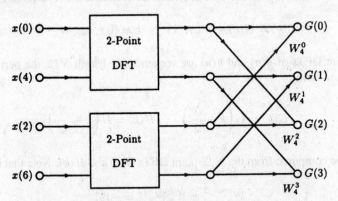

Fig. 7-3. Decimation of the four-point DFT into two two-point DFTs in the decimation-in-time FFT.

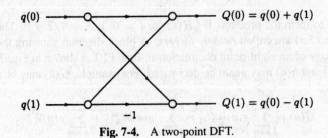

Fig. 7-4. A two-point DFT.

The basic computational unit of the FFT, shown in Fig. 7-5(a), is called a *butterfly*. This structure may be simplified by factoring out a term W_N^r from the lower branch as illustrated in Fig. 7-5(b). The factor that remains is $W_N^{N/2} = -1$. A complete eight-point radix-2 decimation-in-time FFT is shown in Fig. 7-6.

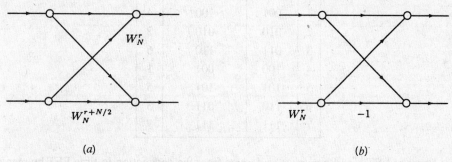

(a)   (b)

Fig. 7-5.   (*a*) The butterfly, which is the basic computational element of the FFT algorithm.
(*b*) A simplified butterfly, with only one complex multiplication.

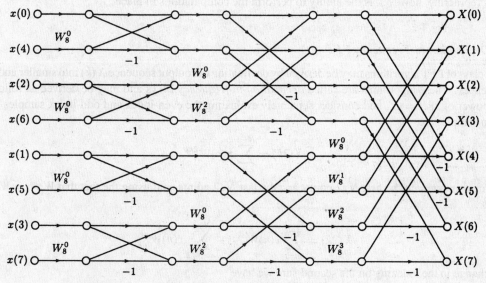

Fig. 7-6.   A complete eight-point radix-2 decimation-in-time FFT.

Computing an N-point DFT using a radix-2 decimation-in-time FFT is much more efficient than calculating the DFT directly. For example, if $N = 2^\nu$, there are $\log_2 N = \nu$ *stages* of computation. Because each stage requires $N/2$ complex multiplies by the twiddle factors W_N^r and N complex additions, there are a total of $\frac{1}{2} N \log_2 N$ complex multiplications[1] and $N \log_2 N$ complex additions.

From the structure of the decimation-in-time FFT algorithm, note that once a butterfly operation has been performed on a pair of complex numbers, there is no need to save the input pair. Therefore, the output pair may be stored in the same registers as the input. Thus, only one array of size N is required, and it is said that the computations may be performed *in place*. To perform the computations in place, however, the input sequence $x(n)$ must be stored (or accessed) in nonsequential order as seen in Fig. 7-6. The *shuffling* of the input sequence that takes place is due to the successive decimations of $x(n)$. The ordering that results corresponds to a bit-reversed indexing of the original sequence. In other words, if the index n is written in binary form, the order in which in the input sequence must be accessed is found by reading the binary representation for n in reverse order as illustrated in the table below for $N = 8$:

---

[1]The number of multiplications is actually a bit less than this because some of the twiddle factors are equal to 1.

n	Binary	Bit-Reversed Binary	n'
0	000	000	0
1	001	100	4
2	010	010	2
3	011	110	6
4	100	001	1
5	101	101	5
6	110	011	3
7	111	111	7

Alternate forms of FFT algorithms may be derived from the decimation-in-time FFT by manipulating the flowgraph and rearranging the order in which the results of each stage of the computation are stored. For example, the nodes of the flowgraph may be rearranged so that the input sequence $x(n)$ is in normal order. What is lost with this reordering, however, is the ability to perform the computations in place.

7.2.2   Decimation-in-Frequency FFT

Another class of FFT algorithms may be derived by decimating the output sequence $X(k)$ into smaller and smaller subsequences. These algorithms are called *decimation-in-frequency* FFTs and may be derived as follows. Let N be a power of 2, $N = 2^\nu$, and consider separately evaluating the even-index and odd-index samples of $X(k)$. The even samples are

$$X(2k) = \sum_{n=0}^{N-1} x(n) W_N^{2nk}$$

Separating this sum into the first $N/2$ points and the last $N/2$ points, and using the fact that $W_N^{2nk} = W_{N/2}^{nk}$, this becomes

$$X(2k) = \sum_{n=0}^{\frac{N}{2}-1} x(n) W_{N/2}^{nk} + \sum_{n=N/2}^{N-1} x(n) W_{N/2}^{nk}$$

With a change in the indexing on the second sum we have

$$X(2k) = \sum_{n=0}^{\frac{N}{2}-1} x(n) W_{N/2}^{nk} + \sum_{n=0}^{\frac{N}{2}-1} x\left(n + \frac{N}{2}\right) W_{N/2}^{(n+\frac{N}{2})k}$$

Finally, because $W_{N/2}^{(n+\frac{N}{2})k} = W_{N/2}^{nk}$,

$$X(2k) = \sum_{n=0}^{\frac{N}{2}-1} \left[x(n) + x\left(n + \frac{N}{2}\right) \right] W_{N/2}^{nk}$$

which is the $N/2$-point DFT of the sequence that is formed by adding the first $N/2$ points of $x(n)$ to the last $N/2$.
   Proceeding in the same way for the odd samples of $X(k)$ leads to

$$X(2k + 1) = \sum_{n=0}^{\frac{N}{2}-1} W_N^n \left[x(n) - x\left(n + \frac{N}{2}\right) \right] W_{N/2}^{nk} \tag{7.4}$$

A flowgraph illustrating this first stage of decimation is shown in Fig. 7-7. As with the decimation-in-time FFT, the decimation may be continued until only two-point DFTs remain. A complete eight-point decimation-in-frequency FFT is shown in Fig. 7-8. The complexity of the decimation-in-frequency FFT is the same as the decimation-in-time, and the computations may be performed in place. Finally, note that although the input sequence $x(n)$ is in normal order, the frequency samples $X(k)$ are in bit-reversed order.

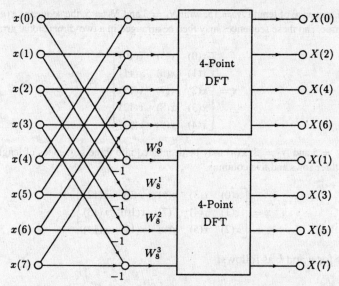

Fig. 7-7. An eight-point decimation-in-frequency FFT algorithm after the first stage of decimation.

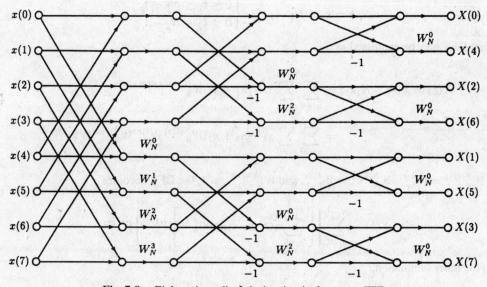

Fig. 7-8. Eight-point radix-2 decimation-in-frequency FFT.

7.3 FFT ALGORITHMS FOR COMPOSITE N

It is not always possible to work with sequences whose length is a power of 2. However, efficient computation of the DFT is still possible if the sequence length may be written as a product of factors. For example, suppose that N may be factored as follows:

$$N = N_1 \cdot N_2$$

We then decompose $x(n)$ into N_2 sequences of length N_1 and arrange these sequences in an array as follows:

$$\mathbf{x} = \begin{bmatrix} x(0) & x(N_2) & \cdots & x(N_2(N_1 - 1)) \\ x(1) & x(N_2 + 1) & \cdots & x(N_2(N_1 - 1) + 1) \\ \vdots & \vdots & & \vdots \\ x(N_2 - 1) & x(2N_2 - 1) & \cdots & x(N_1 N_2 - 1) \end{bmatrix} \tag{7.5}$$

EXAMPLE 7.3.1   For a sequence of length $N = 15$, with $N_1 = 3$ and $N_2 = 5$, the sequence $x(n)$ may be decimated into five sequences of length three, and these sequences may then be arranged in a two-dimensional array as follows:

$$\mathbf{x} = \begin{bmatrix} x(0) & x(5) & x(10) \\ x(1) & x(6) & x(11) \\ x(2) & x(7) & x(12) \\ x(3) & x(8) & x(13) \\ x(4) & x(9) & x(14) \end{bmatrix}$$

Alternatively, if we let $N_1 = 5$ and $N_2 = 3$, $x(n)$ may be decimated into three sequences of length five and arranged in a two-dimensional array of three rows and five columns,

$$\mathbf{x} = \begin{bmatrix} x(0) & x(3) & x(6) & x(9) & x(12) \\ x(1) & x(4) & x(7) & x(10) & x(13) \\ x(2) & x(5) & x(8) & x(11) & x(14) \end{bmatrix}$$

By defining *index maps* for n and k as follows,

$$n = N_2 \cdot n_1 + n_2 \qquad \begin{cases} 0 \le n_1 \le N_1 - 1 \\ 0 \le n_2 \le N_2 - 1 \end{cases}$$

$$k = k_1 + N_1 \cdot k_2 \qquad \begin{cases} 0 \le k_1 \le N_1 - 1 \\ 0 \le k_2 \le N_2 - 1 \end{cases}$$

the N-point DFT may be expressed as

$$X(k) = X(k_1 + N_1 k_2) = \sum_{n_2=0}^{N_2-1} \sum_{n_1=0}^{N_1-1} x(N_2 n_1 + n_2) W_N^{(k_1 + N_1 k_2)(N_2 n_1 + n_2)} \tag{7.6}$$

$$= \sum_{n_2=0}^{N_2-1} \sum_{n_1=0}^{N_1-1} x(N_2 n_1 + n_2) W_N^{N_2 k_1 n_1} W_N^{k_1 n_2} W_N^{N_1 k_2 n_2} W_N^{N_1 N_2 k_2 n_1}$$

Because $W_N^{N_2 k_1 n_1} = W_{N_1}^{k_1 n_1}$, $W_N^{N_1 k_2 n_2} = W_{N_2}^{k_2 n_2}$, and $W_N^{N_1 N_2 k_2 n_1} = 1$, the DFT becomes

$$X(k) = \sum_{n_2=0}^{N_2-1} \left\{ \left[\sum_{n_1=0}^{N_1-1} x(N_2 n_1 + n_2) W_{N_1}^{n_1 k_1} \right] W_N^{k_1 n_2} \right\} W_{N_2}^{k_2 n_2} \tag{7.7}$$

Note that the inner summation,

$$G(n_2, k_1) = \sum_{n_1=0}^{N_1-1} x(N_2 n_1 + n_2) W_{N_1}^{n_1 k_1}$$

is the N_1-point DFT of the sequence $x(N_2 n_1 + n_2)$, which is row n_2 of the two-dimensional array in Eq. (7.5). Computing the N_1-point DFT of each row of the array produces another array,

$$\mathbf{G} = \begin{bmatrix} G(0, 0) & G(0, 1) & \cdots & G(0, N_1 - 1) \\ G(1, 0) & G(1, 1) & \cdots & G(1, N_1 - 1) \\ \vdots & \vdots & & \vdots \\ G(N_2 - 1, 0) & G(N_2 - 1, 1) & \cdots & G(N_2 - 1, N_1 - 1) \end{bmatrix}$$

consisting of the complex numbers $G(n_2, k_1)$. Note that because the data in row n_2 is not needed after the N_1-point DFT of $x(N_2 n_1 + n_2)$ is computed, $G(n_2, k_1)$ may be stored in the same row (i.e., the computations may be performed in place).

The next step in the evaluation of $X(k)$ in Eq. (7.7) is to multiply by the twiddle factors $W_N^{k_1 n_2}$:

$$\tilde{G}(n_2, k_1) = W_N^{k_1 n_2} G(n_2, k_1)$$

The final step is to compute the N_2-point DFT of the columns of the array $\tilde{G}(n_2, k_1)$:

$$X(k_1 + N_1 k_2) = \sum_{n_2=0}^{N_2-1} \tilde{G}(n_2, k_1) W_{N_2}^{k_2 n_2}$$

The DFT coefficients are then read out *row-wise* from the two-dimensional array:

$$X(k) = X(k_1 + N_1 k_2)$$

A pictorial representation of this decomposition is shown in Fig. 7-9 for $N = 15$.

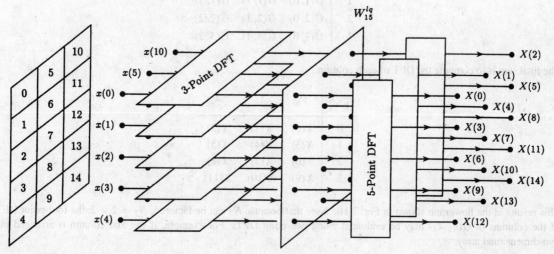

Fig. 7-9. Computation of a 15-point DFT with $N_1 = 3$ and $N_2 = 5$ using 3-point and 5-point DFTs.

EXAMPLE 7.3.2 Suppose that we want to compute the 12-point DFT of $x(n)$. With $N_1 = 3$ and $N_2 = 4$, the first step is to form a two-dimensional array consisting of $N_1 = 3$ columns and $N_2 = 4$ rows,

n_2 \ n_1	0	1	2
0	$x(0)$	$x(4)$	$x(8)$
1	$x(1)$	$x(5)$	$x(9)$
2	$x(2)$	$x(6)$	$x(10)$
3	$x(3)$	$x(7)$	$x(11)$

and compute the DFT of each row,

n_2 \ k_1	0	1	2
0	$G(0, 0)$	$G(0, 1)$	$G(0, 2)$
1	$G(1, 0)$	$G(1, 1)$	$G(1, 2)$
2	$G(2, 0)$	$G(2, 1)$	$G(2, 2)$
3	$G(3, 0)$	$G(3, 1)$	$G(3, 2)$

For example, the DFT of the first row is

$$G(0, k) = x(0) + x(4)W_3^k + x(8)W_3^{2k} \qquad k = 0, 1, 2$$

The next step is to multiply each term by the appropriate twiddle factor. The array of factors is

$$\begin{bmatrix} 1 & 1 & 1 \\ 1 & W_{12} & W_{12}^2 \\ 1 & W_{12}^2 & W_{12}^4 \\ 1 & W_{12}^3 & W_{12}^6 \end{bmatrix}$$

This produces the array $\tilde{G}(n_2, k_1)$:

n_2 \ k_1	0	1	2
0	$\tilde{G}(0,0)$	$\tilde{G}(0,1)$	$\tilde{G}(0,2)$
1	$\tilde{G}(1,0)$	$\tilde{G}(1,1)$	$\tilde{G}(1,2)$
2	$\tilde{G}(2,0)$	$\tilde{G}(2,1)$	$\tilde{G}(2,2)$
3	$\tilde{G}(3,0)$	$\tilde{G}(3,1)$	$\tilde{G}(3,2)$

The final step is to compute the DFT of each column:

k_2 \ k_1	0	1	2
0	$X(0)$	$X(1)$	$X(2)$
1	$X(3)$	$X(4)$	$X(5)$
2	$X(6)$	$X(7)$	$X(8)$
3	$X(9)$	$X(10)$	$X(11)$

This results in the flowgraph shown in Fig. 7-10. Note that because N_2 can be factored, $N_2 = 2 \times 2$, the four-point DFTs of the columns of $\tilde{G}(n_2, k_1)$ may be evaluated using two-point DFTs. For example, if the first column is arranged in a two-dimensional array,

$$\begin{bmatrix} \tilde{G}(0,0) & \tilde{G}(2,0) \\ \tilde{G}(1,0) & \tilde{G}(3,0) \end{bmatrix}$$

after taking the two-point DFTs of the rows, the terms are multiplied by the twiddle factors

$$\begin{bmatrix} 1 & 1 \\ 1 & W_4 \end{bmatrix} = \begin{bmatrix} 1 & 1 \\ 1 & -j \end{bmatrix}$$

and then the two-point DFTs of the columns are computed.

Up to this point, we have only assumed that N could be factored as $N = N_1 \cdot N_2$. It is possible, however, that either or both of these factors could be factored further. What is important for the FFT algorithm to be efficient is that N be a highly composite number:

$$N = N_1 \cdot N_2 \cdots N_\nu$$

In this case, it is possible to define multidimensional index maps for n and k as follows,

$$n = N_\nu n_1 + N_{\nu-1} n_2 + \cdots + n_\nu$$
$$k = k_1 + N_1 k_2 + \cdots + N_\nu k_\nu$$

and the development of the FFT algorithm proceeds as described above. If $N = R^\nu$, the corresponding FFT algorithm is called a *Radix-R algorithm*. If the factors are not equal, the FFT is called a *mixed-radix algorithm*.

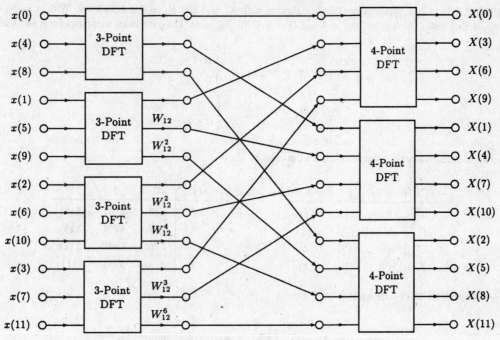

Fig. 7-10. FFT algorithm for $N = 12$.

7.4 PRIME FACTOR FFT

For some values of N, with the appropriate index mapping, it is possible to completely eliminate the twiddle factors. These mapping have the form

$$n = ((An_1 + Bn_2))_N \qquad \begin{cases} 0 \le n_1 \le N_1 - 1 \\ 0 \le n_2 \le N_2 - 1 \end{cases}$$

$$k = ((Ck_1 + Dk_2))_N \qquad \begin{cases} 0 \le k_1 \le N_1 - 1 \\ 0 \le k_2 \le N_2 - 1 \end{cases}$$

where A, B, C, and D are integers, and $((\cdot))_N$ denotes the evaluation of the index modulo N. If $N = N_1 \cdot N_2$, and if N_1 and N_2 are *relatively prime* (i.e., they have no common factors), the twiddle factors may be eliminated with the appropriate values for A, B, C, and D. The requirements on these numbers are as follows:

1. All numbers between 0 and $N - 1$ for n and k must appear uniquely as n_1 and n_2 are varied and as k_1 and k_2 are varied.

2. The numbers A, B, C, and D are such that

$$W_N^{(An_1 + Bn_2)(Ck_1 + Dk_2)} = W_{N_1}^{n_1 k_1} W_{N_2}^{n_2 k_2}$$

The second condition requires that

$$((AC))_N = N_2 \qquad ((BD))_N = N_1 \qquad ((AD))_N = ((BC))_N = 0$$

Finding a set of numbers that satisfies these two conditions falls in the domain of *number theory*, which will not be considered here. However, one set of numbers that satisfies these conditions is

$$A = N_2 \qquad B = N_1$$
$$C = N_2((N_2^{-1}))_{N_1} \qquad D = N_1((N_1^{-1}))_{N_2}$$

where $((N_1^{-1}))_{N_2}$ denotes the *multiplicative inverse* of N_1 modulo N_2. For example, if $N = 12$ with $N_1 = 3$ and $N_2 = 4$, $((4^{-1}))_3 = 1$ because $((4 \cdot 1))_3 = 1$ and $((3^{-1}))_4 = 3$ because $((3 \cdot 3))_3 = 1$.

EXAMPLE 7.4.1   A 12-point prime factor algorithm with $N_1 = 3$ and $N_2 = 4$ is as follows. With $A = N_2 = 4$ and $B = N_1 = 3$, and with $C = N_2((N_2^{-1}))_{N_1} = 4$ and $D = N_1((N_1^{-1}))_{N_2} = 9$. Thus, the index mappings for n and k are

$$n = ((4n_1 + 3n_2))_{12} \quad \begin{cases} 0 \le n_1 \le 2 \\ 0 \le n_2 \le 3 \end{cases}$$

$$k = ((4k_1 + 9k_2))_{12} \quad \begin{cases} 0 \le k_1 \le 2 \\ 0 \le k_2 \le 3 \end{cases}$$

and the two-dimensional array representation for the input is

n_2 \ n_1	0	1	2
0	$x(0)$	$x(4)$	$x(8)$
1	$x(3)$	$x(7)$	$x(11)$
2	$x(6)$	$x(10)$	$x(2)$
3	$x(9)$	$x(1)$	$x(5)$

k_2 \ k_1	0	1	2
0	$X(0)$	$X(4)$	$X(8)$
1	$X(9)$	$X(1)$	$X(5)$
2	$X(6)$	$X(10)$	$X(2)$
3	$X(3)$	$X(7)$	$X(11)$

The representation for $X(k)$ is therefore

$$X((4k_1 + 9k_2))_{12} = \sum_{n_2=0}^{3} \left\{ \left[\sum_{n_1=0}^{2} x((4n_1 + 3n_2))_{12} W_3^{n_1 k_1} \right] \right\} W_4^{n_2 k_2}$$

Thus, the DFT is evaluated by first computing the three-point DFT of each row of the input array, followed by the four-point DFT of each column. The following figure shows how the four-point DFTs are interconnected to the three-point DFTs.

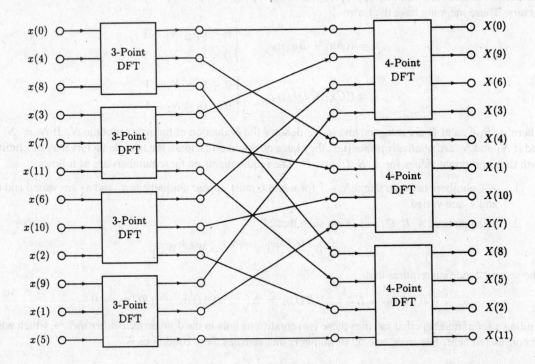

Because a 4-point DFT does not require any multiplications (see Prob. 7.11), and because each 3-point DFT requires only 4 complex multiplications, the 12-point prime factor algorithm requires 16 complex multiplies. For a mixed-radix FFT, there are, in addition, six twiddle factors. The cost for eliminating these six multiplications is an increase in complexity in indexing and in programming.

Solved Problems

Radix-2 FFT Algorithms

7.1 Assume that a complex multiply takes 1 μs and that the amount of time to compute a DFT is determined by the amount of time it takes to perform all of the multiplications.

  (a) How much time does it take to compute a 1024-point DFT directly?

  (b) How much time is required if an FFT is used?

  (c) Repeat parts (a) and (b) for a 4096-point DFT.

  (a) Including possible multiplications by ± 1, computing an N-point DFT directly requires N^2 complex multiplications. If it takes 1 μs per complex multiply, the direct evaluation of a 1024-point DFT requires

$$t_{DFT} = (1024)^2 \cdot 10^{-6}\,s \approx 1.05\,s$$

  (b) With a radix-2 FFT, the number of complex multiplications is approximately $(N/2)\log_2 N$ which, for $N = 1024$, is equal to 5120. Therefore, the amount of time to compute a 1024-point DFT using an FFT is

$$t_{FFT} = 5120 \cdot 10^{-6}\,ms = 5.12\,ms$$

  (c) If the length of the DFT is increased by a factor of 4 to $N = 4096$, the number of multiplications necessary to compute the DFT directly increases by a factor of 16. Therefore, the time required to evaluate the DFT directly is

$$t_{DFT} = 16.78\,s$$

  If, on the other hand, an FFT is used, the number of multiplications is

$$2{,}048 \cdot \log_2 4{,}096 = 24{,}576$$

  and the amount of time to evaluate the DFT is

$$t_{FFT} = 24.576\,ms$$

7.2 A complex-valued sequence $x(n)$ of length $N = 8192$ is to be convolved with a complex-valued sequence $h(n)$ of length $L = 512$.

  (a) Find the number of (complex) multiplications required to perform this convolution directly.

  (b) Repeat part (a) using the overlap-add method with 1024-point radix-2 decimation-in-time FFTs to evaluate the convolutions.

  (a) If $x(n)$ is of length $N = 8192$, and $h(n)$ of length $L = 512$, performing the convolution directly requires

$$512 \cdot 8{,}192 = 4{,}194{,}304$$

  complex multiplications.

  (b) Using the method of overlap-add with 1024-point FFTs, the number of multiplications is as follows. Because $h(n)$ is of length 512, we may segment $x(n)$ into sequences $x_i(n)$ of length $N = 512$ so that the 1024-point circular convolutions of $h(n)$ with $x_i(n)$ will be the same as linear convolutions (although we could use sections of length 513, this does not result in any computational savings). With the length of $x(n)$ being equal to 8192, this means that we will have 16 sequences of length 512. Therefore, to perform the convolution, we must compute 17 DFTs and 16 inverse DFTs. In addition, we must form the products $Y_i(k) = H(k)X_i(k)$ for $i = 1, 2, \ldots, 16$. Thus, the total number of complex multiplications is approximately

$$33 \cdot 512 \log_2(1{,}024) + 16 \cdot 1{,}024 = 185{,}344$$

  which is about 4.5 percent of the number of complex multiplies necessary to perform the convolution directly.

7.3   Speech that is sampled at a rate of 10 kHz is to be processed in real time. Part of the computations required involve collecting blocks of 1024 speech values and computing a 1024-point DFT and a 1024-point inverse DFT. If it takes $1\mu s$ for each real multiply, how much time remains for processing the data after the DFT and the inverse DFT are computed?

With a 10-kHz sampling rate, a block of 1024 samples is collected every 102.4 ms. With a radix-2 FFT, the number of complex multiplications for a 1024-point DFT is approximately $512 \log_2 1024 = 5120$. With a complex multiply consisting of four real multiplies, this means that we have to perform $5,120 \cdot 4 = 20,480$ real multiplies for the DFT and the same number for the inverse DFT. With $1\ \mu s$ per multiply, this will take

$$t = 2 \cdot 20.48 = 40.96 \text{ ms}$$

which leaves 61.44 ms for any additional processing.

7.4   Sampling a continuous-time signal $x_a(t)$ for 1 s generates a sequence of 4096 samples.

   (a)   What is the highest frequency in $x_a(t)$ if it was sampled without aliasing?

   (b)   If a 4096-point DFT of the sampled signal is computed, what is the frequency spacing in hertz between the DFT coefficients?

   (c)   Suppose that we are only interested in the DFT samples that correspond to frequencies in the range $200 \le f \le 300$ Hz. How many complex multiplies are required to evaluate these values computing the DFT directly, and how many are required if a decimation-in-time FFT is used?

   (d)   How many frequency samples would be needed in order for the FFT algorithm to be more efficient than evaluating the DFT directly?

   (a)   Collecting 4096 samples in 1 s means that the sampling frequency is $f_s = 4096$ Hz. If $x_a(t)$ is to be sampled without aliasing, the sampling frequency must be at least twice the highest frequency in $x_a(t)$. Therefore, $x_a(t)$ should have no frequencies above $f_0 = 2048$ Hz.

   (b)   With a 4096-point DFT, we are sampling $X(e^{j\omega})$ at 4096 equally spaced frequencies between 0 and 2π, which corresponds to 4096 frequency samples over the range $0 \le f \le 4096$ Hz. Therefore, the frequency spacing is $\Delta f = 1$ Hz.

   (c)   Over the frequency range from 200 to 300 Hz we have 101 DFT samples. Because it takes 4096 complex multiplies to evaluate each DFT coefficient, the number of multiplies necessary to evaluate only these frequency samples is

$$101 \cdot 4,096 = 413,696$$

On the other hand, the number of multiplications required if an FFT is used is

$$2,048 \log_2 4,096 = 24,576$$

Therefore, even though the FFT generates all of the frequency samples in the range $0 \le f \le 4096$ Hz, it is more efficient than evaluating these 101 samples directly.

   (d)   An N-point FFT requires $\frac{1}{2}N \log_2 N$ complex multiplies, and to evaluate M DFT coefficients directly requires $M \cdot N$ complex multiplications. Therefore, the FFT will be more efficient in finding these M samples if

$$M \cdot N > \tfrac{1}{2}N \log_2 N$$

   or

$$M \ge \tfrac{1}{2} \log_2 N$$

With $N = 4096$, the number of frequency samples is $M = 6$.

7.5   Because some of the $\frac{1}{2}N \log_2 N$ multiplications in the decimation-in-time and decimation-in-frequency FFT algorithms are multiplications by ± 1, it is possible to more efficiently implement these algorithms by writing programs that specifically excluded these multiplications.

   (a)   How many multiplications are there in an eight-point decimation-in-time FFT if we exclude the multiplications by ± 1?

(b) Repeat part (a) for a 16-point decimation-in-time FFT.

(c) Generalize the results in parts (a) and (b) for $N = 2^\nu$.

(a) For an eight-point decimation-in-time FFT, we may count the number of complex multiplications in the flow-graph given in Fig. 7-6. In the first stage of the FFT, there are no complex multiplications, whereas in the second stage, there are two multiplications by W_8^2. Finally, in the third stage there are three multiplications by W_8, W_8^2, and W_8^3. Thus, there are a total of five complex multiplies.

(b) A 16-point DFT is formed from two 8-point DFTs as follows:

$$X(k) = G(k) + W_{16}^k H(k) \qquad k = 0, 1, \ldots, 15$$

where $G(k)$ and $H(k)$ are eight-point DFTs. There are eight butterflies in the last stage that produces $X(k)$ from $G(k)$ and $H(k)$. Because the simplified butterfly in Fig. 7-5(b) only requires only one complex multiply, and noting that one of these is by $W_{16}^0 = 1$, we have a total of seven twiddle factors. In addition, we have two 8-point FFTs, which require five complex multiplies each. Therefore, the total number of multiplies is $2 \cdot 5 + 7 = 17$.

(c) Let $L(\nu)$ be the number of complex multiplies required for a radix-2 FFT when $N = 2^\nu$. From parts (a) and (b) we see that $L(3) = 5$ and $L(4) = 17$. Given that an FFT of length $N = 2^{\nu-1}$ requires $L(\nu - 1)$ multiplies, for an FFT of length $N = 2^\nu$, we have an additional $2^{\nu-1}$ butterflies. Because each butterfly requires one multiply, and because one of these multiplies is by $W_N^0 = 1$, the number of multiplies required for an FFT of length 2^ν is

$$L(\nu) = 2 \cdot L(\nu - 1) + 2^{\nu-1} - 1$$

Solving this recursion for $L(\nu)$, we have the following closed-form expression for $L(\nu)$:

$$L(\nu) = 2^\nu \left[\frac{\nu}{2} - 1 + \left(\frac{1}{2} \right)^\nu \right]$$

7.6 The FFT requires the multiplication of complex numbers:

$$(a_1 + jb_1) \cdot (a_2 + jb_2) = c_1 + jd_1$$

(a) Write out this complex multiplication, and determine how many real multiplies and real adds are required.

(b) Show that the complex multiplication may also be performed as follows:

$$c_1 = (a_1 - b_1) \cdot b_2 + (a_2 - b_2) \cdot a_1$$

$$d_1 = (a_1 - b_1) \cdot b_2 + (a_2 + b_2) \cdot b_1$$

and determine the number of real multiplies and adds required with this method.

(a) The product of two complex number is

$$(a_1 + jb_1) \cdot (a_2 + jb_2) = a_1 a_2 - b_1 b_2 + j(b_1 a_2 + a_1 b_2)$$

which requires four real multiplies and three real adds.

(b) Expanding the expressions for c_1, we have

$$c_1 = (a_1 - b_1) \cdot b_2 + (a_2 - b_2) \cdot a_1 = a_1 b_2 - b_1 b_2 + a_2 a_1 - b_2 a_1 = a_1 a_2 - b_1 b_2$$

as required. Similarly, for d_1 we have

$$d_1 = (a_1 - b_1) \cdot b_2 + (a_2 + b_2) \cdot b_1 = a_1 b_2 - b_1 b_2 + a_2 b_1 + b_2 b_1 = a_1 b_2 + a_2 b_1$$

also as required. This approach only requires three multiplies and four adds.

7.7   The decimation-in-time and decimation-in-frequency FFT algorithms evaluate the DFT of a complex-valued sequence. Show how an N-point FFT program may be used to evaluate the N-point DFT of two *real-valued* sequences.

As we saw in Prob. 6.18, the DFTs of two real-valued sequences may be found from one N-point DFT as follows. First, we form the N-point complex sequence

$$x(n) = x_1(n) + j x_2(n)$$

After finding the N-point DFT of $x(n)$, we extract $X_1(k)$ and $X_2(k)$ from $X(k)$ by exploiting the symmetry of the DFT. Specifically,

$$X_1(k) = \tfrac{1}{2}[X(k) + X^*((N-k))_N]$$

which is the conjugate symmetric part of $X(k)$, and

$$X_2(k) = \tfrac{1}{2}[X(k) - X^*((N-k))_N]$$

which is the conjugate antisymmetric part of $X(k)$.

7.8   Determine how a $2N$-point DFT of a real-valued sequence may be computed using an N-point FFT algorithm.

Let $g(n)$ be a real-valued sequence of length $2N$. From this sequence, we may form two real-valued sequences of length N as follows:

$$x_1(n) = g(2n) \qquad n = 0, 1, \ldots, N-1$$
$$x_2(n) = g(2n+1) \qquad n = 0, 1, \ldots, N-1$$

From these two sequences, we form the complex sequence

$$x(n) = x_1(n) + j x_2(n)$$

Computing the N-point DFT of $x(n)$, we may then extract the N-point DFTs of $x_1(n)$ and $x_2(n)$ as follows (see Prob. 7.7):

$$X_1(k) = \tfrac{1}{2}[X(k) + X^*((N-k))_N]$$
$$X_2(k) = \tfrac{1}{2}[X(k) - X^*((N-k))_N]$$

Now all that is left to do is to relate the $2N$-point DFT of $g(n)$ to the N-point DFTs $X_1(k)$ and $X_2(k)$. Note that

$$G(k) = \sum_{n=0}^{2N-1} g(n) W_{2N}^{nk} = \sum_{n=0}^{N-1} g(2n) W_{2N}^{2nk} + \sum_{n=0}^{N-1} g(2n+1) W_{2N}^{(2n+1)k}$$

$$= \sum_{n=0}^{N-1} x_1(n) W_N^{nk} + W_{2N}^{k} \sum_{n=0}^{N-1} x_2(n) W_N^{nk}$$

Therefore, $$G(k) = X_1(k) + W_{2N}^{k} X_2(k) \qquad k = 0, 1, \ldots, 2N-1$$

where the periodicity of $X_1(k)$ and $X_2(k)$ is used to evaluate $G(k)$ for $N < k < 2N$, that is,

$$X_1(k) = X_1(k+N) \qquad X_2(k) = X_2(k+N)$$

7.9   Given an FFT program to find the N-point DFT of a sequence, how may this program be used to find the inverse DFT?

As we saw in Prob. 6.9, we may find $x(n)$ by first using the DFT program to evaluate the sum

$$g(n) = \sum_{n=0}^{N-1} X^*(n) W_N^{nk}$$

which is the DFT of $X^*(k)$. Then, $x(n)$ may be found from $x(n)$ as follows:

$$x(n) = \frac{1}{N} g^*(n)$$

Alternatively, we may find the DFT of $X(k)$,

$$f(n) = \sum_{n=0}^{N-1} X(n) W_N^{nk}$$

and then extract $x(n)$ as follows:

$$x(n) = \frac{1}{N} f(N-n) = \frac{1}{N} \sum_{k=0}^{N-1} X(k) W_N^{(N-n)k} = \frac{1}{N} \sum_{k=0}^{N-1} X(k) W_N^{-nk}$$

7.10   Let $x(n)$ be a sequence of length N with

$$x(n) = -x\left(n + \frac{N}{2}\right) \qquad n = 0, 1, \ldots, \frac{N}{2} - 1$$

where N is an even integer.

(a)   Show that the N-point DFT of $x(n)$ has only odd harmonics, that is,

$$X(k) = 0 \qquad k \text{ even}$$

(b)   Show how to find the N-point DFT of $x(n)$ by finding the $N/2$-point DFT of an appropriately modified sequence.

(a)   The N-point DFT of $x(n)$ is

$$X(k) = \sum_{n=0}^{N-1} x(n) W_N^{nk} = \sum_{n=0}^{\frac{N}{2}-1} x(n) W_N^{nk} + \sum_{n=N/2}^{N-1} x(n) W_N^{nk}$$

$$= \sum_{n=0}^{\frac{N}{2}-1} x(n) W_N^{nk} + \sum_{n=0}^{\frac{N}{2}-1} x\left(n + \frac{N}{2}\right) W_N^{(n+N/2)k}$$

$$= \sum_{n=0}^{\frac{N}{2}-1} \left[x(n) + (-1)^k x\left(n + \frac{N}{2}\right) \right] W_N^{nk}$$

Because $x(n) = -x(n + N/2)$, if k is even, each term in the sum is zero, and $X(k) = 0$ for $k = 0, 2, 4, \ldots$.

(b)   In the first stage of a decimation-in-frequency FFT algorithm, we separately evaluate the even-index and odd-index samples of $X(k)$. If $X(k)$ has only odd harmonics, the even samples are zero, and we need only evaluate the odd samples. From Eq. (7.4) we see that the odd samples are given by

$$X(2k+1) = \sum_{n=0}^{\frac{N}{2}-1} W_N^n \left[x(n) - x\left(n + \frac{N}{2}\right) \right] W_{N/2}^{nk}$$

With $x(n) = -x(n + N/2)$ this becomes

$$X(2k+1) = \sum_{n=0}^{\frac{N}{2}-1} \left[2 W_N^n x(n) \right] W_{N/2}^{nk}$$

which is the $N/2$-point DFT of the sequence $y(n) = 2W_N^n x(n)$. Therefore, to find the N-point DFT of $x(n)$, we multiply the first $N/2$ points of $x(n)$ by $2W_N^n$,

$$y(n) = 2 W_N^n x(n) \qquad n = 0, 1, 2, \ldots, \frac{N}{2}$$

and then compute the $N/2$-point DFT of $y(n)$. The $N/2$-point DFT of $x(n)$ is then given by

$$X(2k+1) = Y(k) \qquad k = 0, 1, \ldots \frac{(N-2)}{2}$$

$$X(2k) = 0 \qquad k = 0, 1, \ldots \frac{(N-2)}{2}$$

FFT Algorithms for Composite N

7.11 When the number of points in the DFT is a power of 4, we can use a radix-2 FFT algorithm. However, when $N = 4^\nu$, it is more efficient to use a radix-4 FFT algorithm.

(a) Derive the radix-4 decimation-in-time FFT algorithm when $N = 4^\nu$.

(b) Draw the structure for the butterfly in the radix-4 FFT, and compare the number of complex multiplies and adds with a radix-4 FFT to a radix-2 FFT.

(a) To derive a decimation-in-time radix-4 FFT, let $N_1 = N/4$ and $N_2 = 4$, and define the index maps

$$n = 4 \cdot n_1 + n_2 \qquad \begin{cases} 0 \le n_1 \le \dfrac{N}{4} - 1 \\ 0 \le n_2 \le 4 \end{cases}$$

$$k = k_1 + \frac{N}{4} \cdot k_2 \qquad \begin{cases} 0 \le k_1 \le \dfrac{N}{4} - 1 \\ 0 \le k_2 \le 3 \end{cases}$$

We then express $X(k)$ using the decomposition given in Eq. (7.7) with $N_1 = N/4$ and $N_2 = 4$,

$$X(k) = X\left(k_1 + \frac{N}{4}k_2\right) = \sum_{n_2=0}^{3}\left\{\left[\sum_{n_1=0}^{\frac{N}{4}-1} x(4n_1 + n_2)W_{N/4}^{n_1 k_1}\right]W_N^{k_1 n_2}\right\}W_4^{k_2 n_2}$$

The inner summation,

$$G(n_2, k_1) = \sum_{n_1=0}^{\frac{N}{4}-1} x(4n_1 + n_2)W_{N/4}^{n_1 k_1}$$

is the $N/4$-point DFT of the sequence $x(4n_1 + n_2)$, and the outer summation is a 4-point DFT,

$$X\left(k_1 + \frac{N}{4}k_2\right) = \sum_{n_2=0}^{3} \tilde{G}(k_1, n_2)W_4^{k_2 n_2}$$

where
$$\tilde{G}(n_2, k_1) = W_N^{k_1 n_2} G(n_2, k_1)$$

Since $W_4 = -j$, these 4-point transforms have the form

$$X\left(k_1 + \frac{N}{4}k_2\right) = \tilde{G}(k_1, 0) + (-j)^{k_1}\tilde{G}(1, k_1) + (-1)^{k_1}\tilde{G}(2, k_1) + (j)^{k_1}\tilde{G}(3, k_1)$$

for $k_1 = 0, 1, 2, 3$, and $n_2 = 0, 1, \ldots, (N/4) - 1$. If $N_2 = N/4$ is divisible by 4, then the process is repeated. In this way, we generate $\nu = \log_4 N$ stages with $N/4$ butterflies in each stage.

(b) The 4-point butterflies in the radix-4 FFT perform operations of the form

$$F(k_2, k_1) = \sum_{n_2=0}^{3}\left[G(n_2, k_1)W_N^{k_1 n_2}\right]W_4^{k_2 n_2} \qquad k_2 = 0, 1, 2, 3$$

With $W_4^{k_2 n_2} = j^{k_2 n_2}$, the butterflies have the structure shown in the figure below.

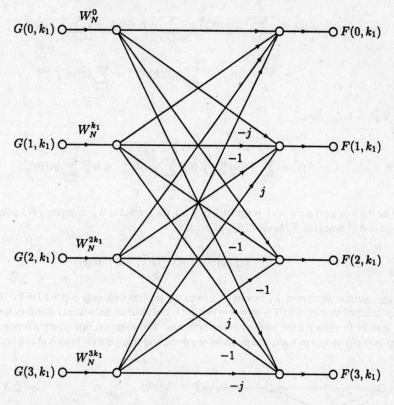

Since multiplications by $\pm j$ only requires interchanging real and imaginary parts and possibly changing a sign bit, then each 4-point butterfly only requires 3 complex multiplications. With $\nu = \log_4 N$ stages, and $N/4$ butterflies per stage, the number of complex multiplies for a DFT of length $N = 4^\nu$ is

$$3 \cdot \frac{N}{4} \log_4 N = \frac{3N}{8} \log_2 N$$

For a radix-2 decimation-in-time FFT, on the other hand, the number of multiplications is

$$\frac{N}{2} \log_2 N$$

Therefore, the number of multiplications in a radix-4 FFT is $\frac{3}{16}$ times the number in a radix-2 FFT.

7.12 Suppose that we would like to find the N-point DFT of a sequence where N is a power of 3, $N = 3^\nu$.

(a) Develop a radix-3 decimation-in-time FFT algorithm, and draw the corresponding flowgraph for $N = 9$.

(b) How many multiplications are required for a radix-3 FFT?

(c) Can the computations be performed in place?

(a) A radix-3 decimation-in-time FFT may be derived in exactly the same way as a radix-2 FFT. First, $x(n)$ is decimated by a factor of 3 to form three sequences of length $N/3$:

$$f(n) = x(3n) \qquad n = 0, 1, \ldots, \frac{N}{3} - 1$$

$$g(n) = x(3n + 1) \qquad n = 0, 1, \ldots, \frac{N}{3} - 1$$

$$h(n) = x(3n + 2) \qquad n = 0, 1, \ldots, \frac{N}{3} - 1$$

Expressing the N-point DFT in terms of these sequences, we have

$$X(k) = \sum_{n=0,3,6,\ldots} x(n)W_N^{nk} + \sum_{n=1,4,5,\ldots} x(n)W_N^{nk} + \sum_{n=2,5,7,\ldots} x(n)W_N^{nk}$$

$$= \sum_{l=0}^{\frac{N}{3}-1} f(l)W_N^{3lk} + \sum_{l=0}^{\frac{N}{3}-1} g(l)W_N^{(3l+1)k} + \sum_{l=0}^{\frac{N}{3}-1} h(l)W_N^{(3l+2)k}$$

Since $W_N^{3lk} = W_{N/3}^{lk}$, then

$$X(k) = \sum_{l=0}^{\frac{N}{3}-1} f(l)W_{N/3}^{lk} + W_N^k \sum_{l=0}^{\frac{N}{3}-1} g(l)W_{N/3}^{lk} + W_N^{2k} \sum_{l=0}^{\frac{N}{3}-1} h(l)W_{N/3}^{lk}$$

Note that the first term is the $N/3$-point DFT of $f(n)$, the second is W_N^k times the $N/3$-point DFT of $g(n)$, and the third is W_N^{2k} times the $N/3$-point DFT of $h(n)$,

$$X(k) = F(k) + W_N^k G(k) + W_N^{2k} H(k)$$

We may continue decimating by factors of 3 until we are left with only 3-point DFTs. The flowgraph for a 9-point decimation-in-time FFT is shown in Fig. 7-11. Only one of the 3-point butterflies is shown in the second stage in order to allow for the labeling of the branches. The complete flowgraph is formed by replicating this 3-point butterfly up by one node, and down by one node, and changing the branch multiplies to their appropriate values.

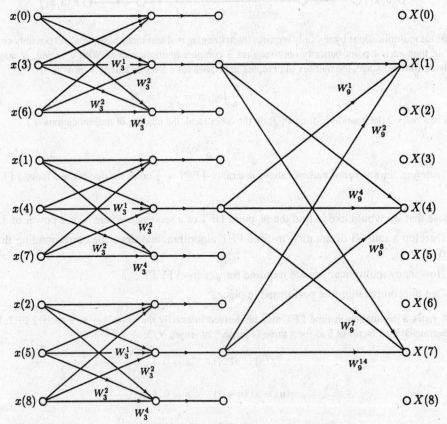

Fig. 7-11.   Flowgraph for a 9-point decimation-in-time FFT (only one butterfly in the second stage is shown).

(b) If $N = 3^\nu$, then there are ν stages in the radix-3 FFT. The general form of each 3-point butterfly, shown in the second stage of the flowgraph in Fig. 7-11, requires six multiplies (some require fewer if we do not consider multiplications by ± 1). Since there are $N/3$ butterflies in each stage, then the total number of multiplications is

$$6N \log_3 N$$

(c) Yes, the computations may be performed in place.

7.13 Derive a radix-3 decimation-in-frequency FFT for $N = 3^\nu$, and draw the corresponding flowgraph for $N = 9$.

As with the radix-2 decimation-in-frequency FFT, with $N = 3^\nu$, we separately evaluate the indices for which $((k))_3 = 0$, $((k))_3 = 1$, and $((k))_3 = 2$. For $((k))_3 = 0$ (i.e., k is a multiple of 3),

$$X(3k) = \sum_{n=0}^{N-1} x(n) W_N^{3nk}$$

Separating this sum into the first $N/3$ points, the second $N/3$ points, and the last $N/3$ points, and using the fact that $W_N^{3nk} = W_{N/3}^{nk}$, this becomes

$$X(3k) = \sum_{n=0}^{\frac{N}{3}-1} x(n) W_{N/3}^{nk} + \sum_{n=\frac{N}{3}}^{\frac{2N}{3}-1} x(n) W_{N/3}^{nk} + \sum_{n=\frac{2N}{3}}^{N-1} x(n) W_{N/3}^{nk}$$

With a change in the indexing in the second and third sums, we have

$$X(3k) = \sum_{n=0}^{\frac{N}{3}-1} x(n) W_{N/3}^{nk} + \sum_{n=0}^{\frac{N}{3}-1} x\left(n + \frac{N}{3}\right) W_{N/3}^{(n+\frac{N}{3})k} + \sum_{n=0}^{\frac{N}{3}-1} x\left(n + \frac{2N}{3}\right) W_{N/3}^{(n+\frac{2N}{3})k}$$

Finally, because $W_{N/3}^{n+\frac{N}{3}} = W_{N/3}^n$, and $W_{N/3}^{n+\frac{2N}{3}} = W_{N/3}^n$,

$$X(3k) = \sum_{n=0}^{\frac{N}{3}-1} \left[x(n) + x\left(n + \frac{N}{3}\right) + x\left(n + \frac{2N}{3}\right) \right] W_{N/3}^{nk}$$

which is the $N/3$-point DFT of the sequence in brackets.

Proceeding in the same way for the samples $X(3k + 1)$, we have

$$X(3k + 1) = \sum_{n=0}^{N-1} x(n) W_N^{n(3k+1)}$$

$$= \sum_{n=0}^{\frac{N}{3}-1} x(n) W_N^{n(3k+1)} + \sum_{n=\frac{N}{3}}^{\frac{2N}{3}-1} x(n) W_N^{n(3k+1)} + \sum_{n=\frac{2N}{3}}^{N-1} x(n) W_N^{n(3k+1)}$$

$$= \sum_{n=0}^{\frac{N}{3}-1} x(n) W_N^{n(3k+1)} + \sum_{n=0}^{\frac{N}{3}-1} x\left(n + \frac{N}{3}\right) W_N^{(n+N/3)(3k+1)} + \sum_{n=0}^{\frac{N}{3}-1} x\left(n + \frac{2N}{3}\right) W_N^{(n+2N/3)(3k+1)}$$

$$= \sum_{n=0}^{\frac{N}{3}-1} \left[x(n) W_N^n + x\left(n + \frac{N}{3}\right) W_N^{n+N/3} + x\left(n + \frac{2N}{3}\right) W_N^{n+2N/3} \right] W_{N/3}^{nk}$$

Finally, for the samples $X(3k + 2)$ we have

$$X(3k + 2) = \sum_{n=0}^{\frac{N}{3}-1} \left[x(n) W_N^{2n} + x\left(n + \frac{N}{3}\right) W_N^{2n+2N/3} + x\left(n + \frac{2N}{3}\right) W_N^{2n+4N/3} \right] W_{N/3}^{nk}$$

The flowgraph for a nine-point decimation-in-frequency FFT is shown below.

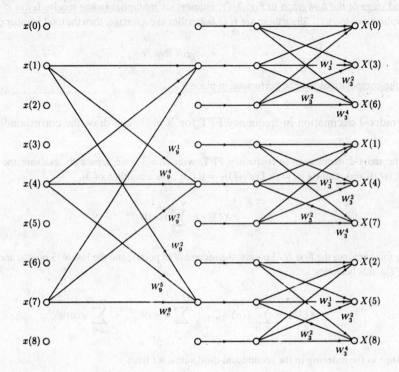

7.14   Suppose that we have a number of eight-point decimation-in-time FFT chips. How could these chips be used to compute a 24-point DFT?

A 24-point DFT is defined by

$$X(k) = \sum_{n=0}^{23} x(n) W_{24}^{nk}$$

Decimating $x(n)$ by a factor of 3, we may decompose this DFT into three 8-point DFTs as follows:

$$X(k) = \sum_{n=0}^{7} x(3n) W_{24}^{3nk} + \sum_{n=0}^{7} x(3n+1) W_{24}^{(3n+1)k} + \sum_{n=0}^{7} x(3n+2) W_{24}^{(3n+2)k}$$

$$= \sum_{n=0}^{7} x(n) W_{8}^{nk} + W_{24}^{k} \sum_{n=0}^{7} x(3n+1) W_{8}^{nk} + W_{24}^{2k} \sum_{n=0}^{7} x(3n+2) W_{8}^{nk}$$

Therefore, if we form the three sequences

$$
\begin{aligned}
f(n) &= x(3n)   &&n = 0, 1, 2, \ldots, 7 \\
g(n) &= x(3n+1) &&n = 0, 1, 2, \ldots, 7 \\
h(n) &= x(3n+2) &&n = 0, 1, 2, \ldots, 7
\end{aligned}
$$

and use the 8-point FFT chips to find the DFTs $F(k)$, $G(k)$, and $H(k)$, the 24-point DFT of $x(n)$ may be found by combining the outputs of the 8-point FFTs as follows:

$$X(k) = F(k) + W_{24}^{k} G(k) + W_{24}^{2k} H(k)$$

Prime Factor FFT

7.15   Find the index maps for a 21-point prime factor FFT with $N_1 = 7$ and $N_2 = 3$. How many multiplications are required compared to a 32-point radix-2 decimation-in-time FFT?

For a 21-point prime factor FFT with $N_1 = 7$ and $N_2 = 3$, we set $A = N_2 = 3$ and $B = N_1 = 7$. Then, with $C = N_2((N_2^{-1}))_{N_1} = 15$ and $D = N_1((N_1^{-1}))_{N_2} = 7$, we have the following index mappings:

$$n = ((3n_1 + 7n_2))_N \qquad \begin{cases} 0 \le n_1 \le 6 \\ 0 \le n_2 \le 2 \end{cases}$$

$$k = ((15k_1 + 7k_2))_N \qquad \begin{cases} 0 \le k_1 \le 6 \\ 0 \le k_2 \le 2 \end{cases}$$

Thus, the two-dimensional array representation for the input is

n_2 \ n_1	0	1	2	3	4	5	6
0	$x(0)$	$x(3)$	$x(6)$	$x(9)$	$x(12)$	$x(15)$	$x(18)$
1	$x(7)$	$x(10)$	$x(13)$	$x(16)$	$x(19)$	$x(1)$	$x(4)$
2	$x(14)$	$x(17)$	$x(20)$	$x(2)$	$x(5)$	$x(8)$	$x(11)$

and the two-dimensional array for the output is

k_2 \ k_1	0	1	2	3	4	5	6
0	$X(0)$	$X(15)$	$X(9)$	$X(3)$	$X(18)$	$X(12)$	$X(6)$
1	$X(7)$	$X(1)$	$X(16)$	$X(10)$	$X(4)$	$X(19)$	$X(13)$
2	$X(14)$	$X(8)$	$X(2)$	$X(17)$	$X(11)$	$X(5)$	$X(20)$

With the prime factor FFT, there are no twiddle factors. Therefore, the only multiplications necessary are those required to compute the three 7-point DFTs, and the seven 3-point DFTs. Because each 3-point DFT requires 6 complex multiplies, and each 7-point DFT requires 42, the number of multiplies for a 21-point prime factor FFT is $(7)(6) + (3)(42) = 168$. For a 32-point radix-2 FFT, on the other hand, we require

$$16 \log_2 32 = 80$$

complex multiplies. Therefore, it would be more efficient to pad a 21-point sequence with zeros and compute a 32-point DFT. The increased efficiency is a result of the fact that $32 = 2^5$ is a much more composite number than $21 = 7 \cdot 3$.

7.16 Suppose that we would like to compute a 15-point DFT of a sequence $x(n)$.

(a) Using a mixed-radix FFT with $N_1 = 5$ and $N_2 = 3$, the DFT is decomposed into two stages, with the first consisting of three 5-point DFTs, and the second stage consisting of five 3-point DFTs. Make a sketch of the connections between the five- and three-point DFTs, indicating any possible twiddle factors, and the order of the inputs and outputs.

(b) Repeat part (a) for the prime factor algorithm with $N_1 = 5$ and $N_2 = 3$, and determine how many complex multiplies are saved with the prime factor algorithm.

(a) Using a mixed-radix FFT with $N_1 = 5$ and $N_2 = 3$, the index mappings for n and k are as follows:

$$n = 3n_1 + n_2 \qquad \begin{cases} 0 \le n_1 \le 4 \\ 0 \le n_2 \le 2 \end{cases}$$

$$k = k_1 + 5k_2 \qquad \begin{cases} 0 \le k_1 \le 4 \\ 0 \le k_2 \le 2 \end{cases}$$

Thus, the two-dimensional array representation for the input is

n_2 \ n_1	0	1	2	3	4
0	$x(0)$	$x(3)$	$x(6)$	$x(9)$	$x(12)$
1	$x(1)$	$x(4)$	$x(7)$	$x(10)$	$x(13)$
2	$x(2)$	$x(5)$	$x(8)$	$x(11)$	$x(14)$

After the five-point DFT of each row in the data array is computed, the resulting complex array is multiplied by the array of twiddle factors:

$$
\begin{bmatrix}
1 & 1 & 1 & 1 & 1 \\
1 & W_{15} & W_{15}^2 & W_{15}^3 & W_{15}^4 \\
1 & W_{15}^2 & W_{15}^4 & W_{15}^6 & W_{15}^8
\end{bmatrix}
$$

The last step then involves computing the three-point DFT of each column. This produces the output array $X(k)$, which is

k_2 \ k_1	0	1	2	3	4
0	$X(0)$	$X(1)$	$X(2)$	$X(3)$	$X(4)$
1	$X(5)$	$X(6)$	$X(7)$	$X(8)$	$X(9)$
2	$X(10)$	$X(11)$	$X(12)$	$X(13)$	$X(14)$

The connections between the three- and five-point DFTs are shown in the following figure, along with the eight twiddle factors:

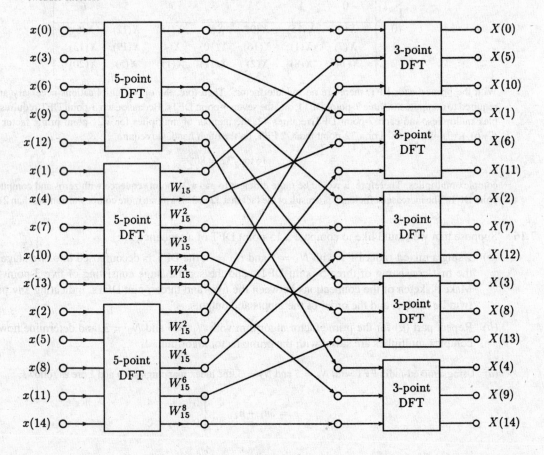

(b) Using the prime factor algorithm with $N_1 = 5$ and $N_2 = 3$, we set $A = N_2 = 3$ and $B = N_1 = 5$. Then, with $C = N_2((N_2^{-1}))_{N_1} = 6$ and $D = N_1((N_1^{-1}))_{N_2} = 10$, we have the following index mappings for n and k:

$$
n = ((3n_1 + 5n_2))_{15} \qquad \begin{cases} 0 \le n_1 \le 4 \\ 0 \le n_2 \le 2 \end{cases}
$$

$$
k = ((6k_1 + 10k_2))_{15} \qquad \begin{cases} 0 \le k_1 \le 4 \\ 0 \le k_2 \le 2 \end{cases}
$$

The two-dimensional array representation for the input is

n_2 \ n_1	0	1	2	3	4
0	$x(0)$	$x(3)$	$x(6)$	$x(9)$	$x(12)$
1	$x(5)$	$x(8)$	$x(11)$	$x(14)$	$x(2)$
2	$x(10)$	$x(13)$	$x(1)$	$x(4)$	$x(7)$

and for the output array we have

k_2 \ k_1	0	1	2	3	4
0	$X(0)$	$X(6)$	$X(12)$	$X(3)$	$X(9)$
1	$X(10)$	$X(1)$	$X(7)$	$X(13)$	$X(4)$
2	$X(5)$	$X(11)$	$X(2)$	$X(8)$	$X(14)$

The interconnections between the five- and three-point DFTs are the same as in the mixed-radix algorithm. However, there are no twiddle factors, and the ordering of the input and output arrays is different. The 15-point prime factor algorithm is diagrammed in the figure below.

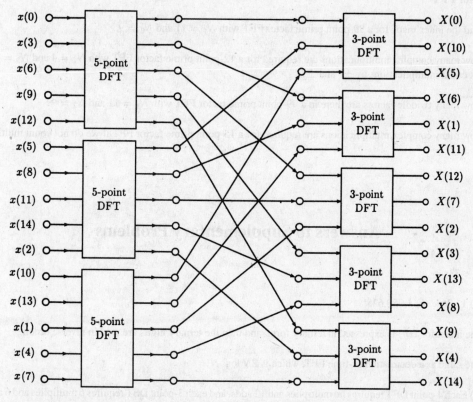

The savings with the prime factor algorithm over the mixed-radix FFT are the eight complex multiplies by the twiddle factors.

Supplementary Problems

Radix-2 FFT Algorithms

7.17   Let $x(n)$ be a sequence of length 1024 that is to be convolved with a sequence $h(n)$ of length L. For what values of L is it more efficient to perform the convolution directly than it is to perform the convolution by taking the inverse DFT of the product $X(k)H(k)$ and evaluating the DFTs using a radix-2 FFT algorithm?

7.18   Suppose that we have a 1025-point data sequence (1 more than $N = 2^{10}$). Instead of discarding the final value, we zero pad the sequence to make it of length $N = 2^{11}$ so that we can use a radix-2 FFT algorithm. (*a*) How many multiplications and additions are required to compute the DFT using a radix-2 FFT algorithm? (*b*) How many multiplications and additions would be required to compute a 1025-point DFT directly?

FFT Algorithms for Composite *N*

7.19   In a radix-3 decimation-in-time FFT, how is the input sequence indexed?

7.20   How many complex multiplications are necessary in a radix-3 decimation-in-frequency FFT?

7.21   Consider the FFT algorithm given in Example 7.3.2. (*a*) How many multiplications and additions are required to compute a 12-point DFT? (*b*) How many multiplications and additions are necessary if the 12-point DFT is computed directly?

Prime Factor FFT

7.22   Find the index maps for a 99-point prime factor FFT with $N_1 = 11$ and $N_2 = 9$.

7.23   How many complex multiplications are required for a 12-point prime factor FFT with $N_1 = 4$ and $N_3 = 3$ if we do not count multiplications by ± 1 and $\pm j$?

7.24   How many twiddle factors are there in a 99-point prime factor FFT with $N_1 = 11$ and $N_2 = 9$?

7.25   How many complex multiplications are required for a 15-point prime factor FFT if we do not count multiplications by ± 1?

Answers to Supplementary Problems

7.17 $L < 33$.

7.18   (*a*) 11,264. (*b*) 1,050,625.

7.19   The index for $x(n)$ is expressed in ternary form, and then the ternary digits are read in reverse order.

7.20   The same as a decimation-in-time FFT, which is $2N \log_3 N$.

7.21   (*a*) Each 4-point DFT requires no multiplies and 12 adds, and each 3-point DFT requires 6 multiplies and 6 adds. With 6 twiddle factors, there are $6 + (4)(6) = 30$ multiplies and $(4)(6) + (3)(12) = 60$ adds. (*b*) 144 multiplies and 132 adds.

7.22 $n = 9n_1 + 11n_2$, and $k = 45k_1 + 55k_2$.

7.23   24.

7.24   None.

7.25   90.

Chapter 8

Implementation of Discrete-Time Systems

8.1 INTRODUCTION

Given a linear shift-invariant system with a rational system function $H(z)$, the input and output are related by a linear constant coefficient difference equation. For example, with a system function

$$H(z) = \frac{b(0) + b(1)z^{-1}}{1 + a(1)z^{-1}}$$

the input $x(n)$ and output $y(n)$ are related by the linear constant coefficient difference equation

$$y(n) = -a(1)y(n-1) + b(0)x(n) + b(1)x(n-1) \qquad (8.1)$$

This difference equation defines a sequence of operations that are to be performed in order to *implement* this system. However, note that this system may also be implemented with the following pair of coupled difference equations:

$$w(n) = -a(1)w(n-1) + x(n)$$
$$y(n) = b(0)w(n) + b(1)w(n-1)$$

With this implementation, it is only necessary to provide one memory location to store $w(n-1)$, whereas Eq. (8.1) requires two memory locations, one to store $y(n-1)$ and one to store $x(n-1)$. This simple example illustrates that there is more than one way to implement a system and that the amount of computation and/or memory required will depend on the implementation. In addition, the implementation may affect the sensitivity of the filter to coefficient quantization, and the amount of round-off noise that appears at the output of the filter.

In this chapter, we look at a number of different ways to implement a linear shift-invariant discrete-time system and look at the effect of finite word lengths on these implementations.

8.2 DIGITAL NETWORKS

For a linear shift-invariant system with a rational system function, the input $x(n)$ and the output $y(n)$ are related by a linear constant coefficient difference equation:

$$y(n) = \sum_{k=0}^{q} b(k)x(n-k) - \sum_{k=1}^{p} a(k)y(n-k)$$

The basic computational elements required to find the output at time n are adders, multipliers, and delays. It is often convenient to use a *block diagram* to illustrate how these adders, multipliers, and delays are interconnected to implement a given system. The notation that is used for these elements is shown in Fig. 8-1. A network is also often represented pictorially using a *signal flowgraph*, which is a network of *directed branches* that are connected at *nodes*. Each branch has an input and an output, with the direction indicated by an arrowhead. The nodes in a flowgraph correspond to either *adders* or *branch points*. Adders correspond to nodes with more than one incoming branch, and branch points are nodes with more than one outgoing branch, as illustrated in Fig. 8-2. With a linear flowgraph, the output of each branch is a linear transformation of the branch input, and the linear operator is indicated next to the arrow. For linear shift-invariant discrete-time filters, these linear operators consist of multiplies and delays. Finally, there are two special types of nodes:

1. *Source nodes.* These are nodes that have no incoming branches and are used for sequences that are input to the filter.

2. *Sink nodes.* These are nodes that have only entering branches and are used to represent output sequences.

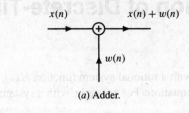

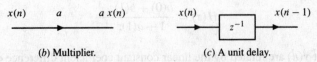

(a) Adder.

(b) Multiplier.   (c) A unit delay.

Fig. 8-1. Notation used for an adder, multiplier, and delay in a digital network.

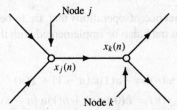

Fig. 8-2. Signal flowgraph consisting of nodes, branches, and node variables. Node j represents an adder, and node k is a branch point.

EXAMPLE 8.2.1   Consider the first-order discrete-time system described by the difference equation

$$y(n) = b(0)x(n) + b(1)x(n-1) + a(1)y(n-1)$$

Shown in the figure below is a block diagram for this system.

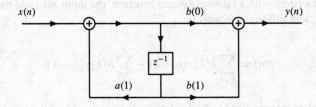

Using a signal flowgraph, this system is represented as follows:

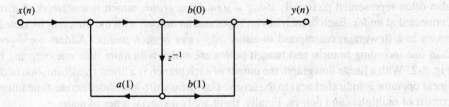

8.3   STRUCTURES FOR FIR SYSTEMS

A causal FIR filter has a system function that is a polynomial in z^{-1}:

$$H(z) = \sum_{n=0}^{N} h(n)z^{-n}$$

For an input $x(n)$, the output is

$$y(n) = \sum_{k=0}^{N} h(k)x(n-k)$$

For each value of n, evaluating this sum requires $(N+1)$ multiplications and N additions. The following subsections describe several different realizations of this system.

8.3.1   Direct Form

The most common way to implement an FIR filter is in *direct form* using a *tapped delay line* as shown in the figure below.

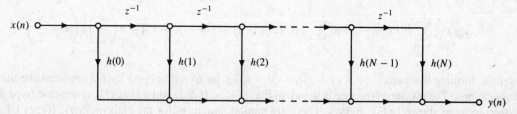

This structure requires $N+1$ multiplications, N additions, and N delays. However, if there are some symmetries in the unit sample response, it may be possible to reduce the number of multiplications (see the section on linear phase filters).

8.3.2   Cascade Form

For a causal FIR filter, the system function may be factored into a product of first-order factors,

$$H(z) = \sum_{n=0}^{N} h(n)z^{-n} = A \prod_{k=1}^{N}(1 - \alpha_k z^{-1})$$

where α_k for $k = 1, \ldots, N$ are the zeros of $H(z)$. If $h(n)$ is real, the complex roots of $H(z)$ occur in complex conjugate pairs, and these conjugate pairs may be combined to form second-order factors with real coefficients,

$$H(z) = A \prod_{k=1}^{N_s}[1 + b_k(1)z^{-1} + b_k(2)z^{-2}]$$

Written in this form, $H(z)$ may be implemented as a cascade of second-order FIR filters as illustrated in Fig. 8-3.

8.3.3   Linear Phase Filters

Linear phase filters have a unit sample response that is either symmetric,

$$h(n) = h(N - n)$$

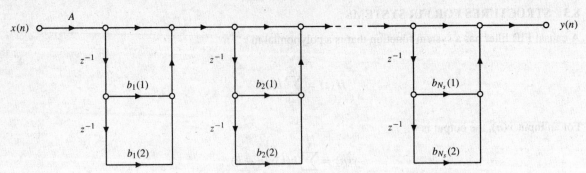

Fig. 8-3. An FIR filter implemented as a cascade of second-order systems.

or antisymmetric (see Sec. 5.3),

$$h(n) = -h(N - n)$$

This symmetry may be exploited to simplify the network structure. For example, if N is even and $h(n)$ is symmetric (type I filter),

$$y(n) = \sum_{k=0}^{N} h(k)x(n - k) = \sum_{k=0}^{\frac{N}{2}-1} h(k)[x(n - k) + x(n - N + k)] + h\left(\frac{N}{2}\right)x\left(n - \frac{N}{2}\right)$$

Therefore, forming the sums $[x(n - k) + x(n - N + k)]$ prior to multiplying by $h(k)$ reduces the number of multiplications. The resulting structure is shown in Fig. 8-4(a). If N is odd and $h(n)$ is symmetric (type II filter), the structure is as shown in Fig. 8-4(b). There are similar structures for the antisymmetric (types III and IV) linear phase filters.

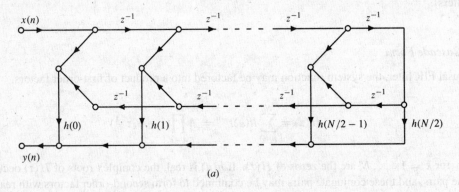

(a)

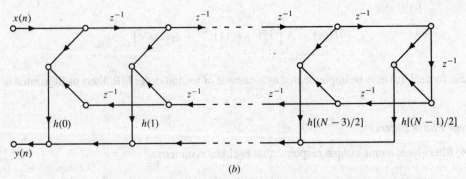

(b)

Fig. 8-4. Direct form implementations for linear phase filters. (a) Type I. (b) Type II.

8.3.4 Frequency Sampling

The frequency sampling structure is an implementation that parameterizes a filter in terms of its DFT coefficients. Specifically, let $H(k)$ be the N-point DFT of an FIR filter with $h(n) = 0$ for $n < 0$ and $n \geq N$.[1] Because the unit sample response of the filter is

$$h(n) = \frac{1}{N} \sum_{k=0}^{N-1} H(k) e^{j2\pi nk/N}$$

the system function may be written as

$$H(z) = \sum_{n=0}^{N-1} h(n)z^{-n} = \sum_{n=0}^{N-1} \left[\frac{1}{N} \sum_{k=0}^{N-1} H(k) e^{j2\pi nk/N} \right] z^{-n}$$

$$= \frac{1}{N} \sum_{k=0}^{N-1} H(k) \sum_{n=0}^{N-1} e^{j2\pi nk/N} z^{-n}$$

Evaluating the sum over n, this becomes

$$H(z) = \frac{1}{N}(1 - z^{-N}) \sum_{k=0}^{N-1} \frac{H(k)}{1 - e^{j2\pi k/N} z^{-1}}$$

which corresponds to a cascade of an FIR filter $\frac{1}{N}(1 - z^{-N})$ with a parallel network of one-pole filters:

$$H_k(z) = \frac{H(k)}{1 - e^{j2\pi k/N} z^{-1}}$$

For a narrowband filter that has most of its DFT coefficients equal to zero, the frequency sampling structure will be an efficient implementation. The frequency sampling structure is shown in Fig. 8-5. If $h(n)$ is real, $H(k) = H^*(N - k)$, and the structure may be simplified. For example, if N is even,

$$H(z) = \frac{1}{N}(1 - z^{-N}) \left[\frac{H(0)}{1 - z^{-1}} + \frac{H(N/2)}{1 + z^{-1}} + \sum_{k=1}^{N/2-1} \frac{A(k) - B(k)z^{-1}}{1 - 2\cos(2\pi k/N) z^{-1} + z^{-2}} \right] \tag{8.2}$$

where
$$A(k) = H(k) + H(N - k)$$
$$B(k) = H(k) e^{-j2\pi k/N} + H(N - k) e^{j2\pi k/N}$$

A similar simplification results when N is odd.

8.4 STRUCTURES FOR IIR SYSTEMS

The input $x(n)$ and output $y(n)$ of a causal IIR filter with a rational system function

$$H(z) = \frac{B(z)}{A(z)} = \frac{\sum_{k=0}^{q} b(k)z^{-k}}{1 + \sum_{k=1}^{p} a(k)z^{-k}}$$

[1] Note that here we are assuming that $h(n)$ is of length N, instead of $N + 1$ as in the previous sections. This is consistent with the convention that the frequency sampling filter is based on an N-point DFT of $h(n)$.

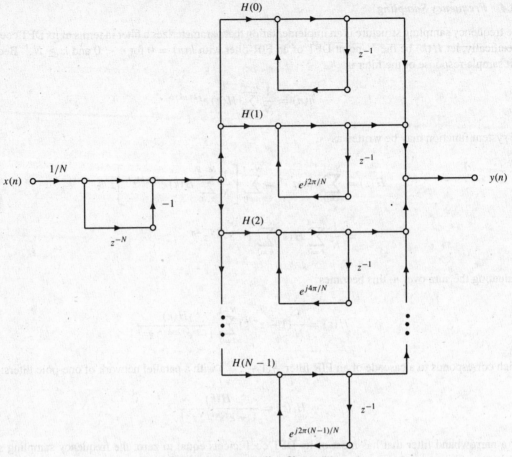

Fig. 8-5. Frequency sampling filter structure.

is described by the linear constant coefficient difference equation

$$y(n) = \sum_{k=0}^{q} b(k)x(n-k) - \sum_{k=1}^{p} a(k)y(n-k) \qquad (8.3)$$

In the following sections, several different implementations of this system are presented, including the direct form structures, the cascade and parallel forms, and the transposed filter structures.

8.4.1 Direct Form

There are two direct form filter structures, referred to as *direct form I* and *direct form II*. The direct form I structure is an implementation that results when Eq. (8.3) is written as a pair of difference equations as follows:

$$w(n) = \sum_{k=0}^{q} b(k)x(n-k)$$

$$y(n) = w(n) - \sum_{k=1}^{p} a(k)y(n-k)$$

The first equation corresponds to an FIR filter with input $x(n)$ and output $w(n)$, and the second equation corresponds to an all-pole filter with input $w(n)$ and output $y(n)$. Therefore, this pair of equations represents a *cascade* of two systems,

$$Y(z) = \frac{1}{A(z)}[B(z)X(z)]$$

as illustrated in Fig. 8-6. The computational requirements for a direct form I structure are as follows:

Number of multiplications: $p + q + 1$ per output sample

Number of additions: $p + q$ per output sample

Number of delays: $p + q$

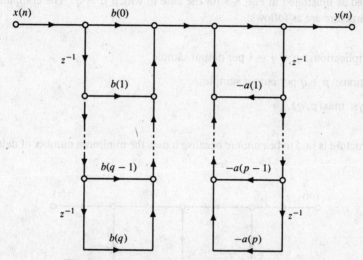

Fig. 8-6. Direct form I realization of an IIR filter.

The *direct form II* structure is obtained by reversing the order of the cascade of $B(z)$ and $1/A(z)$ as illustrated in Fig. 8-7. With this implementation, $x(n)$ is first filtered with the all-pole filter $1/A(z)$ and then with $B(z)$:

$$Y(z) = B(z)\left[\frac{1}{A(z)}X(z)\right]$$

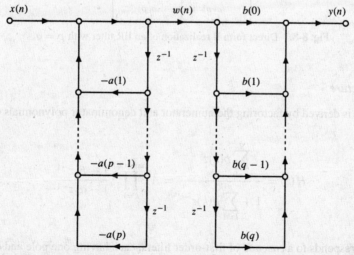

Fig. 8-7. Reversing the order of the cascade in the direct form I filter structure.

If we denote the output of the all-pole filter $1/A(z)$ by $w(n)$, this structure is described by the following pair of coupled difference equations:

$$w(n) = x(n) - \sum_{k=1}^{p} a(k)w(n-k)$$

$$y(n) = \sum_{k=0}^{q} b(k)w(n-k)$$

This structure may be simplified by noting that the two sets of delays are delaying the same sequence. Therefore, they may be combined as illustrated in Fig. 8-8 for the case in which $p = q$. The computational requirements for a direct form II structure are as follows:

Number of multiplications: $p + q + 1$ per output sample

Number of additions: $p + q$ per output sample

Number of delays: $\max(p, q)$

The direct form II structure is said to be *canonic* because it uses the minimum number of delays for a given $H(z)$.

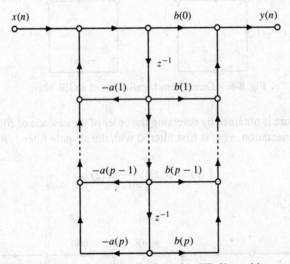

Fig. 8-8. Direct form II realization of an IIR filter with $p = q$.

8.4.2 Cascade Structure

The cascade structure is derived by factoring the numerator and denominator polynomials of $H(z)$:

$$H(z) = \frac{\sum_{k=0}^{q} b(k)z^{-k}}{1 + \sum_{k=1}^{p} a(k)z^{-k}} = A \prod_{k=1}^{\max\{p,q\}} \frac{1 - \beta_k z^{-1}}{1 - \alpha_k z^{-1}}$$

This factorization corresponds to a *cascade* of first-order filters, each having one pole and one zero. In general, the coefficients α_k and β_k will be complex. However, if $h(n)$ is real, the roots of $H(z)$ will occur in complex conjugate pairs, and these complex conjugate factors may be combined to form second-order factors with *real* coefficients:

$$H_k(z) = \frac{1 + \beta_{1k}z^{-1} + \beta_{2k}z^{-2}}{1 + \alpha_{1k}z^{-1} + \alpha_{2k}z^{-2}}$$

A sixth-order IIR filter implemented as a cascade of three second-order systems in direct form II is shown in Fig. 8-9.

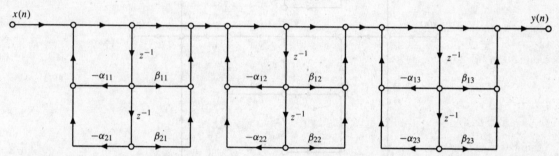

Fig. 8-9. A sixth-order IIR filter implemented as a cascade of three direct form II second-order systems.

There is considerable flexibility in how a system may be implemented in cascade form. For example, there are different *pairings* of the poles and zeros and different ways in which the sections may be *ordered*.

8.4.3 Parallel Structure

An alternative to factoring $H(z)$ is to expand the system function using a partial fraction expansion. For example, with

$$H(z) = \frac{\displaystyle\sum_{k=0}^{q} b(k)z^{-k}}{1 + \displaystyle\sum_{k=1}^{p} a(k)z^{-k}} = A\frac{\displaystyle\prod_{k=1}^{q}(1 - \beta_k z^{-1})}{\displaystyle\prod_{k=1}^{p}(1 - \alpha_k z^{-1})}$$

if $p > q$ and $\alpha_i \neq \alpha_k$ (the roots of the denominator polynomial are distinct), $H(z)$ may be expanded as a sum of p first-order factors as follows:

$$H(z) = \sum_{k=1}^{p} \frac{A_k}{1 - \alpha_k z^{-1}}$$

where the coefficients A_k and α_k are, in general, complex. This expansion corresponds to a sum of p first-order system functions and may be realized by connecting these systems in parallel. If $h(n)$ is real, the poles of $H(z)$ will occur in complex conjugate pairs, and these complex roots in the partial fraction expansion may be combined to form second-order systems with real coefficients:

$$H(z) = \sum_{k=1}^{N_s} \frac{\gamma_{0k} + \gamma_{1k}z^{-1}}{1 + \alpha_{1k}z^{-1} + \alpha_{2k}z^{-2}}$$

Shown in Fig. 8-10 is a sixth-order filter implemented as a parallel connection of three second-order direct form II systems. If $p \leq q$, the partial fraction expansion will also contain a term of the form

$$c_0 + c_1 z^{-1} + \cdots + c_{q-p}z^{-(q-p)}$$

which is an FIR filter that is placed in parallel with the other terms in the expansion of $H(z)$.

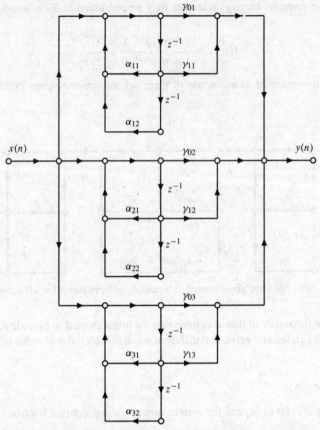

Fig. 8-10. A sixth-order IIR filter implemented as a parallel connection of three second-order direct form II structures.

8.4.4 Transposed Structures

The *transposition theorem* states that the input-output properties of a network remain unchanged after the following sequence of network operations:

1. Reverse the direction of all branches.
2. Change branch points into summing nodes and summing nodes into branch points.
3. Interchange the input and output.

Applying these manipulations to a network results in what is referred to as the *transposed form*. Shown in Fig. 8-11 are second-order transposed direct form I and direct form II filter structures.

8.4.5 Allpass Filters

An allpass filter has a frequency response with a constant magnitude:

$$|H_{ap}(e^{j\omega})| = 1 \qquad \text{all } \omega$$

If the system function of an allpass filter is a rational function of z, it has the form

$$H_{ap}(z) = \prod_{k=1}^{p} \frac{z^{-1} - a_k^*}{1 - a_k z^{-1}}$$

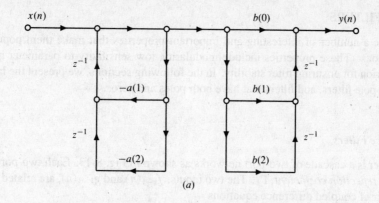

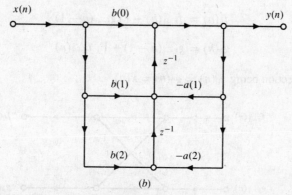

Fig. 8-11. Transposed direct form filter structures. (*a*) Transposed direct form I. (*b*) Transposed direct form II.

If $h(n)$ is real-valued, the complex roots occur in conjugate pairs, and these pairs may be combined to form second-order factors with real coefficients:

$$H_{ap}(z) = \prod_{k=1}^{N_s} \frac{\alpha_k + \beta_k z^{-1} + z^{-2}}{1 + \beta_k z^{-1} + \alpha_k z^{-2}}$$

A direct form II implementation for one of these sections is shown in Fig. 8-12. Because each section only has two distinct coefficients, α_k and β_k, it is possible to implement these sections using as few as two multiplies.

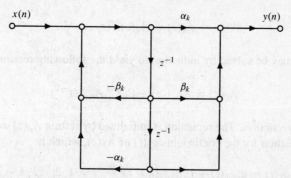

Fig. 8-12. A second-order section of an allpass filter implemented in direct form II.

8.5 LATTICE FILTERS

Lattice filters have a number of interesting and important properties that make them popular in a number of different applications. These properties include modularity, low sensitivity to parameter quantization effects, and a simple criterion for ensuring filter stability. In the following sections, we present the lattice filter structure for FIR filters, all-pole filters, and filters that have both poles and zeros.

8.5.1 FIR Lattice Filters

An FIR lattice filter is a cascade of two-port networks as shown in Fig. 8-13. Each two-port network is defined by the value of its *reflection coefficient*, Γ_k. The two inputs, $f_{k-1}(n)$ and $g_{k-1}(n)$, are related to the outputs $f_k(n)$ and $g_k(n)$ by a pair of coupled difference equations

$$f_k(n) = f_{k-1}(n) + \Gamma_k g_{k-1}(n-1)$$
$$g_k(n) = g_{k-1}(n-1) + \Gamma_k f_{k-1}(n)$$

(8.4)

with the input to the first section being $f_0(n) = g_0(n) = x(n)$.

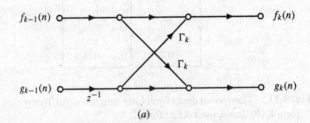

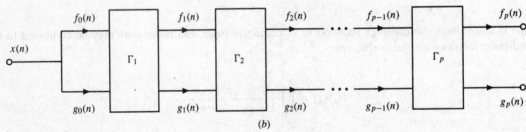

Fig. 8-13. A pth-order FIR lattice filter. (*a*) The two-port network for each lattice filter module. (*b*) A cascade of p lattice filter modules.

With $A_k(z)$ the system function relating the input $x(n)$ to the intermediate output $f_k(n)$,

$$F_k(z) = A_k(z)X(z)$$

these difference equations may be solved by induction to yield the following recurrence formula for $A_k(z)$:

$$A_k(z) = A_{k-1}(z) + \Gamma_k z^{-k} A_{k-1}(z^{-1})$$

(8.5)

which is called the *step-up recursion*. The recursion is initialized by setting $A_0(z) = 1$. This recurrence formula also defines a recurrence relation for the coefficients $a_k(i)$ of $A_k(z)$, which is

$$a_k(i) = a_{k-1}(i) + \Gamma_k a_{k-1}(k-i) \qquad i = 1, 2, \dots, k-1$$
$$a_k(k) = \Gamma_k$$

A simple way to write this recursion is in terms of vectors as follows:

$$
\begin{bmatrix} 1 \\ a_k(1) \\ \vdots \\ a_k(k-1) \\ a_k(k) \end{bmatrix} = \begin{bmatrix} 1 \\ a_{k-1}(1) \\ \vdots \\ a_{k-1}(k-1) \\ 0 \end{bmatrix} + \Gamma_k \begin{bmatrix} 0 \\ a_{k-1}(k-1) \\ \vdots \\ a_{k-1}(1) \\ 1 \end{bmatrix}
$$

EXAMPLE 8.5.1   For a second-order FIR lattice filter with reflection coefficients $\Gamma_1 = \frac{1}{2}$ and $\Gamma_2 = \frac{1}{4}$, the system function relating $x(n)$ to $f_1(n)$ is

$$
A_1(z) = A_0(z) + \Gamma_1 z^{-1} A_0(z^{-1}) = 1 + \tfrac{1}{2} z^{-1}
$$

and the second-order system function relating $x(n)$ to $f_2(n)$ is

$$
\begin{aligned}
A_2(z) &= A_1(z) + \Gamma_2 z^{-2} A_1(z^{-1}) \\
&= \left(1 + \tfrac{1}{2} z^{-1}\right) + \tfrac{1}{4} z^{-2}\left(1 + \tfrac{1}{2} z\right) \\
&= 1 + \tfrac{5}{8} z^{-1} + \tfrac{1}{4} z^{-2}
\end{aligned}
$$

The recurrence formula in Eq. (8.5) provides an algorithm to find the system function $A_p(z)$ from the reflection coefficients Γ_k, $k = 1, 2, \ldots, p$. To find the reflection coefficients Γ_k for a given system function $A_p(z)$, we use the *step-down recursion*, which is given by

$$
A_{k-1}(z) = \frac{1}{1 - \Gamma_k^2}[A_k(z) - \Gamma_k z^{-k} A_k(z^{-1})] \qquad k = p, p-1, \ldots, 1 \tag{8.6}
$$

In terms of the coefficients $a_k(i)$, this recursion is

$$
a_{k-1}(i) = \frac{1}{1 - \Gamma_k^2}[a_k(i) - \Gamma_k a_k(k-i)] \qquad i = 1, 2, \ldots, k-2
$$
$$
a_{k-1}(k-1) = \Gamma_{k-1}
$$

The reflection coefficients are then found from the polynomials $A_k(z)$ by setting $\Gamma_k = a_k(k)$.

EXAMPLE 8.5.2   To find the reflection coefficients Γ_1 and Γ_2 corresponding to the second-order FIR filter $A_2(z) = 1 - \frac{1}{2} z^{-2}$, we begin by setting

$$
\Gamma_2 = a_2(2) = -\tfrac{1}{2}
$$

Next, we find $A_1(z)$ using the step-down recursion,

$$
\begin{aligned}
A_1(z) &= \frac{1}{1 - \Gamma_2^2}[A_2(z) - \Gamma_2 z^{-2} A_2(z^{-1})] \\
&= \tfrac{4}{3}\left[\left(1 - \tfrac{1}{2} z^{-2}\right) + \tfrac{1}{2} z^{-2}\left(1 - \tfrac{1}{2} z^2\right)\right] = 1
\end{aligned}
$$

Because $a_1(1) = 0$, $\Gamma_1 = 0$. Therefore, the reflection coefficients are $\Gamma_1 = 0$ and $\Gamma_2 = -\frac{1}{2}$.

So far, we have only considered the system function relating the input $x(n)$ to the output $f_p(n)$. A similar set of equations relate the input $x(n)$ to the output $g_p(n)$. With

$$
G_p(z) = A_p'(z) X(z)
$$

the relationship between the system function $A_p(z)$ and $A_p'(z)$ is as follows:

$$
A_p'(z) = z^{-p} A_p(z^{-1})
$$

Thus, $f_p(n)$ and $g_p(n)$ are related by an allpass filter, $F_p(z) = H_{ap}(z)G_p(z)$, where

$$H_{ap}(z) = \frac{z^{-p}A_p(z^{-1})}{A_p(z)} \qquad (8.7)$$

An important property of the lattice filter is that the roots of $A_p(z)$ will lie *inside* the unit circle if and only if the reflection coefficients are bounded by 1 in magnitude:

$$|\Gamma_k| < 1 \qquad k = 1, 2, \ldots, p$$

This property is the basis for the *Schur-Cohn* stability test for digital filters. Specifically, a causal filter with a system function

$$H(z) = \frac{B(z)}{A(z)}$$

will be stable if and only if the reflection coefficients associated with $A(z)$ are bounded by 1 in magnitude.

8.5.2   All-Pole Lattice Filters

The structure for an all-pole lattice filter is shown in Fig. 8-14. As with the FIR lattice, a pth-order all-pole filter is a cascade of p stages, with each stage being a two-port network that is parameterized by its reflection coefficient Γ_k. The two inputs, $f_k(n)$ and $g_{k-1}(n)$, are related to the two outputs $f_{k-1}(n)$ and $g_k(n)$ by a pair of coupled difference equations:

$$f_{k-1}(n) = f_k(n) - \Gamma_k g_{k-1}(n-1)$$

$$g_k(n) = g_{k-1}(n-1) + \Gamma_k f_k(n)$$

The system function relating the input $x(n)$ to the output $y(n)$ is

$$H(z) = \frac{1}{A_p(z)}$$

where $A_p(z)$ is the polynomial that is generated by the recursion given in Eq. (8.5). In addition, note that the system function relating $x(n)$ to $w(n)$ is an allpass filter with a system function $H_{ap}(z)$ given in Eq. (8.7).

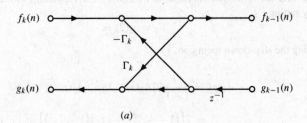

(a)

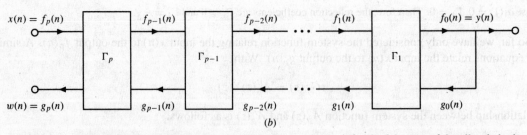

(b)

Fig. 8-14. A pth-order all-pole lattice filter. (a) The two-port network for the kth stage of the all-pole lattice filter. (b) Cascade of p lattice stages.

8.5.3 IIR Lattice Filters

If $H(z)$ is an IIR filter with p poles and q zeros,

$$H(z) = \frac{B_q(z)}{A_p(z)} = \frac{\displaystyle\sum_{k=0}^{q} b_q(k)z^{-k}}{1 + \displaystyle\sum_{k=1}^{p} a_p(k)z^{-k}}$$

with $q \leq p$, a lattice filter implementation of $H(z)$ consists of two components. The first is an all-pole lattice with reflection coefficients $\Gamma_1, \Gamma_2, \ldots, \Gamma_p$ that implements $1/A_p(z)$. The second is a tapped delay line with coefficients $c_q(k)$. The structure is illustrated in Fig. 8-15 for the case in which $p = q$. The relationship between the lattice filter coefficients $c_q(k)$ and the direct form coefficients $b_q(k)$ is given by

$$b_q(k) = \sum_{j=k}^{q} c_q(j)a_j(j-k) \qquad k = 0, 1, \ldots, p \qquad (8.8)$$

Similarly, a recursion that generates the coefficients $c_q(k)$ from the coefficients $b_q(k)$ is

$$c_q(k) = b_q(k) - \sum_{j=k+1}^{q} c_q(j)a_j(j-k) \qquad k = q-1, q-2, \ldots, 0 \qquad (8.9)$$

This recursion is initialized with $c_q(q) = b_q(q)$.

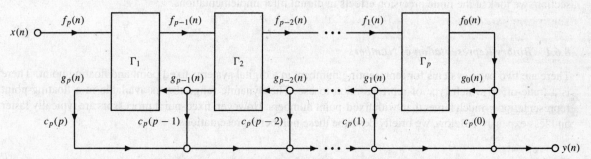

Fig. 8-15. An IIR lattice filter with p poles and p zeros.

EXAMPLE 8.5.3 A third-order low-pass elliptic filter with a cutoff frequency of $\omega_c = 0.5\pi$ has a system function

$$H(z) = \frac{0.2759 + 0.5121z^{-1} + 0.5121z^{-2} + 0.2759z^{-3}}{1 - 0.0010z^{-1} + 0.6546z^{-2} - 0.0775z^{-3}}$$

To implement this filter using a lattice filter structure, we first transform the denominator coefficients into reflection coefficients. Using the step-down recursion, we find

$$\Gamma_1 = 0.0302 \qquad \Gamma_2 = 0.6584 \qquad \Gamma_3 = -0.0775$$

with the second-order system function given by

$$A_2(z) = 1 + 0.0501z^{-1} + 0.6584z^{-2}$$

and the first-order system function

$$A_1(z) = 1 + 0.0302z^{-1}$$

Next, the coefficients $c_3(k)$ are found using the recursion given in Eq. (8.9). Beginning with

$$c_3(3) = b_3(3) = 0.2759$$

we then have

$$c_3(2) = b_3(2) - c_3(3)a_3(1) = 0.5124$$
$$c_3(1) = b_3(1) - c_3(2)a_2(1) - c_3(3)a_3(2) = 0.3058$$
$$c_3(0) = b_3(0) - c_3(1)a_1(1) - c_3(2)a_2(2) - c_3(3)a_3(3) = -0.0493$$

This leads to the lattice filter implementation illustrated below.

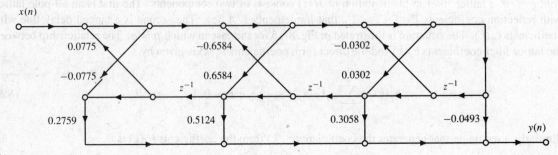

8.6 FINITE WORD-LENGTH EFFECTS

In implementing a discrete-time system in hardware or software, it is important to consider the finite word-length effects. For example, if a filter is to be implemented on a fixed-point processor, the filter coefficients must be quantized to a finite number of bits. This will change the frequency response characteristics of the filter. In this section, we look at the finite precision effects in digital filter implementations.

8.6.1 Binary Representation of Numbers

There are two basic systems for representing numbers in a digital system: fixed point and floating point. There is a trade-off in which type of representation to use. The dynamic range that is available in a floating-point representation is much larger than with fixed-point numbers. However, fixed-point processors are typically faster and less expensive. Below, we briefly describe these number representations.

Fixed Point

In the binary representation of a real number, x, using $B + 1$ bits, there are three commonly used formats: sign magnitude, one's complement, and two's complement, with two's complement being the most common. In these systems, the only difference is in the way that negative numbers are represented.

1. *Sign magnitude*: With a sign-magnitude format, a number x is represented as

$$x = X_m(-1)^{b_0} \cdot \sum_{i=1}^{B} b_i 2^{-i}$$

where X_m is an arbitrary scale factor and where each of the *bits* b_i are either 0 or 1. Thus, b_0 is the sign bit, and the remaining bits represent the magnitude of the fractional number. Bit b_1 is called the *most significant bit (MSB)*, and b_B is called the *least significant bit* (LSB). For example, with $X_m = 1$,

$$x = 0.8125 = 0.11010$$

and

$$-x = -0.8125 = 1.11010$$

2. *One's complement*: In one's complement form, a negative number is represented by complementing all of the bits in the binary representation of the positive number. For example, with $X_m = 1$ and $x = 0.8125 = 0.11010$,

$$-x = -0.8125 = \overline{0.11010} = 1.00101$$

3. *Two's complement*: With a two's complement format, a real number x is represented as

$$x = X_m\left(-b_0 + \sum_{i=1}^{B} b_i 2^{-i}\right)$$

Thus, negative numbers are formed by complementing the bits of the positive number and adding 1 to the least significant bit. For example, with $X_m = 1$, the two's complement representation of $x = -0.8125$ is

$$x = -0.8125 = \overline{0.11010} + 0.00001 = 1.00110$$

Note that with $B + 1$ bits, the smallest difference between two quantized numbers, the *resolution*, is

$$\Delta = X_m 2^{-B}$$

and all quantized numbers lie on the *range* $-X_m \leq x < X_m$.

Floating Point

For a word length of $B + 1$ bits in a fixed-point number system, the resolution is constant over the entire range of numbers, and the resolution decreases (Δ increases) in direct proportion to the dynamic range, $2X_m$. A floating-point number system covers a larger range of numbers at the expense of an overall decrease in resolution, with the resolution varying over the entire range of numbers. The representation used for floating-point numbers is typically of the form

$$x = M \cdot 2^E$$

where M, the *mantissa*, is a signed B_M-bit fractional binary number with $\frac{1}{2} \leq |M| < 1$, and E, the *exponent*, is a B_E-bit signed integer. Because M is a signed fraction, it may be represented using any of the representations described above for fixed-point numbers.

Quantization Errors in Fixed-Point Number Systems

In performing computations within a fixed- or floating-point digital processor, it is necessary to quantize numbers by either truncation or rounding from some level of precision to a lower level. For example, because multiplying two 16-bit fixed-point numbers will produce a product with up to 31 bits of precision, the product will generally need to be quantized back to 16 bits. Truncation and rounding introduce a quantization error

$$e = Q[x] - x$$

where x is the number to be quantized and $Q[x]$ is the quantized number. The characteristics of the error depend upon the number representation that is used. Truncating numbers that are represented in sign-magnitude form result in a quantization error that is negative for positive numbers and positive for negative numbers. Thus, the quantization error is symmetric about zero and falls in the range

$$-\Delta \leq e \leq \Delta$$

where

$$\Delta = X_m 2^{-B}$$

On the other hand, for a two's complement representation, the truncation error is always negative and falls in the range

$$-\Delta \leq e \leq 0$$

With rounding, the quantization error is independent of the type of fixed-point representation and falls in the range

$$-\frac{\Delta}{2} \le e \le \frac{\Delta}{2}$$

For floating-point numbers, the mantissa is either rounded or truncated, and the size of the error depends on the value of the exponent.

8.6.2   Quantization of Filter Coefficients

In order to implement a filter on a digital processor, the filter coefficients must be converted into binary form. This conversion leads to movements in the pole and zero locations and a change in the frequency response of the filter. The accuracy with which the filter coefficients can be specified depends upon the word length of the processor, and the sensitivity of the filter to coefficient quantization depends on the structure of the filter, as well as on the locations of the poles and zeros.

For a second-order section with poles at $z = re^{\pm j\theta}$,

$$A(z) = (1 - re^{j\theta}z^{-1})(1 - re^{-j\theta}z^{-1}) = 1 - 2r\cos\theta z^{-1} + r^2 z^{-2}$$

the filter coefficients in a direct form realization are

$$a(1) = 2r\cos\theta \qquad a(2) = -r^2$$

If $a(1)$ and $a(2)$ are quantized to $B + 1$ bits, the real part of the pole location is restricted to 2^{B+1} possible values, and the radius squared is restricted to 2^B values. The set of allowable pole locations for a 4-bit processor is shown in Fig. 8-16.

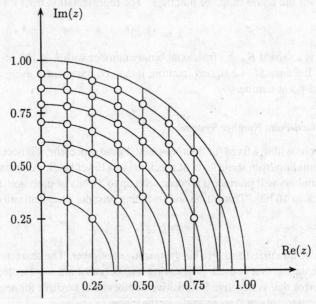

Fig. 8-16.   The set of allowable pole locations in the first quadrant of the z-plane for a second-order IIR filter implemented in direct form using a 4-bit processor.

A general sensitivity analysis of a pth-order polynomial

$$A(z) = 1 + \sum_{k=1}^{p} a(k)z^{-k} = \prod_{k=1}^{p}(1 - \alpha_k z^{-1})$$

shows that the root locations are more sensitive to coefficient quantization errors when the roots are tightly clustered. For example, if the coefficients $a(k)$ are quantized,

$$\hat{A}(z) = 1 + \sum_{k=1}^{p} [a(k) + \Delta a(k)] z^{-k} = \prod_{k=1}^{p} [1 - (\alpha_k + \Delta\alpha_k) z^{-1}]$$

then the sensitivity of the location of the ith pole to changes $\Delta a(k)$ in the coefficients $a(k)$ is approximately

$$\Delta\alpha_i \approx \sum_{k=1}^{p} \frac{\partial \alpha_i}{\partial a(k)} \Delta a(k)$$

With

$$\left(\frac{\partial A(z)}{\partial \alpha_i}\right)_{z=\alpha_i} \frac{\partial \alpha_i}{\partial a(k)} = \left(\frac{\partial A(z)}{\partial a(k)}\right)_{z=\alpha_i}$$

where

$$\frac{\partial A(z)}{\partial a(k)} = z^{-k} \qquad \frac{\partial A(z)}{\partial \alpha_i} = -\prod_{\substack{j=1 \\ j \neq i}}^{p} (1 - \alpha_j z^{-1}) z^{-1}$$

then

$$\frac{\partial \alpha_i}{\partial a(k)} = -\frac{\alpha_i^{p-k}}{\displaystyle\prod_{\substack{j=1 \\ j \neq i}}^{p} (\alpha_i - \alpha_j)}$$

Thus, if the poles are tightly clustered, $|\alpha_i - \alpha_j|$ is small, and small changes in $a(k)$ will result in large changes in the pole locations.

The movement of the poles may be minimized by maximizing the distance between the poles, $|\alpha_i - \alpha_j|$. This may be accomplished by implementing a high-order filter as a combination of first- or second-order systems. For example, with a cascade of second-order sections, each pair of complex conjugate poles and zeros may be realized separately, thereby localizing the coefficient quantization errors to each section.

For an FIR filter,

$$H(z) = \sum_{n=0}^{N} h(n) z^{-n}$$

when the coefficients are quantized, the system function becomes

$$\hat{H}(z) = \sum_{n=0}^{N} \hat{h}(n) z^{-n} = \sum_{n=0}^{N} [h(n) + \Delta h(n)] z^{-n} = H(z) + \Delta H(z)$$

Thus, the quantization errors may be modeled as $H(z)$ in parallel with $\Delta H(z)$ as shown in Fig. 8-17. If we assume that the coefficients $h(n)$ are less than 1 in magnitude, and that the coefficients are rounded to $B + 1$ bits,

$$-2^{-(B+1)} < \Delta h(n) < 2^{-(B+1)}$$

Therefore, a loose bound on the error in the frequency response is

$$|\Delta H(e^{j\omega})| = \left| \sum_{n=0}^{N} \Delta h(n) e^{-jn\omega} \right| \leq \sum_{n=0}^{N} |\Delta h(n)| \leq (N+1) 2^{-(B+1)}$$

As with IIR filters, if the zeros are tightly clustered, the zero locations will be sensitive to coefficient quantization errors. However, FIR filters are commonly implemented in direct form for two reasons:

1.   The zeros of FIR filters are not generally tightly clustered.
2.   In direct form, linear phase is easily preserved.

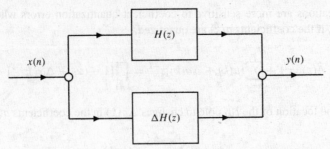

Fig. 8-17. Model for the coefficient quantization error in FIR filters.

8.6.3 Round-Off Noise

Round-off noise is introduced into a digital filter when products or sums of products are quantized. For example, if two $(B + 1)$-bit numbers are multiplied, the product is a $(2B + 1)$-bit number. If the product is to be saved in a $(B + 1)$-bit register or used in a $(B + 1)$-bit adder, it must be quantized to $(B + 1)$-bits, which results in the addition of *round-off noise*. This round-off noise propagates through the filter and appears at the output of the filter as round-off noise. In this section, we illustrate the analysis of round-off noise effects by example.

Consider the second-order IIR filter implemented in direct form I shown in Fig. 8-18(a). The difference equation for this network is

$$y(n) = \sum_{k=0}^{2} b(k)x(n-k) - \sum_{k=1}^{2} a(k)y/n - k$$

If we assume that all numbers are represented by $B + 1$ fixed-point numbers and that the network uses $(B + 1)$-bit adders, each $(2B + 1)$-bit product must be quantized to $B + 1$ bits by either truncation or rounding. Fig. 8-18(b) shows the quantizers explicitly. The difference equation corresponding to this system is the nonlinear equation

$$\hat{y}(n) = \sum_{k=0}^{2} Q[b(k)x(n-k)] - \sum_{k=1}^{2} Q[a(k)\hat{y}(n-k)]$$

If the quantizers are replaced with noise sources that are equal to the quantization error, we have an alternative representation shown in Fig. 8-18(c). This representation is particularly useful when it is assumed that the quantization noise has the following properties:

1. Each quantization noise source is a *wide-sense stationary white noise process*.
2. The probability distribution function of each noise source is uniformly distributed over the quantization interval.
3. Each noise source is uncorrelated with the input to the quantizer, all other noise sources, and the input to the system.

With $B + 1$ bits, and a fractional representation for all numbers, the second property implies that the quantization noise for *rounding* has a zero mean and a variance equal to

$$\sigma_e^2 = \tfrac{1}{12}2^{-2B}$$

To analyze the effect of the round-off noise sources at the output of the filter, it is necessary to know how noise propagates through a filter. If the input to a linear shift-invariant filter with a unit sample response $h(n)$ is wide-sense stationary white noise, $e(n)$, with a mean m_e and a variance σ_e^2, the filtered noise, $f(n) = h(n) * e(n)$, is a wide-sense stationary process with a mean

$$m_f = m_e \sum_{n=-\infty}^{\infty} h(n)$$

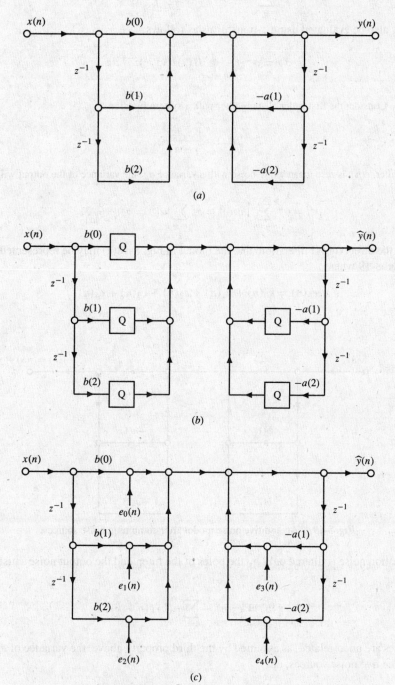

Fig. 8-18. Analysis of round-off noise. (*a*) A second-order direct form I
filter. (*b*) Quantization of products in the filter. (*c*) An additive noise model
for the round-off noise. .

and a variance

$$\sigma_f^2 = \sigma_e^2 \sum_{n=-\infty}^{\infty} |h(n)|^2 = \sigma_e^2 \frac{1}{2\pi} \int_{-\pi}^{\pi} |H(e^{j\omega})|^2 \, d\omega$$

The variance may also be evaluated using z-transforms as follows:

$$\sigma_f^2 = \sigma_e^2 \frac{1}{2\pi j} \oint_C H(z) H(z^{-1}) z^{-1} \, dz$$

EXAMPLE 8.6.1   Consider the first-order all-pole filter with a system function

$$H(z) = \frac{1}{1 - \alpha z^{-1}}$$

If the input to this filter, $e(n)$, is zero mean white noise with a variance σ_e^2, the variance of the output will be

$$\sigma_f^2 = \sigma_e^2 \sum_{n=-\infty}^{\infty} |h(n)|^2 = \sigma_e^2 \sum_{n=0}^{\infty} |\alpha|^{2n} = \sigma_e^2 \frac{1}{1 - |\alpha|^2}$$

Returning to the direct form I filter, note that the model in Fig. 8-18(c) may be represented in the equivalent form shown in Fig. 8-19 where

$$e_a(n) = e_0(n) + e_1(n) + e_2(n) + e_3(n) + e_4(n)$$

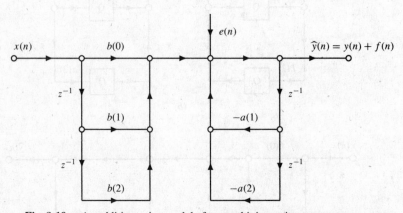

Fig. 8-19.   An additive noise model after combining noise sources.

Thus, the quantization noise is filtered only by the poles of the filter, and the output noise satisfies the difference equation

$$f(n) = e_a(n) - \sum_{k=1}^{2} a(k) f(n - k)$$

If the noise sources are uncorrelated, as assumed by the third property above, the variance of $e_a(n)$ is the sum of the variances of the five noise sources, or

$$\sigma_a^2 = 5\sigma_e^2 = 5 \cdot \frac{2^{-2B}}{12}$$

Assuming that the filter is stable, and that the poles of the filter are complex,

$$\frac{1}{1 + a(1)z^{-1} + a(2)z^{-2}} = \frac{1}{(1 - re^{j\theta}z^{-1})(1 - re^{-j\theta}z^{-1})}$$

the variance of the output noise is

$$\sigma_f^2 = 5\frac{2^{-2B}}{12}\frac{1}{2\pi j}\oint_C \frac{1}{A(z)A(z^{-1})}z^{-1}\,dz$$

$$= 5\frac{2^{-2B}}{12}\frac{1}{2\pi j}\oint_C \frac{z\,dz}{(z-re^{j\theta})(z-re^{-j\theta})(1-re^{j\theta}z)(1-re^{-j\theta}z)}$$

Using Cauchy's residue theorem to evaluate this integral, we find that

$$\sigma_f^2 = 5\frac{2^{-2B}}{12}\left(\frac{1+r^2}{1-r^2}\right)\frac{1}{r^4+1-2r^2\cos 2\theta}$$

Note that as the poles move closer to the unit circle, $r \to 1$, the variance of the output noise increases.

The noise performance of digital filters may be improved by using $(2B + 1)$-bit adders to accumulate sums of products prior to quantization. In this case, the difference equation for the direct form I network becomes

$$\hat{y}(n) = Q\left[\sum_{k=0}^{2}b(k)x(n-k) - \sum_{k=1}^{2}a(k)\hat{y}(n-k)\right]$$

Thus, the sums are accumulated with an accuracy of $2B + 1$ bits, and the sum is then quantized to $B + 1$ bits in order to store $\hat{y}(n-1)$ and $\hat{y}(n-2)$ in $(B+1)$-bit delay registers and to generate the $(B+1)$-bit output $\hat{y}(n)$. Because there is only one quantizer, which quantizes the sum of products, the variance of the noise source in Fig. 8-19 is reduced from $5\sigma_e^2$ to σ_e^2.

8.6.4　Pairing and Ordering

For a filter that is implemented in cascade or parallel form, there is considerable flexibility in terms of selecting which poles are to be paired with which zeros and in selecting the order in which the sections are to be cascaded for a cascade structure. Pairing and ordering may have a significant effect on the shape of the output noise power and on the total output noise variance. The rules that are generally followed for pairing and ordering are as follows:

1. The pole that is closest to the unit circle is paired with the zero that is closest to it in the z-plane, and this pairing is continued until all poles and zeros have been paired.
2. The resulting second-order sections are then ordered in a cascade realization according to the closeness of the poles to the unit circle. The ordering may be done either in terms of increasing closeness to the unit circle or in terms of decreasing closeness to the unit circle. Which ordering is used depends on the consideration of a number of factors, including the shape of the output noise and the output noise variance.

8.6.5　Overflow

Another issue in fixed-point implementations of discrete-time systems is *overflow*. If each fixed-point number is taken to be a fraction that is less than 1 in magnitude, each node variable in the network should be constrained to be less than 1 in magnitude in order to avoid overflow. If we let $h_k(n)$ denote the unit sample response of the system relating the input $x(n)$ to the kth node variable, $w_k(n)$,

$$|w_k(n)| = \left|\sum_{m=-\infty}^{\infty}x(n-m)h_k(m)\right| \le X_{\max}\sum_{m=-\infty}^{\infty}|h_k(m)|$$

where X_{max} is the maximum value of the input $x(n)$. Therefore, a sufficient condition that $|w_k(n)| < 1$ so that no overflow occurs in the network is

$$X_{max} \leq \frac{1}{\sum\limits_{m=-\infty}^{\infty} |h_k(m)|}$$

for all nodes in the network. If this is not satisfied, $x(n)$ may be scaled by a factor s so that

$$sX_{max} \leq \frac{1}{\sum\limits_{m=-\infty}^{\infty} |h_k(m)|} \qquad (8.10)$$

EXAMPLE 8.6.2   In the first-order direct form II network shown below,

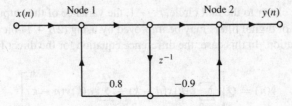

there are two nodes that represent adders, which are labeled "Node 1" and "Node 2." The unit sample response from the input to the first node is

$$h_1(n) = (0.8)^n u(n)$$

Therefore,
$$\sum_{m=-\infty}^{\infty} |h_1(m)| = \frac{1}{1 - (0.8)} = 5$$

The unit sample response from the input to the second node is

$$h_2(n) = (0.8)^n u(n) - 0.9(0.8)^{n-1} u(n-1) = \delta(n) - 0.1(0.8)^{n-1} u(n-1)$$

and
$$\sum_{m=-\infty}^{\infty} |h_2(m)| = 1 + (0.1) \cdot 5 = 1.5$$

Thus, with a fractional representation for $x(n)$, a sufficient condition for no overflow to occur is that $X_{max} \leq 0.2$.

Solved Problems

Structures for FIR Systems

8.1   Find the frequency response of the system defined by the following network:

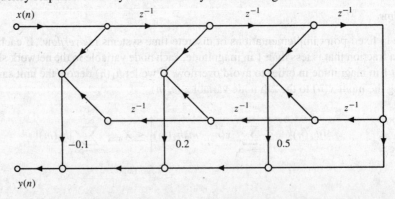

We recognize this structure as a linear phase system with a unit sample response

$$h(n) = -0.1[\delta(n) + \delta(n - 6)] + 0.2[\delta(n - 1) + \delta(n - 5)] + 0.5[\delta(n - 2) + \delta(n - 4)] + \delta(n - 3)$$

Therefore, the frequency response is

$$H(e^{j\omega}) = -0.1[1 + e^{-j6\omega}] + 0.2[e^{-j\omega} + e^{-j5\omega}] + 0.5[e^{-j2\omega} + e^{-j4\omega}] + e^{-j3\omega}$$

$$= e^{-j3\omega}[1 + \cos\omega + 0.4\cos 2\omega - 0.2\cos 3\omega]$$

8.2   A linear shift-invariant system has a unit sample response given by

$$h(0) = -0.01$$
$$h(1) = 0.02$$
$$h(2) = -0.10$$
$$h(3) = 0.40$$
$$h(4) = -0.10$$
$$h(5) = 0.02$$
$$h(6) = -0.01$$

(*a*)   Draw a signal flowgraph for this system that requires the minimum number of multiplications.

(*b*)   If the input to this system is bounded with $|x(n)| < 1$ for all n, what is the maximum value that the output, $y(n)$, can attain?

(*a*)   Because this system is a linear phase filter, it may be implemented with a network that has only four multiplies and six delays as shown in the figure below.

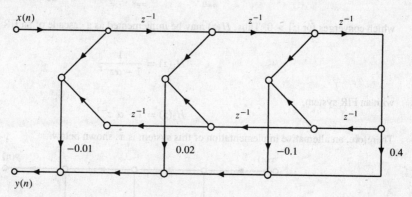

(*b*)   With an input $x(n)$, the output is

$$y(n) = \sum_{k=0}^{6} h(k)x(n - k)$$

Therefore, the magnitude of $y(n)$ is upper bounded by

$$|y(n)| = \left| \sum_{k=0}^{6} h(k)x(n - k) \right| \leq \sum_{k=0}^{6} |h(k)| \, |x(n - k)|$$

With $|x(n)| < 1$ for all n,

$$|y(n)| \leq \sum_{k=0}^{6} |h(k)| = 0.66$$

8.3   The unit sample response of an FIR filter is

$$h(n) = \begin{cases} \alpha^n & 0 \le n \le 6 \\ 0 & \text{otherwise} \end{cases}$$

(a) Draw the direct form implementation of this system.

(b) Show that the corresponding system function is

$$H(z) = \frac{1 - \alpha^7 z^{-7}}{1 - \alpha z^{-1}} \qquad |z| > 0$$

and use this to draw a flowgraph that is a cascade of an FIR system with an IIR system.

(c) For both of these implementations, determine the number of multiplications and additions required to compute each output value and the number of storage registers that are required.

(a) With a unit sample response

$$h(n) = \alpha^n [u(n) - u(n - 7)]$$

the direct form implementation of this system is as shown below.

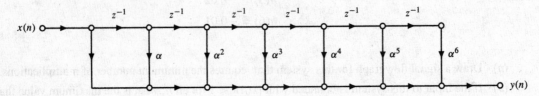

(b) The system function is

$$H(z) = \sum_{n=0}^{6} h(n) z^{-n} = \sum_{n=0}^{6} \alpha^n z^{-n} = \frac{1 - (\alpha z^{-1})^7}{1 - \alpha z^{-1}}$$

which converges for $|z| > 0$. Thus, $H(z)$ may be implemented as a cascade of an IIR system,

$$H_1(z) = \frac{1}{1 - \alpha z^{-1}}$$

with an FIR system,

$$H_2(z) = 1 - \alpha^7 z^{-7}$$

Therefore, an alternative implementation of this system is as shown below.

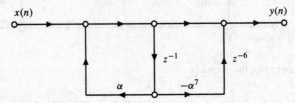

where the branch labeled with z^{-6} represents a delay by 6.

(c) The direct form structure requires six delays, which is the minimum number necessary for this system, six multiplications, and six additions. The cascade, on the other hand, requires one additional delay but only two multiplications and two additions.

8.4   A DSP chip used in real-time signal processing applications has an instruction cycle time of 100 ns. One of the instructions in the instruction set, MACD, will fetch a value from data memory (input signal), fetch another data value from program memory (filter coefficient), multiply the two numbers together, add the product to the accumulator, and then move a number in data memory into the next memory location (this corresponds to a shift or delay of the data sequence). Thus, for an FIR filter of order N, to find the value of the output at time n, we need one instruction to read the new input value, $x(n)$, into the processor, we need $(N + 1)$ MACD instructions to evaluate the sum

$$y(n) = \sum_{k=0}^{N} h(k)x(n - k)$$

and we need one instruction to output the value of $y(n)$. In addition, there are eight other instruction cycles required for each n in order to perform such functions as setting up memory pointers, zeroing the accumulator, and so on.

(a)   With these requirements in mind, determine the maximum bandwidth signal that may be filtered with an FIR filter of order $N = 255$, in real time, using a single DSP chip.

(b)   A speech waveform $x_a(t)$ is sampled at 8 kHz. Determine the maximum length FIR filter that may be used to filter the sampled speech signal in real time.

(a)   For the given DSP chip, we need $N + 11$ instruction cycles to compute a single output value for an FIR filter of order N. Therefore, with $N = 255$, we need 266 cycles, or 266×10^{-7} s to compute each output point. Thus, the signal to be filtered cannot be sampled any faster than

$$f_s = \frac{1}{266 \times 10^{-7}} \text{ Hz} = 37.6 \text{ kHz}$$

Therefore, the bandwidth of the input signal is limited to 18.8 kHz (i.e., $X_a(f)$ must be zero for $|f| > 18.8$ kHz).

(b)   Sampling speech at 8 kHz produces 8000 samples per second. Therefore, we have $T_s = 1/8000 = 0.125$ ms to compute each output. This allows for $M = (0.125 \times 10^{-3})/10^{-7} = 1250$ instruction cycles. Thus, we may implement an FIR filter of order $N = 1250 - 11 = 1239$.

8.5   Find the unit sample response, $h(n)$, of the network drawn below and find the 64-point DFT of $h(n)$.

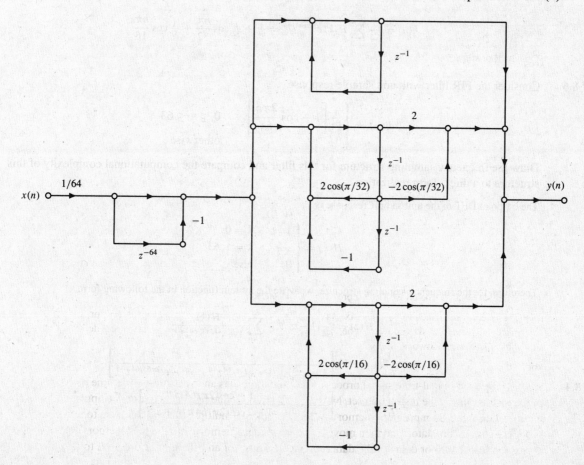

This is a frequency sampling structure for an FIR filter with a unit sample response of length $N = 64$. Because the gain of the first-order section with a pole at $z = 1$ is equal to 1, $H(0) = 1$. With second-order networks of the form

$$G_k(z) = \frac{A(k) - B(k)z^{-1}}{1 - 2\cos(2\pi k/N)\, z^{-1} + z^{-2}}$$

with
$$A(k) = H(k) + H(N - k)$$
$$B(k) = H(k)\, e^{-j2\pi k/N} + H(N - k)\, e^{j2\pi k/N}$$

we see that $H(1)$ and $H(2)$ are nonzero, along with $H(62)$ and $H(63)$. We may therefore solve these equations for $H(1)$, $H(2)$, $H(62)$, and $H(63)$ as follows. Because $A(1) = 2$,

$$H(1) + H(63) = 2$$

and because $B(1) = 2\cos(\pi/32)$,

$$H(1)\, e^{-j\pi/32} + H(63)\, e^{j\pi/32} = 2\cos\frac{\pi}{32}$$

Thus, we have two equations in two unknowns, which may be written in matrix form as follows:

$$\begin{bmatrix} 1 & 1 \\ e^{-j\pi/32} & e^{j\pi/32} \end{bmatrix} \begin{bmatrix} H(1) \\ H(63) \end{bmatrix} = \begin{bmatrix} 2 \\ 2\cos\dfrac{\pi}{32} \end{bmatrix}$$

Solving these equations we find that $H(1) = H(63) = 1$. Similarly, with $A(2) = 2$ and $B(2) = 2\cos(\pi/32)$, we find that $H(2) = H(62) = 1$. Therefore, the 64-point DFT of $h(n)$ is

$$H(k) = \begin{cases} 1 & k = 0, 1, 2, 62, 63 \\ 0 & \text{else} \end{cases}$$

and the unit sample response is

$$h(n) = \frac{1}{64} \sum_{k=0}^{63} H(k)\, e^{j2\pi nk/64} = \frac{1}{64} + \frac{1}{32}\cos\frac{n\pi}{32} + \frac{1}{32}\cos\frac{n\pi}{16}$$

8.6   Consider the FIR filter with unit sample response

$$h(n) = \begin{cases} \dfrac{1}{64}\left[1 - \cos\dfrac{2\pi n}{64}\right] & 0 \le n \le 63 \\ 0 & \text{otherwise} \end{cases}$$

Draw the frequency sampling structure for this filter and compare the computational complexity of this structure to a direct form realization.

The 64-point DFT of the unit sample response is

$$H(k) = \begin{cases} 1 & k = 0 \\ -\dfrac{1}{2} & k = 1, 63 \\ 0 & \text{else} \end{cases}$$

Therefore, for the frequency sampling structure, we write the system function in the following form,

$$H(z) = \frac{1}{64}(1 - z^{-64}) \cdot \sum_{k=0}^{63} \frac{H(k)}{1 - e^{j2\pi k/64}z^{-1}}$$

or
$$H(z) = \frac{1}{64}(1 - z^{-64})\left[\frac{1}{1 - z^{-1}} - \frac{\frac{1}{2}}{1 - e^{j2\pi/64}z^{-1}} - \frac{\frac{1}{2}}{1 - e^{j2\pi 63/64}z^{-1}} \right]$$

$$= \frac{1}{64}(1 - z^{-64})\left[\frac{1}{1 - z^{-1}} - \frac{1 - \cos(2\pi/64)z^{-1}}{1 - 2\cos(2\pi/64)z^{-1} + z^{-2}} \right]$$

This leads to the frequency sampling structure shown below.

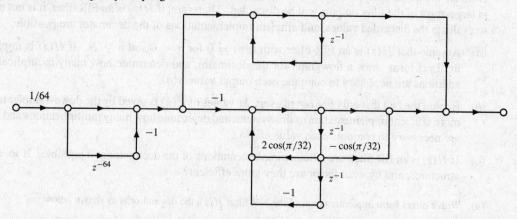

This implementation has 67 delays (4 more than the minimum), and it requires 3 multiplies and 6 adds to evaluate each output $y(n)$. A direct-form realization, on the other hand, has 63 delays and, because $h(n)$ has linear phase, requires 32 multiplies and 63 adds to compute each output value.

8.7 The frequency sampling structure for an FIR filter is based on expressing the system function in the form

$$H(z) = \frac{1}{N}(1 - z^{-N}) \sum_{k=0}^{N-1} \frac{H(k)}{1 - e^{j2\pi k/N}z^{-1}}$$

where $H(k)$ are samples of the frequency response at $\omega_k = 2\pi k/N$. If $h(n)$ is real, the symmetry of the DFT may be used to simplify this structure so that all of the coefficients are real. For example, Eq. (8.2) specifies a structure when N is even. Derive the corresponding structure when N is odd.

If N is odd, we may write $H(z)$ as follows:

$$H(z) = \frac{1}{N}(1 - z^{-N}) \sum_{k=0}^{N-1} \frac{H(k)}{1 - e^{j2\pi k/N}z^{-1}}$$

$$= \frac{1}{N}(1 - z^{-N})\left\{ \frac{H(0)}{1 - z^{-1}} + \sum_{k=1}^{(N-1)/2} \left[\frac{H(k)}{1 - e^{j2\pi k/N}z^{-1}} + \frac{H(N-k)}{1 - e^{j2\pi(N-k)/N}z^{-1}} \right] \right\}$$

$$= \frac{1}{N}(1 - z^{-N})\left[\frac{H(0)}{1 - z^{-1}} + \sum_{k=1}^{(N-1)/2} \frac{H(k) + H(N-k) - H(k)e^{-j2\pi k/N}z^{-1} - H(N-k)e^{j2\pi k/N}z^{-1}}{1 - 2\cos(2\pi k/N)z^{-1} + z^{-2}} \right]$$

or,

$$H(z) = \frac{1}{N}(1 - z^{-N})\left[\frac{H(0)}{1 - z^{-1}} + \sum_{k=1}^{(N-1)/2} \frac{A(k) - B(k)z^{-1}}{1 - 2\cos(2\pi k/N)z^{-1} + z^{-2}} \right]$$

where

$$A(k) = H(k) + H(N-k)$$
$$B(k) = H(k)e^{-j2\pi k/N} + H(N-k)e^{j2\pi k/N}$$

Note that, due to the conjugate symmetry of the DFT, $H(k) = H^*(N - k)$, the coefficients $A(k)$ and $B(k)$ are real.

8.8 As discussed in Chap. 3, sample rate reduction may be realized by cascading a low-pass filter with a down-sampler as shown in the following figure:

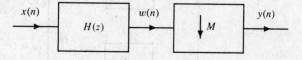

Because the down-sampler only retains one out of every M outputs from the low-pass filter $H(z)$, if M is large, most of the filter outputs will be discarded. Therefore, if $H(z)$ is an FIR filter, it is not necessary to evaluate the discarded values, and efficient implementations of the decimator are possible.

(a) Assume that $H(z)$ is an FIR filter with $h(n) = 0$ for $n < 0$ and $n \geq N$. If $H(z)$ is implemented in direct form, draw a flowgraph for the decimator, and determine how many multiplications and additions are necessary to compute each output value $y(n)$.

(b) Exploit the fact that only one out of every M values of $w(n)$ is saved by the down-sampler to derive a more efficient implementation of this system, and determine how many multiplications and additions are necessary to compute each value of $y(n)$.

(c) If $H(z)$ is an IIR filter, are efficient implementations of the decimator still possible? If so, for which structures, and by what factor are they more efficient?

(a) With a direct form implementation of the FIR filter $H(z)$, the decimator is as shown below.

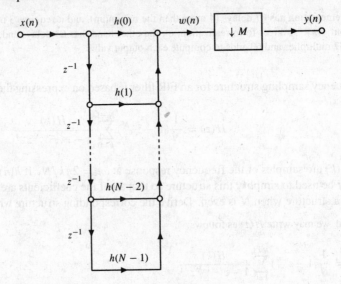

Because we need N multiplies and $N-1$ adds to find each value of $w(n)$, and because only one value of $y(n)$ is computed for every M values of $w(n)$, MN multiplies and $M(N-1)$ adds are performed for each value of $y(n)$.

(b) Because the down-sampler only saves one out of every M values of $w(n)$, the decimator may be implemented more efficiently by only evaluating those values of $w(n)$ that are passed through the down-sampler. This may be accomplished by embedding the down-sampler within the FIR filter as illustrated below.

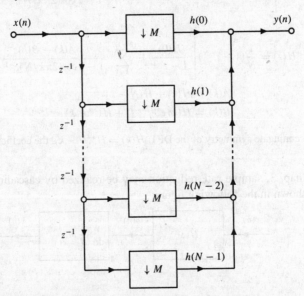

Now, because only one out of every M input samples is multiplied by $h(k)$, this implementation only requires N multiplies and $N - 1$ adds to compute each value of $y(n)$. Thus, the number of multiplies and adds has been reduced by a factor of M.

(c) If $H(z)$ is an IIR filter, it is not possible, in general, to commute the down-sampling operation with branch operations as was done with the FIR filter. For example, if

$$H(z) = \frac{1}{1 + a(1)z^{-1}}$$

we have the system illustrated below.

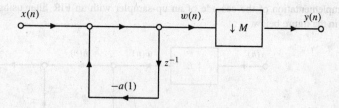

However, in order to evaluate a given value of $w(n)$, the previous value, $w(n - 1)$, must be known. Therefore, the down-sampler cannot be commuted with any branch operations within the filter, because this would discard values of $w(n)$ that are required to compute future values. On the other hand, consider the direct form II implementation of

$$H(z) = \frac{b(0) + b(1)z^{-1}}{1 + a(1)z^{-1}}$$

as illustrated below.

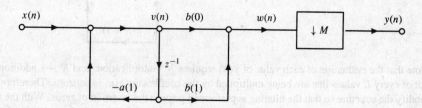

Because $$w(n) = b(0)v(n) + b(1)v(n - 1)$$

and $$y(n) = w(nM) = b(0)v(nM) + b(1)v(nM - 1)$$

the down-sampler may be commuted with the branch operations that form the multiplications by $b(0)$ and $b(1)$ as illustrated in the following figure:

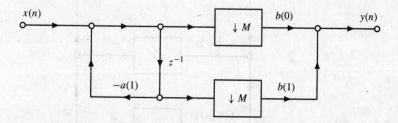

To compute each value of $y(n)$, this structure requires that we find M values of $v(n)$, which requires M multiplies and M adds, and it requires two multiplies and one add to find $y(n)$ from $v(n)$. Thus, the total number of computations is $M + 2$ multiplications and $M + 1$ additions. The direct form II structure is the only one that allows for a savings in computation. For direct form I, transposed direct form I, and transposed direct form II, the down-sampler cannot be commuted with any branch operations.

8.9   The previous problem examined the simplifications that are possible in implementing a decimator. Similar savings are possible for the interpolator shown in the figure below.

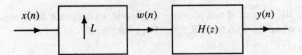

Because the up-sampler inserts $L - 1$ zeros between each sample of $x(n)$, assume that $H(z)$ is the system function of an FIR filter, and use the fact that many of the values of $w(n)$ are equal to zero to derive a more efficient implementation of this system.

A direct implementation of the cascade of an up-sampler with an FIR filter using the transposed direct form is illustrated in the figure below.

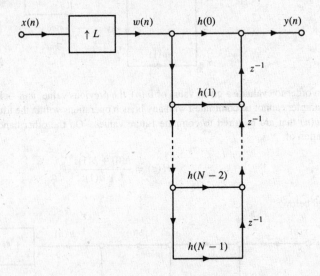

Note that the evaluation of each value of $y(n)$ requires N multiplications and $N - 1$ additions. However, only one out of every L values that are being multiplied by the coefficients $h(n)$ is nonzero. Therefore, it is more efficient to modify the structure so that the filtering is performed prior to the insertion of zeros. With the transposed direct form structure, we may commute the up-sampler with the branch multiplies as illustrated in the following figure:

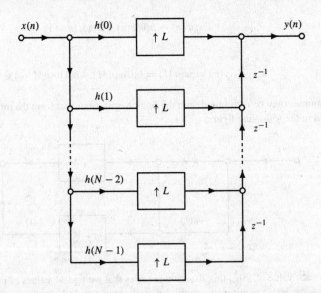

With this simplification, only N multiplies and $N - 1$ adds are required for every L output values.

Structures for IIR Systems

8.10 Consider the causal linear shift-invariant filter with system function

$$H(z) = \frac{1 + 0.875z^{-1}}{(1 + 0.2z^{-1} + 0.9z^{-2})(1 - 0.7z^{-1})}$$

Draw a signal flowgraph for this system using

(a) Direct form I

(b) Direct form II

(c) A cascade of first- and second-order systems realized in direct form II

(d) A cascade of first- and second-order systems realized in transposed direct form II

(e) A parallel connection of first- and second-order systems realized in direct form II

(a) Writing the system function as a ratio of polynomials in z^{-1},

$$H(z) = \frac{1 + 0.875z^{-1}}{1 - 0.5z^{-1} + 0.76z^{-2} - 0.63z^{-3}}$$

it follows that the direct form I realization of $H(z)$ is as follows:

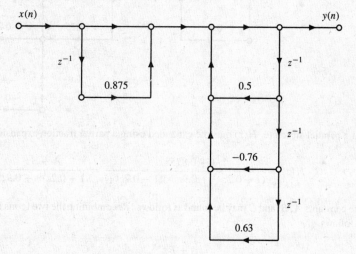

(b) For a direct form II realization of $H(z)$, we have

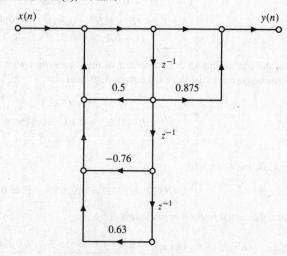

(c) Using a cascade of first- and second-order systems realized in direct form II, we have a choice of either pairing the zero with the first-order factor in the denominator or with the second-order factor. Although it does not make a difference from a computational point of view, because the zero is closer to the pair of complex poles than to the pole at $z = 0.7$, we will pair the zero with the second-order factor. With this pairing, the realization of $H(z)$ is as follows:

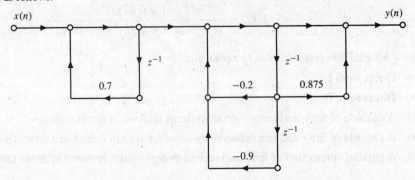

(d) If we change the direct form II systems in part (c) to transposed direct form II, we have the realization shown below.

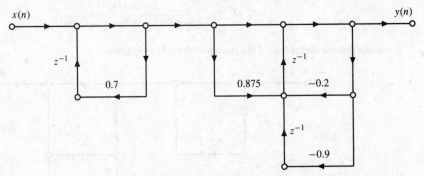

(e) For a parallel structure, $H(z)$ must be expanded using a partial fraction expansion:

$$H(z) = \frac{1 + 0.875z^{-1}}{(1 + 0.2z^{-1} + 0.9z^{-2})(1 - 0.7z^{-1})} = \frac{A + Bz^{-1}}{1 + 0.2z^{-1} + 0.9z^{-2}} + \frac{C}{1 - 0.7z^{-1}}$$

The constants A, B, and C may be found as follows. Recombining the two terms in the partial fraction expansion as follows,

$$H(z) = \frac{A + Bz^{-1}}{1 + 0.2z^{-1} + 0.9z^{-2}} + \frac{C}{1 - 0.7z^{-1}}$$

$$= \frac{(A + C) + (B + 0.2C - 0.7A)z^{-1} + (0.9C - 0.7B)z^{-2}}{(1 + 0.2z^{-1} + 0.9z^{-2})(1 - 0.7z^{-1})}$$

and equating the coefficients in the numerator of this expression with the numerator of $H(z)$, we have the following three equations in the three unknowns A, B, and C:

$$A + C = 1$$
$$B + 0.2C - 0.7A = 0.875$$
$$0.9C - 0.7B = 0$$

Solving for A, B, and C we find

$$A = 0.2794 \qquad B = 0.9265 \qquad C = 0.7206$$

and, therefore, the partial fraction expansion is

$$H(z) = \frac{0.2794 + 0.9265z^{-1}}{1 + 0.2z^{-1} + 0.9z^{-2}} + \frac{0.7206}{1 - 0.7z^{-1}}$$

Thus, a parallel structure for $H(z)$ is shown below.

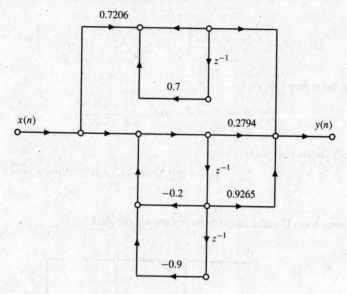

8.11   Consider the filter structure shown in the figure below.

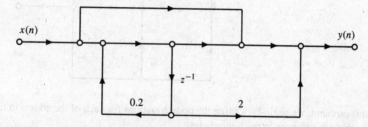

Find the system function and the unit sample response of this system.

For the three nodes labeled in the flowgraph below,

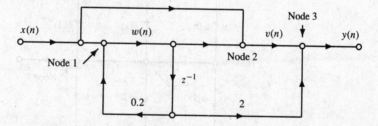

we have the following node equations:

$$w(n) = x(n) + 0.2w(n - 1)$$
$$v(n) = x(n) + w(n)$$
$$y(n) = v(n) + 2w(n - 1)$$

Using z-transforms, the first equation becomes

$$W(z) = \frac{1}{1 - 0.2z^{-1}}X(z)$$

Taking the z-transform of the second equation, and substituting the expression above for $W(z)$, we have

$$V(z) = X(z) + W(z) = X(z) + \frac{1}{1 - 0.2z^{-1}}X(z) = \frac{2 - 0.2z^{-1}}{1 - 0.2z^{-1}}X(z)$$

Finally, taking the z-transform of the last equation, we find

$$Y(z) = V(z) + 2z^{-1}W(z) = \left[\frac{2 - 0.2z^{-1}}{1 - 0.2z^{-1}} + 2z^{-1}\frac{1}{1 - 0.2z^{-1}}\right]X(z) = \frac{2 + 1.8z^{-1}}{1 - 0.2z^{-1}}X(z)$$

Therefore, the system function is

$$H(z) = \frac{2 + 1.8z^{-1}}{1 - 0.2z^{-1}}$$

and the unit sample response is

$$h(n) = 2(0.2)^n u(n) + 1.8(0.2)^{n-1}u(n-1)$$

8.12   Find a direct form II realization for the following network:

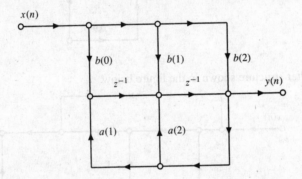

To solve this problem, we begin by writing the node equations for each of the adders in the network. If we label the three nodes that are adders as in the figure below,

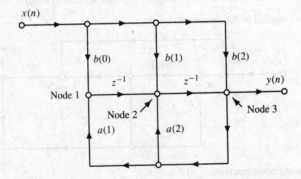

and denote the output of the first node by $w_1(n)$, and the output of the second by $w_2(n)$, we have the following three equations for the three node variables,

$$w_1(n) = b(0)x(n) + a(1)y(n)$$

$$w_2(n) = b(1)x(n) + w_1(n-1) + a(2)y(n)$$

$$y(n) = b(2)x(n) + w_2(n-1)$$

Substituting the first equation into the second, we have

$$w_2(n) = b(1)x(n) + [b(0)x(n-1) + a(1)y(n-1)] + a(2)y(n)$$

Then, substituting this equation into the third equation, we have

$$y(n) = b(2)x(n) + [b(1)x(n-1) + b(0)x(n-2) + a(1)y(n-2) + a(2)y(n-1)]$$

Thus, the direct form II structure for this system is as shown below.

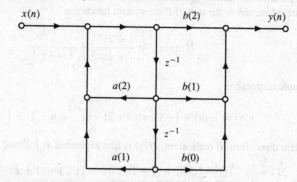

8.13 Find a transposed direct form II realization for the system described by the difference equation

$$y(n) = \frac{3}{4}y(n-1) - \frac{3}{4}y(n-2) + x(n) - \frac{1}{3}x(n-1)$$

and write down the set of difference equations that corresponds to this realization.

The transposed direct form II realization for this system is as follows:

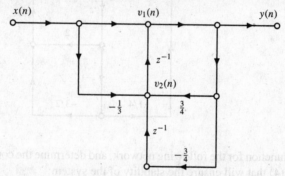

With the node variables $v_1(n)$ and $v_2(n)$ as labeled in the network above, the difference equations that describe this network are as follows:

$$v_1(n) = x(n) + v_2(n-1)$$
$$v_2(n) = -\frac{1}{3}x(n) + \frac{3}{4}v_1(n) - \frac{3}{4}v_1(n-1)$$
$$y(n) = v_1(n)$$

8.14 Find the system function and the unit sample response for the following network, and draw an equivalent direct form II structure:

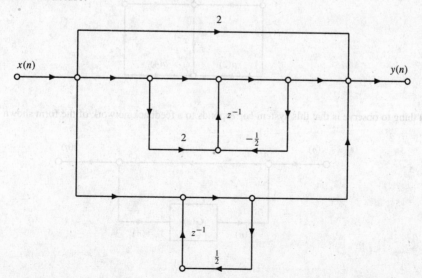

This network is a parallel connection of two first-order systems, plus a feed-through connection with a gain of 2. Therefore, the system function is the sum of three system functions:

$$H(z) = 2 + \frac{1 + 2z^{-1}}{1 + \frac{1}{2}z^{-1}} + \frac{1}{1 - \frac{1}{2}z^{-1}}$$

Thus, the unit sample response is

$$h(n) = 2\delta(n) + \left(-\tfrac{1}{2}\right)^n u(n) + 2\left(-\tfrac{1}{2}\right)^{n-1} u(n-1) + \left(\tfrac{1}{2}\right)^n u(n)$$

To find an equivalent direct form II realization, $H(z)$ is first expressed as follows:

$$H(z) = \frac{2\left(1 + \frac{1}{2}z^{-1}\right)\left(1 - \frac{1}{2}z^{-1}\right) + (1 + 2z^{-1})\left(1 - \frac{1}{2}z^{-1}\right) + \left(1 + \frac{1}{2}z^{-1}\right)}{\left(1 + \frac{1}{2}z^{-1}\right)\left(1 - \frac{1}{2}z^{-1}\right)} = \frac{4 + 2z^{-1} - \frac{3}{2}z^{-2}}{1 - \frac{1}{4}z^{-2}}$$

Therefore, the direct form II structure is as shown in the following figure:

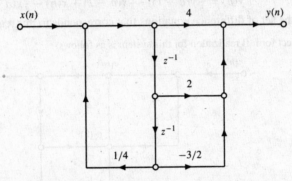

8.15 Find the system function for the following network, and determine the conditions on the coefficients $a(1)$, $a(2)$, $a(3)$, and $a(4)$ that will ensure the stability of the system:

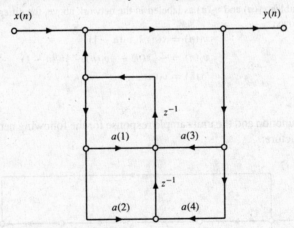

The first thing to observe is that this system corresponds to a feedback network of the form shown in the following figure:

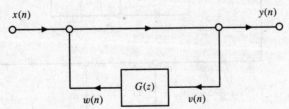

where $G(z)$ is the second-order system shown below.

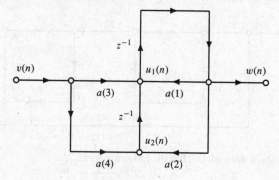

Therefore,
$$Y(z) = X(z) + G(z)Y(z)$$

and
$$H(z) = \frac{1}{1 - G(z)}$$

To find $G(z)$, we begin by writing the node equations for this network:

$$u_1(n) = a(3)v(n) + u_2(n - 1) + a(1)w(n)$$
$$u_2(n) = a(4)v(n) + a(2)w(n)$$
$$w(n) = u_1(n - 1)$$

Taking the z-transform of the first two equations, we have

$$U_1(z) = a(3)V(z) + z^{-1}U_2(z) + a(1)W(z)$$
$$U_2(z) = a(4)V(z) + a(2)W(z)$$

Substituting the second equation into the first gives

$$U_1(z) = a(3)V(z) + z^{-1}[a(4)V(z) + a(2)W(z)] + a(1)W(z)$$

Finally, from the last difference equation, we have

$$W(z) = z^{-1}U_1(z) = z^{-1}\{a(3)V(z) + z^{-1}[a(4)V(z) + a(2)W(z)] + a(1)W(z)\}$$

or
$$W(z)[1 - a(1)z^{-1} - a(2)z^{-2}] = [a(3)z^{-1} + a(4)z^{-2}]V(z)$$

Therefore,
$$G(z) = \frac{W(z)}{V(z)} = \frac{a(3)z^{-1} + a(4)z^{-2}}{1 - a(1)z^{-1} - a(2)z^{-2}}$$

and for $H(z)$ we have

$$H(z) = \frac{1}{1 - G(z)} = \frac{1}{1 - (a(3)z^{-1} + a(4)z^{-2})/(1 - a(1)z^{-1} - a(2)z^{-2})}$$

$$= \frac{1 - a(1)z^{-1} - a(2)z^{-2}}{1 - a(1)z^{-1} - a(2)z^{-2} - a(3)z^{-1} - a(4)z^{-2}}$$

$$= \frac{1 - a(1)z^{-1} - a(2)z^{-2}}{1 - [a(1) + a(3)]z^{-1} - [a(2) + a(4)]z^{-2}}$$

For stability, it is necessary and sufficient that the coefficients $[a(2) + a(4)]$ and $[a(1) + a(3)]$ lie within the stability triangle (see Chap. 5), which requires that

$$|a(2) + a(4)| < 1 \qquad \text{and} \qquad |a(1) + a(3)| < 1 - a(2) - a(4)$$

8.16 Find the system function of the following network:

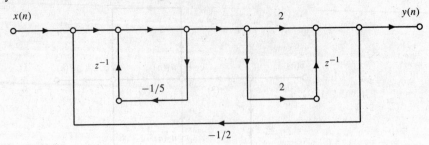

This system is a feedback network that has the following form:

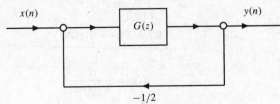

Therefore, the system function is

$$H(z) = \frac{G(z)}{1 + \frac{1}{2}G(z)}$$

With

$$G(z) = \frac{2(1 + z^{-1})}{1 + \frac{1}{5}z^{-1}}$$

we have

$$H(z) = \frac{2\dfrac{1 + z^{-1}}{1 + \frac{1}{5}z^{-1}}}{1 + \dfrac{1}{2}\dfrac{2(1 + z^{-1})}{1 + \frac{1}{5}z^{-1}}} = \frac{1 + z^{-1}}{1 + 0.6z^{-1}}$$

8.17 Find the system function of the following network:

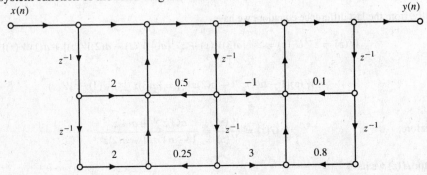

The system function of this network may be found by writing down the difference equations corresponding to each adder and solving these equations using z-transforms. A simpler approach, however, is to redraw the network as follows,

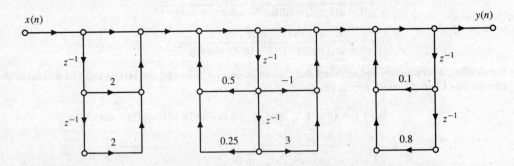

which we recognize as a cascade of three second-order networks. Therefore, the system function is the product of the system functions of each network in the cascade, and we have

$$H(z) = \frac{(1 + 2z^{-1} + 2z^{-2})(1 - z^{-1} + 3z^{-2})}{(1 - 0.5z^{-1} - 0.25z^{-2})(1 - 0.1z^{-1} - 0.8z^{-2})}$$

8.18 Consider the network in the figure below. Redraw the flowgraph as a cascade of second-order sections in transposed direct form II.

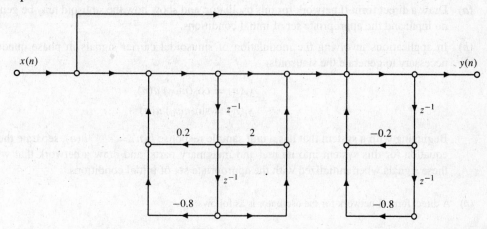

To implement this system as a cascade of second-order transformed direct form II networks, we must first find the system function corresponding to this network. Note that this network is of the form shown in the following figure:

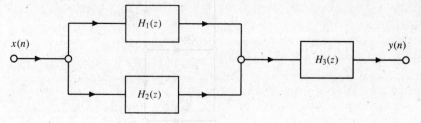

where $H_1(z) = 1$ $H_2(z) = \dfrac{1 + 2z^{-1} + z^{-2}}{1 - 0.2z^{-1} + 0.8z^{-2}}$ $H_3(z) = \dfrac{1}{1 + 0.2z^{-1} + 0.8z^{-2}}$

Therefore, $H(z) = [H_1(z) + H_2(z)]H_3(z)$

or $$H(z) = \left[1 + \frac{1 + 2z^{-1} + z^{-2}}{1 - 0.2z^{-1} + 0.8z^{-2}} \right] \left[\frac{1}{1 + 0.2z^{-1} + 0.8z^{-2}} \right]$$

$$= \frac{2 + 1.8z^{-1} + 1.8z^{-2}}{(1 - 0.2z^{-1} + 0.8z^{-2})(1 + 0.2z^{-1} + 0.8z^{-2})}$$

Therefore, the desired network is as shown in the following figure:

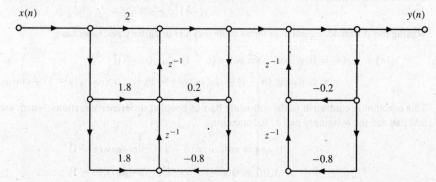

8.19   A digital oscillator has a unit sample response

$$h(n) = \sin[(n+1)\omega_0]\, u(n)$$

The system function of this oscillator is

$$H(z) = \frac{\sin(\omega_0)}{1 - 2\cos(\omega_0)z^{-1} + z^{-2}} \qquad |z| > 1$$

(a)   Draw a direct form II network for this oscillator, and show how the sinusoid may be generated with no input and the appropriate set of initial conditions.

(b)   In applications involving the modulation of sinusoidal carrier signals in phase quadrature, it is necessary to generate the sinusoids

$$y_r(n) = \cos(n\omega_0)\, u(n)$$
$$y_i(n) = \sin(n\omega_0)\, u(n)$$

Beginning with a system that has a unit sample response $h(n) = e^{jn\omega_0} u(n)$, separate the difference equation for this system into its real and imaginary parts, and draw a network that will generate these signals when initialized with the appropriate set of initial conditions.

(a)   A direct form II network for the oscillator is as follows:

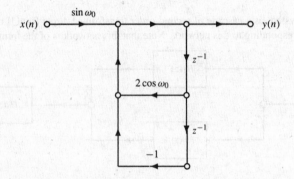

With the input $x(n) = \delta(n)$, the response is $y(n) = \sin[(n+1)\omega_0]$ for $n \geq 0$. Equivalently, if the storage registers corresponding to the delays are initialized so that

$$y(-1) = 0 \qquad y(-2) = -\sin\omega_0$$

the zero-input response will be a sinusoid of frequency ω_0.

(b)   A complex exponential sequence $y(n) = e^{jn\omega_0}\, u(n)$ is generated by the difference equation

$$y(n) = e^{j\omega_0} y(n-1) \qquad n \geq 0$$

with the initial condition

$$y(-1) = e^{-j\omega_0}$$

Writing this difference equation in terms of its real and imaginary parts, we have

$$y_r(n) + jy_i(n) = (\cos\omega_0 + j\sin\omega_0)[y_r(n-1) + jy_i(n-1)]$$
$$= \cos(\omega_0)y_r(n-1) - \sin(\omega_0)y_i(n-1) + j[\sin(\omega_0)y_r(n-1) + \cos(\omega_0)y_i(n-1)]$$

This equation is equivalent to the following pair of coupled difference equations, which are formed from the real part and the imaginary part of the equation:

$$y_r(n) = \cos(\omega_0)y_r(n-1) - \sin(\omega_0)y_i(n-1)$$
$$y_i(n) = \sin(\omega_0)y_r(n-1) + \cos(\omega_0)y_i(n-1)$$

A network that implements this pair of equations is shown below.

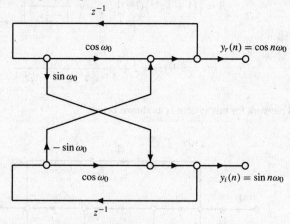

The initial condition required to generate the desired output is $y(-1) = e^{-j\omega_0}$, or

$$y_r(-1) = \cos \omega_0 \qquad y_i(-1) = -\sin \omega_0$$

8.20 Implement the system

$$H(z) = \frac{4 + \frac{9}{4}z^{-1} - \frac{1}{4}z^{-2}}{1 + \frac{1}{4}z^{-1} - \frac{1}{8}z^{-2}}$$

as a parallel network of first-order direct form structures.

Factoring the denominator of the system function, we find

$$H(z) = \frac{4 + \frac{9}{4}z^{-1} - \frac{1}{4}z^{-2}}{\left(1 - \frac{1}{4}z^{-1}\right)\left(1 + \frac{1}{2}z^{-1}\right)}$$

To implement $H(z)$ as a parallel network of first-order filters, we must express $H(z)$ as a sum of first-order factors using a partial fraction expansion. Because the order of the numerator is equal to the order of the denominator, this expansion will contain a constant term,

$$H(z) = C + \frac{A}{1 - \frac{1}{4}z^{-1}} + \frac{B}{1 + \frac{1}{2}z^{-1}}$$

To find the value of C, we divide the numerator polynomial by the denominator as follows:

$$-\tfrac{1}{8}z^{-2} + \tfrac{1}{4}z^{-1} + 1 \overline{\smash{\big)}\ -\tfrac{1}{4}z^{-2} + \tfrac{9}{4}z^{-1} + 4}$$
$$\underline{-\tfrac{1}{4}z^{-2} + \tfrac{1}{2}z^{-1} + 2}$$
$$\tfrac{7}{4}z^{-1} + 2$$

Therefore, $C = 2$, and we may write $H(z)$ as follows:

$$H(z) = 2 + \frac{\frac{7}{4}z^{-1} + 2}{\left(1 - \frac{1}{4}z^{-1}\right)\left(1 + \frac{1}{2}z^{-1}\right)}$$

Finally, with

$$G(z) = \frac{\frac{7}{4}z^{-1} + 2}{\left(1 - \frac{1}{4}z^{-1}\right)\left(1 + \frac{1}{2}z^{-1}\right)}$$

we have, for the coefficients A and B,

$$A = \left[\left(1 - \tfrac{1}{4}z^{-1}\right)G(z)\right]_{z^{-1}=4} = \frac{\frac{7}{4}z^{-1} + 2}{1 + \frac{1}{2}z^{-1}}\bigg|_{z^{-1}=4} = 3$$

and
$$B = \left[\left(1 + \tfrac{1}{2}z^{-1}\right)G(z)\right]_{z^{-1}=-2} = \left.\frac{\tfrac{7}{4}z^{-1}+2}{1 - \tfrac{1}{4}z^{-1}}\right|_{z^{-1}=-2} = -1$$

Thus,
$$H(z) = 2 + \frac{3}{1 - \tfrac{1}{4}z^{-1}} - \frac{1}{1 + \tfrac{1}{2}z^{-1}}$$

and the parallel network for this system is as shown below.

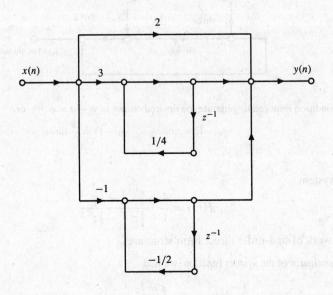

8.21 The system function of a discrete-time system is

$$H(z) = \frac{(1 + z^{-1})^4}{\left(1 - z^{-1} + \tfrac{7}{8}z^{-2}\right)\left(1 + 2z^{-1} + \tfrac{3}{4}z^{-2}\right)}$$

Draw a signal flowgraph of this system using a cascade of second-order systems in direct form II, and write down the set of difference equations that corresponds to this implementation.

Expressing $H(z)$ as a product of two second-order systems, we have

$$H(z) = \frac{1 + 2z^{-1} + z^{-2}}{1 - z^{-1} + \tfrac{7}{8}z^{-2}} \cdot \frac{1 + 2z^{-1} + z^{-2}}{1 + 2z^{-1} + \tfrac{3}{4}z^{-2}}$$

which leads to the following cascade implementation for $H(z)$:

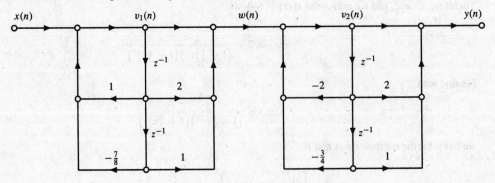

With $w(n)$, $v_1(n)$, and $v_2(n)$ as labeled in the figure above, the set of difference equations for this network is:

$$v_1(n) = x(n) + v_1(n-1) - \tfrac{7}{8}v_1(n-2)$$

$$w(n) = v_1(n) + 2v_1(n-1) + v_1(n-2)$$

$$v_2(n) = w(n) - 2v_2(n-1) - \tfrac{3}{4}v_2(n-2)$$

$$y(n) = v_2(n) + 2v_2(n-1) + v_2(n-2)$$

8.22 Consider the fourth-order *comb filter* that has a system function

$$H(z) = A\,\frac{1 + z^{-4}}{1 + a^4 z^{-4}}$$

where $0 < a < 1$.

(a) Draw a pole-zero diagram for $H(z)$.

(b) Find the value for A so that the peak gain of the filter is equal to 2.

(c) Find a structure for this filter that requires only one multiplier.

(a) The comb filter has four zeros on the unit circle at

$$\beta_k = e^{j(2k+1)\pi/4} \qquad k = 0, 1, 2, 3$$

and four poles at

$$\alpha_k = a e^{j(2k+1)\pi/4} \qquad k = 0, 1, 2, 3$$

A pole-zero diagram for $H(z)$ is shown in the following figure:

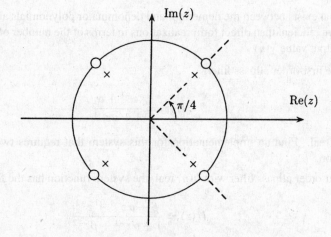

(b) Due to the zeros on the unit circle, the magnitude of the frequency response is zero at $\omega_k = (2k+1)\pi/4$ for $k = 0, 1, 2, 3$, and it increases monotonically until it reaches a maximum value at the frequencies that are midway between the zeros at $\omega_k = k\pi/2$. Therefore, in order for the peak gain to be equal to 2, we want

$$H(z)\big|_{z=1} = 2$$

which implies that

$$A\,\frac{2}{1 + a^4} = 2$$

or $A = 1 + a^4$.

(c) With a system function

$$H(z) = (1 + a^4)\frac{1 + z^{-4}}{1 + a^4 z^{-4}}$$

we may implement this system using two multiplies as shown in the figure below.

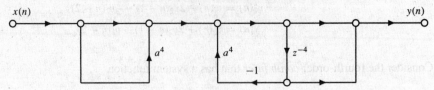

Note, however, that the multiplier may be shared as follows:

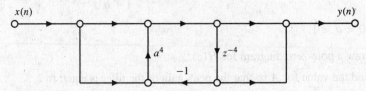

The difference equations for this network are

$$w(n) = x(n) + a^4[x(n) - w(n-4)]$$
$$y(n) = w(n) + w(n-4)$$

8.23   The system function of an allpass filter has the form

$$H(z) = \prod_{k=1}^{N_s} \frac{\alpha_k + \beta_k z^{-1} + z^{-2}}{1 + \beta_k z^{-1} + \alpha_k z^{-2}}$$

The symmetry that exists between the numerator and denominator polynomials allows for special structures that are more efficient than direct form realizations in terms of the number of multiplies required to compute each output value $y(n)$.

(a) Consider the first-order allpass filter,

$$H(z) = \frac{z^{-1} + \alpha}{1 + \alpha z^{-1}}$$

where α is real. Find an implementation for this system that requires two delays but only one multiplication.

(b) For a second-order allpass filter with $h(n)$ real, the system function has the form

$$H(z) = \frac{\beta + \alpha z^{-1} + z^{-2}}{1 + \alpha z^{-1} + \beta z^{-2}}$$

where α and β are real. Derive a structure that implements this system using four delays but only two multiplies.

(a) The direct form realization of a first-order allpass filter requires two multiplies and one delay as shown in the figure below.

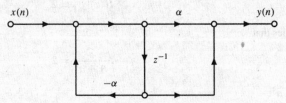

To see how the two multiplies may be combined, consider the difference equation for this system:

$$y(n) = -\alpha y(n-1) + \alpha x(n) + x(n-1)$$
$$= \alpha[x(n) - y(n-1)] + x(n-1)$$

Therefore, only one multiplication is necessary if we form the difference $x(n) - y(n-1)$ prior to multiplying by α. Thus, we have the structure illustrated in the figure below that has two delays but only one multiplication.

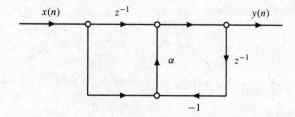

Because this structure requires an extra delay compared to direct form, this structure is not canonic.

(b) As with the first-order allpass filter, we may find a two-multiplier realization of a second-order allpass filter by combining together terms in the difference equation for the allpass filter as follows:

$$y(n) = -\alpha y(n-1) - \beta y(n-2) + \beta x(n) + \alpha x(n-1) + x(n-2)$$
$$= \alpha[x(n-1) - y(n-1)] + \beta[x(n) - y(n-2)] + x(n-2)$$

Thus, only two multiplications are required if we can form the differences $x(n-1) - y(n-1)$ and $x(n) - y(n-2)$ prior to performing any multiplications. A structure that accomplishes this is given in the figure below.

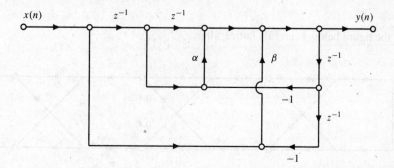

Note that with the additional delays, two multiplications are saved compared to a direct form implementation.

Lattice Filters

8.24 Sketch a lattice filter implementation of the FIR filter

$$H(z) = 8 + 4z^{-1} + 2z^{-2} + z^{-3}$$

To implement this system using a lattice filter structure, we must find the reflection coefficients that generate the polynomial $H(z)$. First, however, it is necessary to normalize $H(z)$ so that the first coefficient is unity:

$$H(z) = 8[1 + 0.5z^{-1} + 0.25z^{-2} + 0.125z^{-3}]$$

Now, with

$$H_3(z) = 1 + 0.5z^{-1} + 0.25z^{-2} + 0.125z^{-3}$$

we see that

$$\Gamma_3 = 0.125$$

Next, we generate the second-order system $H_2(z)$ using the step-down recursion:

$$H_2(z) = \frac{1}{1 - \Gamma_3^2}[H_3(z) - \Gamma_3 z^{-3} H_3(z^{-1})]$$

$$= \frac{1}{1 - (0.125)^2}[1 + 0.5z^{-1} + 0.25z^{-2} + 0.125z^{-3} - 0.125z^{-3}(1 + 0.5z + 0.25z^2 + 0.125z^3)]$$

$$= 1 + 0.4762z^{-1} + 0.1905z^{-2}$$

Therefore, $\Gamma_2 = 0.1905$. Finally, we have

$$H_1(z) = \frac{1}{1 - \Gamma_2^2}[H_2(z) - \Gamma_2 z^{-2} H_2(z^{-1})]$$

$$= \frac{1}{1 - (0.1905)^2}[1 + 0.4762z^{-1} + 0.1905z^{-2} - 0.1905z^{-2}(1 + 0.4762z + 0.1905z^2)]$$

$$= 1 + 0.4z^{-1}$$

and, therefore, $\Gamma_1 = 0.4$. Thus, the lattice filter structure is as shown below.

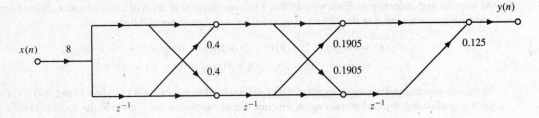

8.25 Shown in the figure below is an FIR lattice filter.

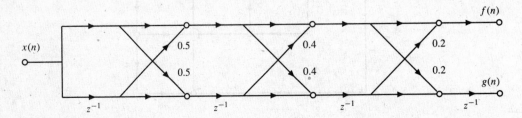

(a) Find the system function $A(z) = F(z)/X(z)$ relating the input $x(n)$ to the output $f(n)$. Does this system have minimum phase?

(b) Repeat part (a) for the system function relating $x(n)$ to $g(n)$.

(a) To find the system function relating $x(n)$ to $f(n)$, we use the step-up recursion. Using the vector form of the recursion, we have for the coefficients $a_1(k)$

$$\begin{bmatrix} 1 \\ a_1(1) \end{bmatrix} = \begin{bmatrix} 1 \\ \Gamma_1 \end{bmatrix} = \begin{bmatrix} 1 \\ 0.5 \end{bmatrix}$$

Then, with $\Gamma_2 = 0.4$, for $a_2(k)$, we have

$$\begin{bmatrix} 1 \\ a_2(1) \\ a_2(2) \end{bmatrix} = \begin{bmatrix} 1 \\ 0.5 \\ 0 \end{bmatrix} + \Gamma_2 \begin{bmatrix} 0 \\ 0.5 \\ 1 \end{bmatrix} = \begin{bmatrix} 1 \\ 0.7 \\ 0.4 \end{bmatrix}$$

Finally, with $\Gamma_3 = 0.2$, we have

$$
\begin{bmatrix} 1 \\ a_3(1) \\ a_3(2) \\ a_3(3) \end{bmatrix} = \begin{bmatrix} 1 \\ 0.7 \\ 0.4 \\ 0 \end{bmatrix} + \Gamma_3 \begin{bmatrix} 0 \\ 0.4 \\ 0.7 \\ 1 \end{bmatrix} = \begin{bmatrix} 1 \\ 0.78 \\ 0.54 \\ 0.2 \end{bmatrix}
$$

Thus, $$A(z) = 1 + 0.78z^{-1} + 0.54z^{-2} + 0.2z^{-3}$$

This system will have minimum phase if the zeros of $A(z)$ are inside this unit circle. Although this could be determined by factoring $A(z)$, because the reflection coefficients used to generate $A(z)$ are bounded by 1 in magnitude, it follows that $A(z)$ has minimum phase.

(b) The system function $A'(z) = G(z)/X(z)$ is related to $X(z)$ as follows:

$$A'(z) = z^{-3}A(z^{-1}) = 0.2 + 0.54z^{-1} + 0.78z^{-2} + z^{-3}$$

Because the zeros of $A'(z)$ are formed by flipping the zeros of $A(z)$ about the unit circle, all of the zeros of $A'(z)$ will be *outside* the unit circle and thus will not have minimum phase.

8.26 Let $A(z)$ be an FIR filter with lattice filter coefficients

$$\Gamma_1 = \tfrac{1}{3} \qquad \Gamma_2 = \tfrac{1}{3} \qquad \Gamma_3 = 1$$

(a) Find the zeros of the system function $A(z)$.

(b) Repeat for the case when $\Gamma_3 = -1$.

(c) Can a general result be proved for lattice filters that have reflection coefficients Γ_j with

$$|\Gamma_j| < 1 \quad \text{for} \quad j = 1, 2, \ldots, p-1$$
$$|\Gamma_p| = 1$$

(a) To find the system function $A(z)$ for a given set of reflection coefficients, we use the step-up recursion. For $a_1(k)$ we have

$$\begin{bmatrix} 1 \\ a_1(1) \end{bmatrix} = \begin{bmatrix} 1 \\ \Gamma_1 \end{bmatrix} = \begin{bmatrix} 1 \\ \tfrac{1}{3} \end{bmatrix}$$

Then, with $\Gamma_2 = \tfrac{1}{3}$ we have

$$\begin{bmatrix} 1 \\ a_2(1) \\ a_2(2) \end{bmatrix} = \begin{bmatrix} 1 \\ \tfrac{1}{3} \\ 0 \end{bmatrix} + \Gamma_2 \begin{bmatrix} 0 \\ \tfrac{1}{3} \\ 1 \end{bmatrix} = \begin{bmatrix} 1 \\ \tfrac{4}{9} \\ \tfrac{1}{3} \end{bmatrix}$$

Finally, with $\Gamma_3 = 1$ we have

$$\begin{bmatrix} 1 \\ a_3(1) \\ a_3(2) \\ a_3(3) \end{bmatrix} = \begin{bmatrix} 1 \\ \tfrac{4}{9} \\ \tfrac{1}{3} \\ 0 \end{bmatrix} + \begin{bmatrix} 0 \\ \tfrac{1}{3} \\ \tfrac{4}{9} \\ 1 \end{bmatrix} = \begin{bmatrix} 1 \\ \tfrac{7}{9} \\ \tfrac{7}{9} \\ 1 \end{bmatrix}$$

Thus, $$A(z) = 1 + \tfrac{7}{9}z^{-1} + \tfrac{7}{9}z^{-2} + z^{-3}$$

The zeros of the system function may be found by factoring $A(z)$. The roots are found to be

$$\beta_1 = -1 \qquad \beta_2 = e^{j0.4646\pi} \qquad \beta_3 = e^{-j0.4646\pi}$$

which are on the unit circle.

(b) If $\Gamma_3 = -1$, the system function may be found by modifying the last step of the step-up recursion in part (a) as follows:

$$\begin{bmatrix} 1 \\ a_3(1) \\ a_3(2) \\ a_3(3) \end{bmatrix} = \begin{bmatrix} 1 \\ \frac{4}{9} \\ \frac{1}{3} \\ 0 \end{bmatrix} - \begin{bmatrix} 0 \\ \frac{1}{3} \\ \frac{4}{9} \\ 1 \end{bmatrix} = \begin{bmatrix} 1 \\ \frac{1}{9} \\ -\frac{1}{9} \\ -1 \end{bmatrix}$$

Thus,

$$A(z) = 1 + \tfrac{1}{9}z^{-1} - \tfrac{1}{9}z^{-2} - z^{-3}$$

and the zeros of $A(z)$ are

$$\beta_1 = 1 \qquad \beta_2 = e^{j0.6875\pi} \qquad \beta_3 = e^{-j0.6875\pi}$$

which are again on the unit circle.

(c) If $A_p(z)$ is a pth-order FIR filter with reflection coefficients Γ_j where $|\Gamma_j| < 1$ for $j = 1, \ldots, p - 1$, and $|\Gamma_p| = 1$,

$$A_p(z) = A_{p-1}(z) \pm z^{-p} A_{p-1}(z^{-1})$$

and it follows that the polynomial $A_p(z)$ is *symmetric* or *antisymmetric*, that is,

$$A_p(z) = \pm z^{-p} A_p(z^{-1})$$

Therefore, $A_p(z)$ has (generalized) linear phase, which implies that all of the zeros of $A_p(z)$ lie on the unit circle or in conjugate reciprocal pairs. However, if $|\Gamma_j| < 1$ for $j = 1, 2, \ldots, p - 1$, the zeros must lie *on* the unit circle. The reason for this is as follows. The Schur-Cohn stability criterion states that none of the roots of $A_p(z)$ may lie *outside* the unit circle if $|\Gamma_j| \leq 1$ for $j = 1, 2, \ldots, p$. Therefore, if $A_p(z)$ has generalized linear phase with no zeros outside the unit circle, then all of the zeros must be *on* the unit circle.

8.27 Draw a lattice filter implementation for the all-pole filter

$$H(z) = \frac{1}{1 - 0.2z^{-1} + 0.4z^{-2} + 0.6z^{-3}}$$

and determine the number of multiplications, additions, and delays required to implement the filter. Compare this structure to a direct form realization of $H(z)$ in terms of the number of multiplies, adds, and delays.

To implement this filter using a lattice filter, we must first derive the reflection coefficients Γ_1, Γ_2, and Γ_3 corresponding to the denominator polynomial. With

$$A_3(z) = 1 - 0.2z^{-1} + 0.4z^{-2} + 0.6z^{-3}$$

it follows that $\Gamma_3 = 0.6$. Next, using the step-down recursion to find $A_2(z)$, we have

$$\begin{aligned} A_2(z) &= \frac{1}{1 - \Gamma_3^2}[A_3(z) - \Gamma_3 z^{-3} A_3(z^{-1})] \\ &= \frac{1}{1 - (0.6)^2}[1 - 0.2z^{-1} + 0.4z^{-2} + 0.6z^{-3} - 0.6z^{-3}(1 - 0.2z + 0.4z^2 + 0.6z^3)] \\ &= 1 - 0.6875z^{-1} + 0.8125z^{-2} \end{aligned}$$

Thus, for Γ_2 we have $\Gamma_2 = 0.8125$. Finally, for $A_1(z)$, we have

$$\begin{aligned} A_1(z) &= \frac{1}{1 - \Gamma_2^2}[A_2(z) - \Gamma_2 z^{-2} A_2(z^{-1})] \\ &= \frac{1}{1 - (0.8125)^2}[1 - 0.6875z^{-1} + 0.8125z^{-2} - 0.8125z^{-2}(1 - 0.6875z + 0.8125z^2)] \\ &= 1 - 0.3793z^{-1} \end{aligned}$$

and, therefore, $\Gamma_1 = -0.3793$. Thus, the structure is as follows:

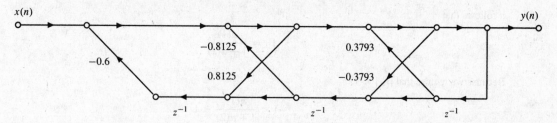

This filter structure has three delays and requires five multiplications and five additions to evaluate each value of the output, $y(n)$. A direct-form structure also requires three delays but only three multiplications and three additions.

8.28   Without factoring any polynomials, determine whether or not the following causal filter is stable:

$$H(z) = \frac{1 + 2z^{-1} + 2z^{-2} + z^{-3}}{1 + 1.58z^{-1} + 1.638z^{-2} + 1.556z^{-3} + 0.4z^{-4}}$$

We may easily check the stability of this filter using the Schur-Cohn stability test, which involves checking to see whether or not the magnitudes of the reflection coefficients corresponding to the denominator polynomial are bounded by 1 in magnitude. With

$$A_4(z) = 1 + 1.58z^{-1} + 1.638z^{-2} + 1.556z^{-3} + 0.4z^{-4}$$

it follows that $|\Gamma_4| = 0.4 < 1$. Using the step-down recursion to find $A_3(z)$, we have

$$A_3(z) = \frac{1}{1 - \Gamma_4^2}[A_4(z) - \Gamma_4 z^{-4} A_4(z^{-1})]$$

$$= \frac{1}{1 - 0.4^2}[1 + 1.58z^{-1} + 1.638z^{-2} + 1.556z^{-3} + 0.4z^{-4}$$

$$- 0.4z^{-4}(1 + 1.58z + 1.638z^2 + 1.556z^3 + 0.4z^4)]$$

$$= 1 + 1.14z^{-1} + 1.17z^{-2} + 1.1z^{-3}$$

Therefore, $|\Gamma_3| = 1.1 > 1$, and it follows that the filter is unstable.

8.29   Use the Schur-Cohn stability test to derive the stability conditions

$$|a(2)| < 1 \qquad |a(1)| < 1 + a(2)$$

for a second-order filter

$$H(z) = \frac{b(0)}{1 + a(1)z^{-1} + a(2)z^{-2}}$$

In order for $H(z)$ to be stable, it is necessary and sufficient for the reflection coefficients Γ_1 and Γ_2 to have a magnitude that is less than 1. In terms of the first two reflection coefficients, the denominator of $H(z)$ is

$$A(z) = 1 + (\Gamma_1 + \Gamma_1\Gamma_2)z^{-1} + \Gamma_2 z^{-2}$$

Therefore, because $a(2) = \Gamma_2$, the constraint that $|\Gamma_2| < 1$ gives us the first condition,

$$|a(2)| < 1$$

Next, with

$$a(1) = \Gamma_1 + \Gamma_1\Gamma_2 = [1 + a(2)]\Gamma_1$$

it follows that

$$\Gamma_1 = \frac{a(1)}{1 + a(2)}$$

Because we require that $|\Gamma_1| < 1$,

$$-1 < \frac{a(1)}{1 + a(2)} < 1$$

or

$$a(1) < 1 + a(2)$$
$$a(1) > -1 - a(2)$$

These two equations are equivalent to

$$|a(1)| < 1 + a(2)$$

as was to be shown.

8.30   Implement the allpass filter

$$H_{ap}(z) = \frac{-0.512 + 0.64z^{-1} - 0.8z^{-2} + z^{-3}}{1 - 0.8z^{-1} + 0.64z^{-2} - 0.512z^{-3}}$$

using a lattice filter structure.

To find the lattice filter structure for this allpass filter, we use the step-down recursion to find the reflection coefficients corresponding to the denominator polynomial,

$$A_3(z) = 1 - 0.8z^{-1} + 0.64z^{-2} - 0.512z^{-3}$$

First, we note that $\Gamma_3 = -0.512$. Then, we find $A_2(z)$ as follows:

$$A_2(z) = \frac{1}{1 - \Gamma_3^2}[A_3(z) - \Gamma_3 z^{-3} A_3(z^{-1})]$$

$$= \frac{1}{1 - (0.512)^2}[1 - 0.8z^{-1} + 0.64z^{-2} - 0.512z^{-3} + 0.512z^{-3}(1 - 0.8z + 0.64z^2 - 0.512z^3)]$$

$$= 1 - 0.6401z^{-1} + 0.3123z^{-2}$$

Thus, the second reflection coefficient is $\Gamma_2 = 0.3123$. Finally, we have

$$A_1(z) = \frac{1}{1 - \Gamma_2^2}[A_2(z) - \Gamma_2 z^{-2} A_2(z^{-1})]$$

$$= \frac{1}{1 - (0.3123)^2}[1 - 0.6401z^{-1} + 0.3123z^{-2} - 0.3123z^{-2}(1 - 0.6401z + 0.3123z^2)]$$

$$= 1 - 0.4878z^{-1}$$

and, therefore, $\Gamma_1 = -0.4878$. Thus, a lattice implementation of this allpass filter is as shown in the figure below.

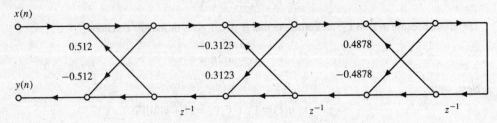

8.31 Find the system function for the lattice filter given in the figure below.

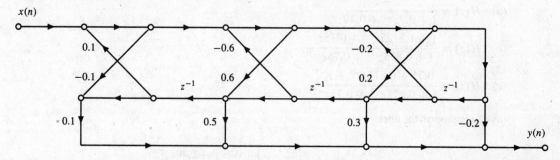

This structure implements a third-order IIR filter with three poles and three zeros:

$$H(z) = \frac{B_3(z)}{A_3(z)}$$

To find the system function, first we use the step-up recursion to find $A_3(z)$ from the reflection coefficients Γ_1, Γ_2, and Γ_3. With $\Gamma_1 = 0.2$, we have

$$\begin{bmatrix} 1 \\ a_1(1) \end{bmatrix} = \begin{bmatrix} 1 \\ \Gamma_1 \end{bmatrix} = \begin{bmatrix} 1 \\ 0.2 \end{bmatrix}$$

Next, for $a_2(k)$ we have

$$\begin{bmatrix} 1 \\ a_2(1) \\ a_2(2) \end{bmatrix} = \begin{bmatrix} 1 \\ 0.2 \\ 0 \end{bmatrix} + \Gamma_2 \begin{bmatrix} 0 \\ 0.2 \\ 1 \end{bmatrix} = \begin{bmatrix} 1 \\ 0.32 \\ 0.6 \end{bmatrix}$$

Finally, for $a_3(k)$ we have

$$\begin{bmatrix} 1 \\ a_3(1) \\ a_3(2) \\ a_3(3) \end{bmatrix} = \begin{bmatrix} 1 \\ 0.32 \\ 0.6 \\ 0 \end{bmatrix} + \Gamma_3 \begin{bmatrix} 0 \\ 0.6 \\ 0.32 \\ 1 \end{bmatrix} = \begin{bmatrix} 1 \\ 0.26 \\ 0.568 \\ -0.1 \end{bmatrix}$$

Therefore, the denominator polynomial is

$$A_3(z) = 1 + 0.26z^{-1} + 0.568z^{-2} - 0.1z^{-3}$$

To find the numerator $B_3(z)$, we use Eq. (8.8),

$$b_q(k) = \sum_{j=k}^{q} c_q(j) a_j(j - k)$$

Thus, we have

$$b_3(3) = c_3(3) a_3(0) = 0.1$$
$$b_3(2) = c_3(2) a_2(0) + c_3(3) a_3(1) = 0.5260$$
$$b_3(1) = c_3(1) a_1(0) + c_3(2) a_2(1) + c_3(3) a_3(2) = 0.5168$$
$$b_3(0) = c_3(0) a_0(0) + c_3(1) a_1(1) + c_3(2) a_2(2) + c_3(3) a_2(2) = 0.15$$

Therefore, the system function is

$$H(z) = \frac{0.15 + 0.5168z^{-1} + 0.5260z^{-2} + 0.1z^{-3}}{1 + 0.26z^{-1} + 0.568z^{-2} - 0.1z^{-3}}$$

8.32   Sketch a lattice filter structure for each of the following system functions:

(a) $H(z) = \dfrac{2 - z^{-1}}{1 + 0.7z^{-1} + 0.49z^{-2}}$

(b) $H(z) = \dfrac{1 + 1.3125z^{-1} + 0.75z^{-2}}{1 + 0.875z^{-1} + 0.75z^{-2}}$

(c) $H(z) = \dfrac{0.75 + 0.875z^{-1} + z^{-2}}{1 + 0.875z^{-1} + 0.75z^{-2}}$

(a)   To implement the filter

$$H(z) = \frac{2 - z^{-1}}{1 + 0.7z^{-1} + 0.49z^{-2}}$$

using a lattice filter structure, we must first find the reflection coefficients corresponding to the denominator polynomial. Using the step-down recursion, we find

$$\Gamma_1 = 0.4698 \qquad \Gamma_2 = 0.49$$

Next, we find the coefficients $c_1(k)$ that produce the numerator polynomial

$$B(z) = 2 - z^{-1}$$

Using the recursion

$$c_q(k) = b_q(k) - \sum_{j=k+1}^{q} c_q(j)a_j(j - k)$$

we have

$$c_1(1) = b_1(1) = -1$$

and

$$c_1(0) = b_1(0) - c_1(1)a_1(1) = 2.4698$$

(note that $a_1(1) = \Gamma_1$). Therefore, the lattice filter structure for this system is as shown in the figure below.

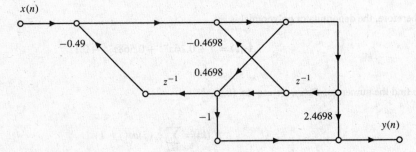

(b)   To find the lattice filter structure for

$$H(z) = \frac{1 + 1.3125z^{-1} + 0.75z^{-2}}{1 + 0.875z^{-1} + 0.75z^{-2}}$$

we first use the step-down recursion to find the reflection coefficients for the denominator, which are

$$\Gamma_1 = 0.5 \qquad \Gamma_2 = 0.75$$

Next, we use the recursion

$$c_q(k) = b_q(k) - \sum_{j=k+1}^{q} c_q(j)a_j(j - k)$$

to find the coefficients $c_3(k)$, which are

$$c_2(2) = b_2(2) = 0.75$$

$$c_2(1) = b_2(1) - c_2(2)a_2(1) = 0.65625$$

$$c_2(0) = b_2(0) - c_2(1)a_1(1) - c_2(2)a_2(2) = 0.109375$$

Thus, the lattice filter structure for this system is as shown in the figure below.

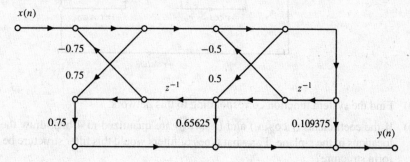

(c)   Note that the system function

$$H(z) = \frac{0.75 + 0.875z^{-1} + z^{-2}}{1 + 0.875z^{-1} + 0.75z^{-2}}$$

is an allpass filter. Because the denominator is the same as the system function in part (b), the reflection coefficients are $\Gamma_1 = 0.5$ and $\Gamma_2 = 0.75$, and the lattice filter is as shown in the figure below.

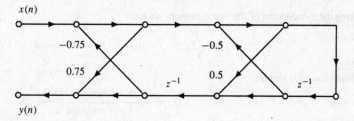

Finite Word-Length Effects

8.33 Express the fractions $\frac{7}{32}$ and $-\frac{7}{32}$ in sign-magnitude, one's complement, and two's complement notation using 6 bits.

With $B + 1 = 6$ bits, 1 bit will be a sign bit, and 5 will be the fractional bits. Because

$$\tfrac{7}{32} = \tfrac{1}{8} + \tfrac{1}{16} + \tfrac{1}{32} = 2^{-3} + 2^{-4} + 2^{-5}$$

the 6-bit representation for $x = \frac{7}{32}$ in all three binary forms is

$$x = 0.00111$$

For $x = -\frac{7}{32}$ we have, in sign-magnitude form,

$$-\tfrac{7}{32} = 1.00111$$

and in one's complement form,

$$x = \overline{0.00111} = 1.11000$$

and in two's complement form

$$x = \overline{0.00111} + 0.00001 = 1.11001$$

8.34 Consider the following implementation of a second-order filter:

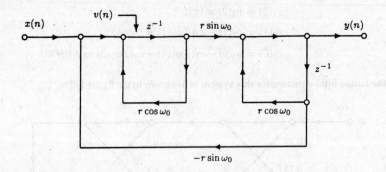

(a) Find the system function corresponding to this network.

(b) If the coefficients $(r \cos \omega_0)$ and $(r \sin \omega_0)$ are quantized to 4 bits, draw the set of allowable pole locations in the z-plane. For what types of filters would this filter structure be preferred over a direct form structure?

(a) This filter structure is called the *coupled form* realization. The system function for this filter may be found as follows. The difference equations relating $x(n)$, $v(n)$, and $y(n)$ are

$$v(n) = x(n) - r \sin(\omega_0)y(n-1) + r \cos(\omega_0)v(n-1)$$
$$y(n) = r \sin(\omega_0)v(n-1) + r \cos(\omega_0)y(n-1)$$

Taking the z-transform of the first equation, we have

$$V(z) = X(z) - r \sin(\omega_0)z^{-1}Y(z) + r \cos(\omega_0)z^{-1}V(z)$$

Solving for $V(z)$ yields

$$V(z) = \frac{X(z) - r \sin(\omega_0)z^{-1}Y(z)}{1 - r \cos(\omega_0)z^{-1}}$$

Substituting this into the z-transform of the second difference equation gives

$$Y(z) = r \sin(\omega_0)z^{-1}\left[\frac{X(z) - r \sin(\omega_0)z^{-1}Y(z)}{1 - r \cos(\omega_0)z^{-1}}\right] + r \cos(\omega_0)z^{-1}Y(z)$$

Solving this equation for $Y(z)$, we have

$$Y(z) = \frac{r \sin(\omega_0)z^{-1}}{1 - 2r \cos(\omega_0)z^{-1} + r^2z^{-2}}X(z)$$

Therefore, the system function is

$$H(z) = \frac{r \sin(\omega_0)z^{-1}}{1 - 2r \cos(\omega_0)z^{-1} + r^2z^{-2}}$$

(b) This filter has poles at

$$z = re^{\pm j\omega_0} = r \cos \omega_0 \pm jr \sin \omega_0$$

Thus, the coefficients in this structure are the real and imaginary parts of the pole locations. Therefore, if the coefficients are quantized to $B + 1$ bits, the poles will lie at the intersections of 2^{B+1} evenly spaced horizontal and vertical lines in the z-plane. These positions are illustrated in the figure below for the first quadrant when $B + 1 = 4$.

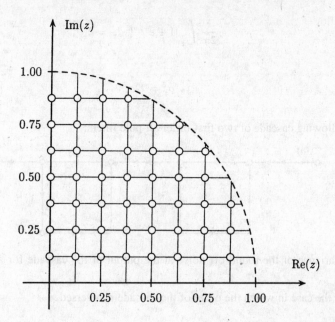

Note that, compared with the direct form structure, the allowable pole locations are uniformly distributed within the unit circle. The cost for this uniform spacing is four multiplies and three additions per output value, compared to only two multiplications and two additions for a direct form implementation. This extra cost may be worthwhile, particularly for low-pass filters that have poles close to the unit circle in the vicinity of $z = 1$, where the density of the allowable pole locations is sparse in the direct form implementation compared to the coupled form implementation.

8.35   A white noise sequence $e(n)$ with variance σ_e^2 is input to a filter with a system function

$$H(z) = \frac{(1 + 2z^{-2})(1 + 3z^{-1})(1 + z^{-1})}{\left(1 + \frac{1}{2}z^{-2}\right)\left(1 + \frac{1}{3}z^{-1}\right)}$$

Find the variance of the output sequence.

The variance of the output sequence is

$$\sigma_f^2 = \sigma_e^2 \cdot \sum_{n=-\infty}^{\infty} |h(n)|^2 = \sigma_e^2 \frac{1}{2\pi} \int_{-\pi}^{\pi} |H(e^{j\omega})|^2 \, d\omega$$

For the given filter, note that

$$H(z) = \frac{1 + 2z^{-2}}{1 + \frac{1}{2}z^{-2}} \cdot \frac{1 + 3z^{-1}}{1 + \frac{1}{3}z^{-1}} \cdot (1 + z^{-1})$$

where the first two terms are allpass filters with

$$\left|\frac{1 + 2e^{-j2\omega}}{1 + \frac{1}{2}e^{-j2\omega}}\right|^2 = 4 \qquad \left|\frac{1 + 3e^{-j\omega}}{1 + \frac{1}{3}e^{-j\omega}}\right|^2 = 9$$

Therefore,

$$\sigma_f^2 = \sigma_e^2 \frac{36}{2\pi} \int_{-\pi}^{\pi} |1 + e^{-j\omega}|^2 \, d\omega$$

Using Parseval's theorem,

$$\sum_{n=-\infty}^{\infty} |h(n)|^2 = \frac{1}{2\pi} \int_{-\pi}^{\pi} |H(e^{j\omega})|^2 \, d\omega$$

we find

$$\frac{1}{2\pi} \int_{-\pi}^{\pi} |1 + e^{-j\omega}|^2 \, d\omega = 2$$

Thus, we have

$$\sigma_f^2 = 72\sigma_e^2$$

8.36 Consider the following cascade of two first-order all-pole filters:

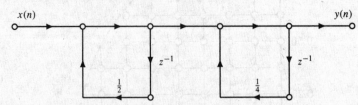

(a) Find the variance of the round-off noise at the output of the cascade for an 8-bit processor with rounding.

(b) Repeat for the case in which the order of the cascade is reversed.

(a) A model for the round-off noise is shown in the following figure:

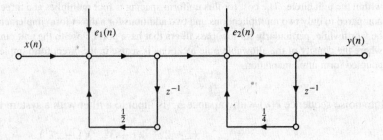

where the variance of each noise source is equal to

$$\sigma_e^2 = \frac{2^{-2B}}{12}$$

The system function of the filter is

$$H(z) = \frac{1}{1 - \frac{1}{2}z^{-1}} \cdot \frac{1}{1 - \frac{1}{4}z^{-1}} = \frac{2}{1 - \frac{1}{2}z^{-1}} - \frac{1}{1 - \frac{1}{4}z^{-1}}$$

and the unit sample response is

$$h(n) = 2\left(\tfrac{1}{2}\right)^n u(n) - \left(\tfrac{1}{4}\right)^n u(n)$$

Note that because $e_1(n)$ is filtered by $h(n)$, and $e_2(n)$ is only filtered by the second filter in the cascade, which has a unit sample response

$$h_2(n) = \left(\tfrac{1}{4}\right)^n u(n)$$

the output noise, $f(n)$, is

$$f(n) = e_1(n) * h(n) + e_2(n) * h_2(n)$$

Therefore, the variance of $f(n)$ is

$$\sigma_f^2 = \sigma_e^2 \cdot \sum_{n=-\infty}^{\infty} |h(n)|^2 + \sigma_e^2 \cdot \sum_{n=-\infty}^{\infty} |h_2(n)|^2$$

With
$$|h(n)|^2 = \left[2\left(\tfrac{1}{2}\right)^n - \left(\tfrac{1}{4}\right)^n\right]^2 = 4\left(\tfrac{1}{2}\right)^{2n} + \left(\tfrac{1}{4}\right)^{2n} - 4\left(\tfrac{1}{8}\right)^n$$

we have
$$\sum_{n=-\infty}^{\infty} |h(n)|^2 = \frac{4}{1-\tfrac{1}{4}} + \frac{1}{1-\tfrac{1}{16}} - \frac{4}{1-\tfrac{1}{8}} = 1.8286$$

Next, we have
$$\sum_{n=0}^{\infty} |h_2(n)|^2 = \sum_{n=0}^{\infty} \left(\tfrac{1}{4}\right)^{2n} = \frac{1}{1-\tfrac{1}{16}} = 1.0667$$

Therefore, the variance of the round-off noise at the output of the filter is
$$\sigma_f^2 = 1.8286\sigma_e^2 + 1.0667\sigma_e^2 = 2.8953\sigma_e^2$$

which, for an 8-bit processor ($B = 7$), is
$$\sigma_f^2 = 2.8953 \frac{2^{-2B}}{12} = 0.2413 \cdot 2^{-14} = 1.4726 \cdot 10^{-5}$$

(b) If the order of the cascade is reversed, we have the following network:

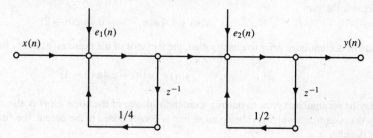

The variance of the round-off noise due to $e_1(n)$ is the same as in part (a), but because the unit sample response of the second system in the cascade is now
$$h_2(n) = \left(\tfrac{1}{2}\right)^n u(n)$$

the variance of the noise due to $e_2(n)$ is
$$\sigma_e^2 \sum_{n=0}^{\infty} |h_2(n)|^2 = \sigma_e^2 \sum_{n=0}^{\infty} \left(\tfrac{1}{2}\right)^{2n} = \sigma_e^2 \cdot \frac{1}{1-\tfrac{1}{4}} = 1.3333\,\sigma_e^2$$

Thus, the variance of the round-off noise at the output of the filter is
$$\sigma_f^2 = 1.8286\sigma_e^2 + 1.3333\sigma_e^2 = 3.1619\sigma_e^2$$

which, for an 8-bit processor is
$$\sigma_f^2 = 3.1619 \frac{2^{-2B}}{12} = 0.2635 \cdot 2^{-14} = 1.6082 \cdot 10^{-5}$$

With this structure, the round-off noise is slightly larger.

8.37 Consider a linear shift-invariant system with a system function
$$H(z) = \frac{1 - 0.4z^{-1}}{(1 - 0.6z^{-1})(1 - 0.8z^{-1})}$$

Suppose that this system is implemented on a 16-bit fixed-point processor and that the sums of products are accumulated prior to quantization. Let σ_e^2 be the variance of the round-off noise.

(a) If the system is implemented in direct form II, find the variance of the round-off noise at the output of the filter.

(b) Repeat part (a) if the system is implemented in parallel form.

(a) The direct form II implementation of this system is shown in the figure below along with the two round-off noise sources.

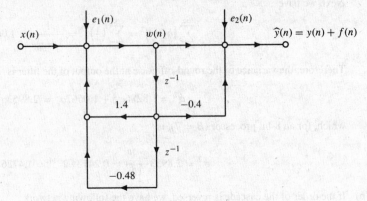

Because the sum

$$x(n) + 1.4w(n-1) - 0.48w(n-2)$$

may be accumulated prior to quantization, the variance of the noise $e_1(n)$ is σ_e^2. Similarly, because the sum

$$y(n) = w(n) - 0.4w(n-1)$$

may be accumulated prior to quantization, the variance of the noise $e_2(n)$ is also σ_e^2. With $e_1(n)$ being filtered by the system and with $e_2(n)$ being noise that is simply added to the output, the quantization noise at the output of the filter is

$$f(n) = h(n) * e_1(n) + e_2(n)$$

which has a variance equal to

$$\sigma_f^2 = \sigma_e^2 \cdot \sum_{n=-\infty}^{\infty} |h(n)|^2 + \sigma_e^2$$

To find the unit sample response of the filter, we expand $H(z)$ in a partial fraction expansion as follows:

$$H(z) = \frac{1 - 0.4z^{-1}}{(1 - 0.6z^{-1})(1 - 0.8z^{-1})} = \frac{-1}{1 - 0.6z^{-1}} + \frac{2}{1 - 0.8z^{-1}}$$

Therefore,

$$h(n) = -(0.6)^n u(n) + 2(0.8)^n u(n)$$

and

$$|h(n)|^2 = [-(0.6)^n + 2(0.8)^n]^2 u(n) = [(0.6)^{2n} - 4(0.48)^n + 4(0.8)^{2n}]u(n)$$

Evaluating the sum of the squares of $h(n)$, we have

$$\sum_{n=-\infty}^{\infty} |h(n)|^2 = \sum_{n=0}^{\infty} [(0.36)^n - 4(0.48)^n + 4(0.64)^n] = \frac{1}{1 - 0.36} - \frac{4}{1 - 0.48} + \frac{4}{1 - 0.64} \approx 5$$

Thus, the variance of the output noise is

$$\sigma_f^2 \approx 6\sigma_e^2$$

(b) Using the partial fraction expansion for $H(z)$ given in part (a), the parallel form implementation of this filter is shown in the following figure:

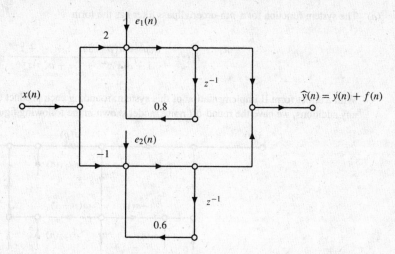

As indicated in the figure, there are two noise sources. The first, $e_1(n)$, is filtered with a first-order all-pole filter that has a unit sample response

$$h_1(n) = (0.8)^n u(n)$$

and the second, $e_2(n)$, is filtered with a first-order all-pole filter that has a unit sample response

$$h_2(n) = (0.6)^n u(n)$$

Because the output noise is

$$f(n) = e_1(n) * h_1(n) + e_2(n) * h_2(n)$$

the variance of $f(n)$ is

$$\sigma_f^2 = \sigma_e^2 \left[\sum_{n=-\infty}^{\infty} |h_1(n)|^2 + \sum_{n=-\infty}^{\infty} |h_2(n)|^2 \right] = \sigma_e^2 \left[\frac{1}{1-(0.8)^2} + \frac{1}{1-(0.6)^2} \right] = 4.34\sigma_e^2$$

8.38 A linear shift-invariant system with a system function of the form

$$H(z) = \prod_{k=1}^{N} \frac{b_k(0) + b_k(1)z^{-1} + b_k(2)z^{-2}}{1 + a_k(1)z^{-1} + a_k(2)z^{-2}}$$

is to be implemented as a cascade of N second-order sections, where each section is realized in either direct form I or II or in their transposed forms. How many different cascaded realizations are possible.

Let us assume that each factor in $H(z)$ is unique, so that there are N different second-order polynomials in the numerator and the same number of polynomials in the denominator. In this case, there are $N!$ different pairings of factors in the numerator with factors in the denominator. In addition, for each of these pairings, there are $N!$ different orderings of these sections. Therefore, there are $(N!)^2$ different pairings and orderings. With four different structures for each section (direct form I, direct form II, transposed direct form I, and transposed direct form II), there are a total of $4^N(N!)^2$ different realizations. For a tenth-order system ($N = 5$), this corresponds to 14,745,600 different structures. This is why general pairing and ordering rules are important.

8.39 Let $H(z)$ be a pth-order allpass filter with a gain of 1 that is implemented in direct form II using a processor with $B + 1$ bits.

(a) If the product of two $(B + 1)$-bit numbers is *rounded* to $B + 1$ bits before any additions are performed, find the variance of the round-off noise at the output of the filter.

(b) Repeat part (a) for the case in which sums of products are accumulated prior to quantization.

(a) The system function for a pth-order allpass filter has the form

$$H(z) = \frac{a(p) + a(p-1)z^{-1} + \cdots + z^{-p}}{1 + a(1)z^{-1} + \cdots + a(p)z^{-p}}$$

With a direct form II implementation of this system, rounding each product to $B + 1$ bits prior to performing any additions, we have the round-off noise model shown in the following figure:

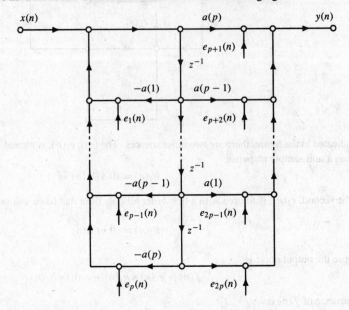

where the variance of each noise source is

$$\sigma_e^2 = \frac{2^{-2B}}{12}$$

Note, however, that this noise model may be simplified as illustrated in the following figure:

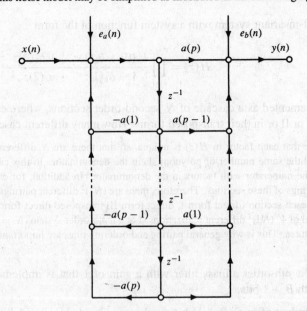

where

$$e_a(n) = e_1(n) + e_2(n) + \cdots + e_p(n)$$
$$e_b(n) = e_{p+1}(n) + e_{p+2}(n) + \cdots + e_{2p}(n)$$

Assuming that each noise source is uncorrelated with the others, the variances of $e_a(n)$ and $e_b(n)$ are

$$\sigma_a^2 = p\sigma_e^2 \qquad \sigma_b^2 = p\sigma_e^2$$

Because the output noise is

$$f(n) = e_a(n) * h(n) + e_b(n)$$

where $h(n)$ is the unit sample response of the allpass filter, the variance of $f(n)$ is

$$\sigma_f^2 = \sigma_a^2 \cdot \sum_{n=-\infty}^{\infty} |h(n)|^2 + \sigma_b^2$$

Equivalently, we may write this using Parseval's theorem as follows:

$$\sigma_f^2 = \sigma_a^2 \frac{1}{2\pi} \int_{-\pi}^{\pi} |H(e^{j\omega})|^2 d\omega + \sigma_b^2$$

Because $H(e^{j\omega})$ is an allpass filter with $|H(e^{j\omega})| = 1$, the variance of the output noise is

$$\sigma_f^2 = \sigma_a^2 + \sigma_b^2 = 2p\sigma_e^2$$

(b) If the products are accumulated prior to quantization, the variances of $e_a(n)$ and $e_b(n)$ in the noise model given in part (a) will be reduced by a factor of p:

$$\sigma_a^2 = \sigma_e^2 \qquad \sigma_b^2 = \sigma_e^2$$

Therefore, the variance of the output noise becomes

$$\sigma_f^2 = 2\sigma_e^2$$

8.40 In the figure below are direct form II and transposed direct form II realizations of the first-order system

$$H(z) = \frac{b(0) + b(1)z^{-1}}{1 - a(1)z^{-1}}$$

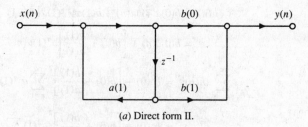

(a) Direct form II.

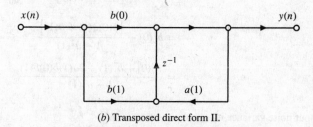

(b) Transposed direct form II.

Assume that both systems are implemented using $(B + 1)$-bit fixed-point arithmetic and that all products are *rounded* to $B + 1$ bits before any additions are performed.

(a) Using a linear noise model for the round-off noise, find the variance of the round-off noise at the output of the direct form II filter.

(b) Repeat part (a) for the transposed direct form II filter.

(c) How would the variance of the output noise change if the sums of products were accumulated prior to quantization?

(a) The linear noise model for round-off noise in the direct form II implementation is shown in the figure below.

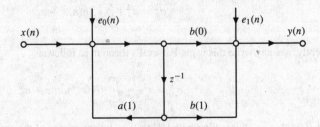

The variance of the noise $e_0(n)$ is σ_e^2 and the variance of $e_1(n)$ is $2\sigma_e^2$, where

$$\sigma_e^2 = \tfrac{1}{12}2^{-2B}$$

Because the output noise is

$$f(n) = e_1(n) + e_0(n) * h(n)$$

where $h(n)$ is the unit sample response of the filter, the variance of the output noise is

$$\sigma_f^2 = 2\sigma_e^2 + \sigma_e^2 \sum_{n=-\infty}^{\infty} |h(n)|^2$$

With

$$H(z) = \frac{b(0) + b(1)z^{-1}}{1 - a(1)z^{-1}}$$

it follows that the unit sample response is

$$h(n) = b(0)a^n(1)u(n) + b(1)a^{n-1}(1)u(n-1)$$

$$= b(0)\delta(n) + \left[b(0) + \frac{b(1)}{a(1)}\right]a^n(1)u(n-1)$$

Therefore,

$$\sum_{n=-\infty}^{\infty} |h(n)|^2 = b^2(0) + \left[b(0) + \frac{b(1)}{a(1)}\right]^2 \sum_{n=1}^{\infty} a^{2n}(1)$$

$$= b^2(0) + \left[b(0) + \frac{b(1)}{a(1)}\right]^2 a^2(1) \sum_{n=0}^{\infty} a^{2n}(1)$$

$$= b^2(0) + \frac{\left[b(0)a(1) + b(1)\right]^2}{1 - a^2(1)}$$

$$= \frac{b^2(0) + b^2(1) + 2a(1)b(0)b(1)}{1 - a^2(1)}$$

and the output noise variance is

$$\sigma_f^2 = 2\sigma_e^2 + \sigma_e^2 \frac{b^2(0) + b^2(1) + 2a(1)b(0)b(1)}{1 - a^2(1)}$$

(b) For the transposed direct form II implementation, the noise model is as follows,

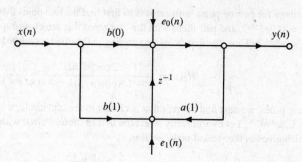

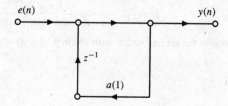

where the variance of $e_0(n)$ is σ_e^2, and the variance of $e_1(n)$ is $2\sigma_e^2$. Note that because neither noise source is filtered by the zeros of the system, an equivalent model for the generation of the filtered noise $f(n)$ is as shown in the figure below,

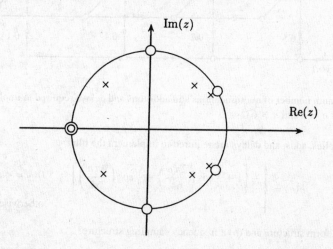

where $e(n) = e_0(n) + e_1(n - 1)$. Because $e_0(n)$ and $e_1(n)$ are uncorrelated, the variance of $e(n)$ is $3\sigma_e^2$, and the variance of the output noise is

$$\sigma_f^2 = 3\sigma_e^2 \sum_{n=0}^{\infty} [a(1)]^2 = 3\sigma_e^2 \frac{1}{1 - a^2(1)}$$

(c) If we accumulate the sums of products prior to quantization, for the direct form II implementation, the variance of the noise $e_1(n)$ would be σ_e^2 instead of $2\sigma_e^2$, and everything else remains the same. For the transposed direct form II structure, on the other hand, the variance of $e_1(n)$ would be σ_e^2 instead of $2\sigma_e^2$, which implies that the variance of $e(n)$ would be $2\sigma_e^2$.

8.41　A sixth-order filter with a system function

$$H(z) =$$

$$\frac{(1 + z^{-2})(1 + z^{-1})^2(1 - 2\cos(\pi/6)z^{-1} + z^{-2})}{(1 - 1.6\cos(\pi/4)z^{-1} + 0.64z^{-2})(1 + 1.6\cos(\pi/4)z^{-1} + 0.64z^{-2})(1 - 1.8\cos(\pi/6)z^{-1} + 0.81z^{-2})}$$

is to be implemented as a cascade of second-order sections. Considering only the effects of round-off noise, determine what the best pole-zero pairing is, and the best ordering of the second-order sections.

Note that all six zeros of this system lie on the unit circle, with two at $z = -1$, a complex pair at $z = \pm j$, and a complex pair at $z = e^{\pm j\pi/6}$. The poles, on the other hand, are at $z = 0.8e^{\pm j\pi/4}$, $z = 0.8e^{\pm j3\pi/4}$, and $z = 0.9e^{\pm j\pi/6}$. A pole-zero diagram showing each of these poles and zeros is given below.

The general strategy for pairing poles with zeros is to first find the two poles that are closest to the unit circle, in this case those at $z = 0.9e^{\pm j\pi/6}$, and pair these with the two zeros that are closest to these poles, which are those on the unit circle at $z = e^{\pm j\pi/6}$. Thus, the first pole-zero pairing yields the second-order section

$$H_1(z) = \frac{1 - 2\cos(\pi/6)z^{-1} + z^{-2}}{1 - 1.8\cos(\pi/6)z^{-1} + 0.81z^{-2}}$$

Of the remaining poles, we next find the pair that is closest to the unit circle, which are either those at $z = 0.8e^{\pm j\pi/4}$ or those at $z = 0.8e^{\pm j3\pi/4}$. Let us arbitrarily select the first of these. Paired with these poles would then be the zeros at $z = \pm j$, which gives us the second-order section

$$H_2(z) = \frac{1 + z^{-2}}{1 - 1.6\cos(\pi/6)z^{-1} + 0.64z^{-2}}$$

Finally, for the last section we have

$$H_3(z) = \frac{(1 + z^{-1})^2}{1 + 1.6\cos(\pi/6)z^{-1} + 0.64z^{-2}}$$

The cascade is then done in the reverse order, with the first second-order section being $H_3(z)$, followed by $H_2(z)$, and then $H_1(z)$.

Supplementary Problems

Structures for FIR Systems

8.42   Find the unit sample response for the following network:

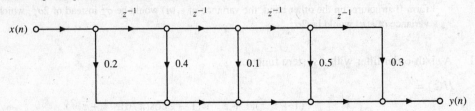

8.43   What is the frequency response of the following network?

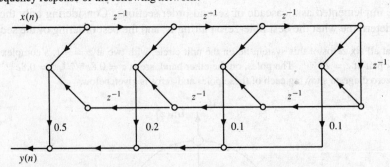

8.44   What is the minimum number of multiplications and additions and delays required to implement a linear phase filter with $h(n) = 0$ for $n < 0$ and $n > 63$?

8.45   How many multiplies, adds, and delays are required to implement the filter

$$h(n) = \begin{cases} \frac{1}{64}\left[1 + \frac{1}{2}\cos\left(\frac{2\pi n}{64}\right) + \frac{1}{4}\cos\left(\frac{4\pi n}{64}\right)\right] & 0 \le n \le 63 \\ 0 & \text{otherwise} \end{cases}$$

using (*a*) a direct form structure and (*b*) a frequency sampling structure?

8.46   Draw a frequency sampling structure for the FIR high-pass filter of length $N = 32$ with

$$H(k) = \begin{cases} \frac{1}{32} & k = 15, 16, 17 \\ 0 & \text{else} \end{cases}$$

Structures for IIR Systems

8.47   Find the system function for the following network, where az^{-1} is a unit delay combined with a multiplication by a:

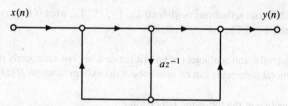

8.48   Find the unit sample response of the following network:

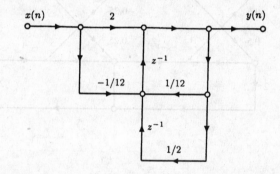

8.49   Find the constant A so that the second-order all-pole filter

$$H(z) = \frac{A}{1 + \alpha_1 z^{-1} + \alpha_2 z^{-2}}$$

has unit gain at $\omega = 0$, and find a structure that only requires two multiplications.

8.50   What is the system function corresponding to the following filter structure?

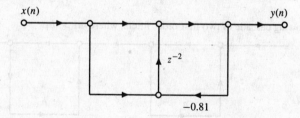

8.51   Find the transposed direct form II realization of the system described by the difference equation

$$y(n) = 0.5y(n-1) - 0.25y(n-2) + x(n) - 2x(n-1) + x(n-2)$$

Lattice Filters

8.52   What is the frequency response of the FIR filter that has reflection coefficients $\Gamma_1 = \Gamma_2 = \cdots = \Gamma_9 = 0$ and $\Gamma_{10} = 1$?

8.53   Draw a lattice filter implementation for the allpass filter

$$H(z) = \frac{0.25 - 0.5z^{-1} + z^{-2}}{1 - 0.5z^{-1} + 0.25z^{-2}}$$

8.54   If the system function of a causal filter is

$$H(z) = \frac{1}{1 + 1.1z^{-1} + 0.9z^{-2} + 1.45z^{-3} + 0.5z^{-4}}$$

is this filter stable?

8.55   If $H(z)$ is an FIR filter with reflection coefficients $\Gamma_1, \Gamma_2, \ldots, \Gamma_p$, what is the system function of the filter $G(z)$ with reflection coefficients $\Gamma_1, 0, \Gamma_2, 0, \Gamma_3, \ldots, 0, \Gamma_p$?

8.56   Suppose that the last reflection coefficient of an FIR lattice filter, not necessarily minimum phase, has unit magnitude $|\Gamma_p| = 1$. What general statements can be made about the system function $H(z)$?

8.57   Find the system function of the following lattice filter:

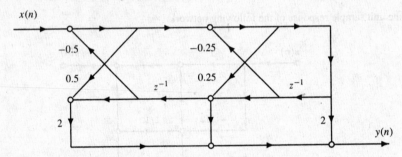

Finite Word-Length Effects

8.58   (a) What fraction does the binary number $x = 1.01101$ represent in one's complement notation? (b) What about two's complement notation?

8.59   White noise with a variance σ_e^2 is input to a linear shift-invariant filter with a system function

$$H(z) = \frac{1 + \frac{5}{6}z^{-1}}{\left(1 - \frac{1}{3}z^{-1}\right)\left(1 + \frac{1}{4}z^{-1}\right)}$$

Find the variance of the noise at the output of the filter.

8.60   Consider the following cascade of two first-order filters, where $|a| > |b|$:

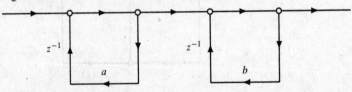

Without explicitly evaluating the variance, determine whether the variance of the round-off noise at the output of this filter will be larger or smaller if the order of the cascade is reversed.

8.61   Consider a linear shift-invariant system with a system function

$$H(z) = \frac{1 + 0.5z^{-1}}{1 - 0.3z^{-1}}$$

Assume that each product is rounded to 16 bits *before* any additions are performed, and let σ_e^2 be the variance of the round-off noise.

(a) If this system is implemented in direct form II, compute the variance of the round-off noise at the output of the filter, σ_f^2, due to all noise sources.

(b) Repeat part (a) if the system is implemented in direct form I.

8.62 Suppose that an FIR filter with a system function

$$H(z) = 1 + 0.2z^{-1} + 0.4z^{-2} - 0.25z^{-3} + 0.1z^{-4}$$

is implemented on a 16-bit fixed-point processor. If sums of products are accumulated prior to rounding, find the variance of the round-off noise at the output of the filter.

8.63 The second-order system

$$H(z) = \frac{1}{1 - 1.2728z^{-1} + 0.81z^{-2}}$$

is implemented in direct form II using 16-bit fixed point arithmetic. Assuming that all sums of products are accumulated prior to rounding, find the quantization noise power at the filter output.

8.64 To minimize the effects of round-off noise, what is the best pairing of poles and zeros into second-order sections for the system

$$H(z) = \frac{(1 + 0.9z^{-2})(1 - 2.4\cos(0.75\pi)z^{-1} + 1.44z^{-2})}{(1 - 1.4\cos(0.25\pi)z^{-1} + 0.49z^{-2})(1 - 1.8\cos(0.9\pi)z^{-1} + 0.81z^{-2})}$$

and what is the best ordering for the second-order sections?

Answers to Supplementary Problems

8.42 $h(n) = 0.2\delta(n) + 0.4\delta(n-1) + 0.1\delta(n-2) + 0.5\delta(n-3) + 0.3\delta(n-4)$.

8.43 $H(e^{j\omega}) = e^{-j3.5\omega}[\cos(3.5\omega) + 0.4\cos(2.5\omega) + 0.2\cos(1.5\omega) + 0.2\cos(0.5\omega)]$.

8.44 63 delays, 63 additions, and 32 multiplications.

8.45 (a) 33 multiplies, 63 adds, and 63 delays. (b) 9 multiplies, 10 adds, and 69 delays.

8.46

8.47 $\dfrac{1 + az^{-1}}{1 - az^{-1}}$.

8.48 $h(n) = \left(-\frac{2}{3}\right)^n u(n) + \left(\frac{3}{4}\right)^n u(n)$.

8.49 $A = 1 + \alpha_1 + \alpha_2$, and the structure is

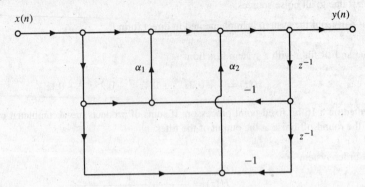

8.50 $H(z) = \dfrac{1 + z^{-2}}{1 + 0.81z^{-2}}$.

8.51

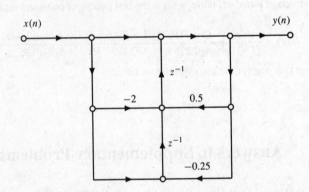

8.52 $H(e^{j\omega}) = 1 + e^{-j10\omega}$.

8.53

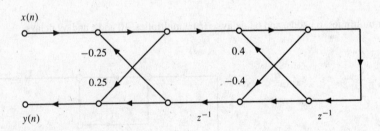

8.54   No, the filter is unstable because $\Gamma_3 = 1.2$.

8.55 $G(z) = H(z^2)$.

8.56   The system function will have generalized linear phase, $H(z) = \pm z^{-p} H(z^{-1})$.

8.57 $H(z) = \dfrac{2 + 1.75z^{-1} + 3.25z^{-2}}{1 + 0.375z^{-1} + 0.5z^{-2}}$.

8.58   (a) $\frac{9}{16}$. (b) $\frac{19}{32}$.

8.59 $\sigma_f^2 = \sigma_e^2 \left[\frac{9}{2} + \frac{16}{15} - \frac{48}{11} \right] = 1.20\sigma_e^2$.

8.60   The variance of the output noise will be larger if the pole closest to the unit circle is the second filter in the cascade. Thus, the output noise variance will be larger if the order of the cascade is reversed.

8.61   (a) $\sigma_f^2 = 2.7\sigma_e^2$. (b) $\sigma_f^2 = 2.2\sigma_e^2$.

8.62   The variance of the output noise is simply the noise variance, $\sigma_e^2 = \frac{1}{12}2^{-2B} = \frac{1}{12}2^{-30}$.

8.63 $\sigma_f^2 = 0.2077\sigma_e^2$ where $\sigma_e^2 = \frac{1}{12}2^{-30}$.

8.64 $H_1(z) = \dfrac{1 + 0.9z^{-2}}{1 - 1.4\cos(0.25\pi)z^{-1} + 0.49z^{-2}}$, and $H_2(z) = \dfrac{1 - 2.4\cos(0.75\pi)z^{-1} + 1.44z^{-2}}{1 - 1.8\cos(0.9\pi)z^{-1} + 0.81z^{-2}}$.

Chapter 9

Filter Design

9.1 INTRODUCTION

This chapter considers the problem of designing a digital filter. The design process begins with the filter specifications, which may include constraints on the magnitude and/or phase of the frequency response, constraints on the unit sample response or step response of the filter, specification of the type of filter (e.g., FIR or IIR), and the filter order. Once the specifications have been defined, the next step is to find a set of filter coefficients that produce an acceptable filter. After the filter has been designed, the last step is to implement the system in hardware or software, quantizing the filter coefficients if necessary, and choosing an appropriate filter structure (Chap. 8).

9.2 FILTER SPECIFICATIONS

Before a filter can be designed, a set of filter specifications must be defined. For example, suppose that we would like to design a low-pass filter with a cutoff frequency ω_c. The frequency response of an ideal low-pass filter with linear phase and a cutoff frequency ω_c is

$$H_d(e^{j\omega}) = \begin{cases} e^{-j\alpha\omega} & |\omega| \leq \omega_c \\ 0 & \omega_c < |\omega| \leq \pi \end{cases}$$

which has a unit sample response

$$h_d(n) = \frac{\sin(n - \alpha)\omega_c}{\pi(n - \alpha)}$$

Because this filter is unrealizable (noncausal and unstable), it is necessary to relax the ideal constraints on the frequency response and allow some deviation from the ideal response. The specifications for a low-pass filter will typically have the form

$$1 - \delta_p < |H(e^{j\omega})| \leq 1 + \delta_p \qquad 0 \leq |\omega| < \omega_p$$
$$|H(e^{j\omega})| \leq \delta_s \qquad \omega_s \leq |\omega| < \pi$$

as illustrated in Fig. 9-1. Thus, the specifications include the passband cutoff frequency, ω_p, the stopband cutoff frequency, ω_s, the passband deviation, δ_p, and the stopband deviation, δ_s. The passband and stopband deviations

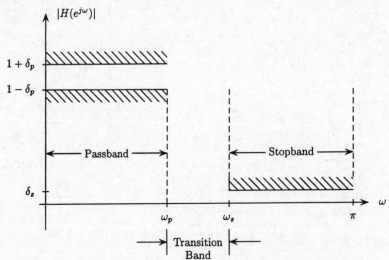

Fig. 9-1. Filter specifications for a low-pass filter.

358

are often given in decibels (dB) as follows:

$$\alpha_p = -20 \log(1 - \delta_p)$$

and

$$\alpha_s = -20 \log(\delta_s)$$

The interval $[\omega_p, \omega_s]$ is called the *transition band*.

Once the filter specifications have been defined, the next step is to design a filter that meets these specifications.

9.3 FIR FILTER DESIGN

The frequency response of an Nth-order causal FIR filter is

$$H(e^{j\omega}) = \sum_{n=0}^{N} h(n) e^{-jn\omega}$$

and the design of an FIR filter involves finding the coefficients $h(n)$ that result in a frequency response that satisfies a given set of filter specifications. FIR filters have two important advantages over IIR filters. First, they are guaranteed to be stable, even after the filter coefficients have been quantized. Second, they may be easily constrained to have (generalized) linear phase. Because FIR filters are generally designed to have linear phase, in the following we consider the design of linear phase FIR filters.

9.3.1 Linear Phase FIR Design Using Windows

Let $h_d(n)$ be the unit sample response of an ideal frequency selective filter with linear phase,

$$H_d(e^{j\omega}) = A(e^{j\omega}) e^{-j(\alpha\omega - \beta)}$$

Because $h_d(n)$ will generally be infinite in length, it is necessary to find an FIR approximation to $H_d(e^{j\omega})$. With the window design method, the filter is designed by windowing the unit sample response,

$$h(n) = h_d(n) w(n)$$

where $w(n)$ is a finite-length window that is equal to zero outside the interval $0 \le n \le N$ and is symmetric about its midpoint:

$$w(n) = w(N - n)$$

The effect of the window on the frequency response may be seen from the complex convolution theorem,

$$H(e^{j\omega}) = \frac{1}{2\pi} H_d(e^{j\omega}) * W(e^{j\omega}) = \frac{1}{2\pi} \int_{-\pi}^{\pi} H_d(e^{j\theta}) W(e^{j(\omega - \theta)}) \, d\theta$$

Thus, the ideal frequency response is *smoothed* by the discrete-time Fourier transform of the window, $W(e^{j\omega})$.

There are many different types of windows that may be used in the window design method, a few of which are listed in Table 9-1.

How well the frequency response of a filter designed with the window design method approximates a desired response, $H_d(e^{j\omega})$, is determined by two factors (see Fig. 9-2):

1. The width of the main lobe of $W(e^{j\omega})$.
2. The peak side-lobe amplitude of $W(e^{j\omega})$.

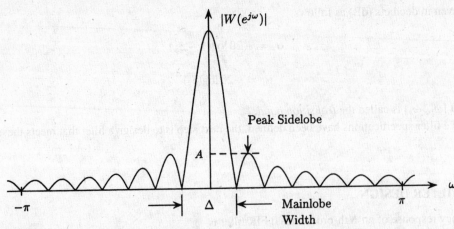

Fig. 9-2. The DTFT of a typical window, which is characterized by the width of its main lobe, Δ, and the peak amplitude of its side lobes, A, relative to the amplitude of $W(e^{j\omega})$ at $\omega = 0$.

Ideally, the main-lobe width should be narrow, and the side-lobe amplitude should be small. However, for a fixed-length window, these cannot be minimized independently. Some general properties of windows are as follows:

1. As the length N of the window increases, the width of the main lobe decreases, which results in a decrease in the transition width between passbands and stopbands. This relationship is given approximately by

$$N \Delta f = c \tag{9.1}$$

where Δf is the transition width, and c is a parameter that depends on the window.

2. The peak side-lobe amplitude of the window is determined by the shape of the window, and it is essentially independent of the window length.

3. If the window shape is changed to decrease the side-lobe amplitude, the width of the main lobe will generally increase.

Listed in Table 9.2 are the side-lobe amplitudes of several windows along with the approximate transition width and stopband attenuation that results when the given window is used to design an Nth-order low-pass filter.

Table 9-1   Some Common Windows

Rectangular	$w(n) = \begin{cases} 1 & 0 \leq n \leq N \\ 0 & \text{else} \end{cases}$		
Hanning[1]	$w(n) = \begin{cases} 0.5 - 0.5\cos\left(\dfrac{2\pi n}{N}\right) & 0 \leq n \leq N \\ 0 & \text{else} \end{cases}$		
Hamming	$w(n) = \begin{cases} 0.54 - 0.46\cos\left(\dfrac{2\pi n}{N}\right) & 0 \leq n \leq N \\ 0 & \text{else} \end{cases}$		
Blackman	$w(n) = \begin{cases} 0.42 - 0.5\cos\left(\dfrac{2\pi n}{N}\right) + 0.08\cos\left(\dfrac{4\pi n}{N}\right) & 0 \leq n \leq N \\ 0 & \text{else} \end{cases}$		

[1]In the literature, this window is also called a Hann window or a von Hann window.

**Table 9-2   The Peak Side-Lobe Amplitude of Some Common Windows and the Approximate
Transition Width and Stopband Attenuation of an Nth-Order Low-Pass Filter
Designed Using the Given Window.**

Window	Side-Lobe Amplitude (dB)	Transition Width (Δf)	Stopband Attenuation (dB)
Rectangular	-13	$0.9/N$	-21
Hanning	-31	$3.1/N$	-44
Hamming	-41	$3.3/N$	-53
Blackman	-57	$5.5/N$	-74

EXAMPLE 9.3.1   Suppose that we would like to design an FIR linear phase low-pass filter according to the following specifications:

$$0.99 \leq |H(e^{j\omega})| \leq 1.01 \qquad 0 \leq |\omega| \leq 0.19\pi$$
$$|H(e^{j\omega})| \leq 0.01 \qquad 0.21\pi \leq |\omega| \leq \pi$$

For a stopband attenuation of $20 \log(0.01) = -40$ dB, we may use a Hanning window. Although we could also use a Hamming or a Blackman window, these windows would overdesign the filter and produce a larger stopband attenuation at the expense of an increase in the transition width. Because the specification calls for a transition width of $\Delta\omega = \omega_s - \omega_p = 0.02\pi$, or $\Delta f = 0.01$, with

$$N \Delta f = 3.1$$

for a Hanning window (see Table 9.2), an estimate of the required filter order is

$$N = \frac{3.1}{\Delta f} = 310$$

The last step is to find the unit sample response of the ideal low-pass filter that is to be windowed. With a cutoff frequency of $\omega_c = (\omega_s + \omega_p)/2 = 0.2\pi$, and a delay of $\alpha = N/2 = 155$, the unit sample response is

$$h_d(n) = \frac{\sin[0.2\pi(n - 155)]}{(n - 155)\pi}$$

In addition to the windows listed in Table 9-1, Kaiser developed a *family* of windows that are defined by

$$w(n) = \frac{I_0[\beta(1 - [(n - \alpha)/\alpha]^2)^{1/2}]}{I_0(\beta)} \qquad 0 \leq n \leq N$$

where $\alpha = N/2$, and $I_0(\cdot)$ is a zeroth-order modified Bessel function of the first kind, which may be easily generated using the power series expansion

$$I_0(x) = 1 + \sum_{k=1}^{\infty} \left[\frac{(x/2)^k}{k!} \right]^2$$

The parameter β determines the shape of the window and thus controls the trade-off between main-lobe width and side-lobe amplitude. A *Kaiser window* is nearly optimum in the sense of having the most energy in its main lobe for a given side-lobe amplitude. Table 9-3 illustrates the effect of changing the parameter β.

There are two empirically derived relationships for the Kaiser window that facilitate the use of these windows to design FIR filters. The first relates the stopband ripple of a low-pass filter, $\alpha_s = -20 \log(\delta_s)$, to the parameter β,

$$\beta = \begin{cases} 0.1102(\alpha_s - 8.7) & \alpha_s > 50 \\ 0.5842(\alpha_s - 21)^{0.4} + 0.07886(\alpha_s - 21) & 21 \leq \alpha_s \leq 50 \\ 0.0 & \alpha_s < 21 \end{cases}$$

Table 9-3   Characteristics of the Kaiser Window as a Function of β

Parameter β	Side Lobe (dB)	Transition Width ($N\Delta f$)	Stopband Attenuation (dB)
2.0	−19	1.5	−29
3.0	−24	2.0	−37
4.0	−30	2.6	−45
5.0	−37	3.2	−54
6.0	−44	3.8	−63
7.0	−51	4.5	−72
8.0	−59	5.1	−81
9.0	−67	5.7	−90
10.0	−74	6.4	−99

The second relates N to the transition width Δf and the stopband attenuation α_s,

$$N = \frac{\alpha_s - 7.95}{14.36\Delta f} \qquad \alpha_s \geq 21 \tag{9.2}$$

Note that if $\alpha_s < 21$ dB, a rectangular window may be used ($\beta = 0$), and $N = 0.9/\Delta f$.

EXAMPLE 9.3.2   Suppose that we would like to design a low-pass filter with a cutoff frequency $\omega_c = \pi/4$, a transition width $\Delta\omega = 0.02\pi$, and a stopband ripple $\delta_s = 0.01$. Because $\alpha_s = -20\log(0.01) = -40$, the Kaiser window parameter is

$$\beta = 0.5842(40 - 21)^{0.4} + 0.07886(40 - 21) = 3.4$$

With $\Delta f = \Delta\omega/2\pi = 0.01$, we have

$$N = \frac{40 - 7.95}{14.36 \cdot (0.01)} = 224$$

Therefore, $$h(n) = h_d(n)w(n)$$

where $$h_d(n) = \frac{\sin[(n - 112)\pi/4]}{(n - 112)\pi}$$

is the unit sample response of the ideal low-pass filter.

Although it is simple to design a filter using the window design method, there are some limitations with this method. First, it is necessary to find a closed-form expression for $h_d(n)$ (or it must be approximated using a very long DFT). Second, for a frequency selective filter, the transition widths between frequency bands, and the ripples within these bands, will be approximately the same. As a result, the window design method requires that the filter be designed to the tightest tolerances in all of the bands by selecting the smallest transition width and the smallest ripple. Finally, window design filters are not, in general, *optimum* in the sense that they do not have the smallest possible ripple for a given filter order and a given set of cutoff frequencies.

9.3.2   Frequency Sampling Filter Design

Another method for FIR filter design is the frequency sampling approach. In this approach, the desired frequency response, $H_d(e^{j\omega})$, is first uniformly sampled at N equally spaced points between 0 and 2π:

$$H(k) = H_d\left(e^{j2\pi k/N}\right) \qquad k = 0, 1, \ldots, N-1$$

These frequency samples constitute an N-point DFT, whose inverse is an FIR filter of order $N - 1$:

$$h(n) = \frac{1}{N} \sum_{k=0}^{N-1} H(k) e^{j2\pi nk/N} \qquad 0 \le n \le N - 1$$

The relationship between $h(n)$ and $h_d(n)$ (see Chap. 3) is

$$h(n) = \sum_{k=-\infty}^{\infty} h_d(n + kN) \qquad 0 \le n \le N - 1$$

Although the frequency samples match the ideal frequency response exactly, there is no control on how the samples are *interpolated* between the samples. Because filters designed with the frequency sampling method are not generally very good, this method is often modified by introducing one or more *transition samples* as illustrated in Fig. 9-3. These transition samples are optimized in an iterative manner to maximize the stopband attenuation or minimize the passband ripple.

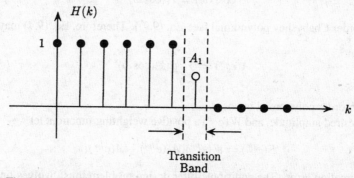

Fig. 9-3.   Introducing a transition sample with an amplitude of A_1 in the frequency sampling method.

9.3.3 Equiripple Linear Phase Filters

The design of an FIR low-pass filter using the window design technique is simple and generally results in a filter with relatively good performance. However, in two respects, these filters are not optimal:

1. First, the passband and stopband deviations, δ_p and δ_s, are approximately equal. Although it is common to require δ_s to be much smaller than δ_p, these parameters cannot be independently controlled in the window design method. Therefore, with the window design method, it is necessary to *overdesign* the filter in the passband in order to satisfy the stricter requirements in the stopband.

2. Second, for most windows, the ripple is not uniform in either the passband or the stopband and generally decreases when moving away from the transition band. Allowing the ripple to be uniformly distributed over the entire band would produce a smaller *peak ripple*.

An equiripple linear phase filter, on the other hand, is optimal in the sense that the magnitude of the ripple is minimized in all bands of interest for a given filter order, N. In the following discussion, we consider the design of a type I linear phase filter. The results may be easily modified to design other types of linear phase filters.

The frequency response of an FIR linear phase filter may be written as

$$H(e^{j\omega}) = A(e^{j\omega})e^{-j\alpha\omega} \tag{9.3}$$

where the amplitude, $A(e^{j\omega})$, is a real-valued function of ω. For a type I linear phase filter,

$$h(n) = h(N - n)$$

where N is an even integer. The symmetry of $h(n)$ allows the frequency response to be expressed as

$$A(e^{j\omega}) = \sum_{k=0}^{L} a(k)\cos(k\omega) \tag{9.4}$$

where $L = N/2$ and

$$a(0) = h\left(\frac{N}{2}\right)$$

$$a(k) = h\left(k + \frac{N}{2}\right) \qquad k = 1, 2, \ldots, \frac{N}{2}$$

The terms $\cos(k\omega)$ may be expressed as a sum of powers of $\cos\omega$ in the form

$$\cos(k\omega) = T_k(\cos\omega)$$

where $T_k(x)$ is a kth-order Chebyshev polynomial [see Eq. (9.9)]. Therefore, Eq. (9.4) may be written as

$$A(e^{j\omega}) = \sum_{k=0}^{L} \alpha(k)(\cos\omega)^k$$

Thus, $A(e^{j\omega})$ is an Lth-order polynomial in $\cos\omega$.

With $A_d(e^{j\omega})$ a desired amplitude, and $W(e^{j\omega})$ a positive weighting function, let

$$E(e^{j\omega}) = W(e^{j\omega})[A_d(e^{j\omega}) - A(e^{j\omega})]$$

be a weighted approximation error. The equiripple filter design problem thus involves finding the coefficients $a(k)$ that minimize the maximum absolute value of $E(e^{j\omega})$ over a set of frequencies, $\mathcal{F}$,

$$\min_{a(k)} \left\{ \max_{\omega \in \mathcal{F}} |E(e^{j\omega})| \right\}$$

For example, to design a low-pass filter, the set $\mathcal{F}$ will be the frequencies in the passband, $[0, \omega_p]$, and the stopband, $[\omega_s, \pi]$, as illustrated in Fig. 9-4. The transition band, (ω_p, ω_s), is a *don't care* region, and it is not

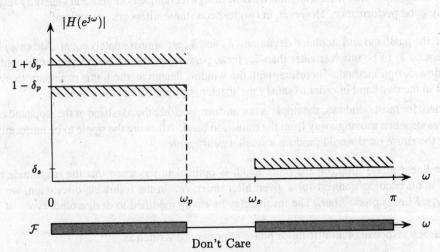

Fig. 9-4. The set R in the equiripple filter design problem, consisting of the passband $[0, \omega_p]$ and the stopband $[\omega_p, \omega_s]$. The transition band (ω_p, ω_s) is a don't care region.

considered in the minimization of the weighted error. The solution to this optimization problem is given in the *alternation theorem*, which is as follows:

Alternation Theorem: Let $\mathcal{F}$ be a union of closed subsets over the interval $[0, \pi]$. For a positive weighting function $W(e^{j\omega})$, a necessary and sufficient condition for

$$A(e^{j\omega}) = \sum_{k=0}^{L} a(k) \cos(k\omega)$$

to be the unique function that minimizes the maximum value of the weighted error $|E(e^{j\omega})|$ over the set $\mathcal{F}$ is that the $E(e^{j\omega})$ have at least $L + 2$ *alternations*. That is to say, there must be at least $L + 2$ *extremal frequencies*,

$$\omega_0 < \omega_1 < \cdots < \omega_{L+1}$$

over the set $\mathcal{F}$ such that

$$E(e^{j\omega_k}) = -E(e^{j\omega_{k+1}}) \qquad k = 0, 1, \ldots, L$$

and

$$|E(e^{j\omega_k})| = \max_{\omega \in \mathcal{F}} |E(e^{j\omega})| \qquad k = 0, 1, \ldots, L + 1$$

Thus, the alternation theorem states that the optimum filter is equiripple. Although the alternation theorem specifies the minimum number of extremal frequencies (or ripples) that the optimum filter must have, it may have more. For example, a low-pass filter may have either $L + 2$ or $L + 3$ extremal frequencies. A low-pass filter with $L + 3$ extrema is called an *extraripple filter*.

From the alternation theorem, it follows that

$$W(e^{j\omega_k})[A_d(e^{j\omega_k}) - A(e^{j\omega_k})] = (-1)^k \epsilon \qquad k = 0, 1, \ldots, L + 1$$

where

$$\epsilon = \pm \max_{\omega \in \mathcal{F}} |E(e^{j\omega})|$$

is the maximum absolute weighted error. These equations may be written in matrix form in terms of the unknowns $a(0), \ldots, a(L)$ and ϵ as follows:

$$\begin{bmatrix} 1 & \cos(\omega_0) & \cdots & \cos(L\omega_0) & 1/W(e^{j\omega_0}) \\ 1 & \cos(\omega_1) & \cdots & \cos(L\omega_1) & -1/W(e^{j\omega_1}) \\ \vdots & \vdots & \vdots & \vdots & \\ 1 & \cos(\omega_L) & \cdots & \cos(L\omega_L) & (-1)^L/W(e^{j\omega_L}) \\ 1 & \cos(\omega_{L+1}) & \cdots & \cos(L\omega_{L+1}) & (-1)^{L+1}/W(e^{j\omega_{L+1}}) \end{bmatrix} \begin{bmatrix} a(0) \\ a(1) \\ \vdots \\ a(L) \\ \epsilon \end{bmatrix} = \begin{bmatrix} A_d(e^{j\omega_0}) \\ A_d(e^{j\omega_1}) \\ \vdots \\ A_d(e^{j\omega_L}) \\ A_d(e^{j\omega_{L+1}}) \end{bmatrix} \qquad (9.5)$$

Given the extremal frequencies, these equations may be solved for $a(0), \ldots, a(L)$ and ϵ. To find the extremal frequencies, there is an efficient iterative procedure known as the Parks-McClellan algorithm, which involves the following steps:

1. Guess an initial set of extremal frequencies.
2. Find ϵ by solving Eq. (9.5). The value of ϵ has been shown to be

$$\epsilon = \frac{\displaystyle\sum_{k=0}^{L+1} b(k) D(e^{j\omega_k})}{\displaystyle\sum_{k=0}^{L+1} (-1)^k b(k)/W(e^{j\omega_k})}$$

where
$$b(k) = \prod_{i=1, i \neq k}^{L+1} \frac{1}{\cos(\omega_k) - \cos(\omega_i)}$$

3. Evaluate the weighted error function over the set $\mathcal{F}$ by interpolating between the extremal frequencies using the Lagrange interpolation formula.

4. Select a new set of extremal frequencies by choosing the $L + 2$ frequencies for which the interpolated error function is maximum.

5. If the extremal frequencies have changed, repeat the iteration from step 2.

A design formula that may be used to estimate the equiripple filter order for a low-pass filter with a transition width Δf, passband ripple δ_p, and stopband ripple δ_s is

$$N = \frac{-10 \log(\delta_p \delta_s) - 13}{14.6 \Delta f} \tag{9.6}$$

EXAMPLE 9.3.3 Suppose that we would like to design an equiripple low-pass filter with a passband cutoff frequency $\omega_p = 0.3\pi$, a stopband cutoff frequency $\omega_s = 0.35\pi$, a passband ripple of $\delta_p = 0.01$, and a stopband ripple of $\delta_s = 0.001$. Estimating the filter using Eq. (9.6), we find

$$N = \frac{-10 \log(\delta_p \delta_s) - 13}{14.6 \Delta f} = 102$$

Because we want the ripple in the stopband to be 10 times smaller than the ripple in the passband, the error must be weighted using the weighting function

$$W(e^{j\omega}) = \begin{cases} 1 & 0 \leq |\omega| \leq 0.3\pi \\ 10 & 0.35\pi \leq |\omega| \leq \pi \end{cases}$$

Using the Parks-McClellan algorithm to design the filter, we obtain a filter with the frequency response magnitude shown below.

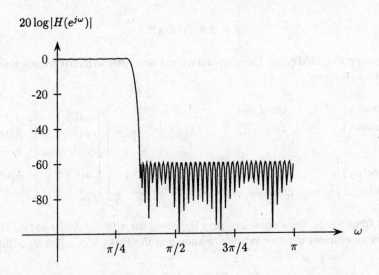

9.4 IIR FILTER DESIGN

There are two general approaches used to design IIR digital filters. The most common is to design an analog IIR filter and then map it into an equivalent digital filter because the art of analog filter design is highly advanced. Therefore, it is prudent to consider optimal ways for mapping these filters into the discrete-time domain. Furthermore, because there are powerful design procedures that facilitate the design of analog filters, this approach to IIR filter design is relatively simple. The second approach to design IIR digital filters is to use an algorithmic design procedure, which generally requires the use of a computer to solve a set of linear or nonlinear equations. These methods may be used to design digital filters with arbitrary frequency response characteristics for which no analog filter prototype exists or to design filters when other types of constraints are imposed on the design.

In this section, we consider the approach of mapping analog filters into digital filters. Initially, the focus will be on the design of digital low-pass filters from analog low-pass filters. Techniques for transforming these designs into more general frequency selective filters will then be discussed.

9.4.1 Analog Low-Pass Filter Prototypes

To design an IIR digital low-pass filter from an analog low-pass filter, we must first know how to design an analog low-pass filter. Historically, most analog filter approximation methods were developed for the design of passive systems having a gain less than or equal to 1. Therefore, a typical set of specifications for these filters is as shown in Fig. 9-5(a), with the passband specifications having the form

$$1 - \delta_p \leq |H_a(j\Omega)| \leq 1$$

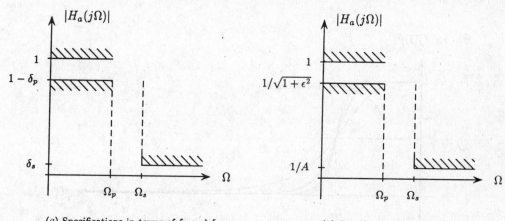

(a) Specifications in terms of δ_p and δ_s.   (b) Specifications in terms of ϵ and A.

Fig. 9-5. Two different conventions for specifying the passband and stopband deviations for an analog low-pass filter.

Another convention that is commonly used is to describe the passband and stopband constraints in terms of the parameters ϵ and A as illustrated in Fig. 9-5(b). Two auxiliary parameters of interest are the *discrimination factor*,

$$d = \left[\frac{(1 - \delta_p)^{-2} - 1}{\delta_s^{-2} - 1} \right]^{1/2} = \frac{\epsilon}{\sqrt{A^2 - 1}}$$

and the *selectivity factor*

$$k = \frac{\Omega_p}{\Omega_s}$$

The three most commonly used analog low-pass filters are the Butterworth, Chebyshev, and elliptic filters. These filters are described below.

Butterworth Filter

A low-pass Butterworth filter is an all-pole filter with a squared magnitude response given by

$$|H_a(j\Omega)|^2 = \frac{1}{1 + (j\Omega/j\Omega_c)^{2N}}$$

The parameter N is the order of the filter (number of poles in the system function), and Ω_c is the 3-dB cutoff frequency. The magnitude of the frequency response may also be written as

$$|H_a(j\Omega)|^2 = \frac{1}{1 + \epsilon^2(j\Omega/j\Omega_p)^{2N}}$$

where

$$\epsilon = \left(\frac{\Omega_p}{\Omega_c}\right)^N$$

The frequency response of the Butterworth filter decreases monotonically with increasing Ω, and as the filter order increases, the transition band becomes narrower. These properties are illustrated in Fig. 9-6, which shows $|H_a(j\Omega)|$ for Butterworth filters of orders $N = 2, 4, 8,$ and 12. Because

$$|H_a(j\Omega)|^2 = H_a(s)H_a(-s)\big|_{s=j\Omega}$$

from the magnitude-squared function, we may write

$$G_a(s) = H_a(s)H_a(-s) = \frac{1}{1 + (s/j\Omega_c)^{2N}}$$

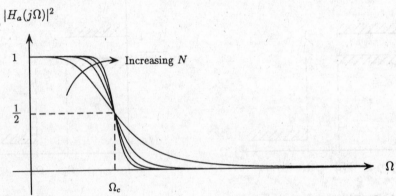

Fig. 9-6. The magnitude of the frequency response for Butterworth filters of orders $N = 2, 4, 8.$

Therefore, the poles of $G_a(s)$ are located at $2N$ equally spaced points around a circle of radius Ω_c,

$$s_k = (-1)^{1/2N}(j\Omega_c) = \Omega_c \exp\left\{ j\frac{(N+1+2k)\pi}{2N}\right\} \qquad k = 0,\ 1,\dots,\ 2N-1 \qquad (9.7)$$

and are symmetrically located about the $j\Omega$-axis. Figure 9-7 shows these pole positions for $N = 6$ and $N = 7$. The system function, $H_a(s)$, is then formed from the N roots of $H_a(s)H_a(-s)$ that lie in the left-half s-plane. For a *normalized* Butterworth filter with $\Omega_c = 1$, the system function has the form

$$H_a(s) = \frac{1}{A_N(s)} = \frac{1}{s^N + a_1 s^{N-1} + \cdots + a_{N-1}s + a_N} \qquad (9.8)$$

Table 9-4 lists the coefficients of $A_N(s)$ for $1 \le N \le 8$. Given $\Omega_p, \Omega_s, \delta_p,$ and δ_s, the steps involved in designing a Butterworth filter are as follows:

1. Find the values for the selectivity factor, k, and the discrimination factor, d, from the filter specifications.

2. Determine the order of the filter required to meet the specifications using the design formula

$$N \ge \frac{\log d}{\log k}$$

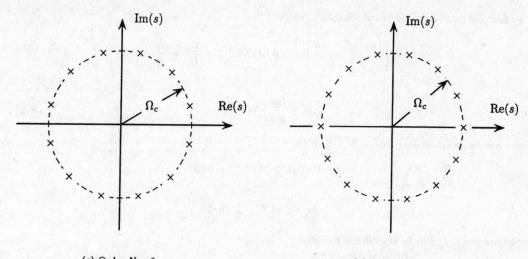

(a) Order $N = 6$.   (b) Order $N = 7$.

Fig. 9-7. The poles of $H_a(a)H_a(-s)$ for a Butterworth filter of order $N = 6$ and $N = 7$.

Table 9-4 The Coefficients in the System Function of a Normalized Butterworth Filter ($\Omega_c = 1$) for Orders $1 \leq N \leq 8$

N	a_1	a_2	a_3	a_4	a_5	a_6	a_7	a_8
1	1.0000							
2	1.4142	1.0000						
3	2.0000	2.0000	1.0000					
4	2.6131	3.4142	2.6131	1.0000				
5	3.2361	5.2361	5.2361	3.2361	1.0000			
6	3.8637	7.4641	9.1416	7.4641	3.8637	1.0000		
7	4.4940	10.0978	14.5918	14.5918	10.0978	4.4940	1.0000	
8	5.1258	13.1371	21.8462	25.6884	21.8462	13.1372	5.1258	1.0000

3. Set the 3-dB cutoff frequency, Ω_c, to any value in the range

$$\Omega_p[(1 - \delta_p)^{-2} - 1]^{-1/2N} \leq \Omega_c \leq \Omega_s[\delta_s^{-2} - 1]^{-1/2N}$$

4. Synthesize the system function of the Butterworth filter from the poles of

$$G_a(s) = H_a(s)H_a(-s) = \frac{1}{1 + (s/j\Omega_c)^{2N}}$$

that lie in the left-half s-plane. Thus,

$$H_a(s) = \prod_{k=0}^{N-1} \frac{-s_k}{s - s_k}$$

where

$$s_k = \Omega_c \exp\left\{ j\frac{(N + 1 + 2k)\pi}{2N} \right\} \qquad k = 0, 1, \ldots, N - 1$$

EXAMPLE 9.4.1   Let us design a low-pass Butterworth filter to meet the following specifications:

$$f_p = 6 \text{ kHz} \qquad f_s = 10 \text{ kHz} \qquad \delta_p = \delta_s = 0.1$$

First, we compute the discrimination and selectivity factors:

$$d = \left[\frac{(1 - \delta_p)^{-2} - 1}{\delta_s^{-2} - 1}\right]^{1/2} = 0.0487 \qquad k = \frac{\Omega_p}{\Omega_s} = \frac{f_p}{f_s} = 0.6$$

Because

$$N \geq \frac{\log d}{\log k} = 5.92$$

it follows that the minimum filter order is $N = 6$. With

$$f_p[(1 - \delta_p)^{-2} - 1]^{-1/2N} = 6770$$

and

$$f_s\left[\delta_s^{-2} - 1\right]^{-1/2N} = 6819$$

the center frequency, f_c, may be any value in the range

$$6770 \leq f_c \leq 6819$$

The system function of the Butterworth filter may then be found using Eq. (9.8) by first constructing a sixth-order normalized Butterworth filter from Table 9-4,

$$H_a(s) = \frac{1}{s^6 + 3.8637s^5 + 7.4641s^4 + 9.1416s^3 + 7.4641s^2 + 3.8637s + 1}$$

and then replacing s with s/Ω_c so that the cutoff frequency is Ω_c instead of unity (see Sec. 9.4.3).

Chebyshev Filters

Chebyshev filters are defined in terms of the Chebyshev polynomials:

$$T_N(x) = \begin{cases} \cos(N\cos^{-1} x) & |x| \leq 1 \\ \cosh(N\cosh^{-1} x) & |x| > 1 \end{cases} \qquad (9.9)$$

These polynomials may be generated recursively as follows,

$$T_{k+1}(x) = 2xT_k(x) - T_{k-1}(x) \qquad k \geq 1$$

with $T_0(x) = 1$ and $T_1(x) = x$. The following properties of the Chebyshev polynomials follow from Eq. (9.9):

1. For $|x| \leq 1$ the polynomials are bounded by 1 in magnitude, $|T_N(x)| \leq 1$, and oscillate between ± 1. For $|x| > 1$, the polynomials increase monotonically with x.
2. $T_N(1) = 1$ for all N.
3. $T_N(0) = \pm 1$ for N even, and $T_N(0) = 0$ for N odd.
4. All of the roots of $T_N(x)$ are in the interval $-1 \leq x \leq 1$.

There are two types of Chebyshev filters. A type I Chebyshev filter is all-pole with an equiripple passband and a monotonically decreasing stopband. The magnitude of the frequency response is

$$|H_a(j\Omega)|^2 = \frac{1}{1 + \epsilon^2 T_N^2(\Omega/\Omega_p)}$$

where N is the order of the filter, Ω_p is the passband cutoff frequency, and ϵ is a parameter that controls the passband ripple amplitude. Because $T_N^2(\Omega/\Omega_p)$ varies between 0 and 1 for $|\Omega| < \Omega_p$, $|H_a(j\Omega)|^2$ oscillates between 1 and $1/(1 + \epsilon^2)$. As the order of the filter increases, the number of oscillations (ripples) in the passband increases, and the transition width between the passband and stopband becomes narrower. Examples are given in Fig. 9-8 for $N = 5, 6$.

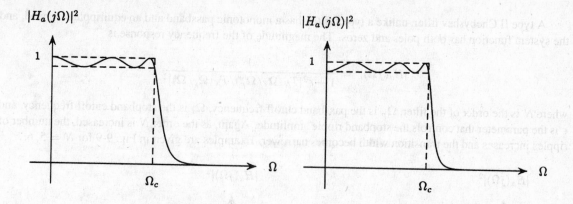

(a) Odd order ($N = 5$).   (b) Even order ($N = 6$).

Fig. 9-8. Frequency response of Chebyshev type I filter for orders $N = 5$ and $N = 6$.

The system function of a type I Chebyshev filter has the form

$$H_a(s) = H_a(0) \prod_{k=0}^{N-1} \frac{-s_k}{s - s_k}$$

where $H_a(0) = (1 - \epsilon^2)^{-1/2}$ if N is even, and $H_a(0) = 1$ if N is odd. Given the passband and stopband cutoff frequencies, Ω_p and Ω_s, and the passband and stopband ripples, δ_p and δ_s (or the parameters ϵ and A), the steps involved in designing a type I Chebyshev filter are as follows:

1. Find the values for the selectivity factor, k, and the discrimination factor, d.
2. Determine the filter order using the formula

$$N \geq \frac{\cosh^{-1}(1/d)}{\cosh^{-1}(1/k)}$$

3. Form the rational function

$$G_a(s) = H_a(s)H_a(-s) = \frac{1}{1 + \epsilon^2 T_N^2(s/j\Omega_p)}$$

where $\epsilon = [(1 - \delta_p)^{-2} - 1]^{1/2}$, and construct the system function $H_a(s)$ by taking the N poles of $G_a(s)$ that lie in the left-half s-plane.

EXAMPLE 9.4.2   If we were to design a low-pass type I Chebyshev filter to meet the specifications given in Example 9.4.1 where we found $d = 0.0487$ and $k = 0.6$, the required filter order would be

$$N \geq \frac{\cosh^{-1}(1/d)}{\cosh^{-1}(1/k)} = 3.38$$

or $N = 4$. Therefore, with

$$\epsilon = [(1 - \delta_p)^{-2} - 1]^{1/2} = 0.4843$$

and

$$T_4(x) = 4x^3 - 4x$$

then

$$|H_a(j\Omega)|^2 = \frac{1}{1 + 3.7527(\Omega/\Omega_p)^2[(\Omega/\Omega_p)^2 - 1]^2}$$

where $\Omega_p = 2\pi(6000)$.

A type II Chebyshev filter, unlike a type I filter, has a monotonic passband and an equiripple stopband, and the system function has both poles and zeros. The magnitude of the frequency response is

$$|H_a(j\Omega)|^2 = \frac{1}{1 + \epsilon^2[T_N(\Omega_s/\Omega_p)/T_N(\Omega_s/\Omega)]^2}$$

where N is the order of the filter, Ω_p is the passband cutoff frequency, Ω_s is the stopband cutoff frequency, and ϵ is the parameter that controls the stopband ripple amplitude. Again, as the order N is increased, the number of ripples increases and the transition width becomes narrower. Examples are given in Fig. 9-9 for $N = 5, 6$.

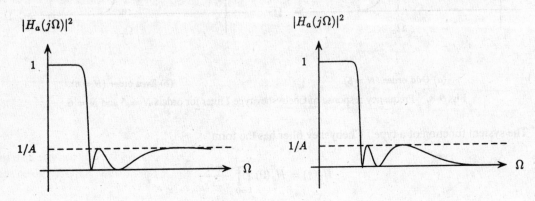

(a) Odd order ($N = 5$).   (b) Even order ($N = 6$).

Fig. 9-9.   Frequency response of a Chebyshev type II filter for orders $N = 5$ and $N = 6$.

The system function of a type II Chebyshev filter has the form

$$H_a(s) = \prod_{k=0}^{N-1} \frac{a_k}{b_k} \frac{s - b_k}{s - a_k}$$

The poles are located at

$$a_k = \frac{\Omega_s^2}{s_k}$$

where s_k for $k = 0, 1, \ldots, N - 1$ are the poles of a type I Chebyshev filter. The zeros b_k lie on the $j\Omega$-axis at the frequencies for which $T_N(\Omega_s/\Omega) = 0$. The procedure for designing a type II Chebyshev filter is the same as for a type I filter, except that

$$\epsilon = \left(\delta_s^{-2} - 1\right)^{-1/2}$$

Elliptic Filter

An elliptic filter has a system function with both poles and zeros. The magnitude of its frequency response is

$$|H_a(\Omega)|^2 = \frac{1}{1 + \epsilon^2 U_N^2(\Omega/\Omega_p)}$$

where $U_N(\Omega/\Omega_p)$ is a Jacobian elliptic function. The Jacobian elliptic function $U_N(x)$ is a rational function of order N with the following property:

$$U_N\left(\frac{1}{\Omega}\right) = \frac{1}{U_N(\Omega)}$$

Elliptic filters have an equiripple passband and an equiripple stopband. Because the ripples are distributed uniformly across both bands (unlike the Butterworth and Chebyshev filters, which have a monotonically decreasing passband and/or stopband), these filters are optimum in the sense of having the smallest transition width for a given filter order, cutoff frequency Ω_p, and passband and stopband ripples. The frequency response for a 4th-order elliptic filter is shown in Fig. 9-10.

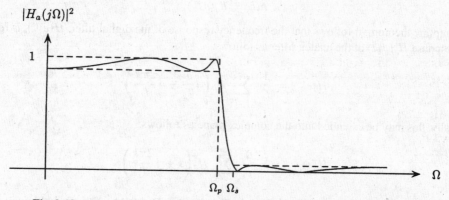

Fig. 9-10.   The magnitude of the frequency response of a sixth-order elliptic filter.

The design of elliptic filters is more difficult than the Butterworth and Chebyshev filters, because their design relies on the use of tables or series expansions. However, the filter order necessary to meet a given set of specifications may be estimated using the formula

$$N \geq \frac{\log(16/d^2)}{\log(1/q)}$$

where d is the discrimination factor, and

$$q = q_0 + 2q_0^5 + 15q_0^9 + 150q_0^{13}$$

where

$$q_0 = \frac{1}{2} \frac{1 - (1 - k^2)^{1/4}}{1 + (1 - k^2)^{1/4}}$$

with k being the selectivity factor.

9.4.2   Design of IIR Filters from Analog Filters

The design of a digital filter from an analog prototype requires that we transform $h_a(t)$ to $h(n)$ or $H_a(s)$ to $H(z)$. A mapping from the s-plane to the z-plane may be written as

$$H(z) = H_a(s)\big|_{s=m(z)}$$

where $s = m(z)$ is the mapping function. In order for this transformation to produce an acceptable digital filter, the mapping $m(z)$ should have the following properties:

1. The mapping from the $j\Omega$-axis to the unit circle, $|z| = 1$, should be one to one and *onto* the unit circle in order to preserve the frequency response characteristics of the analog filter.

2. Points in the left-half s-plane should map to points *inside* the unit circle to preserve the stability of the analog filter.

3. The mapping $m(z)$ should be a rational function of z so that a rational $H_a(s)$ is mapped to a rational $H(z)$.

Described below are two approaches that are commonly used to map analog filters into digital filters.

Impulse Invariance

With the *impulse invariance* method, a digital filter is designed by sampling the impulse response of an analog filter:

$$h(n) = h_a(nT_s)$$

From the sampling theorem, it follows that the frequency response of the digital filter, $H(e^{j\omega})$, is related to the frequency response $H_a(j\Omega)$ of the analog filter as follows:

$$H(e^{j\omega}) = \frac{1}{T_s} \sum_{k=-\infty}^{\infty} H_a\left(j\frac{\omega}{T_s} + j\frac{2\pi k}{T_s}\right)$$

More generally, this may be extended into the complex plane as follows:

$$H(z)\big|_{z=e^{sT_s}} = \frac{1}{T_s} \sum_{k=-\infty}^{\infty} H_a\left(s + j\frac{2\pi k}{T_s}\right)$$

The mapping between the s-plane and the z-plane is illustrated in Fig. 9-11. Note that although the $j\Omega$-axis maps *onto* the unit circle, the mapping is not one to one. In particular, each interval of length $2\pi/T_s$ along the $j\Omega$-axis is mapped onto the unit circle (i.e., the frequency response is aliased). In addition, each point in the left-half s-plane is mapped to a point *inside* the unit circle. Specifically, strips of width $2\pi/T_s$ map *onto* the z-plane. If the frequency response of the analog filter, $H_a(j\Omega)$, is sufficiently bandlimited, then

$$H(e^{j\omega}) \approx \frac{1}{T_s} H_a\left(\frac{j\omega}{T_s}\right)$$

Although the impulse invariance may produce a reasonable design in some cases, this technique is essentially limited to bandlimited analog filters.

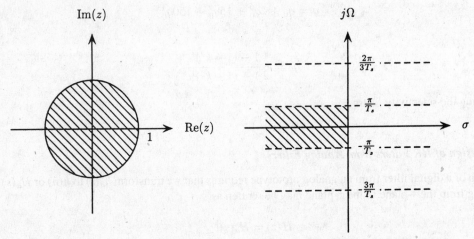

Fig. 9-11.   Properties of the s-plane to z-plane mapping in the impulse invariance method.

To see how poles and zeros of an analog filter are mapped using the impulse invariance method, consider an analog filter that has a system function

$$H_a(s) = \sum_{k=1}^{p} \frac{A_k}{s - s_k}$$

The impulse response, $h_a(t)$, is

$$h_a(t) = \sum_{k=1}^{p} A_k e^{s_k t} u(t)$$

Therefore, the digital filter that is formed using the impulse invariance technique is

$$h(n) = h_a(nT_s) = \sum_{k=1}^{p} A_k e^{s_k n T_s} u(nT_s) = \sum_{k=1}^{p} A_k (e^{s_k T_s})^n u(n)$$

and the system function is

$$H(z) = \sum_{k=1}^{p} \frac{A_k}{1 - e^{s_k T_s} z^{-1}} \qquad (9.10)$$

Thus, a pole at $s = s_k$ in the analog filter is mapped to a pole at $z = e^{s_k T_s}$ in the digital filter,

$$\frac{1}{s - s_k} \implies \frac{1}{1 - e^{s_k T_s} z^{-1}}$$

The zeros, however, do not get mapped in any obvious way.

The Bilinear Transformation

The bilinear transformation is a mapping from the s-plane to the z-plane defined by

$$s = \frac{2}{T_s} \frac{1 - z^{-1}}{1 + z^{-1}} \qquad (9.11)$$

Given an analog filter with a system function $H_a(s)$, the digital filter is designed as follows:

$$H(z) = H_a\left(\frac{2}{T_s} \frac{1 - z^{-1}}{1 + z^{-1}} \right)$$

The bilinear transformation is a rational function that maps the left-half s-plane *inside* the unit circle and maps the $j\Omega$-axis in a one-to-one manner *onto* the unit circle. However, the relationship between the $j\Omega$-axis and the unit circle is highly nonlinear and is given by the *frequency warping function*

$$\omega = 2 \arctan\left(\frac{\Omega T_s}{2} \right) \qquad (9.12)$$

As a result of this warping, the bilinear transformation will only preserve the magnitude response of analog filters that have an ideal response that is piecewise constant. Therefore, the bilinear transformation is generally only used in the design of frequency selective filters.

The parameter T_s in the bilinear transformation is normally included for historical reasons. However, it does not enter into the design process, because it only scales the $j\Omega$-axis in the frequency warping function, and this scaling may be done in the specification of the analog filter. Therefore, T_s may be set to any value to simplify the design procedure. The steps involved in the design of a digital low-pass filter with a passband cutoff frequency ω_p, stopband cutoff frequency ω_s, passband ripple δ_p, and stopband ripple δ_s are as follows:

1. *Prewarp* the passband and stopband cutoff frequencies of the digital filter, ω_p and ω_s, using the inverse of Eq. (9.12) to determine the passband and cutoff frequencies of the analog low-pass filter. With $T_s = 2$, the prewarping function is

$$\Omega = \tan\left(\frac{\omega}{2} \right)$$

2. Design an analog low-pass filter with the cutoff frequencies found in step 1 and a passband and stopband ripple δ_p and δ_s, respectively.

3. Apply the bilinear transformation to the filter designed in step 2.

EXAMPLE 9.4.3   Let us design a first-order digital low-pass filter with a 3-dB cutoff frequency of $\omega_c = 0.25\pi$ by applying the bilinear transformation to the analog Butterworth filter

$$H_a(s) = \frac{1}{1 + s/\Omega_c}$$

Because the 3-dB cutoff frequency of the Butterworth filter is Ω_c, for a cutoff frequency $\omega_c = 0.25\pi$ in the digital filter, we must have

$$\Omega_c = \frac{2}{T_s} \tan\left(\frac{0.25\pi}{2}\right) = \frac{0.828}{T_s}$$

Therefore, the system function of the analog filter is

$$H_a(s) = \frac{1}{1 + sT_s/0.828}$$

Applying the bilinear transformation to the analog filter gives

$$H(z) = H_a(s)\big|_{s = \frac{2}{T_s}\frac{1-z^{-1}}{1+z^{-1}}} = \frac{1}{1 + (2/0.828)[(1 - z^{-1})/(1 + z^{-1})]} = 0.2920\frac{1 + z^{-1}}{1 - 0.4159z^{-1}}$$

Note that the parameter T_s does not enter into the design.

9.4.3 Frequency Transformations

The preceding section considered the design of digital low-pass filters from analog low-pass filters. There are two approaches that may be used to design other types of frequency selective filters, such as high-pass, bandpass, or bandstop filters. The first is to design an analog low-pass filter and then apply a frequency transformation to map the analog filter into the desired frequency selective prototype. This analog prototype is then mapped to a digital filter using a suitable s-plane to z-plane mapping. Table 9-5 provides a list of some analog-to-analog transformations.

Table 9-5   The Transformation of an Analog Low-pass Filter with a 3-dB Cutoff Frequency Ω_p to Other Frequency Selective Filters

Transformation	Mapping	New Cutoff Frequencies
Low-pass	$s \to \dfrac{\Omega_p}{\Omega'_p}s$	Ω'_p
High-pass	$s \to \dfrac{\Omega_p\Omega'_p}{s}$	Ω'_p
Bandpass	$s \to \Omega_p\dfrac{s^2 + \Omega_l\Omega_u}{s(\Omega_u - \Omega_l)}$	Ω_l, Ω_u
Bandstop	$s \to \Omega_p\dfrac{s(\Omega_u - \Omega_l)}{s^2 + \Omega_l\Omega_u}$	Ω_l, Ω_u

The second approach that may be used is to design an analog low-pass filter, map it into a digital filter using a suitable s-plane to z-plane mapping, and then apply an appropriate frequency transformation in the discrete-time domain to produce the desired frequency selective digital filter. Table 9-6 provides a list of some digital-to-digital transformations. The two approaches do not always result in the same design. For example, although the second approach could be used to design a high-pass filter using the impulse invariance technique, with the first approach the design would be unacceptable due to the aliasing that would occur when sampling the analog high-pass filter.

9.5 FILTER DESIGN BASED ON A LEAST SQUARES APPROACH

The design techniques described in the previous section are based on converting an analog filter into a digital filter. It is also possible to perform the design directly in the time domain without any reference to an analog filter. This section describes several methods for designing a digital filter directly.

Table 9-6   The Transformation of a Digital Low-Pass Filter with a Cutoff Frequency ω_c to Other Frequency Selective Filters

Filter Type	Mapping	Design Parameters
Low-pass	$z^{-1} \rightarrow \dfrac{z^{-1} - \alpha}{1 - \alpha z^{-1}}$	$\alpha = \dfrac{\sin[(\omega_c - \omega_c')/2]}{\sin[(\omega_c + \omega_c')/2]}$ $\omega_c' = $ desired cutoff frequency
High-pass	$z^{-1} \rightarrow -\dfrac{z^{-1} + \alpha}{1 + \alpha z^{-1}}$	$\alpha = -\dfrac{\cos[(\omega_c + \omega_c')/2]}{\cos[(\omega_c - \omega_c')/2]}$ $\omega_c' = $ desired cutoff frequency
Bandpass	$z^{-1} \rightarrow -\dfrac{z^{-2} - [2\alpha\beta/(\beta + 1)]z^{-1} + [(\beta - 1)/(\beta + 1)]}{[(\beta - 1)/(\beta + 1)]z^{-2} - [2\alpha\beta/(\beta + 1)]z^{-1} + 1}$	$\alpha = \dfrac{\cos[(\omega_{c2} + \omega_{c1})/2]}{\cos[(\omega_{c2} - \omega_{c1})/2]}$ $\beta = \cot[(\omega_{c2} - \omega_{c1})/2]\tan(\omega_c/2)$ $\omega_{c1} = $ desired lower cutoff frequency $\omega_{c2} = $ desired upper cutoff frequency
Bandstop	$z^{-1} \rightarrow \dfrac{z^{-2} - [2\alpha/(\beta + 1)]z^{-1} + [(1 - \beta)/(1 + \beta)]}{[(1 - \beta)/(1 + \beta)]z^{-2} - [2\alpha/(\beta + 1)]z^{-1} + 1}$	$\alpha = \dfrac{\cos[(\omega_{c1} + \omega_{c2})/2]}{\cos[(\omega_{c1} - \omega_{c2})/2]}$ $\beta = \tan[(\omega_{c2} - \omega_{c1})/2]\tan(\omega_c/2)$ $\omega_{c1} = $ desired lower cutoff frequency $\omega_{c2} = $ desired upper cutoff frequency

9.5.1   Padé Approximation

Let $h_d(n)$ be the unit sample response of an ideal filter that is to be approximated by a causal filter that has a unit sample response, $h(n)$, and a rational system function,

$$H(z) = \sum_{n=0}^{\infty} h(n)z^{-n} = \frac{\displaystyle\sum_{k=0}^{q} b(k)z^{-k}}{1 + \displaystyle\sum_{k=1}^{p} a(k)z^{-k}} \qquad (9.13)$$

Because $H(z)$ has $p + q + 1$ free parameters, it is generally possible to find values for the coefficients $a(k)$ and $b(k)$ so that $h(n) = h_d(n)$ for $n = 0, 1, \ldots, p + q$. The procedure that is used to find these coefficients is to write $H(z) = B(z)/A(z)$ as follows,

$$A(z)H(z) = B(z)$$

and note that, in the time domain, the left-hand side corresponds to a convolution

$$a(n) * h(n) = h(n) + \sum_{k=1}^{p} a(k)h(n - k) = b(n)$$

(note that $b(n)$ is a finite-length sequence that is equal to zero for $n < 0$ and $n > q$). Setting $h(n) = h_d(n)$ for $n = 0, 1, \ldots, p + q$ results in a set of $p + q + 1$ linear equations in $p + q + 1$ unknowns,

$$h_d(n) + \sum_{k=1}^{p} a(k)h_d(n - k) = \begin{cases} b(n) & n = 0, 1, \ldots, q \\ 0 & n = q + 1, \ldots, q + p \end{cases} \qquad (9.14)$$

that may be solved using a two-step approach. In the first step, the coefficients $a(k)$ are found using the last p equations in Eq. (9.14), which may be written in matrix form as

$$
\begin{bmatrix}
h_d(q) & h_d(q-1) & \cdots & h_d(q-p+1) \\
h_d(q+1) & h_d(q) & \cdots & h_d(q-p+2) \\
\vdots & \vdots & & \vdots \\
h_d(q+p-1) & h_d(q+p-2) & \cdots & h_d(q)
\end{bmatrix}
\begin{bmatrix}
a(1) \\
a(2) \\
\vdots \\
a(p)
\end{bmatrix}
= -
\begin{bmatrix}
h_d(q+1) \\
h_d(q+2) \\
\vdots \\
h_d(q+p)
\end{bmatrix}
$$

Assuming that these equations are linearly independent, the coefficients may be uniquely determined. In the second step, the coefficients $b(k)$ are found from the first $q+1$ equations in Eq. (9.14) as follows:

$$
b(n) = h_d(n) + \sum_{k=1}^{p} a(k)h_d(n-k) \qquad n = 0, 1, \ldots, q
$$

Although Padé's method produces an exact match of $h(n)$ to $h_d(n)$ for $n = 0, 1, \ldots, p+q$, because $h(n)$ is unconstrained for $n > p+q$, the Padé method does not generally produce a good approximation to $h_d(n)$ for $n > p+q$.

9.5.2 Prony's Method

With a least-squares approach to filter design, the problem is to find the coefficients $a(k)$ and $b(k)$ that minimize the least-squares error

$$
\mathcal{E} = \sum_{n=0}^{U} |h_d(n) - h(n)|^2 \tag{9.15}
$$

where U is some preselected upper limit. Because $\mathcal{E}$ is a nonlinear function of the coefficients $a(k)$ and $b(k)$, solving this minimization problem is, in general, difficult. With Prony's method, however, an approximate least-squares solution may be found using a two-step procedure as follows. Ideally, because [see Eq. (9.14)]

$$
h_d(n) + \sum_{k=1}^{p} a(k)h_d(n-k) = 0 \qquad n \geq q+1
$$

the first step is to find the coefficients $a(k)$ that minimize

$$
\mathcal{E} = \sum_{n=q+1}^{\infty} e^2(n)
$$

where

$$
e(n) = h_d(n) + \sum_{k=1}^{p} a(k)h_d(n-k)
$$

Once the coefficients $a(k)$ have been determined, the coefficients $b(k)$ are found using the Padé approach of forcing $h(n) = h_d(n)$ for $n = 0, 1, \ldots, q$:

$$
b(n) = \sum_{k=1}^{p} a(k)h_d(n-k) \qquad 0 \leq n \leq q
$$

The coefficients $a(k)$ that minimize $\mathcal{E}$ may be found by setting the partial derivatives of $\mathcal{E}$ equal to zero,

$$
\frac{\partial \mathcal{E}}{\partial a(k)} = 0 \qquad k = 0, 1, \ldots, p
$$

and solving for the unknowns $a(k)$. Setting the derivatives equal to zero produces the following set of linear equations:

$$\begin{bmatrix} r_d(1,1) & r_d(1,2) & \cdots & r_d(1,p) \\ r_d(2,1) & r_d(2,2) & \cdots & r_d(2,p) \\ \vdots & \vdots & & \vdots \\ r_d(p,1) & r_d(p,2) & \cdots & r_d(p,p) \end{bmatrix} \begin{bmatrix} a(1) \\ a(2) \\ \vdots \\ a(p) \end{bmatrix} = - \begin{bmatrix} r_d(1,0) \\ r_d(2,0) \\ \vdots \\ r_d(p,0) \end{bmatrix} \qquad (9.16)$$

where

$$r_d(k,l) = \sum_{n=q+1}^{\infty} h_d(n-l)h_d(n-k)$$

is the correlation of $h_d(n)$.

9.5.3 FIR Least-Squares Inverse

The inverse of a linear shift-invariant system with unit sample response $g(n)$ and system function $G(z)$ is the system that has a unit sample response, $h(n)$, such that

$$h(n) * g(n) = \delta(n)$$

or

$$H(z)G(z) = 1$$

In most applications, the system function $H(z) = 1/G(z)$ is not a viable solution. One of the reasons is that, unless $G(z)$ is minimum phase, $1/G(z)$ cannot be both causal and stable. Another consideration comes from the fact that, in some applications, it may be necessary to constrain $H(z)$ to be an FIR filter. Because $1/G(z)$ will be infinite in length unless $G(z)$ is an all-pole filter, constraining $h(n)$ to be FIR would only be an approximation to the inverse filter.

In the FIR least-squares inverse filter design problem, the goal is to find the FIR filter $h(n)$ of length N such that

$$h(n) * g(n) \approx \delta(n)$$

The filter that minimizes the squared error

$$\mathcal{E} = \sum_{n=0}^{\infty} |e(n)|^2$$

where

$$e(n) = \delta(n) - h(n) * g(n) = \delta(n) - \sum_{l=0}^{N-1} h(l)g(n-l) \qquad (9.17)$$

may be found by solving the linear equations

$$\sum_{l=0}^{N-1} h(l)r_g(k-l) = \begin{cases} g(0) & k = 0 \\ 0 & k = 1,2,\ldots,N-1 \end{cases} \qquad (9.18)$$

where

$$r_g(k) = \sum_{n=0}^{\infty} g(n)g(n-k)$$

In many cases, constraining the least-squares inverse filter to minimize the difference between $h(n) * g(n)$ and $\delta(n)$ is overly restrictive. For example, if a delay may be tolerated, we may consider finding the filter $h(n)$ so that

$$h(n) * g(n) \approx \delta(n - n_0)$$

for some delay n_0. In most cases, a nonzero delay will produce a better approximate inverse filter and, in many cases, the improvement will be substantial. The least-squares inverse filter with delay is found by solving the linear equations

$$\sum_{l=0}^{N-1} h(l)r_g(k-l) = \begin{cases} g(n_0 - k) & k = 0,1,\ldots,n_0 \\ 0 & k = n_0+1,\ldots,N \end{cases} \qquad (9.19)$$

Solved Problems

FIR Filter Design

9.1   Use the window design method to design a linear phase FIR filter of order $N = 24$ to approximate the following ideal frequency response magnitude:

$$|H_d(e^{j\omega})| = \begin{cases} 1 & |\omega| \le 0.2\pi \\ 0 & 0.2\pi < |\omega| \le \pi \end{cases}$$

The ideal filter that we would like to approximate is a low-pass filter with a cutoff frequency $\omega_p = 0.2\pi$. With $N = 24$, the frequency response of the filter that is to be designed has the form

$$H(e^{j\omega}) = \sum_{n=0}^{24} h(n)e^{-jn\omega}$$

Therefore, the delay of $h(n)$ is $\alpha = N/2 = 12$, and the ideal unit sample response that is to be windowed is

$$h_d(n) = \frac{\sin[0.2\pi(n-12)]}{(n-12)\pi}$$

All that is left to do in the design is to select a window. With the length of the window fixed, there is a trade-off between the width of the transition band and the amplitude of the passband and stopband ripple. With a rectangular window, which provides the smallest transition band,

$$\Delta\omega = 2\pi \cdot \frac{0.9}{24} = 0.075\pi$$

and the filter is

$$h(n) = \begin{cases} \dfrac{\sin[0.2\pi(n-12)]}{(n-12)\pi} & 0 \le n \le 24 \\ 0 & \text{otherwise} \end{cases}$$

However, the stopband attenuation is only 21 dB, which is equivalent to a ripple of $\delta_s = 0.089$. With a Hamming window, on the other hand,

$$h(n) = \left[0.54 - 0.46\cos\left(\frac{2\pi n}{24}\right)\right] \cdot \frac{\sin[0.2\pi(n-12)]}{(n-12)\pi} \qquad 0 \le n \le 24$$

and the stopband attenuation is 53 dB, or $\delta_s = 0.0022$. However, the width of the transition band increases to

$$\Delta\omega = 2\pi \cdot \frac{3.3}{24} = 0.275\pi$$

which, for most designs, would be too wide.

9.2   Use the window design method to design a minimum-order high-pass filter with a stopband cutoff frequency $\omega_s = 0.22\pi$, a passband cutoff frequency $\omega_p = 0.28\pi$, and a stopband ripple $\delta_s = 0.003$.

A stopband ripple of $\delta_s = 0.003$ corresponds to a stopband attenuation of $\alpha_s = -20\log\delta_s = 50.46$. For the minimum-order filter, we use a Kaiser window with

$$\beta = 0.1102(\alpha_s - 8.7) = 4.6$$

Because the transition width is $\Delta\omega = 0.06\pi$, or $\Delta f = 0.03$, the required window length is

$$N = \frac{\alpha_s - 7.95}{14.36\Delta f} = 98.67$$

Rounding this up to $N = 99$ results in a type II linear phase filter, which will have a zero in its system function at $z = -1$. Because this produces a null in the frequency response at $\omega = \pi$, this is not acceptable. Therefore, we increase the order by 1 to obtain a type I linear phase filter with $N = 100$.

In order to have a transition band that extends from $\omega_s = 0.22\pi$ to $\omega_p = 0.28\pi$, we set the cutoff frequency of the ideal high-pass filter equal to the midpoint:

$$\omega_c = \frac{\omega_p + \omega_s}{2} = 0.25\pi$$

The unit sample response of an ideal zero-phase high-pass filter with a cutoff frequency $\omega_c = 0.25\pi$ is

$$h_{\text{hp}}(n) = \delta(n) - \frac{\sin(0.25\pi n)}{n\pi}$$

where the second term is a low-pass filter with a cutoff frequency $\omega_c = 0.25\pi$. Delaying $h_{\text{hp}}(n)$ by $N/2 = 50$, we have

$$h_d(n) = \delta(n - 50) - \frac{\sin[0.25\pi(n - 50)]}{(n - 50)\pi}$$

and the resulting FIR high-pass filter is

$$h(n) = h_d(n) \cdot w(n)$$

where $w(n)$ is a Kaiser window with $N = 100$ and $\beta = 4.6$.

9.3 Given a desired frequency response $H_d(e^{j\omega})$, show that the rectangular window design minimizes the least-squares error

$$\mathcal{E}_{LS} = \frac{1}{2\pi} \int_{-\pi}^{\pi} |H_d(e^{j\omega}) - H(e^{j\omega})|^2 d\omega$$

For this problem, we use Parseval's theorem to express the least-squares error $\mathcal{E}_{LS}$ in the time domain:

$$\mathcal{E}_{LS} = \sum_{n=-\infty}^{\infty} |h_d(n) - h(n)|^2$$

If we assume that $h(n)$ is of order N, with $h(n) = 0$ for $n < 0$ and $n > N$,

$$\mathcal{E}_{LS} = \sum_{n=0}^{N} |h_d(n) - h(n)|^2 + \sum_{n=-\infty}^{-1} |h_d(n)|^2 + \sum_{n=N+1}^{\infty} |h_d(n)|^2$$

Because the last two terms are constants that are not affected by the filter $h(n)$, the least-squares error $\mathcal{E}_{LS}$ is minimized by minimizing the first term, which is done by setting $h(n) = h_d(n)$ for $n = 0, 1, \ldots, N$ (i.e., using a rectangular window in the window design method).

9.4 If $h_d(n)$ is the unit sample response of an ideal filter, and $h(n)$ is an Nth-order FIR filter, the least-squares error

$$\mathcal{E}_{LS} = \frac{1}{2\pi} \int_{-\pi}^{\pi} |H_d(e^{j\omega}) - H(e^{j\omega})|^2 d\omega$$

is minimized when $h(n)$ is designed using the rectangular window design method. If $\mathcal{E}_R$ is the squared error using a rectangular window, find the excess squared error that results when a Hanning window is used instead of a rectangular window; that is, find an expression for

$$\mathcal{E}_{\text{ex}} = \mathcal{E}_H - \mathcal{E}_R$$

where $\mathcal{E}_H$ is the squared error using a Hanning window.

Using Parseval's theorem, it is more convenient to express the least-squares error in the time domain as follows:

$$\mathcal{E}_{LS} = \sum_{n=-\infty}^{\infty} |h_d(n) - h(n)|^2$$

Because $e(n) = h_d(n) - h(n) = h_d(n)$ for $n < 0$ and $n > N$,

$$\mathcal{E}_{\text{ex}} = \mathcal{E}_H - \mathcal{E}_R = \sum_{n=0}^{N} |h_d(n) - w_H(n)h_d(n)|^2 - \sum_{n=0}^{N} |h_d(n) - w_R(n)h_d(n)|^2$$

where $w_H(n)$ and $w_R(n)$ are the Hanning and rectangular windows, respectively. However, the second sum is equal to zero. Therefore, the excess squared error is simply

$$\mathcal{E}_{\text{ex}} = \sum_{n=0}^{N} |h_d(n) - w_H(n)h_d(n)|^2 = \sum_{n=0}^{N} |1 - w_H(n)|^2 |h_d(n)|^2$$

$$= \sum_{n=0}^{N} \left| 0.5 + 0.5 \cos\left(\frac{2\pi n}{N}\right) \right|^2 \cdot |h_d(n)|^2$$

which is the desired relationship.

9.5   Consider the following specifications for a low-pass filter:

$$0.99 \leq |H(e^{j\omega})| \leq 1.01 \qquad 0 \leq |\omega| \leq 0.3\pi$$
$$|H(e^{j\omega})| \leq 0.01 \qquad 0.35\pi \leq |\omega| \leq \pi$$

Design a linear phase FIR filter to meet these specifications using the window design method.

Designing a low-pass filter with the window design method generally produces a filter with ripples of the same amplitude in the passband and stopband. Therefore, because the passband and stopband ripples in the filter specifications are the same, we only need to be concerned about the stopband ripple requirement. A stopband ripple of $\delta_s = 0.01$ corresponds to a stopband attenuation of -40 dB. Therefore, from Table 9-2 it follows that we may use a Hanning window, which provides an attenuation of approximately 44 dB. The specification on the transition band is that $\Delta\omega = 0.05\pi$, or $\Delta f = 0.025$. Therefore, the required filter order is

$$N = \frac{3.1}{\Delta f} = 124$$

and we have

$$w(n) = 0.5 - 0.5 \cos\left(\frac{2\pi n}{124}\right) \qquad 0 \leq n \leq 124$$

With an ideal low-pass filter that has a cutoff frequency of $\omega_c = 0.325$ (the midpoint of the transition band), and a delay of $N/2 = 62$ so that $h_d(n)$ is placed symmetrically within the interval $[0, 124]$, we have

$$h_d(n) = \frac{\sin[0.325\pi(n - 62)]}{\pi(n - 62)}$$

Therefore, the filter is

$$h(n) = \left[0.5 - 0.5 \cos\left(\frac{2\pi n}{124}\right)\right] \cdot \frac{\sin[0.325\pi(n - 62)]}{\pi(n - 62)} \qquad 0 \leq n \leq 124$$

Note that if we were to use a Hamming or a Blackman window instead of a Hanning window, the stopband and passband ripple requirements would have been exceeded, and the required filter order would have been larger. With a Blackman window, for example, the filter order required to meet the transition band requirement is

$$N = \frac{5.5}{0.025} = 220$$

9.6   We would like to filter an analog signal $x_a(t)$ with an analog low-pass filter that has a cutoff frequency $f_c = 2$ kHz, a transition width $\Delta f = 500$ Hz, and a stopband attenuation of 50 dB. This filter is to be implemented digitally, as illustrated in the following figure:

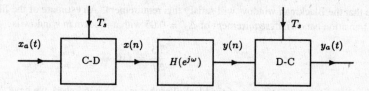

Design a digital filter to meet the analog filter specifications with a sampling frequency $f_s = 10$ kHz.

With a sampling frequency of 10 kHz, the digital filter should have a cutoff frequency $\omega_c = 2\pi f_c/f_s = 0.4\pi$ and a transition bandwidth $\Delta\omega = 2\pi\Delta f/f_s = 0.1\pi$. For a stopband attenuation of 50 dB, we may use a Kaiser window with

$$\beta = 0.1102(50 - 8.7) = 4.55$$

For the length of the window, we have

$$N = \frac{-20\log(\delta_s) - 7.95}{14.36 \cdot \Delta f} = \frac{50 - 7.95}{14.36 \cdot (0.05)} = 58.56$$

or $N = 59$. Finally, the unit sample response of the ideal filter that is to be windowed is a low-pass filter with a cutoff frequency $\omega_c = 0.4\pi$ and a delay $N/2 = 29.5$. Therefore,

$$h(n) = w(n)h_d(n)$$

where $w(n)$ is a Kaiser window with $N = 59$ and $\beta = 4.55$, and

$$h_d(n) = \frac{\sin[0.4\pi(n - 29.5)]}{(n - 29.5)\pi}$$

9.7 Find the Kaiser window parameters, β and N, to design a low-pass filter with a cutoff frequency $\omega_c = \pi/2$, a stopband ripple $\delta_s = 0.002$, and a transition bandwidth no larger than 0.1π.

The parameter β for the Kaiser window depends only on the stopband ripple requirements. With $\delta_s = 0.002$,

$$\alpha_s = -20\log(0.002) = 53.98$$

and we have

$$\beta = 0.1102(\alpha_s - 8.7) = 4.99$$

The window length, N, on the other hand, is determined by the stopband ripple, δ_s, and the transition width as follows:

$$N = \frac{\alpha_s - 7.95}{14.36\Delta f} = 64.1$$

Therefore, the required filter order is $N = 65$.

9.8 Consider the following specifications for a bandpass filter:

$$|H(e^{j\omega})| \leq 0.01 \qquad 0 \leq |\omega| \leq 0.2\pi$$
$$0.95 \leq |H(e^{j\omega})| \leq 1.05 \qquad 0.3\pi \leq |\omega| \leq 0.7\pi$$
$$|H(e^{j\omega})| \leq 0.02 \qquad 0.8\pi \leq |\omega| \leq \pi$$

(a) Design a linear phase FIR filter to meet these specifications using a Blackman window.

(b) Repeat part (a) using a Kaiser window.

(a) For this filter, the width of each transition band is $\Delta\omega = 0.1\pi$. The ripples in the lower stopband, passband, and upper stopband are $\delta_1 = 0.01$, $\delta_2 = 0.05$, and $\delta_3 = 0.02$, respectively, and are all different. Because the ripples produced with the window design method will be approximately the same in all three bands, the filter must be designed so that it has a maximum ripple of $\delta_1 = 0.01$ in all three bands. With

$$-20\log\delta_1 = -40$$

it follows that the Blackman window will satisfy this requirement. An estimate of the filter order necessary to meet the transition bandwidth requirement of $\Delta f = 0.05$ with a Blackman window is

$$N = \frac{5.5}{\Delta f} = 110$$

Finally, for the unit sample response of the ideal filter that is to be windowed, we have

$$h_d(n) = \frac{1}{2\pi} \int_{-\pi}^{\pi} H_d(e^{j\omega}) e^{jn\omega} d\omega$$

where $H_d(e^{j\omega})$ is the frequency response of an ideal bandpass filter. For the cutoff frequencies of $H_d(e^{j\omega})$, we choose the midpoints of the transition bands of $H(e^{j\omega})$. Therefore,

$$|H_d(e^{j\omega})| = \begin{cases} 1 & 0.25\pi \leq |\omega| \leq 0.75\pi \\ 0 & \text{otherwise} \end{cases}$$

Thus, the unit sample response of the ideal bandpass filter with zero phase is

$$h_d(n) = \frac{1}{2\pi} \int_{-\pi}^{\pi} H_d(e^{j\omega}) e^{jn\omega} d\omega$$

$$= \frac{1}{2\pi} \int_{-0.75\pi}^{-0.25\pi} e^{jn\omega} d\omega + \frac{1}{2\pi} \int_{0.25\pi}^{0.75\pi} e^{jn\omega} d\omega$$

and

$$h_d(n) = \frac{1}{2jn\pi} e^{jn\omega} \Big|_{-0.75\pi}^{-0.25\pi} + \frac{1}{2jn\pi} e^{jn\omega} \Big|_{0.25\pi}^{0.75\pi}$$

$$= \frac{1}{2jn\pi} [e^{-j0.25\pi n} - e^{-j0.75\pi n} + e^{j0.75\pi n} - e^{j0.25\pi n}]$$

$$= \frac{\sin(0.75\pi n)}{n\pi} - \frac{\sin(0.25\pi n)}{n\pi}$$

However, we want to delay this filter so that it is centered at $N/2 = 55$. Therefore, the unit sample response of the filter that is to be windowed should be

$$h_d(n) = \frac{\sin[0.75\pi(n - 55)]}{(n - 55)\pi} - \frac{\sin[0.25\pi(n - 55)]}{(n - 55)\pi}$$

(b) For a Kaiser window design, the order of the filter that is required is

$$N = \frac{-20 \log(0.01) - 7.95}{14.36(0.05)} = 44.64$$

Therefore, we set $N = 45$. Next, for the Kaiser window parameter, with an attenuation of 40 dB, we have

$$\beta = 0.5842(40 - 21)^{0.4} + 0.07886(40 - 21) = 3.3953$$

Therefore, the filter is

$$h(n) = w(n) \cdot h_d(n)$$

where

$$h_d(n) = \frac{\sin[0.75\pi(n - 22.5)]}{(n - 22.5)\pi} - \frac{\sin[0.25\pi(n - 22.5)]}{(n - 22.5)\pi}$$

9.9   Suppose that we would like to design a bandstop filter to meet the following specifications:

$$0.95 \leq |H(e^{j\omega})| \leq 1.05 \qquad 0 \leq |\omega| \leq 0.2\pi$$

$$|H(e^{j\omega})| \leq 0.005 \qquad 0.22\pi \leq |\omega| \leq 0.75\pi$$

$$0.95 \leq |H(e^{j\omega})| \leq 1.05 \qquad 0.8\pi \leq |\omega| \leq \pi$$

(a) Design a linear phase FIR filter to meet these filter specifications using the window design method.

(b) What is the approximate order of the equiripple filter that will meet these specifications?

(a) Recall that with the window design method, the ripples in the passbands and stopbands will be approximately the same, along with the widths of the transition bands. Because the smallest ripple occurs in the stopband, we must pick a window that provides a stopband attenuation of

$$\alpha_s = -20\log(0.005) = 46.02$$

Thus, we may use a Hamming window or a Kaiser window with

$$\beta = 0.5842(\alpha_s - 21)^{0.4} + 0.07886(\alpha_s - 21) = 4.09$$

The transition width between the lower stopband and the passband is $\Delta\omega = 0.02\pi$ and between the upper stopband and the passband it is $\Delta\omega = 0.05\pi$. Therefore, we must design the filter to meet the lower transition bandwidth requirement, $\Delta\omega = 0.02\pi$, or $\Delta f = 0.01$. Thus, for a Hamming window, the estimated filter order is

$$N = \frac{3.3}{0.01} = 330$$

For a Kaiser window, on the other hand, the filter order is

$$N = \frac{\alpha_s - 7.95}{14.36\Delta f} = 265.1$$

or $N = 266$.

(b) For an equiripple filter, the filter order may be estimated as follows,

$$N = \frac{-10\log(\delta_p\delta_s) - 13}{14.6\Delta f} = \frac{-10\log(0.05 \cdot 0.005) - 13}{(14.6)(0.01)} = 157.67$$

or $N = 158$.

9.10   Use the window design method to design a type II bandpass filter according to the following specifications:

$$|H(e^{j\omega})| \leq 0.0050 \qquad |\omega| \leq 0.1\pi$$
$$0.995 \leq |H(e^{j\omega})| \leq 1.0050 \qquad 0.25\pi \leq |\omega| \leq 0.6\pi$$
$$|H(e^{j\omega})| \leq 0.0025 \qquad 0.8\pi \leq |\omega|$$

With the window design method, the amplitudes of the ripples in each band of a multiband filter will be approximately equal, and the transition bands will have approximately the same width. Because the requirements on the peak ripple in the three bands of this bandpass filter are not the same, it is necessary to design the filter so that it has the smallest ripple in all three bands, which, in this case, requires that we set $\delta_s = 0.0025$. In addition, because the transition bands do not have the same width, it is necessary to set the desired transition width, $\Delta\omega$, equal to the smaller of the two ($\Delta\omega = 0.15\pi$).

With $\alpha_s = -20\log\delta_s = 52$ dB, it follows that we may use a Hamming window, and with

$$N\Delta f = 3.3$$

this gives an estimated filter order of

$$N = \frac{3.3}{0.075} = 44$$

For a type II filter, however, N must be odd, so we set $N = 45$.

Now we must find the unit sample response of the ideal bandpass filter that is to be windowed. Because the width of both the upper and lower transition bands will be approximately $\Delta\omega = 0.15\pi$, for the ideal filter we set the lower cutoff frequency equal to

$$\omega_1 = 0.25\pi - \frac{\Delta\omega}{2} = 0.175\pi$$

and the upper cutoff frequency equal to

$$\omega_2 = 0.6\pi + \frac{\Delta\omega}{2} = 0.675\pi$$

Therefore, the magnitude of the frequency response of the ideal filter is

$$|H_d(e^{j\omega})| = \begin{cases} 1 & 0.175\pi \leq |\omega| \leq 0.675\pi \\ 0 & \text{otherwise} \end{cases}$$

Repeating the steps in the derivation of the unit sample response of an ideal bandpass filter given in Prob. 9.8, using the given cutoff frequencies and a delay of $N/2 = 22.5$, we have

$$h_d(n) = \frac{\sin[0.675\pi(n - 22.5)]}{(n - 22.5)\pi} - \frac{\sin[0.175\pi(n - 22.5)]}{(n - 22.5)\pi}$$

9.11 Use the window design method to design a multiband filter that meets the following specifications:

$$0.99 \leq |H(e^{j\omega})| \leq 1.01 \qquad 0 \leq \omega \leq 0.3\pi$$
$$|H(e^{j\omega})| \leq 0.01 \qquad 0.35\pi \leq \omega \leq 0.55\pi$$
$$0.49 \leq |H(e^{j\omega})| \leq 0.51 \qquad 0.6\pi \leq \omega \leq \pi$$

To design a multiband filter that meets these specifications using the window design method, we begin by finding the ideal unit sample response. For the frequency response of the ideal filter, we set the cutoff frequencies equal to the midpoint of the transition bands. Therefore, we have

$$|H_d(e^{j\omega})| = \begin{cases} 1 & |\omega| \leq 0.325\pi \\ 0 & 0.325\pi \leq |\omega| \leq 0.575\pi \\ 0.5 & 0.575\pi \leq |\omega| \leq \pi \end{cases}$$

The unit sample response of this ideal filter may be found easily by noting that $H_d(e^{j\omega})$ may be written as an allpass filter with a gain of 0.5 minus a low-pass filter with a gain of 0.5 and a cutoff frequency $\omega_1 = 0.575\pi$ plus a low-pass filter with a gain of 1 and a cutoff frequency of $\omega_2 = 0.325\pi$. Therefore, if we assume that $H_d(e^{j\omega})$ has linear phase with a delay of n_d,

$$h_d(n) = 0.5\delta(n) - 0.5\frac{\sin[0.575\pi(n - n_d)]}{(n - n_d)\pi} + \frac{\sin[0.325\pi(n - n_d)]}{(n - n_d)\pi}$$

Having found the ideal unit sample response, the next step is to choose an appropriate window. When $h_d(n)$ is multiplied by a window $w(n)$, the frequency response is the convolution of the transform of the window $W(e^{j\omega})$ with $H_d(e^{j\omega})$. Assuming that the length of the filter is long compared to the inverse of the transition width, so that the discontinuities between the bands may be treated independently, the ripples in the three bands will be approximately the same as they would be for a low-pass filter, except that they will be scaled by the amplitude of the discontinuities at the band edge. Therefore, if the ripple in the lower passband and the stopband are δ_s, the ripple in the upper passband will be $\delta_s/2$. Consequently, we must use a window that would produce a low-pass filter with a ripple no larger than 0.01. Thus, we may use a Hanning window. Finally, to determine the filter order, note that because the widths of both transition bands are the same, $\Delta\omega = 0.05\pi$, an estimate of the filter order is

$$N = \frac{3.1}{\Delta f} = 124$$

Note that another way to design this filter would have been to design a network of three filters in parallel: a low-pass filter, a bandpass filter, and a high-pass filter. This approach would give greater control over the ripple amplitudes and the transition widths but would require a *trial and error* approach to establish the specifications for the three filters.

9.12 Shown in the following figure is the magnitude of the frequency response of a type I high-pass filter that was designed using the Parks-McClellan algorithm.

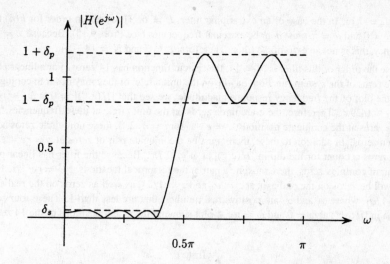

The stopband cutoff frequency is $\omega_s = 0.4\pi$, and the passband cutoff frequency is $\omega_p = 0.5\pi$. In addition, the stopband ripple is $\delta_s = 0.0574$, and the passband ripple is $\delta_p = 0.1722$.

(a) Determine the weighting function, $W(e^{j\omega})$, used to design this filter, and find the length of the unit sample response.

(b) Describe approximately where the zeros of the system function of this filter lie in the z-plane.

(a) To determine the weighting function, we observe that

$$\frac{\delta_p}{\delta_s} = 3$$

Therefore, the weighting function used to design the filter has a value in the stopband that is 3 times larger than the value in the passband. This makes the errors in the stopband more costly and, therefore, smaller by a factor of 3. So, a weighting function that could have been used to design this filter is as follows:

$$W(e^{j\omega}) = \begin{cases} 3 & 0 \le \omega \le 0.4\pi \\ 1 & 0.5\pi \le \omega \le \pi \end{cases}$$

To determine the length of the unit sample response, recall that a type I equiripple high-pass (or low-pass) filter must either have $L + 2$ or $L + 3$ alternations where $L = N/2$. Therefore, the order of the filter may be determined by counting the alternations. For this filter, we have nine alternations, which are labeled in the figure below.

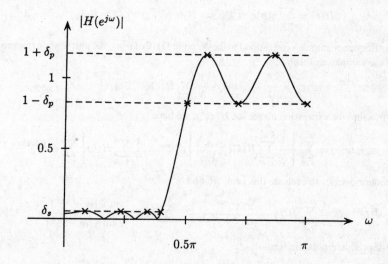

Thus, $L = 7$ or, in the case of an extraripple filter, $L = 6$. However, in order for $h(n)$ to be an extraripple filter, $\omega = 0$ and $\omega = \pi$ must *both* be extremal frequencies (see Prob. 9.15). Because $\omega = 0$ is not an extremal frequency, this is not an extraripple filter. Therefore, $L = 7$ and $N = 14$.

(b) Because the order of this filter is $N = 14$, the system function has 14 zeros. For a linear phase filter, we know that the zeros of the system function may lie on the unit circle, or they may occur in conjugate reciprocal pairs. From the plot of the frequency response magnitude, we see that $|H(e^{j\omega})| = 0$ at $\omega_1 \approx 0.175\pi$, $\omega_2 \approx 0.3\pi$, and $\omega_3 \approx 0.39\pi$. Therefore, there are three zeros on the unit circle at these frequencies. Because there must also be zeros at the conjugate positions, $z = e^{-j\omega_i}$ for $i = 1, 2, 3$, these unit circle zeros account for six of the fourteen zeros. In addition to these, there must be a conjugate pair of zeros at $z = re^{\pm j\omega_4}$, where $\omega_4 \approx 0.7\pi$. These zeros account for the dip in $|H(e^{j\omega})|$ at $\omega = 0.7\pi$. Because the filter has linear phase, in addition to this pair of complex zeros, there must be a pair at the reciprocal locations, $z^{-1} = re^{\pm j\omega_4}$. For the same reason, there will be zeros on the real axis at $z = \alpha_1$ and $z = 1/\alpha_1$, as well as zeros on the real axis at $z = -\alpha_2$ and $z = -1/\alpha_2$, where α_1 and α_2 are positive real numbers that are less than 1. These four zeros account for the minima in $|H(e^{j\omega})|$ at $\omega = 0$ and $\omega = \pi$. A plot showing the actual positions of the 14 zeros of $H(z)$ is given below.

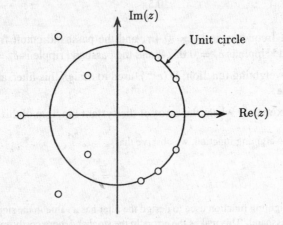

9.13 With the frequency sampling method, the frequency samples match the ideal frequency response exactly. Derive an interpolation formula that shows how the frequency samples $H(k)$ are interpolated.

The frequency response of an FIR filter of length N is

$$H(e^{j\omega}) = \sum_{n=0}^{N-1} h(n)e^{-jn\omega}$$

If $h(n)$ is designed using the frequency sampling method,

$$H(k) = \sum_{n=0}^{N-1} h(n)e^{-j2\pi nk/N} = H_d(e^{j2\pi k/N}) \qquad k = 0, 1, \ldots, N-1$$

Because these frequency samples correspond to the N-point DFT of $h(n)$, the unit sample response may be expressed in terms of these samples as follows:

$$h(n) = \frac{1}{N} \sum_{k=0}^{N-1} H(k)e^{j2\pi nk/N}$$

Substituting this into the expression above for $H(e^{j\omega})$, we have

$$H(e^{j\omega}) = \sum_{n=0}^{N-1} \left\{ \frac{1}{N} \sum_{k=0}^{N-1} H(k)e^{j2\pi nk/N} \right\} e^{-jn\omega} = \frac{1}{N} \sum_{k=0}^{N-1} H(k) \left\{ \sum_{n=0}^{N-1} e^{-jn(\omega - 2\pi k/N)} \right\}$$

Using the geometric series to evaluate this sum, we find

$$H(e^{j\omega}) = \frac{1}{N} \sum_{k=0}^{N-1} H(k) \frac{1 - e^{-j(\omega - 2\pi k/N)N}}{1 - e^{-j(\omega - 2\pi k/N)}} = \sum_{k=0}^{N-1} H(k) \frac{\sin \frac{N}{2}\left(\omega - \frac{2\pi k}{N}\right)}{N \sin \frac{1}{2}\left(\omega - \frac{2\pi k}{N}\right)} e^{-j(\frac{N-1}{2})(\omega - \frac{2\pi k}{N})}$$

which is the desired interpolation formula.

9.14 Given a low-pass filter that has been designed and implemented, either in hardware or software, it may be of interest to try to improve the frequency response characteristics by repetitive use of the filter. Suppose that $h(n)$ is the unit sample response of a zero phase FIR filter with a frequency response that satisfies the following specifications:

$$1 - \delta_p < H(e^{j\omega}) < 1 + \delta_p \qquad 0 \le \omega \le \omega_p$$
$$-\delta_s < H(e^{j\omega}) < \delta_s \qquad \omega_s \le \omega \le \pi$$

(Note that $H(e^{j\omega})$ having zero phase implies that $H(e^{j\omega})$ is real-valued for all ω).

(a) If a new filter is formed by cascading $h(n)$ with itself,

$$g(n) = h(n) * h(n)$$

$G(e^{j\omega})$ satisfies a set of specifications of the form

$$A < G(e^{j\omega}) < B \qquad 0 \le \omega \le \omega_p$$
$$C < G(e^{j\omega}) < D \qquad \omega_s \le \omega \le \pi$$

Find A, B, C, and D in terms of δ_p and δ_s of the low-pass filter $h(n)$.

(b) If $\delta_p \ll 1$ and $\delta_s \ll 1$, what are the approximate passband and stopband ripples of $G(e^{j\omega})$?

(a) With a cascade, $g(n) = h(n) * h(n)$, the frequency response is

$$G(e^{j\omega}) = H^2(e^{j\omega})$$

Therefore, in the passband we have

$$(1 - \delta_p)^2 < G(e^{j\omega}) < (1 + \delta_p)^2 \qquad 0 \le \omega \le \omega_p$$

and in the stopband we have

$$0 \le G(e^{j\omega}) < \delta_s^2 \qquad \omega_s \le \omega \le \pi$$

(b) If we assume that $\delta_p \ll 1$,

$$(1 - \delta_p)^2 = 1 - 2\delta_p + \delta_p^2 \approx 1 - 2\delta_p$$

and

$$(1 + \delta_p)^2 = 1 + 2\delta_p + \delta_p^2 \approx 1 + 2\delta_p$$

Therefore, the passband specifications are approximately

$$1 - 2\delta_p < G(e^{j\omega}) < 1 + 2\delta_p \qquad 0 \le \omega \le \omega_p$$

In other words, the passband ripple is approximately doubled. In the stopband, however, the ripple is much smaller with the cascade. In fact, in decibels, the stopband attenuation is doubled. With other systems built out of interconnections of $h(n)$ it is possible to improve the filter characteristics in both the passband and the stopband.

9.15 Show that a type I equiripple low-pass filter of order N may have either $L+2$ or $L+3$ alternations where $L = N/2$.

For a type I linear phase filter of order N, the frequency response is

$$H(e^{j\omega}) = A(e^{j\omega})e^{-jN\omega/2}$$

where

$$A(e^{j\omega}) = \sum_{k=0}^{L} a(k)\cos k\omega$$

with $L = N/2$. Because the desired response, $A_d(e^{j\omega})$, and the weighting function, $W(e^{j\omega})$, are piecewise constant,

$$\frac{dE(e^{j\omega})}{d\omega} = \frac{d}{d\omega}\{W(e^{j\omega})[A_d(e^{j\omega}) - A(e^{j\omega})]\} = -\frac{dA(e^{j\omega})}{d\omega}$$

However, because $A(e^{j\omega})$ is a trigonometric polynomial of degree $N/2$ in $\cos\omega$,

$$A(e^{j\omega}) = \sum_{k=0}^{L} a(k)\cos k\omega = \sum_{k=0}^{L} \alpha(k)(\cos\omega)^k$$

then $$\frac{dA(e^{j\omega})}{d\omega} = -\sin\omega \sum_{k=0}^{L} k\alpha(k)(\cos\omega)^{k-1} = -\sin\omega \sum_{k=0}^{L-1}(k+1)\alpha(k+1)(\cos\omega)^k$$

Therefore, the derivative of $A(e^{j\omega})$ with respect to ω is always equal to zero at $\omega = 0$ and $\omega = \pi$. In addition, however, the derivative is equal to zero at $L - 1$ other frequencies between 0 and π, which correspond to the roots of the polynomial given by the sum. Therefore, $A(e^{j\omega})$ may have at most $L + 1$ local maxima and minima. In addition, however, the band-edge frequencies, ω_p and ω_s, must also be extremal frequencies. Thus, the maximum number of alternations is $L + 3$. Because the alternation theorem requires a minimum of $L + 2$, the optimum filter may have either $L + 2$ or $L + 3$ alternations.

9.16   Suppose that we would like to design a type I equiripple bandstop filter of order $N = 30$.

(a)   What is the minimum number of alternations that this filter must have?

(b)   What is the maximum number?

(a)   For a type I linear phase FIR filter of order N, the alternation theorem states that the minimum number of alternations is $L + 2$, where $L = N/2$. Therefore, with $N = 30$, the minimum number of alternations is 17.

(b)   As shown in Prob. 9.15, with

$$H(e^{j\omega}) = A(e^{j\omega})e^{-j\omega N/2}$$

$A(e^{j\omega})$, and thus $E(e^{j\omega})$, will have, at most, $L + 1$ local maxima and minima in the interval $[0, \pi]$. In addition, however, there may be alternations at the band-edge frequencies. For a bandstop filter, there are four band edges: the lower passband cutoff frequency, the lower stopband cutoff frequency, the upper stopband cutoff frequency, and the upper passband cutoff frequency. Therefore, the maximum number of alternations is $L + 5$.

9.17   We would like to design a bandstop filter to satisfy the following specifications:

$$0.98 \le |H(e^{j\omega})| \le 1.02 \qquad 0 \le \omega \le 0.2\pi$$
$$|H(e^{j\omega})| < 0.001 \qquad 0.22\pi \le \omega \le 0.78\pi$$
$$0.95 \le |H(e^{j\omega})| \le 1.05 \qquad 0.8\pi \le \omega \le \pi$$

(a)   Estimate the order of the equiripple filter required to meet these specifications.

(b)   What weighting function $W(e^{j\omega})$ should be used to design this filter?

(c)   What is the minimum number of extremal frequencies that the optimal filter must have?

(a)   The design formula used to estimate the order for a low-pass equiripple filter is

$$N = \frac{-10\log(\delta_p\delta_s) - 13}{14.6\Delta f}$$

With the smaller of the two passband ripples being equal to $\delta_p = 0.02$, a stopband ripple of $\delta_s = 0.001$, and a transition width $\Delta\omega = 0.02\pi$, an estimate of the filter order is

$$N = \frac{-10\log(0.02 \cdot 0.001) - 13}{14.6 \cdot (0.01)} = 232$$

However, because this estimate is for a low-pass filter, the actual filter order required is closer to $N = 242$, which may be confirmed by computer.

(b) With a ripple of $\delta_1 = 0.02$ in the lower passband, $\delta_2 = 0.001$ in the stopband, and $\delta_3 = 0.05$ in the upper passband, an appropriate weighting function would be

$$W(e^{j\omega}) = \begin{cases} 1 & 0 \le \omega \le 0.2\pi \\ 20 & 0.22\pi \le \omega \le 0.78\pi \\ 0.4 & 0.8\pi \le \omega \le \pi \end{cases}$$

However, scaling these weights by any constant would not change the design.

(c) Assuming a filter order of $N = 232$, which is a type I design, the amplitude response has the form

$$A(e^{j\omega}) = \sum_{k=0}^{L} a(k)\cos k\omega$$

where $L = N/2 = 116$. Therefore, the minimum number of extremal frequencies is $L + 2 = 118$.

9.18 We would like to design an equiripple high-pass filter of order $N = 64$. The stopband ripple is to be no larger than $\delta_s = 0.001$, and the passband ripple no larger than $\delta_p = 0.01$. If we want a passband cutoff frequency equal to $\omega_p = 0.72\pi$, what will the stopband cutoff frequency be approximately equal to?

For an equiripple low-pass filter, an approximate relation between the filter order N, the passband and stopband ripples, δ_p and δ_s, respectively, and the transition width Δf, is given by

$$N = \frac{-10\log(\delta_p \delta_s) - 13}{14.6\Delta f}$$

Because a high-pass filter may be formed from a low-pass filter as follows,

$$h_{HP}(n) = \delta\left(n - \frac{N}{2}\right) - h_{LP}(n)$$

this formula is also applicable to high-pass filters. With $N = 64$, $\delta_p = 0.01$, and $\delta_s = 0.001$, we find that

$$\Delta f = \frac{37}{(14.6)(64)} = 0.0396$$

Therefore, if the passband cutoff frequency is $\omega_p = 0.72\pi$, the stopband cutoff frequency will be approximately

$$\omega_s = \omega_p - 2\pi\,\Delta f = 0.6408\pi$$

9.19 Suppose that we want to design a low-pass filter of order $N = 63$ with a cutoff frequency $\omega_p = 0.3\pi$ and a stopband cutoff frequency $\omega_s = 0.32\pi$.

(a) What is the approximate stopband attenuation that would obtained if this filter were designed using the window design method with a Kaiser window.

(b) Repeat part (a) for a equiripple filter assuming that we want $\delta_p = \delta_s$.

(a) For a Kaiser window design, the relationship between the filter order N, the stopband attenuation $\alpha_s = -20\log\delta_s$, and the transition width Δf is

$$N = \frac{\alpha_s - 7.95}{14.36\Delta f}$$

Solving this for the stopband attenuation, we have

$$\alpha_s = 14.36 \cdot N\Delta f + 7.95 = 16.99$$

which corresponds to a stopband (and passband) ripple of

$$\delta_s = 10^{-16.99/20} = 0.141$$

(b) For an equiripple filter, the filter order is approximately

$$N = \frac{-10\log(\delta_p\delta_s) - 13}{14.6\Delta f}$$

With $\delta_p = \delta_s$, this becomes

$$N = \frac{\alpha_s - 13}{14.6\Delta f}$$

where $\alpha_s = -20\log\delta_s$. Solving for α_s, we have

$$\alpha_s = 14.6 \cdot N\Delta f + 13 = 22.04$$

The corresponding stopband ripple is

$$\delta_s = 10^{-22.04/20} = 0.079$$

9.20 The linear phase constraint on FIR filters places constraints on the unit sample response and the location of the zeros of the system function. In the table below, indicate with a check which filter types could successfully be used to approximate the given filter type.

	Type I	Type II	Type III	Type IV
Low-pass filter				
High-pass filter				
Bandpass filter				
Bandstop filter				
Differentiator				

A type I linear phase filter has no constraints on the locations of its zeros. Therefore, a type I filter may be used for the design of any type of filter. The type II linear phase filter will always have a zero at $\omega = \pi$. Therefore, these filters should only be used for low-pass and bandpass filters. The type III linear phase filter is constrained to have zeros at $\omega = \pi$ and $\omega = 0$. Therefore, type III filters should only be used for the design of bandpass filters. Finally, because the type IV filters have a zero at $\omega = 0$, they should not be used in the design of low-pass or bandstop filters. These results are summarized in the table below.

	Type I	Type II	Type III	Type IV
Low-pass filter	x	x		
High-pass filter	x			x
Bandpass filter	x	x	x	x
Bandstop filter	x			
Differentiator	x			x

IIR Filter Design

9.21 For historical reasons, the design formulas for analog filters are given assuming a peak gain of 1 in the passband. In terms of the parameters ϵ and A, the filter specifications have the form

$$\frac{1}{\sqrt{1+\epsilon^2}} \le |H_a(j\Omega)| \le 1$$

$$|H_a(j\Omega)| \le \frac{1}{A}$$

Suppose that we would like to use the bilinear transformation to design a discrete-time IIR low-pass filter that satisfies the following frequency response constraints:

$$1 - \delta_p \le |H(e^{j\omega})| \le 1 + \delta_p \qquad 0 \le \omega \le \omega_p$$

$$|H(e^{j\omega})| \le \delta_s \qquad \omega_s \le \omega \le \pi$$

Find the relationship between the parameters δ_p and δ_s for the discrete-time filter and between the parameters ϵ and A for the continuous-time filter.

For a digital low-pass filter with a frequency response magnitude

$$1 - \delta_p \leq |H(e^{j\omega})| \leq 1 + \delta_p$$

dividing by $1 + \delta_p$ this becomes

$$\frac{1 - \delta_p}{1 + \delta_p} \leq |H(e^{j\omega})| \leq 1$$

Setting

$$\frac{1 - \delta_p}{1 + \delta_p} = \frac{1}{\sqrt{1 + \epsilon^2}}$$

we have

$$1 + \epsilon^2 = \left(\frac{1 + \delta_p}{1 - \delta_p}\right)^2$$

and

$$\epsilon^2 = \left(\frac{1 + \delta_p}{1 - \delta_p}\right)^2 - 1 = \frac{4\delta_p}{(1 - \delta_p)^2}$$

With a stopband ripple of δ_s, the normalization of the peak passband gain to 1 produces a peak stopband ripple of

$$\frac{\delta_s}{1 + \delta_p}$$

Therefore

$$A = \delta_s^{-1}(1 + \delta_p)$$

9.22 As the order of an analog Butterworth filter is increased, the slope of $|H_a(j\Omega)|^2$ at the 3-dB cutoff frequency, Ω_c, increases. Derive an expression for the slope of $|H_a(j\Omega)|^2$ at Ω_c as a function of the filter order, N.

The magnitude squared of the Butterworth filter's frequency response is

$$|H_a(j\Omega)|^2 = \frac{1}{1 + (j\Omega/j\Omega_c)^{2N}}$$

To evaluate the slope of $|H_a(j\Omega)|^2$ at $\Omega = \Omega_c$, we may set $\Omega_c = 1$ and evaluate the derivative at $\Omega = 1$. Therefore, with

$$|H_a(j\Omega)|^2 = \frac{1}{1 + \Omega^{2N}}$$

we have

$$\frac{d}{d\Omega}|H_a(j\Omega)|^2 = \frac{-2N\Omega^{2N-1}}{(1 + \Omega^{2N})^2}$$

Evaluating this at $\Omega = 1$, we have

$$\left.\frac{d}{d\Omega}|H_a(j\Omega)|^2\right|_{\Omega=1} = -\frac{N}{2}$$

9.23 Show that the frequency response of an Nth-order low-pass Butterworth filter is *maximally flat* at $\Omega = 0$ in the sense that the first $2N - 1$ derivatives of $|H_a(j\Omega)|^2$ are equal to zero at $\Omega = 0$.

An Nth-order Butterworth filter has a magnitude squared frequency response given by

$$|H_a(j\Omega)|^2 = \frac{1}{1 + (j\Omega/j\Omega_c)^{2N}}$$

Without any loss in generality, we may assume that $\Omega_c = 1$ and evaluate the derivative of the function

$$G(\Omega) = \frac{1}{1 + \Omega^{2N}}$$

at $\Omega = 0$. Multiplying both sides of this equation by $(1 + \Omega^{2N})$, we have

$$G(\Omega)[1 + \Omega^{2N}] = 1$$

Differentiating this equation with respect to Ω yields

$$G'(\Omega)[1 + \Omega^{2N}] + G(\Omega)[2N\Omega^{2N-1}] = 0$$

Setting $\Omega = 0$, we have

$$G'(\Omega)\big|_{\Omega=0} = 0$$

Differentiating a second time gives

$$G''(\Omega)[1 + \Omega^{2N}] + G'(\Omega)[4N\Omega^{2N-1}] + G(\Omega)[2N(2N-1)\Omega^{2N-2}] = 0$$

Again setting $\Omega = 0$, we have

$$G''(\Omega)\big|_{\Omega=0} = 0$$

If we continue to differentiate k times, where $k \leq 2N - 1$, we have an equation of the form

$$G^{(k)}(\Omega)[1 + \Omega^{2N}] + \sum_{i=1}^{k-1} G^{(i)}(\Omega)F_i(\Omega) + G(\Omega)[2N(2N-1)\cdots(2N-k+1)\Omega^{2N-k}] = 0$$

where $G^{(i)}(\Omega)$ is the ith derivative of $G(\Omega)$, and $F(\Omega)$ is a polynomial in Ω. Given that $G^{(i)}(\Omega)$ is equal to zero at $\Omega = 0$ for $i = 1, \ldots, k-1$, it follows that

$$G^{(k)}(\Omega)|_{\Omega=0} = 0$$

Differentiating $2N$ times, however, we have

$$G^{(2N)}(\Omega)[1 + \Omega^{2N}] + \sum_{i=1}^{2N-1} G^{(i)}(\Omega)F_i(\Omega) + G(\Omega) \cdot (2N)! = 0$$

Therefore, $$G^{(2N)}(\Omega)\big|_{\Omega=0} = -G(\Omega)\big|_{\Omega=0} \cdot (2N)! = -(2N)!$$

which is *nonzero*, and the maximally flat property is established.

9.24   Design a low-pass Butterworth filter that has a 3-dB cutoff frequency of 1.5 kHz and an attenuation of 40 dB at 3.0 kHz.

Given the 3-dB cutoff frequency of the Butterworth filter, all that is needed is to find the filter order, N, that will give 40 dB of attenuation at 3 kHz, or $\Omega_s = 2\pi \cdot 3000$. At the stopband cutoff frequency Ω_s, the magnitude of the frequency response squared is

$$\left|H_a(j\Omega)\right|^2_{\Omega=2\pi \cdot 3000} = \left.\frac{1}{1 + (j\Omega/j\Omega_c)^{2N}}\right|_{\Omega=2\pi \cdot 3000} = \frac{1}{1 + 2^{2N}}$$

Therefore, if we want the magnitude of the frequency response to be down 40 dB at $\Omega_s = 2\pi \cdot 3000$, the magnitude *squared* must be no larger than 10^{-4}, or

$$\frac{1}{1 + 2^{2N}} \leq 10^{-4}$$

Thus, we want

$$2N = \frac{\log(10^4 - 1)}{\log 2} = 13.29$$

or $N = 7$. For a seventh-order Butterworth filter, the 14 poles of

$$H_a(s)H_a(-s) = \frac{1}{1 + (s/j\Omega_c)^{2N}}$$

lie on a circle of radius $\Omega_c = 2\pi \cdot 3000$, at angles of

$$\theta_k = \frac{(N + 1 + 2k)\pi}{N} = \frac{(4 + k)\pi}{7} \qquad k = 0, 1, \ldots, 13$$

as illustrated in the following figure:

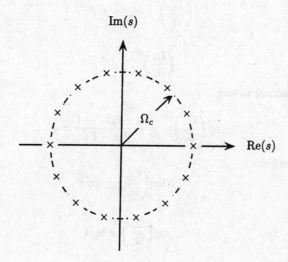

The poles of $H_a(s)$ are the seven poles of $H_a(s)H_a(-s)$ that lie in the left-half s-plane, that is,

$$s_k = -\Omega_c e^{\pm jk\pi/7} \qquad k = 0, 1, 2, 3$$

Except for the isolated pole at $s = -\Omega_c$, the remaining six poles occur in complex conjugate pairs. The conjugate pairs may be combined to form second-order factors with real coefficients to yield factors of the form

$$H_k(s) = \frac{1}{s^2 - 2\Omega_c \cos(k\pi/7)s + \Omega_c^2} \qquad k = 1, 2, 3$$

Thus, the system function of the seventh-order Butterworth filter is

$$H_a(s) = \prod_{k=0}^{N-1} \frac{-s_k}{s - s_k} = \frac{\Omega_c}{s + \Omega_c} \cdot \prod_{k=1}^{3} \frac{\Omega_c^2}{s^2 - 2\Omega_c \cos(k\pi/7)s + \Omega_c^2}$$

9.25 Let Ω_p and Ω_s be the desired passband and stopband cutoff frequencies of an analog low-pass filter, and let δ_p and δ_s be the passband and stopband ripples. Show that the order of the Butterworth filter required to meet these specifications is

$$N \geq \frac{\log d}{\log k}$$

with the 3-dB cutoff frequency Ω_c being any value within the range

$$\Omega_p[(1 - \delta_p)^{-2} - 1]^{-1/2N} \leq \Omega_c \leq \Omega_s[\delta_s^{-2} - 1]^{-1/2N}$$

The squared magnitude of the frequency response of the Butterworth filter is

$$|H_a(j\Omega)|^2 = \frac{1}{1 + (j\Omega/j\Omega_c)^{2N}}$$

Because $|H_a(j\Omega)|$ is monotonically decreasing, the maximum error in the passband and stopband occurs at the band edges, Ω_p and Ω_s, respectively. Therefore, we want

$$|H_a(j\Omega_p)|^2 = \frac{1}{1 + (j\Omega_p/j\Omega_c)^{2N}} \geq (1 - \delta_p)^2$$

and

$$|H_a(j\Omega_s)|^2 = \frac{1}{1 + (j\Omega_s/j\Omega_c)^{2N}} \leq \delta_s^2$$

From the first equation, we have

$$\left(\frac{\Omega_p}{\Omega_c}\right)^{2N} \leq (1 - \delta_p)^{-2} - 1 \qquad (9.20)$$

and from the second,

$$\left(\frac{\Omega_s}{\Omega_c}\right)^{2N} \geq \delta_s^{-2} - 1 \qquad (9.21)$$

Dividing these two equations, we have

$$\left(\frac{\Omega_p}{\Omega_s}\right)^{2N} \leq \frac{(1 - \delta_p)^{-2} - 1}{\delta_s^{-2} - 1} = d^2$$

and taking the logarithm gives

$$N \log\left(\frac{\Omega_p}{\Omega_s}\right) \leq \log d$$

Dividing by

$$\log\left(\frac{\Omega_p}{\Omega_s}\right) = \log k$$

we have

$$N \geq \frac{\log d}{\log k}$$

(note that the inequality is reversed because $\log k < 0$). Because the right side of this equation will not generally be an integer, the order N is taken to be the smallest integer larger than $(\log d)/(\log k)$. Finally, once the order N is fixed, it follows from Eqs. (9.20) and (9.21) that Ω_c may be any value in the range

$$\Omega_p[(1 - \delta_p)^{-2} - 1]^{-1/2N} \leq \Omega_c \leq \Omega_s\left[\delta_s^{-2} - 1\right]^{-1/2N}$$

9.26 Suppose that we would like to design an analog Chebyshev low-pass filter so that

$$1 - \delta_p \leq |H_a(j\Omega)| \leq 1 \qquad |\Omega| \leq \Omega_p$$
$$|H_a(j\Omega)| \leq \delta_s \qquad \Omega_s \leq |\Omega|$$

Find an expression for the required filter order, N, as a function of Ω_p, Ω_s, δ_p, and δ_s.

For a Chebyshev filter, the magnitude of the frequency response squared is

$$|H_a(j\Omega)|^2 = \frac{1}{1 + \epsilon^2 T_N^2(\Omega/\Omega_p)}$$

where

$$T_N(x) = \begin{cases} \cos(N \cos^{-1} x) & |x| \leq 1 \\ \cosh(N \cosh^{-1} x) & |x| > 1 \end{cases}$$

is an Nth-order Chebyshev polynomial. Over the passband, $|\Omega| < \Omega_p$, the magnitude of the frequency response oscillates between 1 and $(1 + \epsilon^2)^{-1/2}$. Therefore, the ripple amplitude, δ_p, is

$$\delta_p = 1 - (1 + \epsilon^2)^{-1/2}$$

or

$$\epsilon^2 = (1 - \delta_p)^{-2} - 1$$

At the stopband frequency Ω_s we have

$$|H_a(j\Omega_s)|^2 = \frac{1}{1 + \epsilon^2 T_N^2(\Omega_s/\Omega_p)}$$

which we want to be less than or equal to δ_s^2:

$$\frac{1}{1 + \epsilon^2 T_N^2(\Omega_s/\Omega_p)} \leq \delta_s^2$$

Therefore,

$$T_N^2\left(\frac{\Omega_s}{\Omega_p}\right) \geq \frac{(\delta_s^{-2} - 1)}{\epsilon^2} = \frac{(\delta_s^{-2} - 1)}{(1 - \delta_p)^{-2} - 1} = \frac{1}{d^2}$$

Because $(\Omega_s/\Omega_p) > 1$, then $T_N(\Omega_s/\Omega_p) = \cosh(N\cosh^{-1}(\Omega_s/\Omega_p))$, and we have

$$\cosh\left(N\cosh^{-1}\frac{\Omega_s}{\Omega_p}\right) \geq \frac{1}{d}$$

or

$$N \geq \frac{\cosh^{-1}(1/d)}{\cosh^{-1}(\Omega_s/\Omega_p)} = \frac{\cosh^{-1}(1/d)}{\cosh^{-1}(1/k)}$$

which is the desired expression.

9.27 If $H_a(s)$ is a third-order type I Chebyshev low-pass filter with a cutoff frequency $\Omega_p = 1$ and $\epsilon = 0.1$, find $H_a(s)H_a(-s)$.

The magnitude of the frequency response squared for an Nth-order type I Chebyshev filter is

$$|H_a(j\Omega)|^2 = \frac{1}{1 + \epsilon^2 T_N^2(\Omega/\Omega_p)}$$

where $T_N(x)$ is an Nth-order Chebyshev polynomial that is defined recursively as follows,

$$T_{k+1}(x) = 2x T_k(x) - T_{k-1}(x) \qquad k \geq 1$$

with $T_0(x) = 1$ and $T_1(x) = x$. Therefore, to find the third-order Chebyshev polynomial, we first find $T_2(x)$ as follows,

$$T_2(x) = 2x T_1(x) - T_0(x) = 2x^2 - 1$$

and then we have

$$T_3(x) = 2x T_2(x) - T_1(x) = 4x^3 - 2x - x = x(4x^2 - 3)$$

Thus, the denominator polynomial in $|H_a(j\Omega)|^2$ is

$$1 + \epsilon^2 T_3^2\left(\frac{\Omega}{\Omega_p}\right) = 1 + 0.01[\Omega(4\Omega^2 - 3)]^2$$

$$= 1 + 0.01\Omega^2(16\Omega^4 - 24\Omega^2 + 9)$$

and we have

$$|H_a(j\Omega)|^2 = \frac{1}{1 + 0.09\Omega^2 - 0.24\Omega^4 + 0.16\Omega^6}$$

Because

$$|H_a(j\Omega)|^2 = [H_a(s)H_a(-s)]_{s=j\Omega}$$

to find the rational function

$$G_a(s) = H_a(s)H_a(-s)$$

we make the substitution $\Omega = s/j$ in $|H_a(j\Omega)|^2$ as follows:

$$G_a(s) = \frac{1}{1 + 0.09(s/j)^2 - 0.24(s/j)^4 + 0.16(s/j)^6} = \frac{1}{1 - 0.09s^2 - 0.24s^4 - 0.16s^6}$$

9.28 Show that the bilinear transformation maps the $j\Omega$-axis in the s-plane *onto* the unit circle, $|z| = 1$, and maps the left-half s-plane, $\text{Re}(s) < 0$ *inside* the unit circle, $|z| < 1$.

To investigate the characteristics of the bilinear transformation, let $z = re^{j\omega}$ and $s = \sigma + j\Omega$. The bilinear transformation may then be written as

$$s = \frac{2}{T_s}\frac{z-1}{z+1} = \frac{2}{T_s}\frac{re^{j\omega}-1}{re^{j\omega}+1}$$

$$= \frac{2}{T_s}\left(\frac{r^2-1}{1+r^2+2r\cos\omega} + j\frac{2r\sin\omega}{1+r^2+2r\cos\omega}\right)$$

Therefore,

$$\sigma = \frac{2}{T_s}\frac{r^2-1}{1+r^2+2r\cos\omega}$$

and

$$\Omega = \frac{2}{T_s}\frac{2r\sin\omega}{1+r^2+2r\cos\omega}$$

Note that if $r < 1$, then $\sigma < 0$, and if $r > 1$, then $\sigma > 0$. Consequently, the left-half s-plane is mapped inside the unit circle, and the right-half s-plane is mapped outside the unit circle. If $r = 1$, then $\sigma = 0$, and

$$\Omega = \frac{2}{T_s}\frac{2\sin\omega}{2+2\cos\omega}$$

Thus, the $j\Omega$-axis is mapped *onto* the unit circle. Using trigonometric identities, this may be written in the equivalent form

$$\Omega = \frac{2}{T_s}\tan\left(\frac{\omega}{2}\right)$$

or

$$\omega = 2\arctan\left(\frac{\Omega T_s}{2}\right)$$

9.29 Let $H_a(s)$ be an all-pole filter with no zeros in the finite s-plane,

$$H_a(s) = A\prod_{k=1}^{p}\frac{1}{s - s_k}$$

If $H_a(s)$ is mapped into a digital filter using the bilinear transformation, will $H(z)$ be an all-pole filter?

With $T_s = 2$, the bilinear transformation is

$$s = \frac{1 - z^{-1}}{1 + z^{-1}}$$

and the system function of the digital filter is

$$H(z) = A\prod_{k=1}^{p}\frac{1}{\dfrac{1 - z^{-1}}{1 + z^{-1}} - s_k} = A\prod_{k=1}^{p}\frac{(1 + z^{-1})}{1 - z^{-1} - s_k(1 + z^{-1})}$$

$$= A\prod_{k=1}^{p}\frac{1 + z^{-1}}{(1 - s_k) - (1 + s_k)z^{-1}}$$

This may be written in the more conventional form as follows,

$$H(z) = A'\prod_{k=1}^{p}\frac{1 + z^{-1}}{1 - \alpha_k z^{-1}}$$

where

$$A' = \prod_{k=1}^{p}\frac{1}{1 - s_k}$$

and
$$\alpha_k = \frac{1 + s_k}{1 - s_k}$$

Therefore, $H(z)$ has p poles (inside the unit circle if $\text{Re}(s_k) < 0$) and p zeros at $z = -1$. Note that these zeros come from the p zeros in $H_a(s)$ at $s = \infty$, which are mapped to $z = -1$ by the bilinear transformation. Thus, $H(z)$ will *not* be an all-pole filter.

9.30   Shown in the figure below is the magnitude of the frequency response of a low-pass filter that was designed by mapping a type I analog Chebyshev filter into a discrete-time filter using the bilinear transformation.

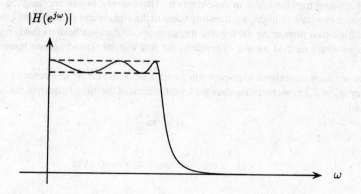

Find the filter order (i.e., the number of poles and zeros in $H(z)$).

The magnitude-squared response of a type I analog Chebyshev filter is

$$|H_a(j\Omega)|^2 = \frac{1}{1 + \epsilon^2 T_N^2(\Omega/\Omega_p)}$$

where
$$T_N(x) = \begin{cases} \cos(N \cos^{-1} x) & |x| \leq 1 \\ \cosh(N \cosh^{-1} x) & |x| > 1 \end{cases}$$

is an Nth-order Chebyshev polynomial. Over the passband, $|\Omega| < \Omega_p$, the magnitude of the frequency response oscillates between 1 and $(1 + \epsilon^2)^{-1/2}$. As the frequency varies from $\Omega = 0$ to $\Omega = \Omega_p$, $\theta = N \cos^{-1}(\Omega/\Omega_p)$ varies from $\theta = N\pi/2$ to $\theta = 0$. Therefore,

$$T_N^2\left(\frac{\Omega}{\Omega_p}\right) = \cos\left[N \cos^{-1}\left(\frac{\Omega}{\Omega_p}\right)\right]$$

oscillates between zero and 1 $N + 1$ times over the interval $[0, \Omega_p]$ [i.e., $T_N^2(\Omega/\Omega_p)$ reaches its maximum or minimum value $N + 1$ times]. The bilinear transformation is a one-to-one mapping of the $j\Omega$-axis onto the unit circle. Therefore, $|H(e^{j\omega})|^2$ will alternate $N + 1$ times between 1 and $1/(1 + \epsilon^2)$ over the interval $[0, \omega_p]$, where

$$\omega_p = \tan\left(\frac{\Omega_p}{2}\right)$$

Because there are six alternations of $|H(e^{j\omega})|^2$ in the passband, $N + 1 = 6$, and $H_a(s)$ is a fifth-order type I Chebyshev filter,

$$H_a(s) = A \prod_{k=1}^{5} \frac{1}{s - s_k}$$

where A is a constant. Applying the bilinear transformation to $H_a(s)$ results in a discrete-time filter with a system function $H(z)$ that has five poles inside the unit circle, and five zeros on the unit circle at $z = -1$ (as shown in Prob. 9.29, the five zeros come from the five zeros in $H_a(s)$ at $s = \infty$).

9.31   Use the bilinear transformation to design a discrete-time Chebyshev high-pass filter with an equiripple
passband with

$$0 \leq |H(e^{j\omega})| \leq 0.1 \qquad 0 \leq \omega \leq 0.1\pi$$

and

$$0.9 \leq |H(e^{j\omega})| \leq 1.0 \qquad 0.3\pi \leq \omega \leq \pi$$

To design a discrete-time high-pass filter, there are two approaches that we may use. We may design an analog type I
Chebyshev low-pass filter, map it into a Chebyshev low-pass filter using the bilinear transformation, and then perform
a low-pass-to-high-pass transformation in the z-domain. Alternatively, before applying the bilinear transformation,
we could perform a low-pass-to-high-pass transformation in the s-plane and then map the analog high-pass filter into
a discrete-time high-pass filter using the bilinear transformation. Because both methods result in the same design,
it does not matter which method we use. Therefore, we will use the second approach, because it is a little easier
algebraically.

Given that we want to design a high-pass filter with a stopband cutoff frequency $\omega_s = 0.1\pi$ and a passband
cutoff frequency $\omega_p = 0.3\pi$, we first transform the specifications of the digital filter into the continuous-time domain.
With $T_s = 2$ and

$$\Omega = \tan \frac{\omega}{2}$$

we have

$$\Omega_s = \tan \frac{\omega_s}{2} = \tan(0.05\pi) = 0.1584$$

and

$$\Omega_p = \tan \frac{\omega_p}{2} = \tan(0.15\pi) = 0.5095$$

Using the transformation $s \longrightarrow 1/s$ to map these high-pass filter cutoff frequencies to low-pass filter cutoff frequencies, we have

$$\Omega_p = \frac{1}{0.5095} = 1.9627$$

and

$$\Omega_s = \frac{1}{0.1584} = 6.3138$$

Therefore, the selectivity factor for the analog Chebyshev filter is

$$k = \frac{\Omega_p}{\Omega_s} = 0.3109$$

With $\delta_p = \delta_s = 0.1$, the discrimination factor is

$$d = \left[\frac{(1 - \delta_p)^{-2} - 1}{\delta_s^{-2} - 1} \right]^{1/2} = 0.04867$$

Thus, the required filter order is

$$N = \frac{\cosh^{-1}(1/d)}{\cosh^{-1}(1/k)} = 2.03$$

Although we should round up to $N = 3$, with a second-order Chebyshev filter we should come close to meeting the
specifications. Therefore, we will use $N = 2$.

The next step is to design a second-order low-pass Chebyshev filter with

$$0.9 \leq |H_a(j\Omega)| \leq 1 \qquad 0 \leq \Omega \leq \Omega_p$$

$$|H_a(j\Omega)| \leq 0.1 \qquad \Omega_s \leq \Omega$$

where $\Omega_p = 1.9627$ and $\Omega_s = 6.3138$. With

$$\frac{1}{\sqrt{1 + \epsilon^2}} = 1 - \delta_p$$

it follows that

$$\epsilon^2 = \frac{1}{(1 - \delta_p)^2} - 1 = 0.2346$$

For a second-order Chebyshev filter, we need to generate a second-order Chebyshev polynomial, which is

$$T_2(x) = 2xT_1(x) - T_0(x) = 2x^2 - 1$$

Squaring $T_2(x)$, we have

$$T_2^2(x) = 4x^4 - 4x^2 + 1$$

and, for the magnitude squared frequency response of the Chebyshev filter, we have

$$|H_a(j\Omega)|^2 = \frac{1}{1 + \epsilon^2 T_N^2(\Omega/\Omega_p)} = \frac{1}{1 + \epsilon^2[4(\Omega/\Omega_p)^4 - 4(\Omega/\Omega_p)^2 + 1]}$$

Substituting for the given values of Ω_p and ϵ, we have

$$|H_a(j\Omega)|^2 = \frac{1}{1.2346 - 0.2436\Omega^2 + 0.0632\Omega^4}$$

Next, we find $H_a(s)H_a(-s)$ with the substitution $\Omega = -js$,

$$H_a(s)H_a(-s) = |H_a(j\Omega)|^2\big|_{\Omega=-js} = \frac{1}{1.2346 + 0.2436\,s^2 + 0.0632\,s^4}$$

Factoring the denominator polynomial, we find that the two roots in the left-half s-plane are at

$$s_1 = -1.1163 + j1.7811 \qquad s_2 = s_1^* = -1.1163 - j1.7811$$

Thus, the second-order Chebyshev filter is

$$H_a(s) = \frac{1}{\sqrt{1 - \epsilon^2}} \frac{s_1 s_1^*}{(s - s_1)(s - s_1^*)} = \frac{3.9778}{s^2 + 2.2327s + 4.4185}$$

Now we transform this low-pass filter into a high-pass filter with the low-pass-to-high-pass transformation $s \longrightarrow 1/s$. This gives

$$H_a(s) = \frac{3.9778s^2}{1 + 2.2327s + 4.4185s^2}$$

Finally, applying the bilinear transformation, we have

$$H(z) = \frac{3.9778\left(\dfrac{1 - z^{-1}}{1 + z^{-1}}\right)^2}{1 + 2.2327\left(\dfrac{1 - z^{-1}}{1 + z^{-1}}\right) + 4.4185\left(\dfrac{1 - z^{-1}}{1 + z^{-1}}\right)^2}$$

Multiplying numerator and denominator by $(1 + z^{-1})^2$ gives

$$H(z) = \frac{3.9778(1 - z^{-1})^2}{(1 + z^{-1})^2 + 2.2327(1 + z^{-1})(1 - z^{-1}) + 4.4185(1 - z^{-1})^2}$$

$$= \frac{0.52(1 - 2z^{-1} + z^{-2})}{1 - 0.894z^{-1} + 0.4164z^{-2}}$$

The magnitude of the frequency response is plotted in the following figure.

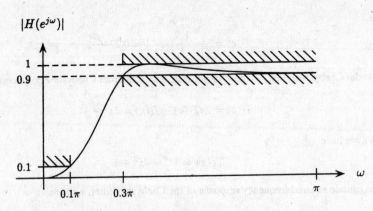

As a check on the design, we may compute the magnitude of the frequency response at $\omega = 0.1\pi$, which is

$$|H(e^{j\omega})|_{\omega=0.1\pi} = 0.1044$$

which comes close to satisfying the stopband specifications. At the passband edge, we have

$$|H(e^{j\omega})|_{\omega=0.3\pi} = 0.9044$$

which does satisfy the constraint.

9.32 We would like to design a digital low-pass filter that has a passband cutoff frequency $\omega_p = 0.375\pi$ with $\delta_p = 0.01$ and a stopband cutoff frequency $\omega_s = 0.5\pi$ with $\delta_s = 0.01$. The filter is to be designed using the bilinear transformation. What order Butterworth, Chebyshev, and elliptic filters are necessary to meet the design specifications?

To find the required filter order, we begin by finding the discrimination factor and the selectivity factor for the analog low-pass filter prototype. With $\delta_p = \delta_s = 0.01$, the discrimination factor is

$$d = \left[\frac{(1-\delta_p)^{-2}-1}{\delta_s^{-2}-1}\right]^{1/2} = \left[\frac{(0.99)^{-2}-1}{(0.01)^{-2}-1}\right]^{1/2} = 1.425 \cdot 10^{-3}$$

For the selectivity factor, we first find the cutoff frequencies for the analog prototype. With $\omega_p = 0.375\pi$ and $\omega_s = 0.5\pi$, we prewarp the frequencies as follows ($T_s = 2$),

$$\Omega_p = \tan\left(\frac{0.375\pi}{2}\right) = 0.6682$$

$$\Omega_s = \tan\left(\frac{0.5\pi}{2}\right) = 1$$

Therefore,

$$k = \frac{\Omega_p}{\Omega_s} = 0.6682$$

For the Butterworth filter, the required filter order is

$$N = \frac{\log d}{\log k} = 16.25$$

or $N = 17$. For the Chebyshev filter,

$$N = \frac{\cosh^{-1}(1/d)}{\cosh^{-1}(1/k)} = 7.55$$

so the minimum order is $N = 8$. Finally, for the elliptic filter, we first evaluate

$$q = q_0 + 2q_0^5 + 15q_0^9 + 150q_0^{13}$$

where

$$q_0 = \frac{1}{2}\frac{1-(1-k^2)^{1/4}}{1+(1-k^2)^{1/4}}$$

With $k = 0.6682$, we have

$$q_0 = \frac{1}{2}\frac{1 - (1 - 0.6682^2)^{1/4}}{1 + (1 - 0.6682^2)^{1/4}} = 0.0369$$

and

$$q = 0.0369$$

Therefore, for the filter order, we find

$$N = \frac{\log(16/d^2)}{\log(1/q)} = 4.81$$

or $N = 5$.

9.33 With impulse invariance, a first-order pole in $H_a(s)$ at $s = s_k$ is mapped to a pole in $H(z)$ at $z = e^{s_k T_s}$:

$$\frac{1}{s - s_k} \implies \frac{1}{1 - e^{s_k T_s} z^{-1}}$$

Determine how a second-order pole is mapped with impulse invariance.

If the system function of a continuous-time filter is

$$H_a(s) = \frac{1}{(s - s_k)^2}$$

the impulse response is

$$h_a(t) = t e^{s_k t} u(t)$$

where $u(t)$ is the unit step function. Sampling $h_a(t)$ with a sampling period T_s, we have

$$h(n) = h_a(nT_s) = nT_s e^{s_k nT_s} u(n)$$

Using the z-transform property

$$nx(n) \overset{z}{\longleftrightarrow} -z\frac{dX(z)}{dz}$$

and the z-transform pair

$$\alpha^n u(n) \overset{z}{\longleftrightarrow} \frac{1}{1 - \alpha z^{-1}}$$

it follows that the z-transform of $h(n)$ is

$$H(z) = -T_s z \frac{d}{dz}\left[\frac{1}{1 - e^{s_k T_s} z^{-1}}\right] = \frac{T_s e^{s_k T_s} z^{-1}}{(1 - e^{s_k T_s} z^{-1})^2}$$

Therefore, for a second-order pole, we have the mapping

$$\frac{1}{(s - s_k)^2} \implies \frac{T_s e^{s_k T_s} z^{-1}}{(1 - e^{s_k T_s} z^{-1})^2}$$

9.34 Suppose that we would like to design and implement a low-pass filter with

$$0.99 \leq |H(e^{j\omega})| \leq 1.01 \qquad 0 \leq \omega \leq 0.40\pi$$

$$|H(e^{j\omega})| \leq 0.001 \qquad 0.42\pi \leq \omega \leq \pi$$

(a) What order FIR equiripple filter is required to satisfy these specifications?

(b) Repeat part (a) for an elliptic filter.

(c) Compare the complexity of the implementations for the equiripple and elliptic filters in terms of the number of coefficients that must be stored, the number of delays that are required, and the number of multiplications necessary to compute each output sample $y(n)$.

(a) With a transition width of $\Delta\omega = 0.02\pi$, an estimate of the required filter order for an FIR equiripple filter is

$$N = \frac{-10\log(\delta_p\delta_s) - 13}{14.6\Delta f} = \frac{-10\log(0.01 \cdot 0.001) - 13}{14.6(0.02\pi)/2\pi} = \frac{50 - 13}{(14.6)(0.01)} = 253.4$$

or $N = 254$.

(b) For an elliptic filter, we have

$$d = \left[\frac{(1 - \delta_p)^{-2} - 1}{\delta_s^{-2} - 1} \right]^{1/2} = 1.42 \cdot 10^{-4}$$

and

$$k = \frac{\Omega_p}{\Omega_s} = \frac{\tan(\omega_p/2)}{\tan(\omega_s/2)} = 0.9367$$

With

$$q_0 = \frac{1}{2} \frac{1 - (1 - k^2)^{1/4}}{1 + (1 - k^2)^{1/4}} = 0.1282$$

and

$$q = q_0 + 2q_0^5 + 15q_0^9 + 150q_0^{13} = 0.1283$$

then

$$N \geq \frac{\log(16/d^2)}{\log(1/q)} = 9.98$$

or $N = 10$.

(c) For an FIR filter of order $N = 254$, the output $y(n)$ is

$$y(n) = \sum_{k=0}^{254} h(k)x(n - k)$$

Therefore, implementing this filter requires $N = 254$ delays. Since this filter has linear phase, exploiting the symmetry of the unit sample response,

$$h(n) = h(254 - n)$$

it follows that we must only provide storage for 128 filter coefficients, $h(0), h(1), \ldots, h(127)$. In addition, we may simplify the evaluation of $y(n)$ as follows,

$$y(n) = \sum_{k=0}^{254} h(k)x(n - k) = \sum_{k=0}^{126} h(k)[x(n - k) + x(n - 254 + k)] + h(127)x(n - 127)$$

Thus, 128 multiplications are required to compute each value of $y(n)$. For a 10th-order elliptic filter,

$$y(n) = \sum_{k=1}^{10} a(k)y(n - k) + \sum_{k=0}^{10} b(k)x(n - k)$$

Therefore, it follows that 21 memory locations are required to store the coefficients $a(k)$ and $b(k)$, and 10 delays are required for a canonic implementation. In addition, we see that 21 multiplications are necessary to evaluate each value of $y(n)$. However, since the zeros of $H(z)$ lie on the unit circle, the coefficients $b(k)$ are symmetric,

$$b(k) = b(10 - k)$$

By exploiting this symmetry, we may eliminate five multiplications per output point and five memory locations.

9.35 The input $x_a(t)$ and output $y_a(t)$ of a continuous-time filter with a rational system function are related by a linear constant coefficient differential equation of the form

$$\sum_{k=0}^{p} a(k) \frac{d^k}{dt^k} y_a(t) = \sum_{k=0}^{q} b(k) \frac{d^k}{dt^k} x_a(t)$$

Suppose that we sample $x_a(t)$ and $y_a(t)$,

$$x(n) = x_a(nT_s) \qquad y(n) = y_a(nT_s)$$

and approximate a first derivative with the *first backward difference*,

$$\frac{d}{dt} x_a(t) \longrightarrow \nabla^{(1)} x(n) = \frac{1}{T_s}[x(n) - x(n - 1)]$$

Approximations to higher-order derivatives are then defined as follows,

$$\frac{d^k}{dt^k} x_a(t) \longrightarrow \nabla^{(k)} x(n) = \nabla[\nabla^{k-1} x(n)]$$

Applying these approximations to the differential equation gives the following approximation to the differential equation:

$$\sum_{k=0}^{p} c(k) \nabla^{(k)} y(n) = \sum_{k=0}^{q} d(k) \nabla^{(k)} x(n)$$

The first backward difference defines a mapping from the s-plane to the z-plane that is given by

$$s = \frac{1 - z^{-1}}{T_s}$$

Determine the characteristics of this mapping, and compare it to the bilinear transformation. Is this a good mapping to use? Explain why or why not.

As with the bilinear transformation, the first backward difference will map a rational function of s into a rational function of z. To see how points in the s-plane map to points in the z-plane, let us write the mapping as follows,

$$z = \frac{1}{1 - sT_s}$$

Note that with $s = \sigma + j\Omega$,

$$|z|^2 = \frac{1}{(1 - \sigma T_s)^2 + (\Omega T_s)^2}$$

and it follows that points in the left-half s-plane ($\sigma < 0$) are mapped to points *inside* the unit circle, $|z| < 1$. Thus, stable analog filters are mapped to stable digital filters.

Now, let us look at how the $j\Omega$-axis is mapped to the z-plane. With $s = j\Omega$, we see that

$$z = \frac{1}{1 - j\Omega T_s} = \frac{1 + j\Omega T_s}{1 + (\Omega T_s)^2}$$

which is an equation for a circle of radius $r = \frac{1}{2}$ centered at $z = \frac{1}{2}$. To see this, note that

$$z - \frac{1}{2} = \frac{1}{1 - j\Omega T_s} - \frac{1}{2} = \frac{1}{2} \cdot \frac{1 + j\Omega T_s}{1 - j\Omega T_s}$$

Thus,

$$\left| z - \frac{1}{2} \right| = \frac{1}{2} \cdot \left| \frac{1 + j\Omega T_s}{1 - j\Omega T_s} \right| = \frac{1}{2}$$

The properties of the mapping are illustrated in the following figure.

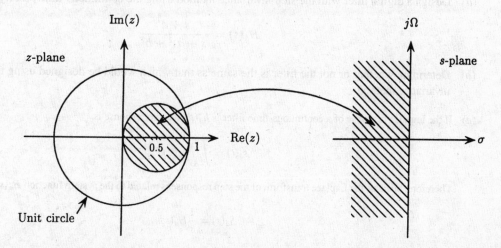

Since the $j\Omega$-axis does *not* map onto the unit circle, the frequency response of the digital filter produced with this mapping will not, in general, be an accurate representation of the frequency response of the analog filter except when ω is close to zero. In other words, the frequency response of a continuous-time filter will be well preserved only for low frequencies.

9.36   Use the impulse invariance method to design a digital filter from an analog prototype that has a system function

$$H_a(s) = \frac{s+a}{(s+a)^2 + b^2}$$

To design a filter using the impulse invariance technique, we first expand $H_a(s)$ in a partial fraction expansion as follows,

$$H_a(s) = \frac{s+a}{(s+a)^2 + b^2} = \frac{A_1}{s+(a+jb)} + \frac{A_2}{s+(a-jb)}$$

where

$$A_1 = [(s+a+jb)H_a(s)]_{s=-a-jb} = \left[\frac{s+a}{s+a-jb}\right]_{s=-a-jb} = \tfrac{1}{2}$$

and

$$A_2 = [(s+a-jb)H_a(s)]_{s=-a+jb} = \left[\frac{s+a}{s+a+jb}\right]_{s=-a+jb} = \tfrac{1}{2}$$

Therefore, with

$$H_a(s) = \frac{\tfrac{1}{2}}{s+a+jb} + \frac{\tfrac{1}{2}}{s+a-jb}$$

using the mapping given in Eq. (9.10), we have

$$H(z) = \frac{\tfrac{1}{2}}{1 - e^{(-a-jb)T_s}z^{-1}} + \frac{\tfrac{1}{2}}{1 - e^{(-a+jb)T_s}z^{-1}} = \frac{1 - e^{-aT_s}\cos(bT_s)z^{-1}}{1 - 2e^{-aT_s}\cos(bT_s)z^{-1} + e^{-2aT_s}z^{-2}}$$

Note that the zero at $s = -a$ is mapped to a zero at $z = e^{-aT_s}\cos(bT_s)$. Thus, the location of the zero in the discrete-time filter depends on the position of the poles as well as the zero in the analog filter.

9.37   With the impulse invariance method, the unit sample response of a digital filter is formed by sampling the impulse response of the continuous-time filter,

$$h(n) = h_a(nT_s)$$

Another approach is to use the *step invariance method* in which the step response of the digital filter is formed by sampling the step response of the continuous-time filter.

(a)   Design a digital filter with the step invariance method using the continuous-time prototype

$$H_a(s) = \frac{s+a}{(s+a)^2 + b^2}$$

(b)   Determine whether or not the filter is the same as that which would be designed using the impulse invariance method.

(a)   If the impulse response of a continuous-time filter is $h_a(t)$, its step response is

$$s_a(t) = \int_{-\infty}^{t} h_a(\tau)\,d\tau$$

Therefore, because the Laplace transform of the step response is related to the system function $H_a(s)$ as follows,

$$S_a(s) = \frac{1}{s}H_a(s)$$

then

$$S_a(s) = \frac{1}{s}\frac{s+a}{(s+a)^2 + b^2}$$

To design a digital filter using step invariance, we first perform a partial fraction expansion of $S_a(s)$,

$$S_a(s) = \frac{A_0}{s} + \frac{A_1}{s+a+jb} + \frac{A_2}{s+a-jb}$$

where

$$A_0 = [sS_a(s)]_{s=0} = \left[\frac{s+a}{(s+a)^2+b^2}\right]_{s=0} = \frac{a}{a^2+b^2}$$

$$A_1 = [(s+a+jb)S_a(s)]_{s=-a-jb} = \left[\frac{s+a}{s(s+a-jb)}\right]_{s=-a-jb} = \frac{-a+jb}{2(a^2+b^2)}$$

and

$$A_2 = [(s+a-jb)S_a(s)]_{s=-a+jb} = \left[\frac{s+a}{s(s+a+jb)}\right]_{s=-a+jb} = \frac{-a-jb}{2(a^2+b^2)}$$

Therefore,

$$S_a(s) = \frac{a}{a^2+b^2}\frac{1}{s} + \frac{-a+jb}{2(a^2+b^2)(s+a+jb)} + \frac{-a-jb}{2(a^2+b^2)(s+a-jb)}$$

Sampling $s_a(t)$,

$$s(n) = s_a(nT_s)$$

and finding the z-transform of $s(n)$ corresponds to the substitution

$$\frac{1}{s-\alpha} \longrightarrow \frac{1}{1-e^{-\alpha T_s}z^{-1}}$$

Thus, the z-transform of the step response is

$$S(z) = \frac{a}{a^2+b^2}\frac{1}{1-z^{-1}} + \frac{-a+jb}{2(a^2+b^2)}\frac{1}{1-e^{(-a-jb)T_s}z^{-1}} + \frac{-a-jb}{2(a^2+b^2)}\frac{1}{1-e^{(-a+jb)T_s}z^{-1}}$$

$$= \frac{a}{a^2+b^2}\frac{1}{1-z^{-1}} + \frac{1}{a^2+b^2}\frac{-a+\{a\cos(bT_s)+b\sin(bT_s)\}e^{-aT_s}z^{-1}}{1-2\cos(bT_s)e^{-aT_s}z^{-1}+e^{-2aT_s}z^{-2}}$$

The system function of the digital filter is then

$$H(z) = (1-z^{-1})S(z) = \frac{a}{a^2+b^2} + \frac{(1-z^{-1})}{(a^2+b^2)}\cdot\frac{-a+\{a\cos(bT_s)+b\sin(bT_s)\}e^{-aT_s}z^{-1}}{1-2\cos(bT_s)e^{-aT_s}z^{-1}+e^{-2aT_s}z^{-2}}$$

(b) Using impulse invariance, we see from Prob. 9.36 that the system function is

$$H(z) = \frac{1/2}{1-e^{(-a-jb)T_s}z^{-1}} + \frac{1/2}{1-e^{(-a+jb)T_s}z^{-1}} = \frac{1-e^{-aT_s}\cos(bT_s)z^{-1}}{1-2e^{-aT_s}\cos(bT_s)z^{-1}+e^{-2aT_s}z^{-2}}$$

Note that although $H(z)$ has the same poles as the filter designed using step invariance, the system functions are not the same. Therefore, the two designs are not equivalent.

9.38 Suppose that we would like to design a discrete-time low-pass filter by applying the impulse invariance method to a continuous-time Butterworth filter that has a magnitude-squared function

$$|H_a(j\Omega)|^2 = \frac{1}{1-(j\Omega/j\Omega_c)^{2N}}$$

The specifications for the discrete-time filter are

$$1 - \delta_p \leq |H(e^{j\omega})| \leq 1 \qquad 0 \leq \omega \leq \omega_p$$

$$|H(e^{j\omega})| \leq \delta_s \qquad \omega_s \leq \omega \leq \pi$$

Show that the design is not affected by the value of the sampling period that is used in the impulse invariance technique.

In the absence of aliasing, the impulse invariance method is a linear mapping from $H_a(j\Omega)$ to $H(e^{j\omega})$ for $|\omega| \leq \pi$. This mapping is

$$H(e^{j\omega}) = H_a(j\Omega)\big|_{\omega = \Omega T_s} \qquad |\omega| \leq \pi$$

Let us assume that there is no aliasing (this will be approximately true if the filter order is large enough). The required filter order is then

$$N \geq \frac{\log d}{\log k}$$

where d, the discrimination factor, is

$$d = \left[\frac{(1 - \delta_p)^{-2} - 1}{\delta_s^{-2} - 1} \right]^{1/2}$$

and k, the selectivity factor, is

$$k = \frac{\Omega_p}{\Omega_s}$$

Clearly, the discrimination factor does not depend on the sampling period T_s. In addition, with $\omega_p = \Omega_p T_s$ and $\omega_s = \Omega_s T_s$, it follows that

$$k = \frac{\omega_p / T_s}{\omega_s / T_s} = \frac{\omega_p}{\omega_s}$$

which does not depend on the sampling period. Therefore, the required filter order is independent of T_s. Next, if we expand the system function of the Butterworth filter in a partial fraction expansion, we have

$$H_a(s) = \sum_{k=1}^{N} \frac{A_k}{s - s_k}$$

where the poles, s_k, are

$$s_k = \Omega_c \exp\left\{ j \frac{(N + 1 + 2k)\pi}{2N} \right\} \qquad k = 0, 1, \ldots, N - 1$$

With impulse invariance, the system function of the discrete-time filter becomes

$$H(z) = \sum_{k=1}^{N} \frac{A_k}{1 - e^{s_k T_s} z^{-1}}$$

and it follows that the poles of $H(z)$ are at

$$z = \exp\{s_k T_s\} = \exp\{\Omega_c T_s \theta_k\}$$

where

$$\theta_k = \exp\left\{ j(N + 1 + 2k)\frac{\pi}{2N} \right\}$$

Because $\omega_c = \Omega_c T_s$ is the 3-dB cutoff frequency of the low-pass filter in the discrete-time domain, it is fixed by the filter specifications. Therefore, the poles of $H(z)$ will not be affected by the sampling period T_s. For example, if we try to decrease T_s to reduce aliasing, this would require an increase in Ω_c to preserve the cutoff frequency. Thus, it follows that the design is not affected by T_s.

9.39   Use the impulse invariance method to design a low-pass digital Butterworth filter to meet the following specifications:

$$0.9 \leq |H(e^{j\omega})| \leq 1 \qquad |\omega| \leq 0.2\pi$$

$$|H(e^{j\omega})| \leq 0.2 \qquad 0.3\pi \leq \omega \leq \pi$$

In the absence of aliasing, the impulse invariance method is a linear mapping from $H_a(j\Omega)$ to $H(e^{j\omega})$ for $|\omega| \leq \pi$, which is given by

$$H(e^{j\omega}) = H_a(j\Omega)\big|_{\omega = \Omega T_s}, \qquad |\omega| \leq \pi$$

Therefore, in order to simplify the design, we will assume that there is no aliasing and then, after the design is completed, check to see that the filter satisfies the given specifications. Because the parameter T_s does not enter into the design using the impulse invariance method (see Prob. 9.38), for convenience we will set $T_s = 1$.

The first step, then, is to design an analog Butterworth filter according to the following specifications:

$$0.9 \leq |H_a(j\Omega)| \leq 1 \qquad 0 \leq |\Omega| \leq 0.2\pi$$
$$|H_a(j\Omega)| \leq 0.2 \qquad 0.3\pi \leq \Omega$$

To determine the filter order, we compute the discrimination factor,

$$d = \left[\frac{(1 - \delta_p)^{-2} - 1}{\delta_s^{-2} - 1} \right]^{1/2} = 0.0989$$

and the selectivity factor,

$$k = \frac{0.2\pi}{0.3\pi} = 0.667$$

Thus, we have

$$N = \frac{\log d}{\log k} = 5.71$$

which, when rounded up, gives $N = 6$.

For the 3-dB cutoff frequency of the Butterworth filter, we will select Ω_c so that

$$|H_a(j\Omega)|^2_{\Omega = 0.2\pi} = (0.9)^2$$

that is, so that the Butterworth filter satisfies the passband specifications exactly (this will provide for some allowance for aliasing in the stopband). With

$$|H_a(j\Omega)|^2_{\Omega = 0.2\pi} = \frac{1}{1 + (0.2\pi / \Omega_c)^{12}} = 0.81$$

we have

$$1 = 0.81 \left[1 + \left(\frac{0.2\pi}{\Omega_c} \right)^{12} \right]$$

or

$$0.19 = 0.81 \left(\frac{0.2\pi}{\Omega_c} \right)^{12}$$

which gives

$$\Omega_c = 0.2\pi \left(\frac{0.81}{0.19} \right)^{1/12} = 0.7090$$

Therefore, the magnitude of the frequency response squared is

$$|H_a(j\Omega)|^2 = \frac{1}{1 + (\Omega/0.7090)^{12}}$$

and the 12 poles of

$$H_a(s)H_a(-s) = \frac{1}{1 + (s/j\Omega_c)^{12}}$$

lie on a circle of radius $\Omega_c = 0.7090$, at angles

$$\theta_k = \frac{(2k + 1)\pi}{12} \qquad k = 0, 1, 2, \ldots, 11$$

as illustrated in the following figure:

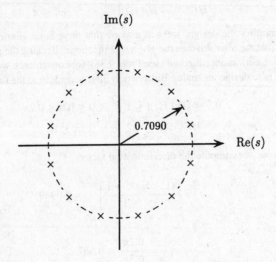

Thus, the poles of the Butterworth filter are the three complex conjugate pole pairs of $H_a(s)H_a(-s)$ that are in the left-half s-plane:

$$s_0 = s_1^* = 0.7090e^{j7\pi/12}$$

$$s_2 = s_3^* = 0.7090e^{j9\pi/12}$$

$$s_4 = s_5^* = 0.7090e^{j11\pi/12}$$

Therefore, with

$$H_a(s) = \prod_{k=0}^{5} \frac{-s_k}{s - s_k}$$

forming second-order polynomials from each conjugate pole pair, we have

$$H_a(s) = \frac{(0.7090)^6}{(s^2 + 0.3670s + 0.5027)(s^2 + 1.0027s + 0.5027)(s^2 + 1.3697s + 0.5027)}$$

The next steps, which are algebraically very tedious, are to perform a partial fraction expansion of $H_a(s)$,

$$H_a(s) = \sum_{k=1}^{6} \frac{A_k}{s - 0.7090e^{j(2k+1)\pi/12}}$$

perform the transformation

$$\frac{1}{s - s_k} \longrightarrow \frac{1}{1 - e^{s_k}z^{-1}}$$

and then recombine the terms. The result is

$$H(z) = \frac{0.0007z^{-1} + 0.0105z^{-2} + 0.0168z^{-3} + 0.0042z^{-4} + 0.0001z^{-5}}{1 - 3.3431z^{-1} + 5.0150z^{-2} - 4.2153z^{-3} + 2.0703z^{-4} - 0.5593z^{-5} + 0.0646z^{-6}}$$

The magnitude of the frequency response in decibels is plotted in the following figure.

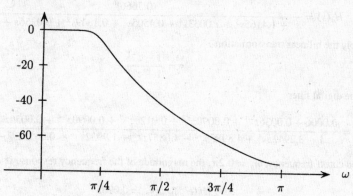

$20 \log |H(e^{j\omega})|$

As a final check on the design, evaluating $H(e^{j\omega})$ at $\omega = 0.2\pi$ and $\omega = 0.3\pi$, we find that

$$|H(e^{j\omega})|_{\omega=0.2\pi} = 0.9219$$

and

$$|H(e^{j\omega})|_{\omega=0.3\pi} = 0.2045$$

Therefore, the filter exceeds the passband specifications and comes close to meeting the stopband specifications. If this filter is unacceptable, the design could be modified by decreasing Ω_c to improve the stopband performance.

9.40   Repeat Prob. 9.39 using the bilinear transformation.

Using the bilinear transformation to design a Butterworth filter according to the specifications given in Prob. 9.39, we first use the transformation

$$\Omega = \frac{2}{T_s} \tan\left(\frac{\omega}{2}\right)$$

to map the passband and stopband frequencies of the digital filter to the cutoff frequencies of the analog filter. With $T_s = 2$, we have

$$\Omega_p = \tan\left(\frac{\omega_p}{2}\right) = \tan(0.1\pi) = 0.3249$$

and

$$\Omega_s = \tan\left(\frac{\omega_s}{2}\right) = \tan(0.15\pi) = 0.5095$$

As we found in Prob. 9.39, the required filter order is $N = 6$. For the 3-dB cutoff frequency of the analog Butterworth filter, we may choose any frequency in the range

$$\Omega_p[(1 - \delta_p)^{-2} - 1]^{-1/2N} \le \Omega_c \le \Omega_s\left(\delta_s^{-2} - 1\right)^{-1/2N}$$

or

$$0.3667 \le \Omega_c \le 0.3910$$

If we select $\Omega_c = 0.3667$, the passband specifications will be met exactly, and the stopband specifications will be exceeded. Conversely, if we set $\Omega_c = 0.3910$, the stopband specifications will be met exactly, and the passband specifications exceeded. Picking a frequency between the two extremes will produce an improvement in both bands. Because the stopband deviation is twice that of the deviation in the passband, we will set $\Omega_c = 0.3667$ in order to improve the stopband performance. From Table 9-4, we find the coefficients in the system function of a sixth-order normalized ($\Omega_c = 1$) Butterworth filter to be

$$H_a(s) = \frac{1}{s^6 + 3.8637s^5 + 7.4641s^4 + 9.1416s^3 + 7.4641s^2 + 3.8637s + 1}$$

To obtain a Butterworth filter with a cutoff frequency $\Omega_c = 0.3666$, we perform the low-pass-to-low-pass transformation

$$s \longrightarrow \frac{s}{0.3667}$$

This gives

$$H_a(s) = \frac{(0.3666)^6}{s^6 + 1.4165s^5 + 1.0033s^4 + 0.4505s^3 + 0.1349s^2 + 0.0256s + 0.0024}$$

Finally, we apply the bilinear transformation

$$s = \frac{1 - z^{-1}}{1 + z^{-1}}$$

which yields the digital filter

$$H(z) = \frac{0.0006 + 0.0036z^{-1} + 0.0090z^{-2} + 0.0120z^{-3} + 0.0090z^{-4} + 0.0036z^{-5} + 0.0006z^{-6}}{1 - 3.2942z^{-1} + 4.8985z^{-2} - 4.0857z^{-3} + 1.9932z^{-4} - 0.5353z^{-5} + 0.0615z^{-6}}$$

At the passband cutoff frequency, $\omega_p = 0.2\pi$, the magnitude of the frequency response is

$$|H(e^{j\omega})|_{\omega=0.2\pi} = 0.0905$$

and at the stopband cutoff frequency, $\omega_s = 0.3\pi$, the magnitude of the frequency response is

$$|H(e^{j\omega})|_{\omega=0.3\pi} = 0.1392$$

Therefore, this filter meets the given specifications.

9.41 Use the bilinear transformation to design a first-order low-pass Butterworth filter that has a 3-dB cutoff frequency $\omega_c = 0.2\pi$.

If a digital low-pass filter is to have a 3-dB cutoff frequency at $\omega_c = 0.2\pi$, the analog Butterworth filter should have a 3-dB cutoff frequency

$$\Omega_c = \tan\left(\frac{\omega_c}{2}\right) = \tan(0.1\pi) = 0.3249$$

For a first-order Butterworth filter,

$$H_a(s)H_a(-s) = \frac{1}{1 + (s/j\Omega_c)^2} = \frac{\Omega_c^2}{\Omega_c^2 - s^2} = \frac{\Omega_c}{\Omega_c + s} \cdot \frac{\Omega_c}{\Omega_c - s}$$

Therefore, the system function is

$$H_a(s) = \frac{\Omega_c}{s + \Omega_c}$$

With $\Omega_c = 0.3249$, applying the bilinear transformation

$$s = \frac{1 - z^{-1}}{1 + z^{-1}}$$

we have $$H(z) = \frac{0.3249}{\dfrac{1 - z^{-1}}{1 + z^{-1}} + 0.3249} = \frac{0.3249(1 + z^{-1})}{(1 - z^{-1}) + 0.3249(1 + z^{-1})} = \frac{0.2452(1 + z^{-1})}{1 - 0.5095z^{-1}}$$

9.42 A second-order continuous-time filter has a system function

$$H_a(s) = \frac{1}{s - a} + \frac{1}{s - b}$$

where $a < 0$ and $b < 0$ are real.

(a) Determine the locations of the poles and zeros of $H(z)$ if the filter is designed using the bilinear transformation with $T_s = 2$.

(b) Repeat part (a) for the impulse invariance technique, again with $T_s = 2$.

(a) The bilinear transformation is defined by the mapping

$$s = \frac{1 - z^{-1}}{1 + z^{-1}}$$

Therefore, for the given filter, we have

$$H(z) = \frac{1}{\dfrac{1 - z^{-1}}{1 + z^{-1}} - a} + \frac{1}{\dfrac{1 - z^{-1}}{1 + z^{-1}} - b} = \frac{1 + z^{-1}}{(1 - a) - z^{-1}} + \frac{1 + z^{-1}}{(1 - b) - z^{-1}}$$

which has poles at

$$z_1 = \frac{1}{1 - a} \qquad \text{and} \qquad z_2 = \frac{1}{1 - b}$$

To find the zeros, it is necessary to combine the two terms in the system function over a common denominator. Doing this, we have

$$H(z) = \frac{[(1 - b) - z^{-1}](1 + z^{-1}) + [(1 - a) - z^{-1}](1 + z^{-1})}{[(1 - a) - z^{-1}][(1 - b) - z^{-1}]}$$

$$= \frac{(2 - a - b) - (a + b)z^{-1} - 2z^{-2}}{[(1 - a) - z^{-1}][(1 - b) - z^{-1}]}$$

Finding the roots of the numerator may be facilitated by noting that $H_a(s)$ has a zero at $s = \infty$, which gets mapped to $z = -1$ with the bilinear transformation. Therefore, one of the factors of the numerator is $(1 + z^{-1})$. Dividing the numerator by this factor, we obtain the second factor, which is $[(2 - a - b) - 2z^{-1}]$. Thus, $H(z)$ has zeros at

$$z_1 = -1 \qquad \text{and} \qquad z_2 = \frac{2}{2 - a - b}$$

(b) With the impulse invariance technique, for first-order poles, the mapping is

$$\frac{1}{s - s_k} \Longrightarrow \frac{1}{1 - e^{s_k T_s} z^{-1}}$$

Therefore, with $T_s = 2$ we have

$$H(z) = \frac{1}{1 - e^{2a} z^{-1}} + \frac{1}{1 - e^{2b} z^{-1}} = \frac{2 - e^{2a} z^{-1} - e^{2b} z^{-1}}{(1 - e^{2a} z^{-1})(1 - e^{2b} z^{-1})}$$

which has two poles at

$$z_1 = e^{2a} \qquad \text{and} \qquad z_2 = e^{2b}$$

and only one zero, which is located at

$$z_0 = \tfrac{1}{2}(e^{2a} + e^{2b})$$

9.43 The system function of a digital filter is

$$H(z) = \sum_{k=1}^{p} \frac{A_k}{1 - \alpha_k z^{-1}}$$

(a) If this filter was designed using impulse invariance with $T_s = 2$, find the system function, $H_a(s)$, of an analog filter that could have been the analog filter prototype. Is your answer unique?

(b) Repeat part (a) assuming that the bilinear transformation was used with $T_s = 2$.

(a) Because $H(z)$ is expanded in a partial fraction expansion, the poles at $z = \alpha_k$ in $H(z)$ are mapped from poles in $H_a(s)$ according to the mapping

$$\alpha_k = e^{s_k T_s}$$

Therefore, if $T_s = 2$,

$$s_k = \tfrac{1}{2} \ln \alpha_k$$

and one possible analog filter prototype is

$$H_a(s) = \sum_{k=1}^{p} \frac{A_k}{s - \frac{1}{2}\ln\alpha_k}$$

Because the mapping from the s-plane to the z-plane is not one to one, this answer is not unique. Specifically, note that we may also write

$$\alpha_k = e^{s_k T_s + j2\pi}$$

Therefore, with $T_s = 2$, we may also have

$$s_k = \frac{1}{2}\ln\alpha_k + j\pi$$

and another possible analog filter prototype is

$$H_a(s) = \sum_{k=1}^{p} \frac{A_k}{s - \left(\frac{1}{2}\ln\alpha_k - j\pi\right)}$$

(b) With the bilinear transformation, because the mapping from the s-plane to the z-plane is a one-to-one mapping, with $T_s = 2$,

$$z = \frac{1+s}{1-s}$$

and the analog filter prototype that is mapped to $H(z)$ is unique and given by

$$H_a(s) = \sum_{k=1}^{p} \frac{A_k}{1 - \alpha_k \dfrac{1-s}{1+s}} = \sum_{k=1}^{p} \frac{A_k(1+s)}{(1-\alpha_k) + (1+\alpha_k)s}$$

9.44   A continuous-time system is called an *integrator* if the response of the system $y_a(t)$ to an input $x_a(t)$ is

$$y_a(t) = \int_{-\infty}^{t} x_a(\tau)d\tau$$

The system function for an integrator is

$$H_a(s) = \frac{1}{s}$$

(a) Design a discrete-time "integrator" using the bilinear transformation, and find the difference equation relating the input $x(n)$ to the output $y(n)$ of the discrete-time system.

(b) Find the frequency response of the discrete-time integrator found in part (a), and determine whether or not this system is a good approximation to the continuous-time system.

(a) With a system function $H_a(s) = 1/s$, the bilinear transformation produces a discrete-time filter with a system function

$$H(z) = \frac{T_s}{2} \frac{1+z^{-1}}{1-z^{-1}}$$

The unit sample response of this filter is

$$h(n) = \frac{T_s}{2}[u(n) + u(n-1)]$$

and the difference equation relating the output $y(n)$ to the input $x(n)$ is

$$y(n) = y(n-1) + \frac{T_s}{2}[x(n) + x(n-1)]$$

(b) Because the frequency response of the continuous-time system, $H_a(j\Omega) = 1/j\Omega$, is related to the discrete-time filter through the mapping

$$\Omega = \frac{2}{T_s} \tan\left(\frac{\omega}{2}\right)$$

the frequency response of the discrete-time system is

$$H(e^{j\omega}) = H_a(j\Omega)\Big|_{\Omega = \frac{2}{T_s}\tan(\frac{\omega}{2})} = \frac{T_s}{2j}\cot\left(\frac{\omega}{2}\right)$$

Note that because $H(e^{j\omega})$ goes to zero at $\omega = \pi$, then $H(e^{j\omega})$ will not be a good approximation to $H_a(j\Omega) = 1/j\Omega$ except for low frequencies. However, if $\omega \ll 1$, using the expansion

$$\cos x \approx 1 - \tfrac{1}{2}x^2 \qquad x \ll 1$$

and

$$\sin x \approx x \qquad x \ll 1$$

we have

$$\cot\left(\frac{\omega}{2}\right) = \frac{\cos(\omega/2)}{\sin(\omega/2)} \approx \frac{1 - \frac{1}{2}(\omega/2)^2}{\omega/2}$$

and we have, for the frequency response,

$$H(e^{j\omega}) = \frac{T_s}{2j}\frac{\cos(\omega/2)}{\sin(\omega/2)} \approx \frac{T_s}{j\omega}\left(1 - \tfrac{1}{8}\omega^2\right)$$

Therefore, for small ω

$$H(e^{j\omega}) \approx T_s H_a(j\Omega)$$

9.45 Let $H_a(j\Omega)$ be an Nth-order low-pass Butterworth filter with a 3-dB cutoff frequency Ω_c.

(a) Show that $H_a(s)$ may be transformed into an Nth-order high-pass Butterworth filter by adding N zeros at $s = 0$ and scaling the gain.

(b) What is the relationship between the corresponding digital low-pass and high-pass Butterworth filters that are designed using the bilinear transformation?

(a) For an Nth-order low-pass Butterworth with a system function $H_a(s)$,

$$H_a(s)H_a(-s) = \frac{1}{1 + (s/j\Omega_c)^{2N}}$$

Adding N zeros to $H_a(s)$ at $s = 0$, we have

$$H_a(s)H_a(-s) = \frac{(-1)^N s^{2N}}{1 + (s/j\Omega_c)^{2N}}$$

Multiplying numerator and denominator by $(j\Omega_c/s)^{2N}$ yields

$$H_a(s)H_a(-s) = \frac{(j\Omega_c)^{2N}}{1 + (j\Omega_c/s)^{2N}}$$

which corresponds to a magnitude-squared frequency response

$$|H_a(j\Omega)|^2 = \frac{\Omega_c^{2N}}{1 + (\Omega_c/\Omega)^{2N}}$$

Scaling $H_a(j\Omega)$ by Ω^{-N} results in a filter that has a frequency response $G_a(j\Omega)$ with a squared magnitude

$$|G_a(j\Omega)|^2 = \frac{1}{\Omega_c^{2N}}|H_a(j\Omega)|^2 = \frac{1}{1 + (\Omega_c/\Omega)^{2N}}$$

which is a high-pass filter with a cutoff frequency Ω_c. Specifically, note that $|G_a(j\Omega)|^2$ is equal to zero at $\Omega = 0$, and that $|G_a(j\Omega)|^2 \to 1$ as $\Omega \to \infty$,

(b) Applying the bilinear transformation to a low-pass Butterworth filter, we have

$$H(z)H(z^{-1}) = H_a(s)H_a(-s)|_{s=(1-z^{-1})/(1+z^{-1})}$$

$$= \cfrac{1}{1 + (j\Omega_c)^{-2N}\cfrac{1-z^{-1}}{1+z^{-1}}}$$

$$= \frac{(1+z^{-1})^{2N}(j\Omega_c)^{2N}}{(j\Omega_c)^{2N}(1+z^{-1})^{2N} + (1-z^{-1})^{2N}}$$

For the high-pass filter, on the other hand, we have

$$G(z)G(z^{-1}) = G_a(s)G_a(-s)|_{s=(1-z^{-1})/(1+z^{-1})}$$

$$= \cfrac{1}{1 + (j\Omega_c)^{2N}\left(\cfrac{1+z^{-1}}{1+z^{-1}}\right)^{2N}}$$

$$= \frac{(1-z^{-1})^{2N}}{(1-z^{-1})^{2N} + (j\Omega_c)^{2N}(1+z^{-1})^{2N}}$$

Therefore, we see that the poles of the low-pass digital Butterworth filter are the same as those of the high-pass digital Butterworth filter. The zeros, however, which are at $z = -1$ in the case of the low-pass filter, are at $z = 1$ in the high-pass filter. Thus, except for a difference in the gain, the high-pass digital Butterworth filter may be derived from the low-pass filter by flipping the N zeros in $H(z)$ at $z = -1$ to $z = 1$.

9.46 The impulse invariance method and the bilinear transformation are two filter design techniques that preserve stability of the analog filter by mapping poles in the left-half s-plane to poles inside the unit circle in the z-plane. An analog filter is minimum phase if all of its poles and zeros are in the left-half s-plane.

(a) Determine whether or not a minimum phase analog filter is mapped to a minimum phase discrete-time system using the impulse invariance method.

(b) Repeat part (a) for the bilinear transformation.

(a) With impulse invariance, an analog filter with a system function

$$H_a(s) = \sum_{k=1}^{p} \frac{A_k}{s - s_k}$$

will be mapped to a digital filter with a system function

$$H(z) = \sum_{k=1}^{p} \frac{A_k}{1 - e^{s_k T_s}z^{-1}}$$

Rewriting this system function as a ratio of polynomials, it follows that the locations of the zeros of $H(z)$ will depend on the locations of poles as well as the zeros of $H_a(s)$, and there is no way to guarantee that the zeros lie inside the unit circle. A simple example showing that a minimum phase continuous-time filter will not necessarily be mapped to a minimum phase discrete-time filter is the following:

$$H_a(s) = \frac{s+8}{(s+1)(s+2)} = \frac{7}{s+1} - \frac{6}{s+2}$$

Using the impulse invariance method with $T_s = 1$, we have

$$H(z) = \frac{7}{1 - e^{-1}z^{-1}} - \frac{6}{1 - e^{-2}z^{-1}} = \frac{1 + (6e^{-1} - 7e^{-2})z^{-1}}{(1 - e^{-1}z^{-1})(1 - e^{-2}z^{-1})}$$

which has a zero at

$$z = -(6e^{-1} - 7e^{-2}) \approx -1.256$$

Therefore, although $H_a(s)$ is minimum phase, $H(z)$ is not.

(b) The mapping between the s-plane and the z-plane with the bilinear transformation is defined by

$$z = \frac{1 + (T_s/2)s}{1 - (T_s/2)s}$$

Therefore, a pole or a zero at $s = s_k$ becomes a pole or a zero at

$$z_k = \frac{1 + (T_s/2)s_k}{1 - (T_s/2)s_k}$$

If $H_a(s)$ is minimum phase, the poles and zeros of $H_a(s)$ are in the left-half s-plane. In other words, if $H_a(s)$ has a pole or a zero at $s = s_k$, where $s_k = \sigma_k + j\Omega_k$,

$$\sigma_k < 0$$

Therefore, $$|z_k|^2 = \frac{|1 + (T_s/2)s_k|^2}{|1 - (T_s/2)s_k|^2} = \frac{|(2/T_s) + s_k|^2}{|(2/T_s) - s_k|^2} = \frac{[(2/T_s) + \sigma_k]^2 + \Omega_k^2}{[(2/T_s) - \sigma_k]^2 + \Omega_k^2} < 1$$

and it follows that a pole or a zero in the left-half s-plane will be mapped to a pole or a zero inside the unit circle in the z-plane (i.e., $H(z)$ is minimum phase).

9.47 The system function of a continuous-time filter $H_a(s)$ of order $N \geq 2$ may be expressed as a cascade of two lower-order systems:

$$H_a(s) = H_{a1}(s)H_{a2}(s)$$

Therefore, a digital filter may either be designed by applying a transformation directly to $H_a(s)$ or by individually transforming $H_{a1}(s)$ and $H_{a2}(s)$ into $H_1(z)$ and $H_2(z)$, respectively, and then realizing $H(z)$ as the cascade:

$$H(z) = H_1(z)H_2(z)$$

(a) If $H_1(z)$ and $H_2(z)$ are designed from $H_{a1}(s)$ and $H_{a2}(s)$ using the impulse invariance technique, compare the cascade $H(z) = H_1(z)H_2(z)$ with the filter that is designed by using the impulse invariance technique directly on $H_a(s)$.

(b) Repeat part (a) for the bilinear transformation.

(a) Due to sampling, aliasing occurs when designing a digital filter using the impulse invariance method. Because the operations of sampling and convolution do not commute, the filter designed by using impulse invariance on $H_a(s)$ will not be the same as the filter designed by cascading the two filters that are designed using impulse invariance on $H_{a1}(s)$ and $H_{a2}(s)$. In other words, if

$$h_a(t) = h_{a1}(t) * h_{a2}(t)$$

then $$h(n) \neq h_1(n) * h_2(n)$$

where $h(n) = h_a(nT_s)$, $h_1(n) = h_{a1}(nT_s)$, and $h_2(n) = h_{a2}(nT_s)$. As an example, consider the continuous-time filter that has a system function

$$H_a(s) = \frac{1}{(s+1)(s+2)} = \frac{1}{s+1} - \frac{1}{s+2}$$

Using the impulse invariance technique on $H_a(s)$ with $T_s = 1$, we have

$$H(z) = \frac{1}{1 - e^{-1}z^{-1}} - \frac{1}{1 - e^{-2}z^{-1}} = \frac{(e^{-1} - e^{-2})z^{-1}}{(1 - e^{-1}z^{-1})(1 - e^{-2}z^{-1})}$$

On the other hand, writing $H_a(s)$ as a cascade of two first-order systems,

$$H_a(s) = \frac{1}{s+1} \cdot \frac{1}{s+2}$$

and using the impulse invariance method on each of these systems with $T_s = 1$, we have

$$H(z) = \frac{1}{1 - e^{-1}z^{-1}} \cdot \frac{1}{1 - e^{-2}z^{-1}}$$

which is not the same as the previous filter.

(b) With the bilinear transformation, $(T_s = 2)$

$$H(z) = H_a\left(\frac{1 - z^{-1}}{1 + z^{-1}}\right) = H_{a1}\left(\frac{1 - z^{-1}}{1 + z^{-1}}\right)H_{a2}\left(\frac{1 - z^{-1}}{1 + z^{-1}}\right) = H_1(z)H_2(z)$$

and the two designs are the same.

9.48   What are the properties of the s-plane-to-z-plane mapping defined by

$$s = \frac{1 + z^{-1}}{1 - z^{-1}}$$

and what might this mapping be used for?

This mapping is very similar to the bilinear transformation which, with $T_s = 2$, is

$$s = \frac{1 - z^{-1}}{1 + z^{-1}}$$

In fact, this mapping may be considered to be a cascade of two mappings. The first is the bilinear transformation, and the second is one that replaces z with $-z$,

$$z' = -z$$

This mapping reflects points in the z-plane about the origin and, for points on the unit circle, corresponds to a shift of 180°:

$$H(z')|_{z'=e^{j\omega}} = H(-z)|_{z=e^{j\omega}} = H(-e^{j\omega}) = H\left(e^{j(\omega+\pi)}\right)$$

Therefore, this mapping has the same properties as the bilinear transformation, except that the $j\Omega$ axis is mapped onto the unit circle by the mapping

$$\omega = 2\arctan\left(\frac{\Omega T_s}{2}\right) + \pi$$

Because the unit circle is rotated by 180°, this mapping may be used to map low-pass analog filters into high-pass digital filters, and high-pass analog filters into low-pass digital filters.

Least-Squares Filter Design

9.49   Suppose that the desired unit sample response of a linear shift-invariant system is

$$h_d(n) = 3\left(\tfrac{1}{2}\right)^n u(n)$$

Use the Padé approximation method to find the parameters of a filter with a system function

$$H(z) = \frac{b(0) + b(1)z^{-1}}{1 + a(1)z^{-1}}$$

that approximates this unit sample response.

Using the Padé approximation method, with $p = q = 1$, we want to solve the following set of linear equations for $b(0)$, $b(1)$, and $a(1)$:

$$\begin{bmatrix} h_d(0) & 0 \\ h_d(1) & h_d(0) \\ h_d(2) & h_d(1) \end{bmatrix} \begin{bmatrix} 1 \\ a(1) \end{bmatrix} = \begin{bmatrix} b(0) \\ b(1) \\ 0 \end{bmatrix}$$

Substituting the given values for $h_d(n)$, we have

$$\begin{bmatrix} 3 & 0 \\ 3/2 & 3 \\ 3/4 & 3/2 \end{bmatrix} \begin{bmatrix} 1 \\ a(1) \end{bmatrix} = \begin{bmatrix} b(0) \\ b(1) \\ 0 \end{bmatrix}$$

Using the last equation, we may easily solve for $a(1)$,

$$\tfrac{3}{4} + \tfrac{3}{2}a(1) = 0$$

or

$$a(1) = -\tfrac{1}{2}$$

Having found $a(1)$, we may solve for $b(0)$ and $b(1)$ using the first two equations

$$\begin{bmatrix} 3 & 0 \\ 3/2 & 3 \end{bmatrix} \begin{bmatrix} 1 \\ -1/2 \end{bmatrix} = \begin{bmatrix} b(0) \\ b(1) \end{bmatrix}$$

or

$$b(0) = 3 \qquad b(1) = 0$$

Therefore, we have

$$H(z) = \frac{3}{1 - 0.5z^{-1}}$$

Notice that the unit sample response corresponding to this system exactly matches the given unit sample response. In general, however, this will not be true. A perfect match depends on $h_d(n)$ being the inverse z-transform of a rational function of z, and it depends upon an appropriate choice for the order of the Padé approximation (the number of poles and zeros).

9.50 Let the first three values of the unit sample response of a desired causal filter be $h_d(0) = 3$, $h_d(1) = \tfrac{1}{4}$, and $h_d(2) = \tfrac{1}{16}$.

(a) Using the Padé approximation method, find the coefficients of a second-order all-pole filter that has a unit sample response $h(n)$, such that $h(n) = h_d(n)$ for $n = 0, 1, 2$.

(b) Repeat part (a) for a filter that has one pole and one zero.

(c) Repeat part (a) for an FIR filter that has two zeros.

(a) For a second-order all-pole filter, the equations for the Padé approximation are

$$\begin{bmatrix} h_d(0) & 0 & 0 \\ h_d(1) & h_d(0) & 0 \\ h_d(2) & h_d(1) & h_d(0) \end{bmatrix} \begin{bmatrix} 1 \\ a(1) \\ a(2) \end{bmatrix} = \begin{bmatrix} b(0) \\ 0 \\ 0 \end{bmatrix}$$

which, with the given values for $h_d(n)$ become

$$\begin{bmatrix} 3 & 0 & 0 \\ \tfrac{1}{4} & 3 & 0 \\ \tfrac{1}{16} & \tfrac{1}{4} & 3 \end{bmatrix} \begin{bmatrix} 1 \\ a(1) \\ a(2) \end{bmatrix} = \begin{bmatrix} b(0) \\ 0 \\ 0 \end{bmatrix}$$

From the last two equations, we have

$$\begin{bmatrix} 3 & 0 \\ \tfrac{1}{4} & 3 \end{bmatrix} \begin{bmatrix} a(1) \\ a(2) \end{bmatrix} = - \begin{bmatrix} \tfrac{1}{4} \\ \tfrac{1}{16} \end{bmatrix}$$

Solving for $a(1)$ and $a(2)$, we find

$$a(1) = -\tfrac{1}{12} \qquad a(2) = -\tfrac{1}{72}$$

Then, using the first equation, we have

$$b(0) = 3$$

Thus, the system function of the filter is

$$H(z) = \frac{3}{1 - \tfrac{1}{12}z^{-1} - \tfrac{1}{72}z^{-2}}$$

(b) Using a first-order system to match the given values of $h_d(n)$,

$$H(z) = \frac{b(0) + b(1)z^{-1}}{1 + a(1)z^{-1}}$$

the equations that we must solve are as follows,

$$\begin{bmatrix} h_d(0) & 0 \\ h_d(1) & h_d(0) \\ h_d(2) & h_d(1) \end{bmatrix} \begin{bmatrix} 1 \\ a(1) \end{bmatrix} = \begin{bmatrix} b(0) \\ b(1) \\ 0 \end{bmatrix}$$

or, using the given values for $h_d(n)$,

$$\begin{bmatrix} 3 & 0 \\ \frac{1}{4} & 3 \\ \frac{1}{16} & \frac{1}{4} \end{bmatrix} \begin{bmatrix} 1 \\ a(1) \end{bmatrix} = \begin{bmatrix} b(0) \\ b(1) \\ 0 \end{bmatrix}$$

We may solve for $a(1)$ using the last equation,

$$\frac{1}{16} + \frac{1}{4}a(1) = 0$$

or

$$a(1) = -\frac{1}{4}$$

Next, we solve for $b(0)$ and $b(1)$ using the first two equations,

$$\begin{bmatrix} 3 & 0 \\ \frac{1}{4} & 3 \end{bmatrix} \begin{bmatrix} 1 \\ a(1) \end{bmatrix} = \begin{bmatrix} b(0) \\ b(1) \end{bmatrix}$$

which gives

$$\begin{bmatrix} b(0) \\ b(1) \end{bmatrix} = \begin{bmatrix} 3 & 0 \\ \frac{1}{4} & 3 \end{bmatrix} \begin{bmatrix} 1 \\ -\frac{1}{4} \end{bmatrix} = \begin{bmatrix} 3 \\ -\frac{1}{2} \end{bmatrix}$$

Thus, the system function is

$$H(z) = \frac{3 - \frac{1}{2}z^{-1}}{1 - \frac{1}{4}z^{-1}}$$

(c) For an FIR filter, the solution is trivial:

$$\begin{aligned} H(z) &= b(0) + b(1)z^{-1} + b(2)z^{-2} \\ &= h_d(0) + h_d(1)z^{-1} + h_d(2)z^{-2} \\ &= 3 + \frac{1}{4}z^{-1} + \frac{1}{16}z^{-2} \end{aligned}$$

9.51 Find the least-squares FIR inverse filter of length 3 for the system that has a unit sample response

$$g(n) = \begin{cases} 2 & n = 0 \\ 1 & n = 1 \\ 0 & \text{else} \end{cases}$$

Also, find the least-squares error,

$$\mathcal{E} = \sum_{n=0}^{\infty} e^2(n)$$

for this least-squares inverse filter.

To find the least-squares inverse, we need to solve the linear equations

$$\sum_{l=0}^{N-1} h(l) r_g(k - l) = \begin{cases} g(0) & k = 0 \\ 0 & k = 1, 2, \ldots, N-1 \end{cases}$$

where

$$r_g(k) = \sum_{n=0}^{\infty} g(n)g(n-k) = g(n) * g(-n)$$

is the deterministic autocorrelation of $g(n)$. With $N = 3$, these equations may be written in matrix form as follows,

$$\begin{bmatrix} r_g(0) & r_g(1) & r_g(2) \\ r_g(1) & r_g(0) & r_g(1) \\ r_g(2) & r_g(1) & r_g(0) \end{bmatrix} \begin{bmatrix} h(0) \\ h(1) \\ h(2) \end{bmatrix} = \begin{bmatrix} g(0) \\ 0 \\ 0 \end{bmatrix}$$

For the given sequence $g(n)$, we compute the autocorrelation sequence as follows,

$$r_g(0) = \sum_{n=0}^{\infty} g^2(n) = 5$$

$$r_g(1) = \sum_{n=0}^{\infty} g(n)g(n-1) = 2$$

$$r_g(2) = \sum_{n=0}^{\infty} g(n)g(n-2) = 0$$

Therefore, the linear equations become

$$\begin{bmatrix} 5 & 2 & 0 \\ 2 & 5 & 2 \\ 0 & 2 & 5 \end{bmatrix} \begin{bmatrix} h(0) \\ h(1) \\ h(2) \end{bmatrix} = \begin{bmatrix} 2 \\ 0 \\ 0 \end{bmatrix}$$

and the solution is

$$h(n) = \begin{cases} 0.494 & n = 0 \\ -0.235 & n = 1 \\ 0.094 & n = 2 \\ 0 & \text{else} \end{cases}$$

Performing the convolution of $h(n)$ with $g(n)$, we have

$$g(n) * h(n) = \begin{cases} 0.988 & n = 0 \\ 0.023 & n = 1 \\ -0.047 & n = 2 \\ 0.094 & n = 3 \end{cases}$$

From this sequence, we may evaluate the squared error,

$$\mathcal{E} = \sum_{n=0}^{\infty} [g(n) * h(n) - \delta(n)]^2 = 0.0118$$

9.52 Find the FIR least-squares inverse filter of length N for the system having a unit sample response

$$g(n) = \delta(n) - \alpha\delta(n-1)$$

where α is an arbitrary real number.

Before we begin, note that if $|\alpha| > 1$, $G(z)$ has a zero that is outside the unit circle. In this case, $G(z)$ is not minimum phase, and the inverse filter $1/G(z)$ cannot be both causal and stable. However, if $|\alpha| < 1$,

$$G^{-1}(z) = \frac{1}{G(z)} = \frac{1}{1 - \alpha z^{-1}}$$

and the inverse filter is

$$g^{-1}(n) = \alpha^n u(n)$$

We begin by finding the least-squares inverse of length $N = 2$. The autocorrelation sequence $r_g(k)$ is

$$r_g(k) = \begin{cases} 1 + \alpha^2 & k = 0 \\ -\alpha & k = \pm 1 \\ 0 & \text{else} \end{cases}$$

Therefore, the linear equations that we must solve are

$$\begin{bmatrix} 1 + \alpha^2 & -\alpha \\ -\alpha & 1 + \alpha^2 \end{bmatrix} \begin{bmatrix} h(0) \\ h(1) \end{bmatrix} = \begin{bmatrix} 1 \\ 0 \end{bmatrix}$$

The solution for $h(0)$ and $h(1)$ is easily seen to be

$$h(0) = \frac{1 + \alpha^2}{1 + \alpha^2 + \alpha^4}$$

$$h(1) = \frac{\alpha}{1 + \alpha^2 + \alpha^4}$$

The system function of this least-squares inverse filter is

$$H(z) = \frac{1 + \alpha^2}{1 + \alpha^2 + \alpha^4} + \frac{\alpha}{1 + \alpha^2 + \alpha^4} z^{-1} = \frac{1 + \alpha^2}{1 + \alpha^2 + \alpha^4} \left(1 + \frac{\alpha}{1 + \alpha^2} z^{-1} \right)$$

which has a zero at

$$z_0 = -\frac{\alpha}{1 + \alpha^2}$$

Note that because

$$|z_0| = \left| \frac{\alpha}{1 + \alpha^2} \right| = \left| \frac{1}{\alpha + \alpha^{-1}} \right| < 1$$

the zero of $H(z)$ is inside the unit circle, and $H(z)$ is minimum phase, regardless of whether the zero of $G(z)$ is inside or outside the unit circle.

Let us now look at the least-squares inverse, $h_N(n)$, of length N. In this case, the linear equations have the form

$$\begin{bmatrix} 1 + \alpha^2 & -\alpha & 0 & \cdots & 0 \\ -\alpha & 1 + \alpha^2 & -\alpha & \cdots & 0 \\ 0 & -\alpha & 1 + \alpha^2 & \cdots & 0 \\ \vdots & \vdots & \vdots & & \vdots \\ 0 & 0 & 0 & \cdots & 1 + \alpha^2 \end{bmatrix} \begin{bmatrix} h_N(0) \\ h_N(1) \\ h_N(2) \\ \vdots \\ h_N(N-1) \end{bmatrix} = \begin{bmatrix} 1 \\ 0 \\ 0 \\ \vdots \\ 0 \end{bmatrix} \qquad (9.22)$$

Solving these equations for arbitrary α and N may be accomplished as follows. For $n = 1, 2, \ldots, N - 2$ these equations may be represented by the homogeneous difference equation,

$$-\alpha h_N(n-1) + (1 + \alpha^2) h_N(n) - \alpha h_N(n+1) = 0$$

The general solution to this equation is of the form

$$h_N(n) = c_1 \alpha^n + c_2 \alpha^{-n} \qquad (9.23)$$

where c_1 and c_2 are constants that are determined by the boundary conditions at $n = 0$ and $n = N - 1$ [the first and last equations in Eq. (9.22)]:

$$(1 + \alpha^2) h_N(0) - \alpha h_N(1) = 1$$

$$-\alpha h_N(N-2) + (1 + \alpha^2) h_N(N-1) = 0 \qquad (9.24)$$

Substituting Eq. (9.23) into Eq. (9.24), we have

$$(1 + \alpha^2)[c_1 + c_2] - \alpha[c_1 \alpha + c_2 \alpha^{-1}] = 1$$

$$-\alpha[c_1 \alpha^{N-2} + c_2 \alpha^{-(N-2)}] + (1 + \alpha^2)[c_1 \alpha^{N-1} + c_2 \alpha^{-(N-1)}] = 0$$

which, after canceling common terms, may be simplified to

$$c_1 + \alpha^2 c_2 = 1$$

$$\alpha^{N+1} c_1 + \alpha^{-(N-1)} c_2 = 0$$

or

$$\begin{bmatrix} 1 & \alpha^2 \\ \alpha^{N+1} & \alpha^{-(N-1)} \end{bmatrix} \begin{bmatrix} c_1 \\ c_2 \end{bmatrix} = \begin{bmatrix} 1 \\ 0 \end{bmatrix}$$

The solution for c_1 and c_2 is

$$\begin{bmatrix} c_1 \\ c_2 \end{bmatrix} = \frac{1}{\alpha^{-(N-1)} - \alpha^{N+3}} \begin{bmatrix} \alpha^{-(N-1)} \\ -\alpha^{N+1} \end{bmatrix}$$

Therefore, $h_N(n)$ is

$$h_N(n) = \begin{cases} \dfrac{\alpha^{n-N} - \alpha^{N-n}}{\alpha^{-N} - \alpha^{N+2}} & 0 \leq n \leq N-1 \\ 0 & \text{else} \end{cases}$$

Let us now look at what happens asymptotically as $N \rightarrow \infty$. If $|\alpha| < 1$,

$$\lim_{N \rightarrow \infty} h_N(n) = \frac{\alpha^{n-N}}{\alpha^{-N}} = \alpha^n \qquad n \geq 0$$

which is the inverse filter, that is,

$$\lim_{N \rightarrow \infty} h_N(n) = \alpha^n u(n) = g^{-1}(n)$$

and

$$\lim_{N \rightarrow \infty} H_N(z) = \frac{1}{1 - \alpha z^{-1}}$$

However, if $|\alpha| > 1$,

$$\lim_{N \rightarrow \infty} h_N(n) = \frac{\alpha^{N-n}}{\alpha^{N+2}} = \alpha^{-n-2} \qquad n \geq 0$$

and

$$\lim_{N \rightarrow \infty} H_N(z) = \frac{\alpha^{-2}}{1 - \alpha^{-1} z^{-1}}$$

which is *not* the inverse filter. Note that although $\hat{d}(n) = h_N(n) * g(n)$ does not converge to $\delta(n)$ as $N \rightarrow \infty$, taking the limit of $\hat{D}_N(z)$ as $N \rightarrow \infty$, we have

$$\lim_{N \rightarrow \infty} \hat{D}_N(z) = \lim_{N \rightarrow \infty} \hat{H}_N(z) G(z) = \frac{1}{\alpha} \left(\frac{1 - \alpha z^{-1}}{\alpha - z^{-1}} \right)$$

which is an all-pass filter, that is,

$$|\hat{D}_N(e^{j\omega})| = \frac{1}{\alpha}$$

9.53   The first five samples of the unit sample response of a causal filter are

$$h(0) = 3 \qquad h(1) = -1 \qquad h(2) = 1 \qquad h(3) = 2 \qquad h(4) = 0$$

If it is known that the system function has two zeros and two poles, determine whether or not the filter is stable.

The system function of this filter has the form

$$H(z) = \frac{b(0) + b(1)z^{-1} + b(2)z^{-2}}{1 + a(1)z^{-1} + a(2)z^{-2}}$$

To determine whether or not this system is stable, it is necessary to find the denominator polynomial,

$$A(z) = 1 + a(1)z^{-1} + a(2)z^{-2}$$

and check to see whether or not the roots of $A(z)$ lie inside the unit circle. Given that $H(z)$ has two poles and two zeros, we may use the Padé approximation method to find the denominator coefficients:

$$\begin{bmatrix} h(0) & 0 & 0 \\ h(1) & h(0) & 0 \\ h(2) & h(1) & h(0) \\ h(3) & h(2) & h(1) \\ h(4) & h(3) & h(2) \end{bmatrix} \begin{bmatrix} 1 \\ a(1) \\ a(2) \end{bmatrix} = \begin{bmatrix} b(0) \\ b(1) \\ b(2) \\ 0 \\ 0 \end{bmatrix}$$

Using the last two equations, we have

$$\begin{bmatrix} h(3) & h(2) & h(1) \\ h(4) & h(3) & h(2) \end{bmatrix} \begin{bmatrix} 1 \\ a(1) \\ a(2) \end{bmatrix} = \begin{bmatrix} 0 \\ 0 \end{bmatrix}$$

which become

$$\begin{bmatrix} h(2) & h(1) \\ h(3) & h(2) \end{bmatrix} \begin{bmatrix} a(1) \\ a(2) \end{bmatrix} = - \begin{bmatrix} h(3) \\ h(4) \end{bmatrix}$$

Substituting the given values for $h(n)$, we have

$$\begin{bmatrix} 1 & -1 \\ 2 & 1 \end{bmatrix} \begin{bmatrix} a(1) \\ a(2) \end{bmatrix} = - \begin{bmatrix} 2 \\ 0 \end{bmatrix}$$

The solution is

$$a(1) = -\tfrac{2}{3} \qquad a(2) = \tfrac{4}{3}$$

and the denominator polynomial is

$$A(z) = 1 - \tfrac{2}{3}z^{-1} + \tfrac{4}{3}z^{-2}$$

Because the roots of this polynomial are not inside the unit circle, the filter is unstable.

Supplementary Problems

FIR Filter Design

9.54   What type of window(s) may be used to design a low-pass filter with a passband cutoff frequency $\omega_p = 0.35\pi$, a transition width $\Delta\omega = 0.025\pi$, and a maximum stopband deviation of $\delta_s = 0.003$?

9.55   Use the window design method to design a minimum-order low-pass filter with a passband cutoff frequency $\omega_s = 0.45\pi$, a stopband cutoff frequency $\omega_s = 0.5\pi$, and a maximum stopband deviation $\delta_s = 0.005$.

9.56   We would like to design a bandstop filter to satisfy the following specifications:

$$0.95 \le |H(e^{j\omega})| \le 1.05 \qquad 0 \le \omega \le 0.3\pi$$
$$|H(e^{j\omega})| < 0.01 \qquad 0.35\pi \le \omega \le 0.8\pi$$
$$0.95 \le |H(e^{j\omega})| \le 1.05 \qquad 0.85\pi \le \omega \le \pi$$

(a)   What weighting function $W(e^{j\omega})$ should be used to design this filter?

(b) What are the minimum and maximum numbers of extremal frequencies that a type I filter of order $N = 128$ must have?

9.57 Suppose that we would like to design a low-pass filter of order $N = 128$ with a passband cutoff frequency $\omega_p = 0.48\pi$ and a stopband cutoff frequency of $\omega_s = 0.52\pi$.

(a) Find the approximate passband and stopband ripple if we were to use a Kaiser window design.

(b) If an equiripple filter were designed so that it had a passband ripple equal to that of the Kaiser window design found in part (a), how small would the stopband ripple be?

9.58 We would like to design an equiripple low-pass filter of order $N = 30$. For a type I filter of order N, what is the minimum number of alternations that this filter may have, and what is the maximum number?

9.59 For a low-pass filter with $\delta_p = \delta_s$, what is the difference in the stopband attenuation in decibels between a Kaiser window design and an equiripple filter if both filters have the same transition width?

IIR Filter Design

9.60 Find the minimum order and the 3-dB cutoff frequency of a continuous-time Butterworth filter that will satisfy the following frequency response constraints:

$$|H_a(j\Omega)| = 0.95 \qquad \Omega = 16,000\pi$$
$$|H_a(j\Omega)| \le 0.1 \qquad \Omega > 24,000\pi$$

9.61 Use the bilinear transformation to design a first-order low-pass Butterworth filter that has a 3-dB cutoff frequency $\omega_c = 0.5\pi$.

9.62 Use the bilinear transformation to design a second-order bandpass Butterworth filter that has 3-dB cutoff frequencies $\omega_l = 0.4\pi$ and $\omega_u = 0.6\pi$.

9.63 If the specifications for an analog low-pass filter are to have a 1-dB cutoff frequency of 1 kHz and a maximum stopband ripple $\delta_s = 0.01$ for $|f| > 5$ kHz, determine the required filter order for the following:

(a) Butterworth filter

(b) Type I Chebyshev filter

(c) Type II Chebyshev filter

(d) Elliptic filter

9.64 Let $H_a(j\Omega)$ be an analog filter with

$$H_a(j\Omega)|_{\Omega=0} = 1$$

(a) If a discrete-time filter is designed using the impulse invariance method, is it necessarily true that

$$H(e^{j\omega})|_{\omega=0} = 1$$

(b) Repeat part (a) for the bilinear transformation.

9.65 Consider a causal and stable continuous-time filter that has a system function

$$H_a(s) = \frac{s+1}{(s+2)^2}$$

If a discrete-time filter is designed using impulse invariance with $T_s = 1$, find $H(z)$.

9.66 The system function of a digital filter is

$$H(z) = \frac{2}{1 - 0.5z^{-1}} - \frac{1}{1 - 0.25z^{-1}}$$

(a) Assuming that this filter was designed using impulse invariance with $T_s = 2$, find the system function of two different analog filters that could have been the analog filter prototype.

(b) If this filter was designed using the bilinear transformation with $T_s = 2$, find the analog filter that was used as the prototype.

9.67 Determine the characteristics of the s-plane-to-z-plane mapping

$$s = \frac{1 - z^{-2}}{1 + z^{-2}}$$

9.68 The system function of an analog filter $H_a(s)$ may be expressed as a parallel connection of two lower-order systems

$$H_a(s) = H_{a1}(s) + H_{a2}(s)$$

If $H_a(s)$, $H_{a1}(s)$, and $H_{a2}(s)$ are mapped into digital filters using the impulse invariance technique, will it be true that

$$H(z) = H_1(z) + H_2(z)$$

What about with the bilinear transformation?

9.69 If an analog filter has an equiripple passband, will the digital filter designed using the impulse invariance method have an equiripple passband? Will it have an equiripple passband if the bilinear transformation is used?

9.70 Can an analog allpass filter be mapped to a digital allpass filter using the bilinear transformation?

9.71 An IIR low-pass digital filter is to be designed to meet the following specifications:

Passband cutoff frequency of 0.22π with a passband ripple less than 0.01

Stopband cutoff frequency of 0.24π with a stopband attenuation greater than 40 dB

(a) Determine the filter order required to meet these specifications if a digital Butterworth filter is designed using the bilinear transformation.

(b) Repeat for a digital Chebyshev filter.

(c) Compare the number of multiplications required to compute each output value using these filters, and compare them to an equiripple linear phase filter.

Least-Squares Filter Design

9.72 Suppose that the desired unit sample response of a linear shift-invariant system is

$$h_d(n) = \delta(n) + 2\left(\tfrac{1}{2}\right)^n u(n - 1)$$

Use the Padé approximation method to find the parameters of a filter with a system function

$$H(z) = \frac{b(0) + b(1)z^{-1}}{1 + a(1)z^{-1}}$$

that approximates this unit sample response.

9.73 The first five samples of the unit sample response of a causal filter are

$$h(0) = 0.2000 \qquad h(1) = 0.7560 \qquad h(2) = 1.0737 \qquad h(3) = -0.8410 \qquad h(4) = -0.6739$$

If it is known that the system function has two zeros and two poles, determine whether or not the filter is stable.

Answers to Supplementary Problems

9.54 A Hamming or a Blackman window or a Kaiser window with $\beta = 4.6$.

9.55 $h(n) = w(n)h_d(n)$, where $w(n)$ is a Kaiser window with $\beta = 4.09$ and $N = 107$, and

$$h_d(n) = \frac{\sin[0.475\pi(n - 53.5)]}{(n - 53.5)\pi}$$

9.56 (a) $W(e^{j\omega}) = \begin{cases} 1 & 0 \le \omega \le 0.3\pi \\ 5 & 0.35\pi \le \omega \le 0.8\pi \\ 1 & 0.85\pi \le \omega \le \pi \end{cases}$

(b) The minimum is 66 and the maximum is 69.

9.57 (a) $\delta_p \approx \delta_s \approx 0.0058$. (b) $\delta_s \approx 0.0016$.

9.58 The minimum number is 17 and the maximum is 18.

9.59 5 dB.

9.60 $N = 9$ and $\Omega_c = 17,342\pi$.

9.61 $H(z) = \frac{1}{2}(1 + z^{-1})$.

9.62 $H(z) = \frac{0.65(1 - z^{-2})}{2.65 + 1.35z^{-2}}$.

9.63 (a) $N = 4$. (b) $N = 3$. (c) $N = 3$. (d) $N = 3$.

9.64 (a) No. (b) Yes.

9.65 $H(z) = \frac{1 - 2e^{-2}z^{-1}}{(1 - e^{-2}z^{-1})^2}$.

9.66 (a) One possible filter has a system function

$$H_a(s) = \frac{2}{s - \frac{1}{2}\ln(0.5)} - \frac{1}{s - \frac{1}{2}\ln(0.25)}$$

and another is

$$H_a(s) = \frac{2}{s - \frac{1}{2}\ln(0.5) + j\pi} - \frac{1}{s - \frac{1}{2}\ln(0.25) + j\pi}$$

Note, however, that the second filter has a complex-valued impulse response.

(b) This filter is unique and has a system function

$$H_a(s) = \frac{4(1 + s)}{1 + 3s} - \frac{4(1 + s)}{3 + 5s}$$

9.67 This is a cascade of two mappings. The first is the bilinear transformation, and the second is the mapping $z \to z^2$, which compresses the frequency axis by a factor of 2. Thus, a low-pass filter is mapped into a bandstop filter, and a high-pass filter is mapped into a bandpass filter.

9.68 True for both methods.

9.69 The digital filter will have an equiripple passband with the bilinear transformation but not with the impulse invariance method.

9.70   Yes.

9.71   (a) Butterworth filter order is $N = 69$.

(b) Chebyshev filter order is $N = 17$.

(c) For an equiripple filter, we require $N = 185$, which requires 185 delays. In addition, 93 multiplications are needed to evaluate each value of $y(n)$. The Butterworth and Chebyshev filters require 69 and 17 delays, respectively, and approximately twice this number of multiplications to evaluate each value of $y(n)$.

9.72   Padé gives $b(0) = 1$, $b(1) = 0.5$, and $a(1) = -0.5$, or

$$H(z) = \frac{1 + 0.5z^{-1}}{1 - 0.5z^{-1}}$$

9.73   Padé's method with $p = q = 2$ gives

$$H(z) = \frac{0.2 + 0.8z^{-1} + 1.4z^{-2}}{1 + 0.22z^{-1} + 0.8z^{-2}}$$

Because the roots of the denominator lie inside the unit circle, this filter is stable.

Index